AF869460

Transport Phenomena in Porous Media

Yasuaki Ichikawa • A.P.S. Selvadurai

Transport Phenomena in Porous Media

Aspects of Micro/Macro Behaviour

Springer

Yasuaki Ichikawa
Okayama University
Okayama
Japan

A.P.S. Selvadurai
McGill University
Montréal
Canada

ISBN 978-3-642-25332-4 e-ISBN 978-3-642-25333-1
DOI 10.1007/978-3-642-25333-1
Springer Heidelberg Dordrecht London New York

Library of Congress Control Number: 2012930123

© Springer-Verlag Berlin Heidelberg 2012
This work is subject to copyright. All rights are reserved, whether the whole or part of the material is concerned, specifically the rights of translation, reprinting, reuse of illustrations, recitation, broadcasting, reproduction on microfilm or in any other way, and storage in data banks. Duplication of this publication or parts thereof is permitted only under the provisions of the German Copyright Law of September 9, 1965, in its current version, and permission for use must always be obtained from Springer. Violations are liable to prosecution under the German Copyright Law.
The use of general descriptive names, registered names, trademarks, etc. in this publication does not imply, even in the absence of a specific statement, that such names are exempt from the relevant protective laws and regulations and therefore free for general use.

Printed on acid-free paper

Springer is part of Springer Science+Business Media (www.springer.com)

To Maria and Sally

Note: Sections designated by a symbol ♣ show some advanced topics, which beginners might like to skip.

Contents

1 Introduction 1
1.1 Transport Phenomena in Porous Media and Modeling 1
1.2 Molecular Dynamics Simulation and Homogenization Analysis 2
1.3 Underground Disposal of HLW's and Bentonite 5

2 Introduction to Continuum Mechanics 9
2.1 Newtonian Mechanics 9
2.2 Deformation Kinematics 11
2.2.1 Motion and Configuration 12
2.2.2 Changes of Frame and Frame Indifference ♣ 16
2.2.3 Motion in a Non-inertial System ♣ 17
2.2.4 Deformation Gradient, Strain and Strain Rate 19
2.2.5 Transport Theorems and Jump Condition 25
2.3 Mass Conservation Law 29
2.4 Law of Conservation of Linear Momentum and Stress 30
2.4.1 Eulerian Descriptions 30
2.4.2 Lagrangian Descriptions ♣ 33
2.5 Conservation of Moment of Linear Momentum and Symmetry of Stress 35
2.6 Incremental Forms of the Equation of Equilibrium ♣ 36
2.6.1 Total Lagrangian Form 36
2.6.2 Updated Lagrangian Form 37
2.7 Specific Description of the Equation of Motion ♣ 38
2.7.1 Eulerian Equation of Motion 38
2.7.2 Lagrangian Equation of Motion 39
2.7.3 Incremental Form of the Total Lagrangian Equation of Motion 39
2.7.4 Incremental Form of the Updated Lagrangian Equation of Motion 39

2.8 Response of Materials: Constitutive Theory 40
2.8.1 Fundamental Principles of Material Response 41
2.8.2 Convected Derivative, Corotational Derivative and Frame Indifference ♣ 42
2.8.3 Invariants of Stress and Strain and Isotropic Elastic Solids 46
2.8.4 Newtonian Fluid 53
2.9 Small Strain Viscoelasticity Theory 56
2.9.1 Boltzmann Integral and Excitation-response Theory 56
2.9.2 Stress Relaxation and the Relaxation Spectra: Generalized Maxwell Model 58
2.9.3 Creep and the Retardation Spectra: Generalized Kelvin-Voigt Model 61
2.9.4 Relaxation and Retardation Spectra and Their Asymptotic Expansion 63
2.9.5 Experiments for Determining Viscous Properties 64
2.10 Small Strain Plasticity: Flow Theory 67
2.10.1 Yield Function and Hardening Law 68
2.10.2 Prager's Consistency Condition 71
2.10.3 Flow Rule and Incremental Constitutive Law 72

3 Non-equilibrium Thermodynamics 77
3.1 Review of Classical Thermodynamics 77
3.1.1 An Application of Classical Thermodynamics for a Newtonian Fluid 78
3.1.2 The Role of the Second Law of Thermodynamics in Classical Theory 79
3.1.3 The Entropy Inequality in Classical Continuum Mechanics 80
3.1.4 Note on the Proposed Framework 81
3.2 Conservation of Energy: The First Law of Thermodynamics 82
3.2.1 Stokes' Power Formulation as Mechanical Conservation of Energy in a Continuum: Eulerian Description 82
3.2.2 Generalized Strain Measure and its Conjugate Stress in a Continuum ♣ 86
3.2.3 Stokes' Power Formula in a Continuum: Lagrangian Description ♣ 87
3.2.4 First Law of Thermodynamics in a Thermo-mechanical Continuum: Conservation of Mechanical and Thermal Energies 88
3.2.5 First Law of Thermodynamics in a Thermo-mechanical Continuum: Eulerian Description 89
3.2.6 First Law of Thermodynamics in a Thermo-mechanical Continuum: Lagrangian Description ♣ 90

3.3 Second Law of Thermodynamics 90
3.3.1 Thermal Energy and the Existence of Entropy: The First Part of the Second Law of Thermodynamics ... 91
3.3.2 Entropy Inequality: Second Part of the Second Law of Thermodynamics 92
3.3.3 Second Law of Thermodynamics in a Thermo-mechanical Continuum: Lagrangian Description ♣ 97
3.4 Thermodynamic Functions 97
3.4.1 Legendre Transformation and Convex Functions 98
3.4.2 Thermodynamic Functions in a Thermo-mechanical Field: Solids with Small Strain 99
3.4.3 Thermodynamic Functions in a Thermo-mechanical Field: A Finitely Strained Solid ♣............ 105
3.4.4 Thermodynamic Functions in a Thermo-mechanical Field: A Fluid 108
3.5 Chemical Process and Thermodynamics 111
3.5.1 Thermodynamic Variables in a Thermo-mechanochemical Field 111
3.5.2 Thermodynamic Functions in a Thermo-mechanochemical Field: A Small Strain Solid 113
3.5.3 Thermodynamic Functions in a Thermo-mechanochemical Field: A Finitely Strained Solid ♣ 118
3.5.4 Thermodynamic Functions in a Thermo-mechanochemical Field: A Fluid 121
3.6 Mixture Theory for a Multi-component Solution 122
3.6.1 Mass Conservation Law 124
3.6.2 Conservations of Linear Momentum and Moment of Momentum 128
3.6.3 First Law of Thermodynamics in a Thermo-mechanochemical Continuum: Conservation of Energy 130
3.6.4 Entropy Inequality in a Thermo-mechanochemical Continuum 132
3.7 Thermodynamics Laws and Constitutive Theory ♣ 133
3.7.1 Constitutive Theory of a Solid with Chemical Processes in the Small Strain Field 134
3.8 Summary of the Framework of Non-equilibrium Thermodynamics ♣ 135

4 Virtual Work Equation, Variational Methods and Energy Principles 139
4.1 Variational Method for a One-dimensional Elastic Problem 139
4.1.1 Strong Form 139
4.1.2 Weak Form and the Virtual Work Equation 140
4.1.3 Principle of the Energy Minimization and a Variational Method 141

4.2 Variational Method for a Three-dimensional Elasticity Problem ... 143
4.2.1 Strong Form ... 143
4.2.2 Weak Form and the Virtual Work Equation ... 144
4.2.3 Principle of the Energy Minimization and a Variational Method ... 145
4.3 The Penalty Method and the Lagrangian Multiplier Method ... 145
4.4 Convolution Integral and Energy Forms for Parabolic and Hyperbolic PDEs ... 147
4.4.1 Energy Form of a Parabolic PDE ... 149
4.4.2 Energy Form of a Hyperbolic PDE ... 150
4.5 Interpolation, Approximation and Galerkin Method ... 151

5 Classical Theory of Diffusion and Seepage Problems in Porous Media ... 157
5.1 Representative Elementary Volume and Averaging ... 157
5.2 Diffusion and Seepage Problem for a Multi-component Solution in Saturated Porous Media ... 159
5.2.1 Mass Conservation for the Fluid Phase ... 160
5.2.2 Mass Conservation in the Solid Phase ... 163
5.2.3 Seepage Theory for the Incompressible Fluid ... 166
5.2.4 Seepage Theory for the Compressible Fluid ... 169
5.3 Navier-Stokes Equation and the Classical Permeability Theory ... 169
5.3.1 Pipe Flow: Hagen-Poiseulle Flow ... 170
5.3.2 Flow Through an Assembly of Pipes and Its 'Permeability' ... 172
5.3.3 Flow in a Tank Filled with Solid Particles and Its Permeability ... 173
5.4 Fick's Law and Evaluation of the Diffusion Coefficient ... 174
5.5 Adsorption Isotherm and the Distribution Coefficient ... 177
5.5.1 Langmuir's Equilibrium Adsorption Isotherm ... 178
5.5.2 Freundlich's Equilibrium Adsorption Isotherm ... 181
5.5.3 Temkin's Equilibrium Adsorption Isotherm ... 181
5.5.4 Langmuir's Nonequilibrium Adsorption Isotherm ... 182
5.5.5 Freundlich's Nonequilibrium Adsorption Isotherm ... 182
5.6 Transport Equations and Similitude Laws ... 182

6 Classical Theory of Consolidation for Saturated Porous Media ... 185
6.1 Mass Conservation Law and Seepage Equation ... 185
6.2 Conservation of Linear Momentum, Effective Stress and Biot's Consolidation Theory ... 186
6.3 Finite Strain Theory of Consolidation ♣ ... 187
6.3.1 Seepage Equation of Consolidation in a Lagrangian Form ... 188
6.3.2 Lagrangian Equation of Equilibrium ... 189
6.3.3 Incremental Form of the Equation of Equilibrium ... 190

6.4 A Weak Form of Biot's Consolidation Equations and Finite Element Analysis 191
6.4.1 Strong Form 191
6.4.2 Weak Form 192
6.4.3 Finite Element Analysis 193
6.5 The Cam Clay Model 196
6.5.1 Normally Consolidated Clay 196
6.5.2 Over-consolidated Clay 197
6.5.3 The Original Cam Clay Model 198
6.5.4 Modified Cam Clay Model 204
6.5.5 Elasto-plastic Constitutive Law 205

7 Introduction to Homogenization Analysis 207
7.1 One-dimensional Problem of an Elastic Bar 207
7.2 Micro/Macro Coordinates 207
7.3 Microscale and Macroscale Problems 208

8 Homogenization Analysis and Permeability of Porous Media 213
8.1 Micro-inhomogeneous Porous Media and Stokes' Equation 214
8.2 Seepage Theory for Two-scale Porous Media 214
8.2.1 Homogenization Analysis and Seepage Problem of Porous Media 215
8.2.2 Analytical Solution for a Microscale Poiseulle Flow 219
8.2.3 Finite Element Analysis for the Mass Conservation Equation in the Micro-domain 220
8.2.4 Numerical Results of Seepage Analysis for Pure Smectitic Clay 223
8.2.5 Three-dimensional Seepage Analysis of Sand 227
8.3 A Permeability Theory for Multiscale Porous Media 230
8.3.1 Multiscale Permeability Theory 231
8.3.2 Seepage Analysis of Bentonite 238

9 Homogenization Analysis of Diffusion in Porous Media 241
9.1 Micro-inhomogeneous Porous Media and Diffusion Problems 241
9.2 Diffusion Theory for Two-scale Porous Media 244
9.2.1 HA for Diffusion Problems in Porous Media 244
9.2.2 Simulation of a Through-diffusion Test 248
9.2.3 HA Diffusion Equation with Higher Order Derivatives 250
9.3 Diffusion Problem for Multiscale Porous Media 253
9.3.1 Multiscale HA for Diffusion Problems in Porous Media 253
9.4 Diffusivity of Compacted Bentonite Considering Micro-structure 257
9.4.1 Experimental Data of Bentonite Diffusivity 257
9.4.2 Microscale Problem of HA for Bentonite 259
9.5 HA and Similitude for the Seepage/Diffusion Problem 264

10 **Long-Term Consolidation of Bentonite and a Homogenization Analysis of the Flow Field** 267
10.1 Long-term Consolidation Test on Bentonite 268
10.2 MD/HA Seepage Analysis and 1D Consolidation Theory 270
10.2.1 1D Finite Strain Consolidation Theory 271
10.2.2 Weak Form of the 1D Finite Strain Consolidation Equation 274
10.2.3 FEM for 1D Finite Strain Consolidation Equation 274
10.2.4 Relation Between Permeability and Void Ratio for Compacted Bentonite 276
10.2.5 Consolidation Experiment and Inhomogeneous 1D Analysis 279

A **Introduction to Vectors** 283
A.1 Vectors in $\mathbb{R}^3$ 283
A.2 Inner Product and the Length of Vectors 284
A.3 Coordinate Transformation 285
A.4 Outer Product 287

B **Partial Differentiation and Integral Theorems** 289
B.1 Calculus of Partial Differentials 289
B.2 Gauss-Green Theorem 291
B.3 Stokes' Theorem and Exact Differentiability 293

C **A Summary of Linear Vector Spaces** 299
C.1 Algebraic System 299
C.2 A Linear Vector Space $\mathcal{V} = \{X, +, *\}$ 300
C.3 Basis and Dimension 301
C.4 Topological Spaces 302
C.5 Cauchy Sequence and Complete Space 303
C.6 Functionals and Dual Space 303
C.7 Tensor Product and Tensor Space 304
C.8 On Completeness of Function Spaces 306
C.8.1 Sequence Space l^p 306
C.8.2 Completeness of l^p 309
C.8.3 Completeness of C$[a, b]$ Under the Norm $\| x \|_\infty$ 310
C.8.4 Bessel's Inequality and Parseval's Equality 311
C.8.5 Completeness of Fourier Basis $\{\phi_n\}$ 313

D **Classical and Chemical Thermodynamics** 317
D.1 Energy Conservation Law and Thermodynamic Systems 317
D.2 Existence of Entropy and Thermal Energy: First Part of the Second Law of Thermodynamics 318
D.2.1 Carnot Cycle 319
D.2.2 Entropy as a State Variable 322

D.3 Interpretation of Entropy within the Framework of Statistical Thermodynamics: Boltzmann's Theory ♣ 323
D.3.1 Statistical Representation of Molecular States and Boltzmann Distribution 324
D.3.2 Molecular Partition Function q 325
D.3.3 Internal Energy and Entropy: Boltzmann's Formula 327
D.4 Entropy Production: Second Part of the Second Law of Thermodynamics 331
D.4.1 Second Law of Thermodynamics for the Irreversible Process 331
D.4.2 Entropy Production 332
D.4.3 Second Law of Thermodynamics and the Minimum Energy Principle 336
D.4.4 Second Law of Thermodynamics in a Thermo-mechanical Continuum: Eulerian Description 336

E Chemical Processes and Classical Thermodynamics 339
E.1 Molar Description of Thermodynamic Functions 339
E.2 Heat of Reaction and Change of Enthalpy 342
E.3 Change of Entropy 345
E.3.1 Progress of Chemical Reaction and Change of Entropy 345
E.3.2 Affinity and Direction of Reaction 346
E.3.3 Change of Entropy and the Diffusion Process 348
E.3.4 Changes of Temperature/Pressure/Phase and Entropy 348
E.4 Change of Gibbs Free Energy and Chemical Potential 350
E.5 Thermodynamics of Gas 351
E.6 Diffusion Behavior of Solutions without Inter-molecular Interaction 353
E.6.1 Chemical Potential 353
E.6.2 Diffusion Coefficient in an Ideal Solution: Stokes–Einstein Equation 354
E.6.3 Diffusion Coefficient of Solute without Inter-molecular Interaction 355
E.7 Diffusion Process in an Electrolyte Solution 356
E.7.1 Chemical Potential of Electrolyte Solute 356
E.7.2 Electrostatic Field due to Distributed Ions and Activity Coefficient: Debye–Hückel Theory 358
E.7.3 Diffusion of Ionic Species in an Electrolyte Solution 361
E.7.4 Electric Conduction in an Electrolyte Solution 362
E.8 Chemical Equilibrium and the Equilibrium Constant 364
E.9 Phenomenological Theory of Non-equilibrium Chemical Reaction Processes 366
E.9.1 Reaction Velocity and Order of Reaction 366
E.9.2 Elementary Reaction and Complex Reaction 367

E.9.3 Temperature Dependence of Reaction Rate: Arrhenius Equation 370
E.9.4 Transition State Theory 371

References 373

Index 379

Preface

Clay is a commonly occurring porous material that is frequently used in engineering endeavors, but there are also other important porous materials, such as those used in some types of electrodes and as catalysts. Porous media should be examined giving due regard to both the physics and chemistry of the constituents. The purpose of this book is to present an approach that examines porous media from both these aspects, outlining a procedure that combines microscale (or even nanoscale) characteristics with macroscale behavior.

A fundamental outline of mechanics (or physics) and chemistry, including thermodynamics, is provided in Chap. 2, *Introduction to Continuum Mechanics*; Chap. 3, *Non-equilibrium Thermodynamics* and Chap. 4, *Virtual Work Equation, Variational Methods and Energy Principles*.

In Chap. 5, *Classical Theory of Diffusion and Seepage Problems in Porous Media* and Chap. 6, *Classical Theory of Consolidation for Saturated Porous Media* we review and re-organize the classical soil mechanics in terms of modern continuum mechanics; looking at the mass conservation law for a multi-component solution, we show that the diffusion field is strictly connected with the seepage field.

The central results for saturated porous media are given in the following chapters: In Chap. 7, *Introduction to Homogenization Analysis*, a one-dimensional (1D) elastic problem is used and the fundamental notion of the homogenization analysis (HA) is outlined, giving a unified procedure for treating a micro-inhomogeneous material, not only in the micro-domain but also in the macro-domain. In Chap. 8, *Homogenization Analysis and Permeability of Porous Media*, we use Stokes' equation, and then apply the HA scheme to obtain the seepage equation together with the HA-based Darcy's law. Using this procedure we can obtain the true velocity and pressure fields in the micro-domain. A relationship between the HA seepage theory and the conventional theory is discussed. Note that the distribution of water viscosity in the interlayer space between clay minerals is obtained using a molecular dynamics (MD) simulation. A diffusion problem in bentonite, including adsorption at the clay edges is discussed in Chap. 9, *Homogenization Analysis for Diffusion Problem in Porous Media*. The distribution of the diffusion coefficient in the micro-domain (i.e., the interlayer space between clay minerals) is calculated using MD, and the macroscale diffusivity obtained by HA is significantly similar to that obtained experimentally. The purpose of Chap. 10, *Long-term Consolidation of Bentonite and Homogenization Analysis of the Flow Field*, is to analyze the problem of secondary consolidation of bentonite, which is frequently observed in long-term experiments. This phenomenon was considered to be the result of time-dependent deformation such as creep of the solid skeleton of bentonite. However, we show that the secondary consolidation is caused by a non-homogeneous distribution of permeability, which occurs as a result of a change in the crystalline structure during the consolidation process. This result was verified by a series of X-ray diffraction analyses, that continuously measured the crystalline structure at each point of a bentonite specimen.

In the Appendices we outline the essential mathematics and classical thermodynamics, including some chemistry. The section on thermodynamics is referred to in Chap. 3.

We acknowledge our sincere gratitude to our wives, Maria and Sally. We are indeed indebted to their dedication, collaboration and encouragement. We should like also to thank Professor Katsuyuki Kawamura of Okayama University: The MD results shown in the last three chapters are due to his work and excellent collaboration. The first author expresses his deep gratitude to Professor Bogdan Raniecki of the Polish Academy of Sciences, who introduced him to the greater realm of mechanics. He is grateful to Dr. Kazumi Kitayama (Nuclear Waste Management Organization of Japan; NUMO), Professor Masashi Nakano (Professor Emeritus of University of Tokyo), Tokyo Electric Power Company (TEPCO), Japan Atomic Energy Agency (JAEA) and NUMO: They have supported his research over many years. Among others, he thanks Prof. Shunsuke Baba (Okayama University), Prof. Kazushi Kimoto (Okayama University), Drs. Hiroyuki Umeki (JAEA) and Kozo Sugihara (JAEA). The works in the last chapters were obtained by a collaboration with his doctral students. He is grateful to Dr. Naoki Fujii (Radioactive Waste Management Funding and Research Center; RWMC), Professors Jian-Guo Wang (University of Western Australia), Gyo-Cheol Jeong (Andong National University, Korea), Yong-Seok Seo (Chungbuk National University, Korea), Drs. Byung-Gon Chae (Korea Institute of Geoscience and Mineral Resources; KIGAM) and Jung-Hae Choi (KIGAM). The second author wishes to thank the Max Plank Gesellschaft for providing a Fellowship that enabled the initiation of this work and to McGill University for research support in the form of the James McGill Professorship. The research support provided by the Natural Sciences and Engineering Research Council of Canada in the form of a Discovery Grant is also acknowledged. Finally, the hospitality of Okayama University during his many visits to the Faculty of Graduate School of Environmental Science is gratefully acknowledged. The authors are indebted to Mrs. Sally Selvadurai, Assist-Ed Editorial Services, Montréal, QC, Canada for her expert editing of the final manuscript.

List of Notations

$\Rightarrow$	'imply/implies'	
$\forall$	'for all'	
$\exists$	'there exist(s)'	
$\in$	'belong(s) to'	
FEM	'finite element method'	
HA	'homogenization analysis'	
iff	'if and only if'	
l.h.s.	'left-hand side'	
MD	'molecular dynamics'	
PDE	Partial differential equation	
r.h.s.	'right-hand side'	
s.t.	'such that'	
w.r.t.	'with respect to'	
$!\alpha$	Summation convention is exempted for α	9
$\boldsymbol{A} : \boldsymbol{B}$	Inner product of tensors $\boldsymbol{A}$, $\boldsymbol{B}$	48
$[\![\cdot]\!]$	Jump	28
A	Affinity	347
$A_r = -\sum_\alpha c_\alpha \nu_{\alpha r}^\dagger \mu_\alpha$	Affinity at reaction stage r	132
$\boldsymbol{a}$	Acceleration	14
a_α	Activity of species α	354
$\boldsymbol{B} = \boldsymbol{F}\boldsymbol{F}^T$	Left Cauchy-Green tensor	20
$\boldsymbol{b}$	Body force per unit mass	30
$\boldsymbol{b}^*$	Diffusive body force	130
$\boldsymbol{C} = \boldsymbol{F}^T\boldsymbol{F}$	Right Cauchy-Green tensor	20
c	Mass-molar concentration	111
c_α	Mass concentration of species α	112
$d\mathcal{C}$	Mass-energy flux of the system	131, 317

C_p	Heat capacity under constant pressure	104, 349
C_v	Heat capacity under constant volume	104, 349
$d\mathcal{C}$	Mass energy flux of system	317
$D_{\alpha\beta}$	Diffusion coefficient	166, 175
D_a	Apparent diffusivity	257
D_e	Effective diffusivity	257
$D^e_{\alpha\beta}$	Effective diffusivity for species α and β	175
$\boldsymbol{D}$	Stretch tensor	23
$d\mathcal{D} = T\, d\mathcal{S} - d\mathcal{Q}$	Internal dissipation	334
$\mathfrak{D}_Q(\cdot)/\mathfrak{D}t$	Corotational derivative due to $\boldsymbol{Q}$	43
$\mathfrak{D}(\cdot)/\mathfrak{D}t$	Zaremba-Jaumann rate	44
$\mathfrak{D}_E(\cdot)/\mathfrak{D}t = \overset{\nabla}{(\cdot)}$	Corotational derivatiove due to $\boldsymbol{\Omega}^E$	46
$\mathfrak{D}_R(\cdot)/\mathfrak{D}t = \overset{\circ}{(\cdot)}$	Corotational derivatiove due to $\boldsymbol{\omega}^R$	46
$e = n/(1-n)$	Void ratio (porous media)	71, 196
de	Increment of void ratio (porous media)	200
de^e	Elastic increment of void ratio (porous media)	200
de^p	Plastic increment of void ratio (porous media)	71, 200
E	Elastic constant	51
$\boldsymbol{E} = (\boldsymbol{C} - \boldsymbol{I})/2$	Lagrangian or Green strain	20
$\mathfrak{E} = \sum_i f(\lambda_i)\, \boldsymbol{N}_i \otimes \boldsymbol{N}_i$	Generalized Lagrangian strain measure	22
$\boldsymbol{E}(n)$	n-th order family of Lagrangian strain measure	22
$\boldsymbol{e} = (\boldsymbol{i} - \boldsymbol{B}^{-1})/2$	Eulerian or Almansi strain	20
$\mathfrak{e} = \boldsymbol{R}\, \mathfrak{E}\, \boldsymbol{R}^T$	Generalized Eulerian strain measure	22
$\boldsymbol{e}(n) = \boldsymbol{R}\, \boldsymbol{E}(n)\, \boldsymbol{R}^T$	n-th order family of Eulerian strain measure	22
$d\boldsymbol{e}^p$	Deviatoric plastic strain increment tensor	68
de^p	Norm of deviatoric plastic strain increment	68
$\mathcal{E}$	Total energy of the system	317
$E(t)$	Relaxation function (viscoelasticity)	58
$\{\boldsymbol{e}_i\}$	Basis for the current configuration	13
$\{\boldsymbol{E}_I\}$	Basis for the reference configuration	13
$f(\boldsymbol{\sigma}, \kappa)$	Yield function	68
$f = \gamma P$	Fugacity	353
$F(V, T, n_\alpha)$	Helmholtz free energy: molar description	340
$f(\boldsymbol{\varepsilon}^e, T)$	Helmholtz free energy: small strain	102
$f(\boldsymbol{\varepsilon}^e, T, c_\alpha)$	Helmholtz free energy: small strain	113
$f(\boldsymbol{E}^e, \mathsf{T})$	Helmholtz free energy: Lagrangian form	106
$f(\mathfrak{e}^e, T)$	Helmholtz free energy: Eulerian form	107
$f(\boldsymbol{E}^e, \mathsf{T}, \mathsf{c}_\alpha)$	Helmholtz free energy: Lagrangian form	119
$\boldsymbol{F} = \mathrm{Grad}\, \boldsymbol{x}$	Deformation gradient	19
G	Shear mudulus	51

g	Gravity acceleration	31
$G(P,T,n_\alpha)$	Gibbs free energy: molar description	340, 350
$g(\boldsymbol{\sigma}^\ddagger,T)$	Gibbs free energy: small strain	102
$g(\boldsymbol{\sigma}^\ddagger,T,c_\alpha)$	Gibbs free energy: small strain	114
$g(\boldsymbol{T}^\ddagger,\mathsf{T})$	Gibbs free energy: Lagrangian form	106
$g(\boldsymbol{T}^\ddagger,\mathsf{T},\mathsf{c}_\alpha)$	Gibbs free energy: Lagrangian form	119
$\Delta G^\ominus$	Change of standard Gibbs free energy	350
h	Hardening parameter	73
$H(P,S^e,n_\alpha)$	Enthalpy: molar description	340
$H_{m\alpha}$	Partial molar enthalpy: molar description	351
$h(\boldsymbol{\sigma}^\ddagger,s)$	Enthalpy: small strain	102
$h(\boldsymbol{\sigma}^\ddagger,s,c_\alpha)$	Enthalpy: small strain	114
$h(\boldsymbol{T}^\ddagger,\mathsf{s})$	Enthalpy: Lagrangian form	106
$h(\boldsymbol{T}^\ddagger,\mathsf{s},\mathsf{c}_\alpha)$	Enthalpy: Lagrangian form	119
$\Delta H^\ominus$	Change of standard enthalpy	343
$H(\tau)$	Relaxation spectrum (viscoelasticity)	59
$\boldsymbol{i}=\delta_{ij}\,\boldsymbol{e}_i\otimes\boldsymbol{e}_j$	Unit tensor in Ω	20
$\boldsymbol{I}=\delta_{IJ}\,\boldsymbol{E}_I\otimes\boldsymbol{E}_J$	Unit tensor in Ω_0	20
$I_1^\sigma=\mathrm{tr}\,(\boldsymbol{\sigma})=I_1^\sigma/\sqrt{3}$	First invariant of the stress	50
$J=\det\boldsymbol{F}=\rho_0/\rho$	Jacobian	25
$J(t)$	Creep function (viscoelasticity)	61
$J_2^\sigma=\boldsymbol{\sigma}':\boldsymbol{\sigma}'/2$	Second invariant of the deviatoric stress	50
$J_3^\sigma=\det(\boldsymbol{\sigma}')$	Third invariant of the deviatoric stress	50
K	Bulk modulus	51
$\mathcal{K}$	Kinematical energy of the system	83, 317
$\boldsymbol{k}$	Thermal conductivity tensor	335
k	Thermal conductivity	335
$\boldsymbol{k}$	Hydraulic conductivity tensor	167, 188
$k_B=R/N_A$ $=1.380658\times10^{-23}\mathrm{J\,K^{-1}}$	Boltzmann constant	324, 351
K_α^d	Distribution coefficient	165
$\boldsymbol{L}=\dot{\boldsymbol{F}}\boldsymbol{F}^{-1}=\boldsymbol{D}+\boldsymbol{W}$	Velocity gradient	23
$L(\tau)$	Retardation spectrum (viscoelasticity)	61
$\mathcal{M}$	Mass of the system	29
m	Molar averaged molecular weight	112
m_α	Molecular weight of species α	111
$n=e/(1+e)$	Porosity (porous media)	157
$\boldsymbol{n}$	Unit outward normal in Ω	26
$\boldsymbol{N}$	Unit outward normal in Ω_0	26
n^*	Total molar amount of substance of system	111
n_α	Molar amount of α-th component	111

$\boldsymbol{n}_i$	Eulerian triads	22
$\boldsymbol{N}_i$	Lagrangian triads	21
$N_A = 6.0221367 \times 10^{23}$	Avogadro's number	351
p	Pore pressure (porous media)	186
$p^k(\boldsymbol{x}^1)$	Characteristic function of pressure	217
$P_\alpha = x_\alpha P$	Partial pressure	353
q	Molecular partition function	325
$\boldsymbol{Q}$	Coordinate transformation tensor	16
$\mathcal{Q}$	Heat supply of the system	88
$d\mathcal{Q}$	Heat supply rate of the system: heat flux	80, 317
$\boldsymbol{q}$	Heat flux	80, 88
$\boldsymbol{q}_\alpha$	Heat flux of species α	131
$\boldsymbol{q}^m$	Total mass flux	125
$\boldsymbol{q}_\alpha^m$	Mass flux of species α	126, 135, 165
$R = N_A k_B$	Gas constant	351
$\boldsymbol{R}$	Rotation tensor	20
$\mathbb{R}^n$	n-dimensional real number space	299
R_α^d	Retardation factor	166, 181
r	Heat of reaction	346
r	Heat source per unit mass	89
r_α	Heat source for species α	131
$\mathcal{S}$	Entropy of system	322
$\mathcal{S} = k_B \ln W$	Entropy of system: Boltzmann's principle	324
$d\mathcal{S}$	Entropy change of the system	322
$d\mathcal{S}^e$	Entropy change exchangeable with surrounding	332
$d\mathcal{S}^i$	Entropy production in the system	332
S	Entropy: molar description	345
ΔS	Entropy change: molar description	345
ΔS^e	Entropy change exchangeable with surrounding: molar description	339, 345
ΔS^i	Entropy production in the system: molar description	345
$(\Delta S^e)^\ominus$	Standard entropy change: molar description	350
s	Entropy: per unit mass	336
ds	Entropy change: per unit mass	92, 336
ds^e	Entropy change exchangeable with surrounding: per unit mass	336
ds^i	Entropy production: per unit mass	336
$\boldsymbol{s}$	Deviatoric stress (porous media)	68
$\boldsymbol{S} = \boldsymbol{\Pi}^T$	Nominal stress	34
T	Temperature	102

$\boldsymbol{t}$	Traction in Ω	30,31
$\boldsymbol{T} = J\boldsymbol{F}^{-1}\boldsymbol{\sigma}\boldsymbol{F}^{-T}$	Second Piola-Kirchhoff stress	34
$\boldsymbol{T}^{\ddagger} = \boldsymbol{T}/\rho_0$	Specific Second Piola-Kirchhoff stress	39
$\boldsymbol{T}^{\ddagger}(n)$	Stress work-conjugate with $\boldsymbol{E}(n)$	87
$\boldsymbol{t}^{E\ddagger} = \boldsymbol{R}\boldsymbol{T}^{\ddagger}\boldsymbol{R}^T$	Corotational Eulerian stress: per unit mass	87
$\mathcal{U}$	Internal energy of the system	88, 317
$\mathcal{U}_\sigma$	Internal work due to stress	84
$d\mathcal{U}$	Increment of internal energy	88
$U(V, S^e, n_\alpha)$	Internal energy: molar description	340
u	Internal energy per unit mass	89
u_σ	Internal energy per unit mass due to stress	84
$u(\boldsymbol{\varepsilon}^e, s, c_\alpha)$	Internal energy per unit mass: small strain	113
$u(\boldsymbol{E}^e, \mathsf{s})$	Internal energy per unit mass: Lagrangian form	105
$u(\boldsymbol{E}^e, \mathsf{s}, \mathsf{c}_\alpha)$	Internal energy per unit mass including chemical effect: Lagrangian form	118
$\boldsymbol{U}$	Right stretch tensor	20
$\boldsymbol{v}$	Velocity of the material point	12
$\boldsymbol{v}$	Mass-average velocity for mixture	125, 161
$\boldsymbol{v}_\alpha$	Particle velocity of species α in fluid	124, 159
$\overline{\boldsymbol{v}}_\alpha = \boldsymbol{v}_\alpha - \boldsymbol{v}$	diffusion velocity	125
$\boldsymbol{v}^*$	Mass-average velocity of solid part	164
$\boldsymbol{v}^*_\alpha$	Particle velocity of species α in solid	159
$\tilde{\boldsymbol{v}} = n(\boldsymbol{v} - \boldsymbol{v}^*)$	Seepage velocity	167
$\boldsymbol{V}$	Right stretch tensor	20
$\boldsymbol{W}$	Spin tensor	23
W	Thermodynamic weight	324
$\mathcal{W}$	Work for the system	83
$d\mathcal{W}$	Power input to the system	83, 317
x_α	Molar fraction of species α	112
α	Thermal expansion coefficient	52
$\delta^c(\cdot)/\delta t,\ \delta_c(\cdot)/\delta t$	Convected derivative	42
$\boldsymbol{\varepsilon}$	Small strain	21
$\bar{\boldsymbol{\varepsilon}} = I_1^\varepsilon \boldsymbol{I}^\varepsilon/3$	Volumetric strain	50
$\bar{\varepsilon} = I_1^\varepsilon/\sqrt{3}$	Magnitude of volumetric strain	50
$\boldsymbol{\varepsilon}' = \boldsymbol{\varepsilon} - \bar{\boldsymbol{\varepsilon}}$	Deviatoric strain	50
$\varepsilon_\theta = \frac{1}{3}\cos^{-1}\{3\sqrt{3}J_3^\varepsilon/2(J_2^\varepsilon)^{3/2}\}$	Strain Lode angle	51
$d\boldsymbol{\varepsilon}^e$	Elastic strain increment	2.10
$d\boldsymbol{\varepsilon}^p$	Plastic strain increment	2.10

$d\bar{\varepsilon}^p = \mathrm{tr}\,(d\boldsymbol{\varepsilon}^p)/\sqrt{3}$	Magnitude of volumetric plastic strain increment	2.10
η	Viscous constant (viscoelasticity)	60
γ_α	Mass source per unit mass	160
$\kappa(\boldsymbol{x};t) = \boldsymbol{v}\cdot\boldsymbol{v}/2$	Kinetic energy per unit mass: Eulerian form	84
κ	Hardening parameter	68
$\lambda,\ \mu$	Lamé's constant for solid	51
$\lambda,\ \mu$	Viscosities for fluid	54
$\lambda^\ddagger = \lambda/\rho,\ \mu^\ddagger = \mu/\rho$	Kinematic viscosities for fluid	108
λ_i	Principal stretch	21
$\lambda,\ \kappa$	Compression and swelling indexes	200
$\mu(t)$	Response function (viscoelasticity)	58
μ	Chemical potential per unit mass of total system	111
μ_α	Chemical potential per unit mass of species α	112
ν	Poisson's ratio	51
ν_α	Stoichiometry coefficient	345
$\boldsymbol{\omega}$	Rotation vector	18
ω	Sum of volume molar fraction	111
ω_α	Volume molar fraction of species α (vol. concentration)	111
$\boldsymbol{\omega}^R = \dot{\boldsymbol{R}}\boldsymbol{R}^T$	Eulerian spin	45
$\boldsymbol{\omega}^{RR} = \boldsymbol{R}^T\boldsymbol{\omega}^R\boldsymbol{R}$	Lagrangian spin	45
$\boldsymbol{\Omega}^E = \dot{\boldsymbol{n}}_i \otimes \boldsymbol{n}_i$	Spin of Eulerian triads	44
$\boldsymbol{\Omega}^L = \dot{\boldsymbol{N}}_i \otimes \boldsymbol{N}_i$	Spin of Lagrangian triads	45
$\phi(t)$	After-effect function (viscoelasticity)	57
ϕ	Total potential (total head)	167, 186, 219
$\Phi(V, T, \mu_\alpha)$	Grand potential: molar description	340
$\phi(\boldsymbol{\varepsilon}, T, \mu_\alpha^\ddagger)$	Grand potential: small strain	114
$\phi(\boldsymbol{E}^e, \mathrm{T}, \mu_\alpha^*)$	Grand potential: Lagrangian form	120
$\boldsymbol{\Pi} = J\boldsymbol{F}^{-1}\boldsymbol{\sigma}$	First Piola-Kirchhoff stress	33
$\boldsymbol{\Pi}^\ddagger = \boldsymbol{\Pi}/\rho$	Specific first Piola-Kirchhoff stress	39
$\overset{\circ}{\boldsymbol{\Pi}} = J^{-1}\boldsymbol{F}\dot{\boldsymbol{\Pi}}$	Nominal stress rate	37
$\overset{\circ}{\boldsymbol{\Pi}}{}^\ddagger = \boldsymbol{F}\dot{\boldsymbol{\Pi}}^\ddagger$	Specific nominal stress rate	40
ρ	Mass density in Ω	29

ρ	Total mass density of multi-component fluid	112, 161
ρ^*	Total mass density of multi-component solid	164
ρ_0	Mass density in Ω_0	29
ρ_α	Mass density of species α	111
$\boldsymbol{\sigma}$	Cauchy stress	32
$\boldsymbol{\sigma}^\ddagger = \boldsymbol{\sigma}/\rho$	Kirchhoff energy stress	78
$\bar{\boldsymbol{\sigma}} = I_1^\sigma \boldsymbol{I}/3$	Volumetric or mean stress	50
$\bar{\sigma} = I_1^\sigma/\sqrt{3}$	Magnitude of the volumetric stress	50
$\boldsymbol{\sigma}' = \boldsymbol{\sigma} - \bar{\boldsymbol{\sigma}}$	Deviatoric stress	50
$\sigma_\theta = \frac{1}{3}\cos^{-1}\{3\sqrt{3}J_3^\sigma/2(J_2^\sigma)^{3/2}\}$	Stress Lode angle	50
$\boldsymbol{\sigma}^*$	Diffusive stress	130
$\boldsymbol{\sigma}' = \boldsymbol{\sigma} + p\boldsymbol{i}$	Effective stress (porous media)	187
$\tau = \eta/E$	Relaxation time of Maxwell model	60
$\tau = \eta J$	Relaxation time of Voigt model	62
τ	Tortuosity	175, 259
$\boldsymbol{\tau} = \boldsymbol{R}\boldsymbol{T}\boldsymbol{R}^T$	Euler stress	34
$\boldsymbol{\tau}^\ddagger = \boldsymbol{R}\boldsymbol{T}^\ddagger\boldsymbol{R}^T$	Specific Euler stress	39
θ	Adsorption ratio	178
ξ	Reaction extent	346
ζ	Height potential	167, 185

Chapter 1
Introduction

1.1 Transport Phenomena in Porous Media and Modeling

Porous materials can be both natural and artificial. Such materials are encountered in a wide range of problems dealing with engineering, agriculture and science: Natural materials include soil and rock, and artificial materials can include membrane materials and catalysts used in chemical engineering, etc. We have applied the *principles of continuum mechanics* to analyze the behavior of these materials.

In continuum mechanics we describe the averaged behavior of a large ensemble of atoms/molecules. We replace the set of discrete particles of molecules by a continuum subset of the n-dimensional real number system $\mathbb{R}^n$, and apply physical laws pertaining to conservation of mass, linear momentum, moment of linear momentum and energy. For example, one mole of carbon (^{12}C) corresponding to 12 g consists of 6.0223×10^{23} atoms, thus it is rational to introduce a continuum hypothesis for describing these materials. We derive a system of governing equations by assuming that the rate of change of these conservation processes (e.g., mass, linear momentum, moment of linear momentum and energy) is equivalent to a flux, which flows into the system (e.g., mass flux, force, moment and heat flux) and a source, which is generated in the system.

The material parameters encountered in these governing equations are determined through experiments. However, more recently, molecular simulation-based computational methods have been applied; even for cases with extremely difficult physical and chemical configurations, we can obtain a set of material properties if we have the correct molecular model. For example, we can determine the material properties (at each point, if desired) for a micron-order ($1\,\mu\text{m} = 10^{-6}\,\text{m}$) or nano-order ($1\,\text{nm} = 10^{-9}\,\text{m}$) material, which is referred to as a *micro-inhomogeneous material*. Then we can introduce a mathematical procedure that gives a perspective of the microscale and macroscale characteristics.

The detailed procedure will be discussed in subsequent chapters; however it is important to follow these steps:

Y. Ichikawa and A.P.S. Selvadurai, *Transport Phenomena in Porous Media*,
DOI 10.1007/978-3-642-25333-1_1, © Springer-Verlag Berlin Heidelberg 2012

- Correctly prescribe the conservation laws.
- Represent the conservation law in a mathematically correct manner.
- Introduce accurate experimental or simulation methods in order to determine the material parameters.
- Solve the equations using, e.g., the finite element method, taking into account the convergence and stability of the scheme employed.

1.2 Molecular Dynamics Simulation and Homogenization Analysis

In the framework of continuum mechanics, the material properties have mainly been determined through experiments using specimens of sufficiently large dimensions in comparison to the dimensions of any inherent fabric. This is particularly true for micro-inhomogeneous materials where the specimen is large compared to the size of this local structure. This procedure, which provides the system of governing equations and the experiment-based material properties, is referred to as the *macro-phenomenological scheme*.

If we use the macro-phenomenological approach, we do not include the intrinsic properties, which represent the movement at a molecular-level. In this sense it is a difficult procedure to establish a correct system of governing and constitutive equations. For example, if we consider the experimental results for a micro-inhomogeneous material, we frequently observe differences if the specimen size is changed. The experimental results are only intrinsically true for the size range of that experiment. In this sense, the macro-phenomenological scheme is interpolation-based.

The high-level radioactive wastes (HLW's) produced at nuclear power plants must be isolated from the population for an extremely long time, i.e., more than 10,000 years. There are plans to construct HLW disposal facilities in deep underground repositories. Thus, the safety of the barrier system must be evaluated for time scales in excess of 10,000 years; this is beyond the scope of any conventional scientific experiment-based approach. We need to use an alternative method that is based on a macro-phenomenological approach.

All materials consist of particles, i.e., atoms and/or molecules. It is possible to determine the forces that act on these particles by using the modern scientific techniques of quantum mechanics and chemical-bond models. Molecular simulation methods provide material properties as a set of particle behaviors under the above chemical-bond forces (Allen and Tildesley 1987; Ueda 1990; Kawamura 1990): The *Molecular Dynamics method* (MD; Fig. 1.1) solves the equation of motion directly in a finite difference scheme, using a very short time step, i.e., less than 1 fs (femtosecond; 1 fs = 10^{-15} s). The *Monte Carlo Method* (MC) calculates a probability of occurrence of the particle configuration. Note that since the *Molecular Mechanics Method* (MM) does not treat the behavior of a molecular group, we exclude MM from the molecular simulation methods.

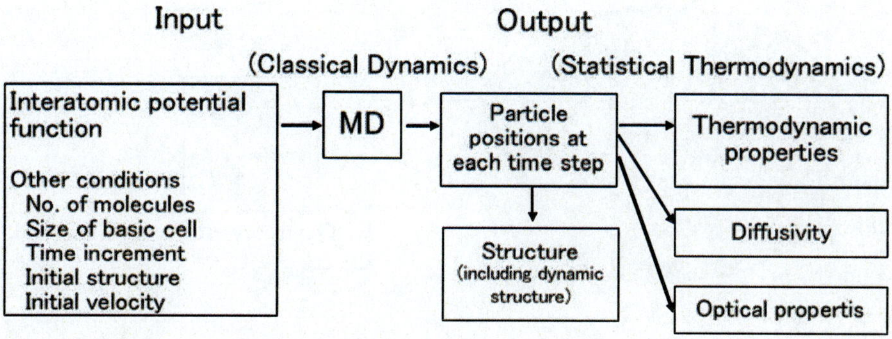

Fig. 1.1 Concept of MD simulation (Ueda, 1990)

Molecular simulations allow scientists to predict future occurrences with some accuracy. However in the foreseeable future it is virtually impossible to consider the sample size of the bentonite buffer that will be used in the HLW repositories using molecular simulation methods. One cc of water has about 0.3×10^{23} molecules and bentonite can absorb more water than the volume of the solid crystals. Even the most-advanced parallel processor computer can only simulate about a hundred million particles. The elapsed time is also limited since we can obtain the results for at most 1 ns ($=10^{-9}$ s). Thus, the use of molecular simulations to forecast events that might 'occur' in, for example, nuclear waste repositories is limited by computer resources.

A mathematical scheme that can treat a micro-inhomogeneous material uniformly at the microscale and the macroscale is referred to as *Homogenization Analysis* (HA) (see Sanchez-Palencia 1980; Bakhvalov & Panasenko 1984). In the HA method, we introduce a perturbation scheme by using both a macroscale coordinate system and a microscale one, and derive a microscale equation, which represents the geometry and material properties in the micro-domain. Then, using the solution of the microscale equation, we determine the macroscale equation (Fig. 1.2). However, since the HA method is implemented within a framework of continuum mechanics, it also experiences difficulties when the material properties of micro-inhomogeneous materials need to be obtained.

We have recently developed a new scheme that combines the MD simulation and the HA method to account for the behavior of bentonite clay, which is a nano-scale micro-inhomogeneous material (Ichikawa et al. 1998; Fig. 1.3).

In the MD/HA method, we obtain a local distribution of material properties using MD simulation, and introduce these material properties into the HA procedure. The MD/HA method overcomes the shortcomings that are inherent in both the MD and HA methods. The characteristic abilities of the MD/HA method are as follows:

1. We can simulate the micro-/macro-behaviors of micro-inhomogeneous materials on the solid basis of physical and chemical laws.

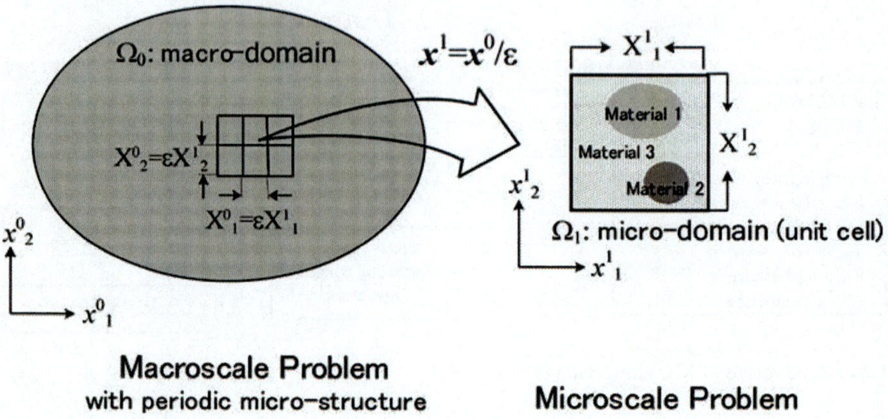

Fig. 1.2 Concept of the homogenization analysis (HA)

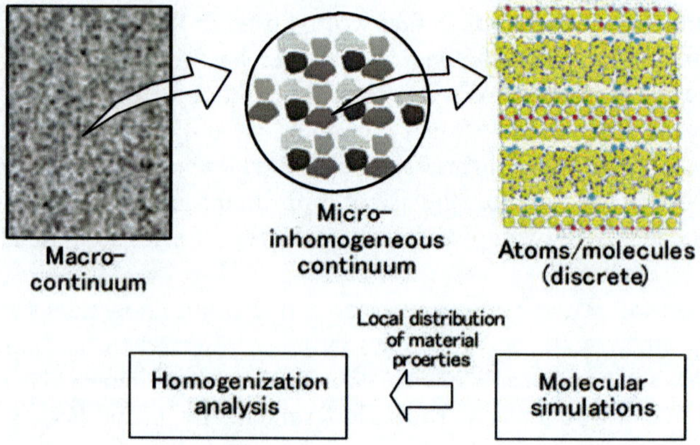

Fig. 1.3 Schematical diagram of the MD/HA method

2. We can predict the material behavior under extreme environments, such as high temperature, high pressure and long elapsed time, which are difficult to assess if we use only a classical macro-phenomenological theory.
3. Using this analytical scheme we may be able to find the true physical/chemical mechanism of complex phenomena, such as water movement in clay under heating.
4. We may be able to find the most appropriate conditions for an engineering process through parametric studies.

We have applied the MD/HA method to bentonite, and have uncovered the mechanisms involved in the extremely low permeability, delayed diffusion and secondary consolidation (Ichikawa et al. 2002, 2004); water molecules close to the clay mineral surface are strongly constrained due to the charged state of the mineral, and the clay minerals themselves form a nano-order of stacked structure.

1.3 Underground Disposal of HLW's and Bentonite

There are plans to dispose of HLW's produced from nuclear power plants in deep underground repositories sited several hundred meters deep (Fig. 1.4). The HLW's

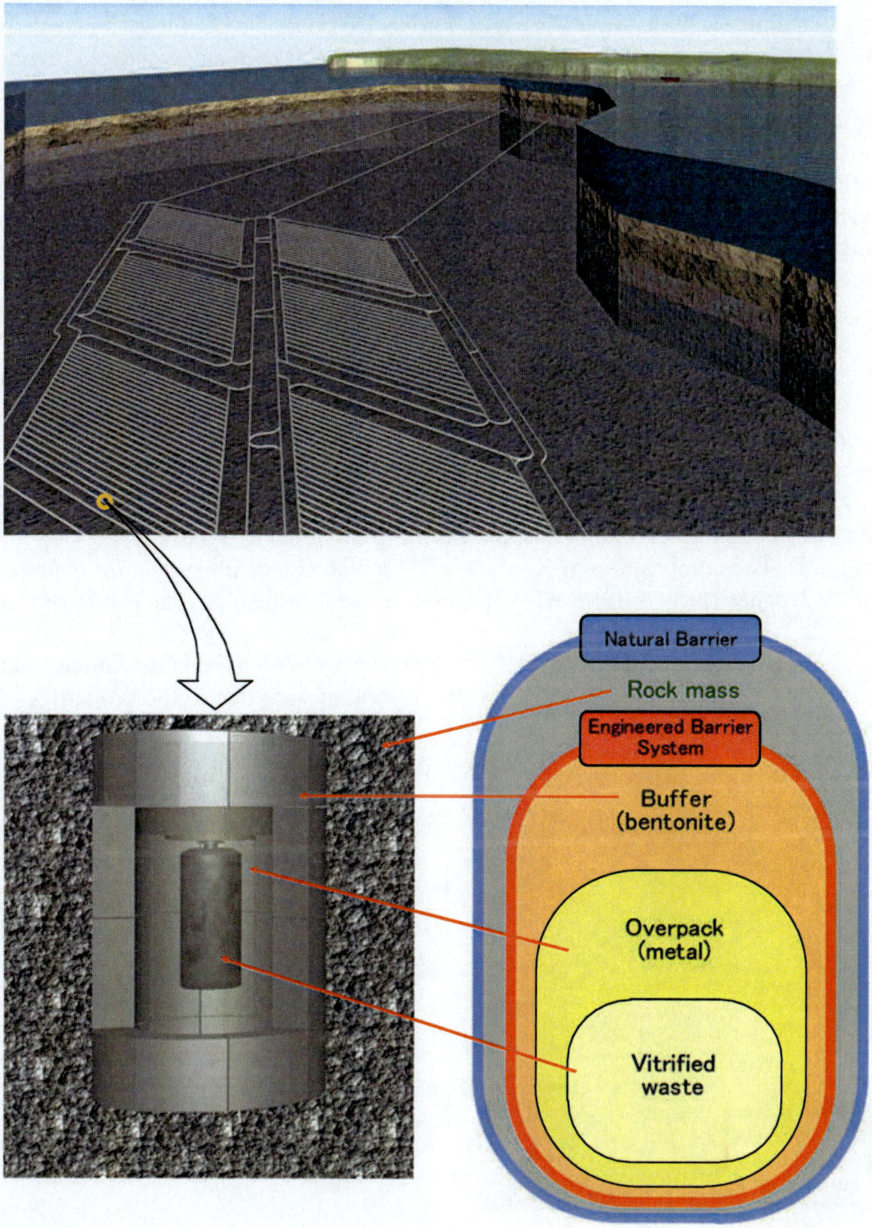

Fig. 1.4 A concept for HLW disposal

are classified into two groups; (1) spent fuel and (2) reprocessed effluent and its vitrified solid (i.e., vitrified waste). The fundamental plan of the Japanese government is to reprocess spent fuel to achieve efficient use of the uranium resources. The vitrified wastes are then kept in a surface facility for about 30 $\sim$ 50 years to allow cooling and the reduction in the radioactivity. Each waste form is then encapsulated in a metal overpack (e.g., steel) and is placed in rock surrounded by a bentonite buffer as shown in Fig. 1.4. Thus the complete barrier system, which prevents radionuclide migration for a long period, forms both an engineered barrier system (EBS, consisting of the vitrified waste, a metal overpack and bentonite buffer) and a natural barrier, i.e., the rock mass.

The radioactivity of HLW's lasts for several thousands years before its radioactivity reduces to the level of natural uranium ore, because it includes long-lived fission products (FP's) and transuranium nuclides (TRU's). The radionuclides released from the stored waste can be transported via ground water flow from the repository. Scientifically, it is difficult to prove the safety of a HLW disposal facility because the long time scale involved is beyond human experience.

Bentonite is the key material of the EBS, which retards the transport of radionuclides. This is because (1) it is an extremely low permeable material, which prevents movement of water and eventually radionuclides, (2) it seals cracks in the rock mass because of its swelling properties, and (3) it shows a high ability for cation adsorption, therefore it can retard the migration of radionuclides.

At the microscale level bentonite is a micro-inhomogeneous material, which consists of smectic clay minerals and macro-grains, mainly quartz, water and air (Fig. 1.5). The composition of Kunigel V1®, which is a candidate buffer material for the Japanese proposals for HLW disposal, and its purified Kunipia F®, is shown in Table 1.1.

The main component of bentonite is smectic clay, such as montmorillonite and beidellite, and the unique properties mentioned above are due to the properties of these smectic clays (Pusch et al. 1990).

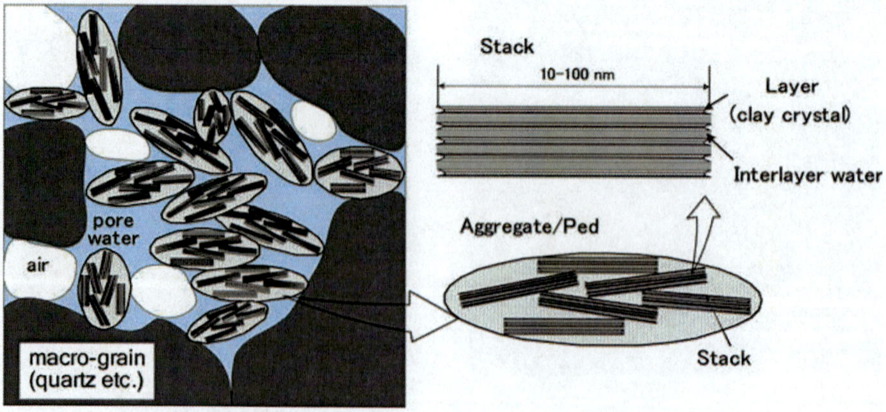

Fig. 1.5 Schematic diagram of micro-structure of bentonite

Table 1.1 Compositions of bentonite (Ito et al. 1993)

	Montmorillonite	Quartz·Chalcedony	Feldspar	Calcite	Dolomite	Analcite	Pyrite	Organics
Kunigel V1®	46 ~ 49	29 ~ 38	2.7 ~ 5.5	2.1 ~ 2.6	2.0 ~ 2.8	3.0 ~ 3.5	0.5 ~ 0.7	0.31 ~ 0.34
Kunipia F®	98 ~ 99	<1	–	<1	–	–	–	–

Unit [wt%]

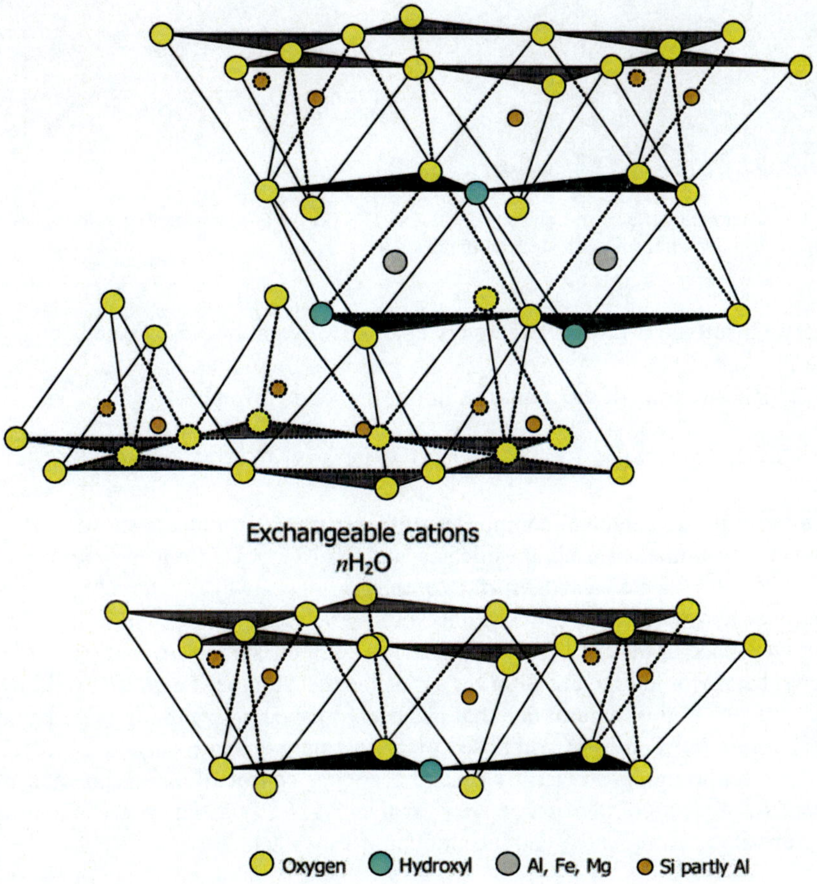

Fig. 1.6 Crystalline structure of montmorillonite

A schematic diagram of the crystalline structure of montmorillonite is shown in Fig. 1.6. The platelet crystal is formed by a sandwich structure consisting of two SiO tetrahedral sheets and one AlO/AlOH octahedral sheet. The Al atom of the octahedral sheet is frequently replaced by other elements, such as Mg, and therefore

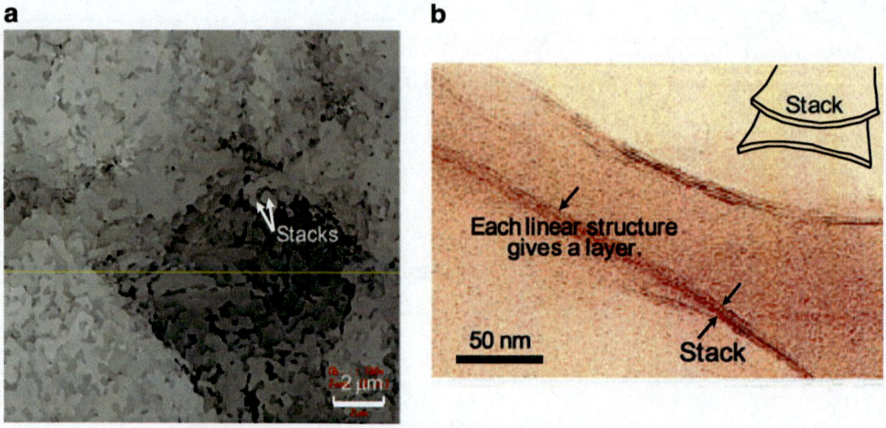

Fig. 1.7 Microphotographs of compacted Kunipia F® (**a**) by a laser scanning confocal microscope, and (**b**) by a transmission electron microscope

the crystal surface is always negatively charged. Several of these crystals combine to form a stack.

An ideal structure of the montmorillonite crystal is given by

$$(\mathrm{Na, K, Ca, Mg})_{0.33}(\mathrm{Al}_{1.67}\mathrm{Mg}_{0.33})\mathrm{Si}_4\mathrm{O}_{10}(\mathrm{OH}_2).$$

In general, the chemical composition of smectic clay minerals shows that they belong to the aqueous aluminum silicates, and the Al atom is frequently replaced by Mg, Fe, Na, Ca, K, etc., and the surface of the crystal is always negatively charged as mentioned above. Exchangeable cations such as Na^+, Mg^{2+}, Ca^{2+}, etc., exist in the interlayer space of the crystal stack in order to compensate for the charge; bentonite that has the major interlayer cation Na^+ is called the Na-type. There are two kinds of bentonite; the almost natural one that has been crushed after mining (e.g., Kunigel V1®), and a purified form composed of almost pure smectic clays (e.g., Kunipia F®). Microphotographs taken by a laser scanning confocal microscope and by a transmission electron microscope are shown in Fig. 1.7; we can observe the microstructure of the Kunipia F® and the amount of clay stacks.

Chapter 2
Introduction to Continuum Mechanics

The mechanics of a deformable body treated here is based on Newton's laws of motion and the laws of thermodynamics. In this Chapter we present the fundamental concepts of continuum mechanics, and, for conciseness, the material is presented in Cartesian tensor formulation with the implicit assumption of Einstein's summation convention. Where this convention is exempted we shall denote the index thus: $(!\alpha)$.

2.1 Newtonian Mechanics

Newtonian mechanics consists of

1. The first law (The law of inertia),
2. The second law (The law of conservation of linear momentum), and
3. The third law (The law of action and reaction).

Newton's first law defines inertial frames of reference. In general it can be described as follows; "if no force acts on a body, it remains immobilized or in a state of constant motion". However, if this is true, the first law is obviously induced by the second law,[1] and the laws might be incomplete as a physical system. We must rephrase the first law as follows:

The First Law: If no force acts on a body, there exist frames of reference, referred to as *inertial frames*, in which we can observe that the body is either stationary or moves at a constant velocity.

[1]In the second law (2.2) if we set $\boldsymbol{f} = \boldsymbol{0}$ and solve the differential equation, we obtain $\boldsymbol{v} = \text{constant}$ since $m = \text{constant}$, which suggests that the first law is included in the second law. This apparent contradiction results from the misinterpretation of the first law.

Y. Ichikawa and A.P.S. Selvadurai, *Transport Phenomena in Porous Media*,
DOI 10.1007/978-3-642-25333-1_2, © Springer-Verlag Berlin Heidelberg 2012

We now understand that the first law assures the existence of inertial frames, and the second and third laws are valid in the inertial frames. This is the essence of Newtonian mechanics.

Let m and $\boldsymbol{v}$ be the mass and velocity of a body, respectively, then the momentum $\boldsymbol{p}$ is given by

$$\boldsymbol{p} = m\boldsymbol{v}. \tag{2.1}$$

We now have the following:

The Second Law: In an inertial frame, if a force $\boldsymbol{f}$ acts on a body, the following law of conservation of linear momentum is valid:

$$\frac{d\boldsymbol{p}}{dt} = \frac{d(m\boldsymbol{v})}{dt} = \boldsymbol{f} \tag{2.2}$$

Since the mass m is conserved in Newtonian mechanics, (2.2) can be rewritten as

$$m\frac{d\boldsymbol{v}}{dt} = \boldsymbol{f}. \tag{2.3}$$

Let us consider a material point which is represented by two different frames as $(\boldsymbol{x}, t)$ and $(\boldsymbol{x}^*, t^*)$ where $\boldsymbol{x}$, $\boldsymbol{x}^*$ denote positions and t, t^* denote time. Under the Galilean transformation given by

$$\boldsymbol{x}^* = \boldsymbol{Q}\boldsymbol{x} + \boldsymbol{V}t, \qquad t^* = t + a, \tag{2.4}$$

both frames form the equivalent inertial frames (Fig. 2.1). This is referred to as the Galilean principle of relativity. Here $\boldsymbol{Q}$ is a time-independent transformation tensor (cf. Appendix A.3), $\boldsymbol{V}$ is a constant vector and a is a scalar constant.

The Galilean transformation (2.4) gives, in fact, the condition which results in the law of conservation of linear momentum (2.2) in both the frames $(\boldsymbol{x}, t)$ and $(\boldsymbol{x}^*, t^*)$. That is, differentiating (2.4) yields

$$\boldsymbol{v}^* = \frac{d\boldsymbol{x}^*}{dt^*} = \boldsymbol{Q}\frac{d\boldsymbol{x}}{dt} + \boldsymbol{V} = \boldsymbol{Q}\boldsymbol{v} + \boldsymbol{V}$$

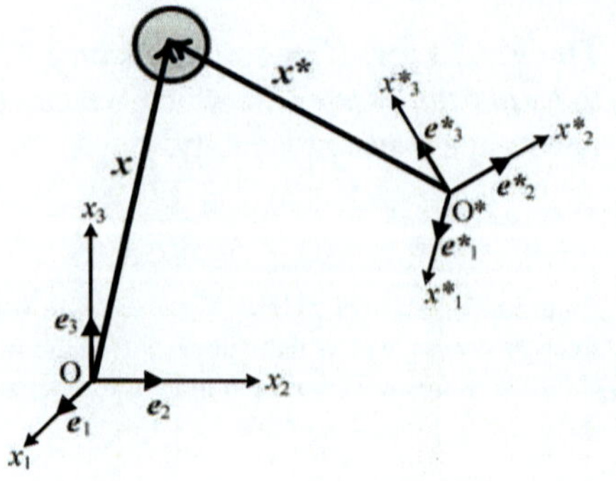

Fig. 2.1 Galilean transformation

and the force vector $\boldsymbol{f}$ is transformed in the same manner:

$$\boldsymbol{f}^* = \boldsymbol{Q}\,\boldsymbol{f}\,. \tag{2.5}$$

The mass conservation law is $m^* = m$, and (2.5) suggests that the law of conservation of linear momentum is satisfied in the inertial frames in the following manner:

$$\boldsymbol{f}^* = m^* \frac{d\boldsymbol{v}^*}{dt^*} = m \frac{d(\boldsymbol{Q}\,\boldsymbol{v} + \boldsymbol{V})}{dt} = \boldsymbol{Q}\left(m \frac{d\boldsymbol{v}}{dt}\right) = \boldsymbol{Q}\,\boldsymbol{f}\,. \tag{2.6}$$

Note that the force vector represented in the form of (2.5) is a fundamental hypothesis of Newtonian mechanics (i.e., the frame indifference of a force vector; see Sect. 2.2.2). The third law gives the interacting forces for a two-body problem, and this will not be treated here.

Note 2.1 (Inertial frame and the relativity principle). The first law (i.e., the law of inertia) gives a condition that there are multiple frames of equivalent inertia that are moving under a relative velocity $\boldsymbol{V}$. Therefore, the first law guarantees that the second law (i.e., the law of conservation of linear momentum) is realized in any inertial frame. This implies that the second law is not always appropriate in different frames of reference, which are moving under a general relative velocity. In fact, in an accelerated frame that includes rotational motion, such as on the surface of Earth, there is a centrifugal force and a Coriolis effect. In Sect. 2.2.2 we will treat a general law of change of frame, which is related to a constitutive theory that describes the material response (see Sect. 2.8).

If the relative velocity $\boldsymbol{V}$ of the Galilean transformation (2.4) approaches the speed of light c, the uniformity of time is not applicable, and the Newtonian framework is no longer valid. That is, the Galilean transformation is changed into the Lorentz transformation under invariance of Maxwell's electromagnetic equations, and the equation of motion is now described in relation to Einstein's theory of relativity. ■

2.2 Deformation Kinematics

When we consider the motion of a material body, constituent atoms and molecules are not directly taken into consideration since this will require an inordinate amount of analysis. Therefore we represent the real material by an equivalent shape of a subdomain of the n-dimensional real number space $\mathbb{R}^n$, and apply the Newtonian principles to this image. This procedure leads to *Continuum Mechanics*; the term 'continuum' is a result of the continuity properties of the n-dimensional real number space $\mathbb{R}^n$.

2.2.1 *Motion and Configuration*

Consider a material point X in the body $\mathfrak{B}$, and we have an image in the n-dimensional real number space $\mathbb{R}^n$ that is referred as a *configuration*.[2] We choose the configuration, $\boldsymbol{X}(X) \in \Omega_0 \subset \mathbb{R}^n$, at the time $t = t_0$ as a *reference configuration*, and treat the subsequent deformation and motion of the *current configuration*, $\boldsymbol{x} \in \Omega \subset \mathbb{R}^n$, at the time $t = t$ with respect to the reference configuration (Fig. 2.2).

Assume that a reference point $\boldsymbol{X} \in \Omega_0$ moves to the current point $\boldsymbol{x} \in \Omega$. The motion is represented as

$$\boldsymbol{x} = \boldsymbol{x}(\boldsymbol{X}, t) \tag{2.7}$$

Note that the time change $\boldsymbol{x}(\boldsymbol{X}, t)$ of a specific point $\boldsymbol{X}$ gives a trajectory. Then the velocity of the material point is calculated by

$$\boldsymbol{v}(\boldsymbol{X}, t) = \frac{d\boldsymbol{x}}{dt}(\boldsymbol{X}, t) \tag{2.8}$$

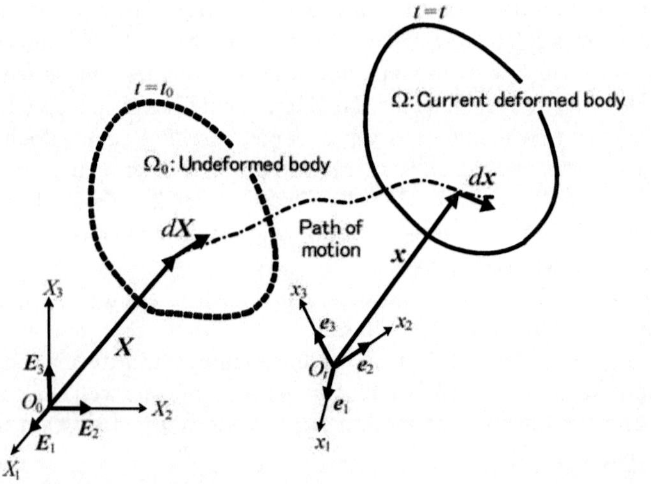

Fig. 2.2 Motion and deformation of a material body

---

[2] 'Configuration' is defined as an invertible continuous function that maps every material point $X \in \mathfrak{B}$ to a point z in a subset of the n-dimensional real number space $\mathbb{R}^n$. A time-dependent motion is considered; therefore the configuration is a function of the material point X and time t. The configuration at a given time t_0 is set as a reference configuration κ, and the point $\boldsymbol{X} \in \mathbb{R}^n$ corresponding to a material point X is written as $\boldsymbol{X} = \kappa(X)$, $X = \kappa^{-1}(\boldsymbol{X})$ where κ^{-1} is an inverse mapping of κ. The current configuration χ at time t maps X to $\boldsymbol{x} \in \mathbb{R}^n$ as $\boldsymbol{x} = \chi(X, t)$, $X = \chi^{-1}(\boldsymbol{X}, t)$. The composite function $\chi_\kappa = \chi \circ \kappa^{-1}$ is introduced as $\boldsymbol{x} = \chi(\kappa^{-1}(\boldsymbol{X}), t) = \chi_\kappa = \chi \circ \kappa^{-1}(\boldsymbol{X}, t) = \chi_\kappa(\boldsymbol{X}, t)$. The function χ_κ gives a mapping between the position vector $\boldsymbol{X}$ in the reference configuration and the position vector $\boldsymbol{x}$ in the current configuration. Since this formal procedure is complicated, the above simplified descriptions are employed.

We should pay attention to the fact that the velocity $\boldsymbol{v}$ has a simple differential form of (2.8) with respect to t, since $\boldsymbol{x}(\boldsymbol{X},t)$ is a function both of $\boldsymbol{X}$ and of t but $\boldsymbol{X}$ is a fixed frame of reference.

Two material points can never occupy the same position, therefore (2.7) has a unique inverse:

$$\boldsymbol{X} = \boldsymbol{X}(\boldsymbol{x},t) \tag{2.9}$$

This relation permits the definition of the velocity $\boldsymbol{v}(\boldsymbol{X},t)$; this can be written as a function of $\boldsymbol{x}$ and t:

$$\boldsymbol{v}(\boldsymbol{X}(\boldsymbol{x},t),t) = \boldsymbol{v}(\boldsymbol{x},t). \tag{2.10}$$

We can generalize this procedure: If a function ϕ is represented in terms of $(\boldsymbol{X},t)$, we call it the *Lagrangian description*, and if ϕ is given in terms of $(\boldsymbol{x},t)$, it is the *Eulerian description*. The choice of the form is arbitrary but will be influenced by any advantage of a problem formulation in either description. For example, in solid mechanics, the Lagrangian description is commonly used, while in fluid mechanics the Eulerian description is popular. This is because in solid mechanics we can attach labels (e.g., visualize 'strain gauges' at various points) on the surface of a solid body, and each material point can be easily traced from the reference state to the current state. On the other hand for a fluid we measure the velocity $\boldsymbol{v}$ or pressure p at the current position $\boldsymbol{x}$, therefore the Eulerian description better represents the fluid (note that for a fluid it is difficult to know the exact reference point $\boldsymbol{X}$ corresponding to all the current points $\boldsymbol{x}$).

The coordinate system with the basis $\{\boldsymbol{E}_I\}$ $(I=1,2,3)$ for the undeformed body Ω_0 is usually different from the coordinate system with the basis $\{\boldsymbol{e}_i\}$ $(i=1,2,3)$ that describes the deformed body Ω in the current configuration (Fig. 2.2):

$$\boldsymbol{X} = X_I \boldsymbol{E}_I, \qquad \boldsymbol{x} = x_i \boldsymbol{e}_i. \tag{2.11}$$

$\{\boldsymbol{E}_I\}$, $\{\boldsymbol{e}_i\}$ represent the orthogonal coordinate systems (i.e., Cartesian); we will employ the same orthogonal coordinate systems $\{\boldsymbol{E}_I\}=\{\boldsymbol{e}_i\}$ for simplicity unless otherwise mentioned. Then, the position vectors $\boldsymbol{X}$ and $\boldsymbol{x}$ can be written as

$$\boldsymbol{X} = X_i \boldsymbol{e}_i, \qquad \boldsymbol{x} = x_i \boldsymbol{e}_i$$

Let $\phi(\boldsymbol{X},t)$ be a Lagrangian function. Since $\boldsymbol{X}$ is time-independent, the time differentiation $\dot{\phi}$ is simply given by

$$\dot{\phi}(\boldsymbol{X},t) = \frac{d\phi}{dt}. \tag{2.12}$$

Here $\dot{\phi} = d\phi/dt$ is referred to as the *material time derivative* of $\phi(\boldsymbol{X},t)$. On the other hand, for an Eulerian function $\phi(\boldsymbol{x},t)$ the total differential $d\phi(\boldsymbol{x},t)$ can be written as

$$d\phi(\boldsymbol{x},t) = \frac{\partial\phi}{\partial t}dt + \frac{\partial\phi}{\partial x_i}dx_i,$$

and $v_i = dx_i/dt$, and therefore we have the following relationship between the material time derivative and Eulerian time derivative

$$\frac{d\phi}{dt} = \frac{\partial \phi}{\partial t} + \boldsymbol{v} \cdot \nabla \phi. \tag{2.13}$$

Here the symbol ∇ (called 'nabla') implies[3]

$$\nabla = \boldsymbol{e}_i \frac{\partial}{\partial x_i}. \tag{2.14}$$

The second term of the r.h.s. of (2.13) gives a convective term.

The velocity $\boldsymbol{v}$ is written either in the Lagrangian form or in the Eulerian form; therefore the acceleration $\boldsymbol{a}$ can be represented in either form:

$$\boldsymbol{a}(\boldsymbol{X}, t) = \frac{d\boldsymbol{v}}{dt} = \frac{\partial \boldsymbol{v}}{\partial t} + \boldsymbol{v} \cdot \nabla \boldsymbol{v} = \boldsymbol{a}(\boldsymbol{x}, t) \tag{2.15}$$

In indicial notation (2.15) is given by

$$a_i = \frac{dv_i}{dt} = \frac{\partial v_i}{\partial t} + v_j \frac{\partial v_i}{\partial v_j}. \tag{2.16}$$

Note 2.2 (Differentiation of vector and tensor valued functions). Let $\boldsymbol{u} = u_i \boldsymbol{e}_i$ and $\boldsymbol{T} = T_{ij} \boldsymbol{e}_i \otimes \boldsymbol{e}_j$ be vector and tensor valued functions. It is understood that the gradients $\mathrm{grad}\, \boldsymbol{u}$, $\mathrm{grad}\, \boldsymbol{T}$ can be of two forms:

$$\mathrm{grad}\, \boldsymbol{u} = \begin{cases} \boldsymbol{u} \otimes \nabla = \dfrac{\partial u_i}{\partial x_j} \boldsymbol{e}_i \otimes \boldsymbol{e}_j, \\ \nabla \otimes \boldsymbol{u} = \dfrac{\partial u_i}{\partial x_j} \boldsymbol{e}_j \otimes \boldsymbol{e}_i, \end{cases} \qquad \mathrm{grad}\, \boldsymbol{T} = \begin{cases} \boldsymbol{T} \otimes \nabla = \dfrac{\partial T_{ij}}{\partial x_k} \boldsymbol{e}_i \otimes \boldsymbol{e}_j \otimes \boldsymbol{e}_k, \\ \nabla \otimes \boldsymbol{T} = \dfrac{\partial T_{jk}}{\partial x_i} \boldsymbol{e}_i \otimes \boldsymbol{e}_j \otimes \boldsymbol{e}_k. \end{cases}$$

The former are called the *right form*, and the latter are the *left form*.

[3] If the gradient is used with respect to Eulerian coordinates with the basis $\{\boldsymbol{e}_i\}$, it is denoted as (2.14). If we explicitly explain the gradient with respect to the Eulerian system, it is denoted as

$$\mathrm{grad} = \nabla_x = \boldsymbol{e}_i \frac{\partial}{\partial x_i}.$$

If the gradient is operated with respect to Lagrangian coordinates $\{\boldsymbol{E}_I\}$, it is represented as

$$\mathrm{Grad} = \nabla_X = \boldsymbol{E}_I \frac{\partial}{\partial X_I}.$$

Right and left forms of divergence $\operatorname{div} \boldsymbol{u}$, $\operatorname{div} \boldsymbol{T}$ and rotation $\operatorname{rot} \boldsymbol{u}$, $\operatorname{rot} \boldsymbol{T}$ are given by

$$\operatorname{div} \boldsymbol{u} = \begin{cases} \boldsymbol{u} \cdot \nabla = \dfrac{\partial u_i}{\partial x_i}, \\ \nabla \cdot \boldsymbol{u} = \dfrac{\partial u_i}{\partial x_i}, \end{cases} \qquad \operatorname{div} \boldsymbol{T} = \begin{cases} \boldsymbol{T} \cdot \nabla = \dfrac{\partial T_{ij}}{\partial x_j} \boldsymbol{e}_i, \\ \nabla \cdot \boldsymbol{T} = \dfrac{\partial T_{ji}}{\partial x_j} \boldsymbol{e}_i, \end{cases}$$

$$\operatorname{rot} \boldsymbol{u} = \begin{cases} \boldsymbol{u} \wedge \nabla = e_{ijk} \dfrac{\partial u_i}{\partial x_j} \boldsymbol{e}_k, \\ \nabla \wedge \boldsymbol{u} = e_{jik} \dfrac{\partial u_i}{\partial x_j} \boldsymbol{e}_k, \end{cases} \qquad \operatorname{rot} \boldsymbol{T} = \begin{cases} \boldsymbol{T} \wedge \nabla = e_{jkl} \dfrac{\partial T_{ik}}{\partial x_l} \boldsymbol{E}_I \otimes \boldsymbol{e}_j, \\ \nabla \wedge \boldsymbol{T} = e_{ikl} \dfrac{\partial T_{lj}}{\partial x_k} \boldsymbol{E}_I \otimes \boldsymbol{e}_j, \end{cases}$$

In this volume we will predominantly use the right form, and we symbolically write the right forms in the same manner as the left form: e.g.

$$\operatorname{grad} \boldsymbol{u} = \frac{\partial u_i}{\partial x_j} \boldsymbol{e}_i \otimes \boldsymbol{e}_j, \qquad \operatorname{grad} \boldsymbol{T} = \frac{\partial T_{ij}}{\partial x_k} \boldsymbol{e}_i \otimes \boldsymbol{e}_j \otimes \boldsymbol{e}_k,$$

$$\operatorname{div} \boldsymbol{u} = \frac{\partial u_i}{\partial x_i}, \qquad \operatorname{div} \boldsymbol{T} = \frac{\partial T_{ij}}{\partial x_j} \boldsymbol{e}_i,$$

$$\operatorname{rot} \boldsymbol{u} = e_{ijk} \frac{\partial u_i}{\partial x_j} \boldsymbol{e}_k, \qquad \operatorname{rot} \boldsymbol{T} = e_{jkl} \frac{\partial T_{ik}}{\partial x_l} \boldsymbol{e}_i \otimes \boldsymbol{e}_j$$

In addition, if the right-divergence form of the second-order tensor $\boldsymbol{T} = T_{ij} \boldsymbol{e}_i \otimes \boldsymbol{e}_j$ is as given above, the divergence theorem (B.6) given in Appendix B.2 can be written as

$$\int_G \operatorname{div} \boldsymbol{T}\, dv = \int_G \nabla \cdot \boldsymbol{T}\, dv = \int_{\partial G} \boldsymbol{T} \boldsymbol{n}\, ds.$$

If $\boldsymbol{u}$ and $\boldsymbol{T}$ are given as functions of the reference basis such that $\boldsymbol{u} = u_I(\boldsymbol{X}) \boldsymbol{E}_I$, $\boldsymbol{T} = T_{IJ}(\boldsymbol{X}) \boldsymbol{E}_I \otimes \boldsymbol{E}_J$, we have

$$\operatorname{Grad} \boldsymbol{u} = \frac{\partial u_I}{\partial X_J} \boldsymbol{E}_I \otimes \boldsymbol{E}_J, \qquad \operatorname{Grad} \boldsymbol{T} = \frac{\partial T_{IJ}}{\partial X_K} \boldsymbol{E}_I \otimes \boldsymbol{E}_J \otimes \boldsymbol{E}_K,$$

$$\operatorname{Div} \boldsymbol{u} = \frac{\partial u_I}{\partial X_I}, \qquad \operatorname{Div} \boldsymbol{T} = \frac{\partial T_{IJ}}{\partial X_J} \boldsymbol{E}_I,$$

$$\operatorname{Rot} \boldsymbol{u} = e_{IJK} \frac{\partial u_I}{\partial X_J} \boldsymbol{E}_K, \qquad \operatorname{Rot} \boldsymbol{T} = e_{JKL} \frac{\partial T_{IK}}{\partial X_L} \boldsymbol{E}_I \otimes \boldsymbol{E}_J. \quad \blacksquare$$

2.2.2 Changes of Frame and Frame Indifference ♣

Let us consider two different points $\boldsymbol{x}$ and $\boldsymbol{x}_0$ in the current body, and introduce a two-point vector $\boldsymbol{u}=\boldsymbol{x}-\boldsymbol{x}_0$ ($\boldsymbol{u}$ does not imply displacement). Then, it is easily observed that the length $|\boldsymbol{u}|=(\boldsymbol{u}\cdot\boldsymbol{u})^{1/2}$ is an invariant; i.e., it has the same value in any coordinate system. We call this property of $|\boldsymbol{u}|$ the *principle of frame indifference*. The concept of frame indifference is a fundamental requirement for a constitutive theory between, for example, stress and strain (see Sect. 2.8).

The frame indifference of the two-point vector $\boldsymbol{u}$ is proved as follows: Let us introduce two coordinate systems, System 1 and System 2, which have bases $\{\boldsymbol{e}_i\}$ and $\{\boldsymbol{e}_i^*\}$, respectively (Fig. 2.3). In the framework of Newtonian mechanics, the two-point vector $\boldsymbol{x}-\boldsymbol{x}_0$ and time t of System 1 can be related to $\boldsymbol{x}^*-\boldsymbol{x}_0^*$ and t^* of System 2 by

$$\boldsymbol{x}^*-\boldsymbol{x}_0^*=\boldsymbol{Q}(\boldsymbol{x}-\boldsymbol{x}_0) \tag{2.17}$$

$$t^*=t-a \tag{2.18}$$

where $\boldsymbol{Q}=Q_{ij}\boldsymbol{e}_i^*\otimes\boldsymbol{e}_j$ is the following coordinate transformation tensor:

$$\boldsymbol{e}_i^*=\boldsymbol{Q}\boldsymbol{e}_i. \tag{2.19}$$

The tensor $\boldsymbol{Q}$ is orthonormal:

$$\boldsymbol{Q}\,\boldsymbol{Q}^T=\boldsymbol{I}^*,\qquad \boldsymbol{Q}^T\boldsymbol{Q}=\boldsymbol{I} \tag{2.20}$$

where $\boldsymbol{I}$ is the unit tensor defined on System 1, and $\boldsymbol{I}^*$ is that defined on System 2. Note that a in (2.18) is a constant.[4] It is understood that the length $|\boldsymbol{x}-\boldsymbol{x}_0|$ is invariant under the change of frame since from (2.17) to (2.20) we have

$$|\boldsymbol{x}^*-\boldsymbol{x}_0^*|^2=(\boldsymbol{x}-\boldsymbol{x}_0)\cdot\boldsymbol{Q}^T\boldsymbol{Q}(\boldsymbol{x}-\boldsymbol{x}_0)=|\boldsymbol{x}-\boldsymbol{x}_0|^2. \tag{2.21}$$

In general, a frame indifferent scalar function f is defined by

$$f^*(\boldsymbol{x}^*)=f(\boldsymbol{x}). \tag{2.22}$$

The frame indifferent vector function $\boldsymbol{u}$ is given by

$$\boldsymbol{u}^*(\boldsymbol{x}^*)=\boldsymbol{Q}\,\boldsymbol{u}(\boldsymbol{x}),\qquad \boldsymbol{u}(\boldsymbol{x})=\boldsymbol{Q}^T\boldsymbol{u}^*(\boldsymbol{x}^*). \tag{2.23}$$

---

[4] Here we deal with a general case in which two coordinate systems may not be inertial systems. If both are inertial systems, $\boldsymbol{Q}$ is time-independent as given by (2.4). Therefore, we have

$$\boldsymbol{x}^*=\boldsymbol{Q}\,\boldsymbol{x}+\boldsymbol{V}t,\quad \boldsymbol{x}_0^*=\boldsymbol{Q}\,\boldsymbol{x}_0+\boldsymbol{V}t\quad\Rightarrow\quad \boldsymbol{u}=\boldsymbol{x}^*-\boldsymbol{x}_0^*=\boldsymbol{Q}(\boldsymbol{x}-\boldsymbol{x}_0),$$

which shows that the two-point vector $\boldsymbol{u}$ is frame indifferent.

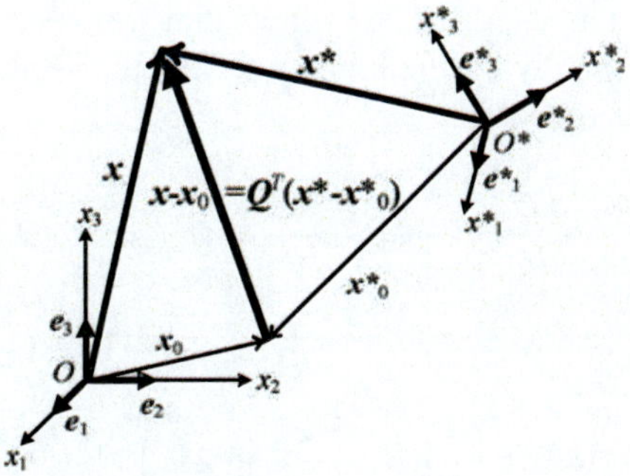

Fig. 2.3 Coordinate transformation and frame indifferent vector

A frame indifferent second-order tensor function $\boldsymbol{T}$ is characterized by the frame indifference of a transformed vector $\boldsymbol{v} = \boldsymbol{T}\boldsymbol{u}$. That is, let $\boldsymbol{u}$ be a frame indifferent vector ($\boldsymbol{u}^* = \boldsymbol{Q}\,\boldsymbol{u}$), and $\boldsymbol{v}$ and $\boldsymbol{v}^*$ be the vectors corresponding to $\boldsymbol{u}$ and $\boldsymbol{u}^*$ transformed by $\boldsymbol{T}$ and $\boldsymbol{T}^*$ such that

$$\boldsymbol{v} = \boldsymbol{T}\,\boldsymbol{u}, \qquad \boldsymbol{v}^* = \boldsymbol{T}^*\boldsymbol{u}^*.$$

Then, if $\boldsymbol{v}^* = \boldsymbol{Q}\,\boldsymbol{v}$, $\boldsymbol{T}$ is said to be frame indifferent. Now we have

$$\boldsymbol{v}^* = \boldsymbol{Q}^*\boldsymbol{v} = \boldsymbol{T}^*\boldsymbol{u}^* = \boldsymbol{T}^*\boldsymbol{Q}\,\boldsymbol{u} \qquad \Rightarrow \qquad \boldsymbol{v} = \boldsymbol{Q}^T\boldsymbol{T}^*\boldsymbol{Q}\,\boldsymbol{u}.$$

Therefore the requirement for frame indifference of a second-order tensor $\boldsymbol{T}$ is defined by

$$\boldsymbol{T} = \boldsymbol{Q}^T\boldsymbol{T}^*\boldsymbol{Q}, \qquad \boldsymbol{T}^* = \boldsymbol{Q}\,\boldsymbol{T}\,\boldsymbol{Q}^T. \tag{2.24}$$

2.2.3 Motion in a Non-inertial System ♣

We consider an equation of motion for the case where System 1 is inertial while System 2 is non-inertial. Since the two-point vector $\boldsymbol{x} - \boldsymbol{x}_0$ should be frame indifferent, satisfying (2.17) and (2.18), a position vector $\boldsymbol{x}^*(t)$ in System 2 is given by

$$\boldsymbol{x}^*(t) = \boldsymbol{x}_0^*(t) + \boldsymbol{Q}(t)(\boldsymbol{x} - \boldsymbol{x}_0) \tag{2.25}$$

where we can regard $\boldsymbol{x}_0$ as the origin of the coordinate system O, and $\boldsymbol{x}_0^*(t)$ as the origin of the coordinate system O^* (see Fig. 2.3). Taking a material time derivative of (2.25) and substituting the inverse relation of (2.17) gives the velocity:

$$\boldsymbol{v}^* = \frac{d\boldsymbol{x}_0^*}{dt} + \boldsymbol{\Omega}(\boldsymbol{x}^* - \boldsymbol{x}_0^*) + \boldsymbol{Q}\boldsymbol{v}, \tag{2.26}$$

$$\boldsymbol{\Omega} = \frac{d\boldsymbol{Q}}{dt}\boldsymbol{Q}^T \tag{2.27}$$

The result (2.26) shows that the velocity vector $\boldsymbol{v}$ is not frame indifferent. Note that by differentiating $(2.20)_1$ the second-order tensor $\boldsymbol{\Omega}$ is understood to be anti-symmetric:

$$\frac{d\boldsymbol{Q}}{dt}\boldsymbol{Q}^T + \boldsymbol{Q}\frac{d\boldsymbol{Q}^T}{dt} = \boldsymbol{0} \quad \Rightarrow \quad \boldsymbol{\Omega} = \frac{d\boldsymbol{Q}}{dt}\boldsymbol{Q}^T = -\boldsymbol{Q}\frac{d\boldsymbol{Q}^T}{dt} = -\boldsymbol{\Omega}^T. \tag{2.28}$$

The acceleration vector is given by a material time derivative of (2.26):

$$\boldsymbol{a}^* = \frac{d\boldsymbol{v}^*}{dt} = \frac{d^2\boldsymbol{x}_0^*}{dt^2} + 2\boldsymbol{\Omega}\left(\boldsymbol{v}^* - \frac{d\boldsymbol{x}_0^*}{dt}\right) + \left(\frac{d\boldsymbol{\Omega}}{dt} - \boldsymbol{\Omega}^2\right)(\boldsymbol{x}^* - \boldsymbol{x}_0^*) + \boldsymbol{Q}\frac{d\boldsymbol{v}}{dt} \tag{2.29}$$

This suggests that the acceleration is not frame indifferent.

Now we define a rotation vector $\boldsymbol{\omega}$ with respect to the original coordinate system by

$$\omega_i = -\frac{1}{2}e_{ijk}\Omega_{kj} = \frac{1}{2}e_{ijk}\Omega_{jk} \tag{2.30}$$

then for any vector $\boldsymbol{a}$ we have

$$\boldsymbol{\Omega}\boldsymbol{a} = \boldsymbol{\omega} \wedge \boldsymbol{a} \tag{2.31}$$

Thus, (2.29) can be written as

$$\frac{d\boldsymbol{v}^*}{dt} = \frac{d^2\boldsymbol{x}_0^*}{dt^2} + \frac{d\boldsymbol{\Omega}}{dt}\boldsymbol{Q}(\boldsymbol{x}-\boldsymbol{x}_0) + \boldsymbol{\omega}\wedge[\boldsymbol{\omega} \wedge \boldsymbol{Q}(\boldsymbol{x}-\boldsymbol{x}_0)] + 2\boldsymbol{\omega}\wedge\boldsymbol{Q}\boldsymbol{v} + \boldsymbol{Q}\frac{d\boldsymbol{v}}{dt} \tag{2.32}$$

The third term of the r.h.s. of this equation gives a centrifugal force and the fourth is the Coriolis force.

As shown by (2.5), a fundamental hypothesis of Newtonian mechanics is that the force $\boldsymbol{f}$ is frame indifferent ($\boldsymbol{f}^* = \boldsymbol{Q}\,\boldsymbol{f}$). Therefore, referring to (2.32) and (2.27), the equation of motion in System 2, which is non-inertial, can be written as

$$m^*\frac{d\boldsymbol{v}^*}{dt^*} = \boldsymbol{f}^* + \boldsymbol{f}^{\text{ai}}, \tag{2.33}$$

$$\begin{aligned} \boldsymbol{f}^{\text{ai}} = m\Bigg[&\frac{d^2\boldsymbol{x}_0^*}{dt^2} + \frac{d\boldsymbol{\Omega}}{dt}(\boldsymbol{x}^* - \boldsymbol{x}_0^*) + \boldsymbol{\omega}\wedge\boldsymbol{\omega}\wedge(\boldsymbol{x}^* - \boldsymbol{x}_0^*) \\ &+2\boldsymbol{\omega}\wedge\left(\boldsymbol{v}^* - \frac{d\boldsymbol{x}_0^*}{dt} - \boldsymbol{\Omega}(\boldsymbol{x}^* - \boldsymbol{x}_0^*)\right)\Bigg] \end{aligned} \tag{2.34}$$

where $\boldsymbol{f}^{\text{ai}}$ is an apparent inertial force observed in the non-inertial system.

2.2.4 Deformation Gradient, Strain and Strain Rate

We distinguish between the coordinate system $\{\boldsymbol{E}_I\}$ of the Lagrangian description and the coordinate system $\{\boldsymbol{e}_i\}$ of the Eulerian description in order to understand the relationship between both systems.

As shown in Fig. 2.2, an increment vector $d\boldsymbol{x} = dx_i \boldsymbol{e}_i$ of a point $\boldsymbol{x} \in \Omega$ in the deformed body Ω is related to the increment $d\boldsymbol{X} = dX_I \boldsymbol{E}_I$ of the corresponding point $\boldsymbol{X} \in \Omega_0$ in the undeformed body Ω_0 by

$$d\boldsymbol{x} = \boldsymbol{F}\, d\boldsymbol{X} \tag{2.35}$$

where the second-order tensor $\boldsymbol{F}$ is the *deformation gradient*;

$$\boldsymbol{F} = \text{Grad}\, \boldsymbol{x} = \nabla_X \boldsymbol{x} = F_{iI}\, \boldsymbol{e}_i \otimes \boldsymbol{E}_I, \tag{2.36}$$

$$F_{iI} = \frac{\partial x_i}{\partial X_I}, \qquad \text{Grad} = \nabla_X = \boldsymbol{E}_I \frac{\partial}{\partial X_I} \tag{2.37}$$

which relates the infinitesimal line segment $d\boldsymbol{X} \in \Omega_0$ to the corresponding segment $d\boldsymbol{x} \in \Omega$. As observed in (2.36), the deformation gradient $\boldsymbol{F}$ plays a role in the coordinate transformation from $\boldsymbol{E}_I$ to $\boldsymbol{e}_i$. Since no material point vanishes, there is an inverse relation of $\boldsymbol{F}$:

$$\boldsymbol{F}^{-1} = \text{grad}\, \boldsymbol{X} = \nabla_x \boldsymbol{X} = F^{-1}_{Ii}\, \boldsymbol{E}_I \otimes \boldsymbol{e}_i, \tag{2.38}$$

$$F^{-1}_{Ii} = \frac{\partial X_I}{\partial x_i}, \qquad \text{grad} = \nabla_x = \boldsymbol{e}_i \frac{\partial}{\partial x_i}. \tag{2.39}$$

If the right form (see Note 2.2) is employed, the related forms of deformation gradient can be written as follows:

$$\boldsymbol{F} = \frac{\partial x_i}{\partial X_I} \boldsymbol{e}_i \otimes \boldsymbol{E}_I, \qquad \boldsymbol{F}^T = \frac{\partial x_i}{\partial X_I} \boldsymbol{E}_I \otimes \boldsymbol{e}_i, \tag{2.40}$$

$$\boldsymbol{F}^{-1} = \frac{\partial X_I}{\partial x_i} \boldsymbol{E}_I \otimes \boldsymbol{e}_i, \qquad \boldsymbol{F}^{-T} = \frac{\partial X_I}{\partial x_i} \boldsymbol{e}_i \otimes \boldsymbol{E}_I. \tag{2.41}$$

Here $\boldsymbol{F}^{-T}$ implies $(\boldsymbol{F}^{-1})^T$.

Now we need to measure the extent of deformation of an elemental length located at a material point. To do so, we compare the length $|d\boldsymbol{x}|$ with its original length $|d\boldsymbol{X}|$ (see Fig. 2.2) by comparing the difference of both lengths as a squared measure:

$$|\, d\boldsymbol{x}\,|^2 - |\, d\boldsymbol{X}\,|^2 = d\boldsymbol{x} \cdot d\boldsymbol{x} - d\boldsymbol{X} \cdot d\boldsymbol{X}$$

Substituting the deformation gradient yields

$$d\boldsymbol{x} \cdot d\boldsymbol{x} = d\boldsymbol{X} \cdot \boldsymbol{C} d\boldsymbol{X}, \qquad d\boldsymbol{X} \cdot d\boldsymbol{X} = d\boldsymbol{x} \cdot \boldsymbol{B}^{-1} d\boldsymbol{x}$$

where

$$\boldsymbol{C} = \boldsymbol{F}^T \boldsymbol{F} = F_{kI} F_{kJ} \, \boldsymbol{E}_I \otimes \boldsymbol{E}_J, \qquad \boldsymbol{B} = \boldsymbol{F} \boldsymbol{F}^T = F_{iI} F_{jI} \, \boldsymbol{e}_i \otimes \boldsymbol{e}_j \tag{2.42}$$

$\boldsymbol{C}$ is referred to as the *right Cauchy-Green tensor* and $\boldsymbol{B}$ the *left Cauchy-Green tensor*. Then the deformation measure can be written as

$$|\, d\boldsymbol{x} \,|^2 - |\, d\boldsymbol{X} \,|^2 = d\boldsymbol{X} \cdot 2\boldsymbol{E} \, d\boldsymbol{X} = d\boldsymbol{x} \cdot 2\boldsymbol{e} \, d\boldsymbol{x} \tag{2.43}$$

where we set

$$\boldsymbol{E} = \frac{1}{2} (\boldsymbol{C} - \boldsymbol{I}) = \frac{1}{2} (C_{IJ} - \delta_{IJ}) \, \boldsymbol{E}_I \otimes \boldsymbol{E}_J, \tag{2.44}$$

$$\boldsymbol{e} = \frac{1}{2} \left(\boldsymbol{i} - \boldsymbol{B}^{-1} \right) = \frac{1}{2} \left(\delta_{ij} - B_{ij}^{-1} \right) \boldsymbol{e}_i \otimes \boldsymbol{e}_j \tag{2.45}$$

Note that $\boldsymbol{I} = \delta_{IJ} \, \boldsymbol{E}_I \otimes \boldsymbol{E}_J$ and $\boldsymbol{i} = \delta_{ij} \, \boldsymbol{e}_i \otimes \boldsymbol{e}_j$ are the unit tensors in the reference and current configurations, respectively. $\boldsymbol{E}$ is referred to as the *Lagrangian strain* or *Green strain*, and $\boldsymbol{e}$ the *Eulerian strain* or *Almansi strain* (see, e.g., Malvern 1969; Spencer 2004).

Since the deformation gradient $\boldsymbol{F}$ is invertible and positive definite ($\det \boldsymbol{F} > 0$), we can introduce the following polar decomposition:

$$\boldsymbol{F} = \boldsymbol{RU} = \boldsymbol{VR} \tag{2.46}$$

where

$$\boldsymbol{R} = R_{iI} \, \boldsymbol{e}_i \otimes \boldsymbol{E}_I, \qquad \boldsymbol{U} = U_{IJ} \, \boldsymbol{E}_I \otimes \boldsymbol{E}_J, \qquad \boldsymbol{V} = V_{ij} \, \boldsymbol{e}_i \otimes \boldsymbol{e}_j \tag{2.47}$$

$\boldsymbol{R}$ is referred to as the *rotation tensor*, $\boldsymbol{U}$ the *right stretch tensor*, $\boldsymbol{V}$ the *left stretch tensor*. $\boldsymbol{R}$ is orthonormal ($\boldsymbol{R}^T \boldsymbol{R} = \boldsymbol{I}, \ \boldsymbol{R} \boldsymbol{R}^T = \boldsymbol{i}$), which gives the rotation of $\boldsymbol{C}$ and $\boldsymbol{B}^{-1}$ to their principal axes. Under the polar decomposition we have

$$\boldsymbol{C} = \boldsymbol{U}^2, \qquad \boldsymbol{B} = \boldsymbol{V}^2 \tag{2.48}$$

As understood from (2.48), $\boldsymbol{U}$ and $\boldsymbol{V}$ are symmetric and positive definite.

Note 2.3 (Small strain theory). In most textbooks on elasticity theory a displacement vector is defined as $\boldsymbol{u} = \boldsymbol{x} - \boldsymbol{X}$. However we know that $\boldsymbol{x} = x_i \boldsymbol{e}_i$, $\boldsymbol{X} = X_I \boldsymbol{E}_I$ and the transformation of both bases are locally defined by the deformation gradient $\boldsymbol{F}$; therefore it is difficult to introduce the globally defined displacement vector $\boldsymbol{u}$ unless a common rectangular Cartesian coordinate system is used.

Now let us use a common basis $\boldsymbol{e}_i$ and introduce an incremental form as

$$d\boldsymbol{u} = d\boldsymbol{x} - d\boldsymbol{X} = (\boldsymbol{F} - \boldsymbol{i}) \, d\boldsymbol{X} = \boldsymbol{H} \, d\boldsymbol{X} \tag{2.49}$$

where

$$H = \frac{\partial \boldsymbol{u}}{\partial \boldsymbol{X}} = \boldsymbol{F} - \boldsymbol{i} = (F_{ij} - \delta_{ij})\, \boldsymbol{e}_i \otimes \boldsymbol{e}_j \tag{2.50}$$

Then the Green strain is given by

$$\boldsymbol{E} = \frac{1}{2}\left(\boldsymbol{H} + \boldsymbol{H}^T + \boldsymbol{H}^T\boldsymbol{H}\right) = \frac{1}{2}\left(\frac{\partial u_i}{\partial X_j} + \frac{\partial u_j}{\partial X_i} + \frac{\partial u_k}{\partial X_i}\frac{\partial u_k}{\partial X_j}\right) \boldsymbol{e}_i \otimes \boldsymbol{e}_j \tag{2.51}$$

Since the third term of the r.h.s. is second-order infinitesimal, the small strain tensor is given by

$$\varepsilon_{ij} = \frac{1}{2}\left(\frac{\partial u_i}{\partial x_j} + \frac{\partial u_j}{\partial x_i}\right) \tag{2.52}$$

where we identify the coordinates X_i with x_i (Little 1973; Davis and Selvadurai 1996; Barber 2002).

The strain with components given by (2.52) is referred to as the tensorial strain, while the strain in which the shearing components are changed into

$$\gamma_{yz} = \frac{\partial u_y}{\partial z} + \frac{\partial u_z}{\partial y}, \qquad \gamma_{zx} = \frac{\partial u_z}{\partial x} + \frac{\partial u_x}{\partial z}, \qquad \gamma_{xy} = \frac{\partial u_x}{\partial y} + \frac{\partial u_y}{\partial x}$$

which can be denoted by the vector

$$\boldsymbol{\varepsilon} = \left[\varepsilon_{xx}, \varepsilon_{yy}, \varepsilon_{zz}, \gamma_{xy}, \gamma_{yz}, \gamma_{zx}\right]^T \tag{2.53}$$

that is referred to as the engineering strain. ■

Note 2.4 (Generalized strain measure (Hill 1978)♣). Since the right Cauchy-Green tensor $\boldsymbol{C} = \boldsymbol{F}^T\boldsymbol{F}$ is symmetric and the components are real numbers, there are three real eigenvalues that are set as λ_i^2 $(i = 1, 2, 3)$ and the corresponding eigenvectors are given by $\boldsymbol{N}_i$; then we have

$$\left(\boldsymbol{F}^T\boldsymbol{F}\right)\boldsymbol{N}_i - \lambda_i^2\boldsymbol{N}_i = \boldsymbol{0}, \qquad \boldsymbol{N}_i \cdot \boldsymbol{N}_j = \delta_{ij} \qquad (i: \text{not summed}). \tag{2.54}$$

Let us set a material fiber along $\boldsymbol{N}_i$ as $d\boldsymbol{X}_i$, therefore we can write

$$d\boldsymbol{X}_i = dX_i\, \boldsymbol{N}_i \quad \Rightarrow \quad \left(\boldsymbol{F}^T\boldsymbol{F}\right) d\boldsymbol{X}_i = \lambda_i^2\, d\boldsymbol{X}_i \qquad (i: \text{not summed}) \tag{2.55}$$

where λ_i is referred to as the principal stretch , and $\boldsymbol{N}_i$ $(i = 1, 2, 3)$ form Lagrangian triads. Referring $(2.48)_1$, the right stretch tensor $\boldsymbol{U}$ can be written as

$$\boldsymbol{U} = \sum_i \lambda_i\, \boldsymbol{N}_i \otimes \boldsymbol{N}_i. \tag{2.56}$$

The rotation tensor $\boldsymbol{R}$ given by $(2.47)_1$ transforms the Lagrangian triads $\boldsymbol{N}_i$ into the Eulerian triads $\boldsymbol{n}_i$:

$$\boldsymbol{n}_i = \boldsymbol{R}\,\boldsymbol{N}_i, \qquad \boldsymbol{R} = \sum_i \boldsymbol{n}_i \otimes \boldsymbol{N}_i. \tag{2.57}$$

The left stretch tensor $\boldsymbol{V}$ given by $(2.47)_3$ can now be written as

$$\boldsymbol{V} = \sum_i \lambda_i\, \boldsymbol{n}_i \otimes \boldsymbol{n}_i. \tag{2.58}$$

Following Hill (1978) the generalized Lagrangian strain measure is defined by

$$\mathfrak{E} = \sum_i f(\lambda_i)\, \boldsymbol{N}_i \otimes \boldsymbol{N}_i. \tag{2.59}$$

Here $f(\lambda_i)$ is a scale function which satisfies the conditions

$$f(1) = 0, \qquad f'(1) = 1.$$

Consider the following example:

$$f(z) = \frac{z^{2n} - 1}{2n}. \tag{2.60}$$

We can introduce the following family of n-th order Lagrangian strain measures:

$$\boldsymbol{E}\,(n) = \sum_i \frac{(\lambda_i)^{2n} - 1}{2n}\, \boldsymbol{N}_i \otimes \boldsymbol{N}_i. \tag{2.61}$$

The Green strain given by (2.44) corresponds to $\boldsymbol{E}\,(1)$. Furthermore from (2.60) we have

$$\lim_{n \to 0} \frac{(\lambda_i)^{2n} - 1}{2n} = \ln \lambda_i.$$

As $n \to 0$, the logarithmic strain $\boldsymbol{E}\,(0)$ is given by

$$\boldsymbol{E}\,(0) = \ln \boldsymbol{U} = \sum_i \ln(\lambda_i)\, \boldsymbol{N}_i \otimes \boldsymbol{N}_i. \tag{2.62}$$

The generalized Eulerian strain measure is given by

$$\mathfrak{e} = \sum_i f(\lambda_i)\, \boldsymbol{n}_i \otimes \boldsymbol{n}_i = \boldsymbol{R}\, \mathfrak{E}\, \boldsymbol{R}^T, \tag{2.63}$$

and the family of n-th order Eulerian strain measures is introduced by

$$\boldsymbol{e}(n) = \sum_i \frac{(\lambda_i)^{2n} - 1}{2n} \boldsymbol{n}_i \otimes \boldsymbol{n}_i, \qquad \boldsymbol{e}(0) = \ln \boldsymbol{V} = \sum_i \ln(\lambda_i)\, \boldsymbol{n}_i \otimes \boldsymbol{n}_i. \quad (2.64)$$

It should be noted that $\boldsymbol{E}(-1) \neq \boldsymbol{e}(1)$. ■

Recall the definition of the deformation gradient $d\boldsymbol{x} = \boldsymbol{F}\, d\boldsymbol{X}$; its material time-derivative defines the following *velocity gradient*:

$$\dot{\overline{d\boldsymbol{x}}} = \dot{\boldsymbol{F}}\, d\boldsymbol{X} = \boldsymbol{L}\, d\boldsymbol{x} \quad \Rightarrow \quad \boldsymbol{L} \equiv \dot{\boldsymbol{F}}\, \boldsymbol{F}^{-1}. \quad (2.65)$$

Since the direct forms of $\dot{\boldsymbol{F}}$, $\boldsymbol{F}^{-1}$ are given by

$$\dot{\boldsymbol{F}} = \dot{F}_{iI}\, \boldsymbol{e}_i \otimes \boldsymbol{E}_I = \frac{\partial v_i}{\partial X_I} \boldsymbol{e}_i \otimes \boldsymbol{E}_I, \qquad \boldsymbol{F}^{-1} = \frac{\partial X_I}{\partial x_i} \boldsymbol{E}_I \otimes \boldsymbol{e}_i \quad (2.66)$$

$\boldsymbol{L}$ can be written as

$$\boldsymbol{L} = \operatorname{grad} \boldsymbol{v} = \frac{\partial v_i}{\partial x_k} \boldsymbol{e}_i \otimes \boldsymbol{e}_k \quad (2.67)$$

On the other hand, $\boldsymbol{F}\, \boldsymbol{F}^{-1} = \boldsymbol{i}$, therefore taking its material time differential with (2.65) $\dot{\boldsymbol{F}} = \boldsymbol{L}\boldsymbol{F}$ yields the inverse of $\dot{\boldsymbol{F}}$:

$$\dot{\overline{\boldsymbol{F}^{-1}}} = -\boldsymbol{F}^{-1}\boldsymbol{L}. \quad (2.68)$$

The velocity gradient $\boldsymbol{L}$ is decomposed into its symmetric part $\boldsymbol{D}$, called the *stretch tensor* or *rate-of-deformation tensor*, and its anti-symmetric part $\boldsymbol{W}$, called the *spin tensor*:

$$\boldsymbol{L} = \dot{\boldsymbol{F}}\, \boldsymbol{F}^{-1} = \boldsymbol{D} + \boldsymbol{W}, \quad (2.69)$$

$$\boldsymbol{D} = \frac{1}{2}\left(\boldsymbol{L} + \boldsymbol{L}^T\right), \qquad \boldsymbol{W} = \frac{1}{2}\left(\boldsymbol{L} - \boldsymbol{L}^T\right). \quad (2.70)$$

The material time differentiation of the Green strain $\boldsymbol{E}$ given by (2.44) is

$$\dot{\boldsymbol{E}} = \frac{1}{2}\left(\dot{\boldsymbol{F}}^T \boldsymbol{F} + \boldsymbol{F}^T \dot{\boldsymbol{F}}\right) = \boldsymbol{F}^T \boldsymbol{D} \boldsymbol{F} \quad \Rightarrow \quad \boldsymbol{D} = \boldsymbol{F}^{-T} \dot{\boldsymbol{E}}\, \boldsymbol{F}^{-1} \quad (2.71)$$

where (2.65) is used.

Note 2.5 (Embedded coordinates ♣). In solids we can trace each material point $\boldsymbol{X}$ by attaching labels on its surface during deformation. Then, as shown in Fig. 2.4, it is easy to introduce a coordinate system in which the label values of the coordinates are not changed ($x^i = X^i$) but the reference basis $\boldsymbol{G}_i$ (before deformation) is changed into the current basis $\boldsymbol{g}_i$ (after deformation). This is referred to as the *embedded coordinate system*.

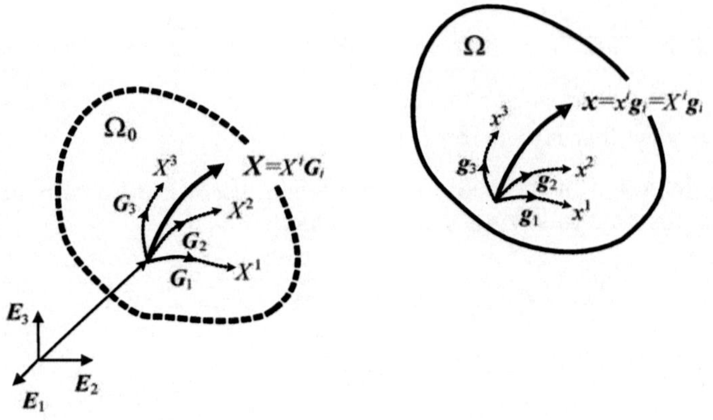

Fig. 2.4 Embedded coordinate system

In the embedded coordinate system, the transformation rule of the base vectors is given by

$$\boldsymbol{g}_i(t) = \boldsymbol{F}(t)\boldsymbol{G}_i \qquad \Rightarrow \qquad \boldsymbol{F}(t) = \boldsymbol{g}_i(t) \otimes \boldsymbol{G}^i, \tag{2.72}$$

and the following relationships are obtained:

$$\boldsymbol{F}^{-1} = \boldsymbol{G}_i \otimes \boldsymbol{g}^i, \qquad \boldsymbol{G}_i = \boldsymbol{F}^{-1}\boldsymbol{g}_i, \tag{2.73}$$

$$\boldsymbol{F}^{-T} = \boldsymbol{g}^i \otimes \boldsymbol{G}_i, \qquad \boldsymbol{g}^i = \boldsymbol{F}^{-T}\boldsymbol{G}^i. \tag{2.74}$$

The unit tensor in the current deformed body is written as $\boldsymbol{i} = \delta_i^{\,j}\boldsymbol{g}^i \otimes \boldsymbol{g}_j = \delta^i{}_j\,\boldsymbol{g}_i \otimes \boldsymbol{g}^j = g_{ij}\,\boldsymbol{g}^i \otimes \boldsymbol{g}^j = g^{ij}\,\boldsymbol{g}_i \otimes \boldsymbol{g}_j$, and the right Cauchy-Green tensor $\boldsymbol{C}$ is given by

$$\boldsymbol{C} = \boldsymbol{F}^T\boldsymbol{F} = (\boldsymbol{G}^i \otimes \boldsymbol{g}_i)(\boldsymbol{g}_j \otimes \boldsymbol{G}^j) = g_{ij}\,\boldsymbol{G}^i \otimes \boldsymbol{G}^j. \tag{2.75}$$

Let an arbitrary second-order tensor $\boldsymbol{K}$ in the current body be written as

$$\boldsymbol{K} = K^{ij}\boldsymbol{g}_i \otimes \boldsymbol{g}_j = K^i{}_j\,\boldsymbol{g}_i \otimes \boldsymbol{g}^j = K_i^{\,j}\,\boldsymbol{g}^i \otimes \boldsymbol{g}_j = K_{ij}\,\boldsymbol{g}^i \otimes \boldsymbol{g}^j, \tag{2.76}$$

then the second-order tensors that have the same components in the undeformed body are given by

$$\begin{aligned}
\boldsymbol{K}^{(I)} &= K^{ij}\,\boldsymbol{G}_i \otimes \boldsymbol{G}_j &&= \boldsymbol{F}^{-1}\boldsymbol{K}\boldsymbol{F}^{-T}\\
\boldsymbol{K}^{(II)} &= K^i{}_j\,\boldsymbol{G}_i \otimes \boldsymbol{G}^j &&= \boldsymbol{F}^{-1}\boldsymbol{K}\boldsymbol{F}\\
\boldsymbol{K}^{(III)} &= K_i^{\,j}\,\boldsymbol{G}^i \otimes \boldsymbol{G}_j &&= \boldsymbol{F}^T\boldsymbol{K}\boldsymbol{F}^{-T}\\
\boldsymbol{K}^{(IV)} &= K_{ij}\,\boldsymbol{G}^i \otimes \boldsymbol{G}^j &&= \boldsymbol{F}^T\boldsymbol{K}\boldsymbol{F}.
\end{aligned} \tag{2.77}$$

Material time differentiation of (2.72) yields $\dot{\boldsymbol{F}} = \dot{\boldsymbol{g}}_i(t) \otimes \boldsymbol{G}^i$ (note $\boldsymbol{G}^i =$ constant). As shown in (2.65), $\dot{\boldsymbol{F}} = \boldsymbol{L}\boldsymbol{F}$, therefore the velocity gradient $\boldsymbol{L}$ in the embedded coordinates is

$$\dot{\boldsymbol{g}}_i(t) = \boldsymbol{L}\boldsymbol{g}_i(t) \qquad \Rightarrow \qquad \boldsymbol{L}(t) = \dot{\boldsymbol{g}}_i(t) \otimes \boldsymbol{g}^i(t). \tag{2.78}$$

The stretch tensor $\boldsymbol{D}$ and spin tensor $\boldsymbol{W}$ are now given by

$$\boldsymbol{D} = \frac{1}{2}\left(\dot{\boldsymbol{g}}_i \otimes \boldsymbol{g}^i + \boldsymbol{g}^i \otimes \dot{\boldsymbol{g}}_i\right) = \frac{1}{2}\left(\dot{g}_{ij}\, \boldsymbol{g}^i \otimes \boldsymbol{g}^j\right), \tag{2.79}$$

$$\boldsymbol{W} = \frac{1}{2}\left(\dot{\boldsymbol{g}}_i \otimes \boldsymbol{g}^i - \boldsymbol{g}^i \otimes \dot{\boldsymbol{g}}_i\right) \tag{2.80}$$

■

2.2.5 *Transport Theorems and Jump Condition*

A volume element dV in the undeformed body Ω_0 is related to the volume element dv in the deformed body Ω through the determinant, J, of the deformation gradient $\boldsymbol{F}$ by

$$dv = J\, dV, \tag{2.81}$$

$$J = \det \boldsymbol{F} = \left|\frac{\partial x_i}{\partial X_I}\right| = e_{IJK} \frac{\partial x_1}{\partial X_I}\frac{\partial x_2}{\partial X_J}\frac{\partial x_3}{\partial X_K} \tag{2.82}$$

where J is referred to as the *Jacobian*. If we recall that

$$e_{IJK} \frac{\partial x_1}{\partial X_I}\frac{\partial x_1}{\partial X_J}\frac{\partial x_2}{\partial X_K} = 0, \quad e_{IJK} \frac{\partial x_2}{\partial X_I}\frac{\partial x_2}{\partial X_J}\frac{\partial x_3}{\partial X_K} = 0, \quad \cdots$$

the material time derivative of the Jacobian is given by

$$\dot{J} = J\frac{\partial v_k}{\partial x_k} \qquad \Rightarrow \qquad \dot{J} = J\, \nabla \cdot \boldsymbol{v} = J \operatorname{tr} \boldsymbol{L} = J \operatorname{tr} \boldsymbol{D} \tag{2.83}$$

Under this relation the material time derivative of the integral of an arbitrary function ϕ can be calculated as

$$\frac{d}{dt}\int_\Omega \phi\, dv = \frac{d}{dt}\int_{\Omega_0} \phi\, J\, dV = \int_\Omega \left(\frac{d\phi}{dt} + \phi \frac{\partial v_i}{\partial x_i}\right) dv$$

We now have the following *Reynolds' transport theorem*:

$$\frac{d}{dt}\int_\Omega \phi\, dv = \int_\Omega \left(\frac{d\phi}{dt} + \phi\nabla\cdot\boldsymbol{v}\right) dv = \int_\Omega \left(\frac{d\phi}{dt} + \phi\,\mathrm{tr}\,\boldsymbol{L}\right) dv$$
$$= \int_\Omega \left(\frac{\partial\phi}{\partial t} + \nabla\cdot(\phi\boldsymbol{v})\right) dv = \int_\Omega \frac{\partial\phi}{\partial t} dv + \int_{\partial\Omega} \phi\boldsymbol{v}\cdot\boldsymbol{n}\, ds. \tag{2.84}$$

Next we give the transport theorem for a surface integral: The determinant of a (3×3)-matrix $\boldsymbol{A}$ is given by

$$e_{rst}\det\boldsymbol{A} = e_{ijk}A_{ir}A_{js}A_{kt}.$$

Therefore (2.81) can be written as

$$e_{ijk}\, J^{-1} = e_{IJK}\frac{\partial X_I}{\partial x_i}\frac{\partial X_J}{\partial x_j}\frac{\partial X_K}{\partial x_k}, \qquad e_{IJK}\, J = e_{ijk}\frac{\partial x_i}{\partial X_I}\frac{\partial x_j}{\partial X_J}\frac{\partial x_k}{\partial X_K}$$

A surface element $d\boldsymbol{S} = d\boldsymbol{X}\wedge\delta\boldsymbol{X}$ consists of line elements $d\boldsymbol{X}$ and $\delta\boldsymbol{X}$ in the undeformed body and the corresponding surface element $d\boldsymbol{s} = d\boldsymbol{x}\wedge\delta\boldsymbol{x}$ consists of line elements $d\boldsymbol{x}$ and $\delta\boldsymbol{x}$ in the deformed body

$$d\boldsymbol{S} = \boldsymbol{N}\,dS = d\boldsymbol{X}\wedge\delta\boldsymbol{X}, \qquad d\boldsymbol{s} = \boldsymbol{n}\,ds = d\boldsymbol{x}\wedge\delta\boldsymbol{x} \tag{2.85}$$

where $\boldsymbol{N}$ and $\boldsymbol{n}$ are outward normals of $d\boldsymbol{S}$ and $d\boldsymbol{s}$, respectively (Fig. 2.5). If $d\boldsymbol{S}$ is deformed into $d\boldsymbol{s}$, $(2.85)_1$ implies

$$N_I\, dS = e_{IJK}\, dX_J\, dX_K = e_{IJK}\frac{\partial X_J}{\partial x_j}\frac{\partial X_K}{\partial x_k} dx_j\, dx_k,$$

and we have

$$\frac{\partial X_I}{\partial x_i} N_I\, dS = e_{IJK}\frac{\partial X_I}{\partial x_i}\frac{\partial X_J}{\partial x_j}\frac{\partial X_K}{\partial x_k} dx_j dx_k = J^{-1} n_i\, ds \;\Rightarrow\; n_i\, ds = J\frac{\partial X_I}{\partial x_i} N_I\, dS$$

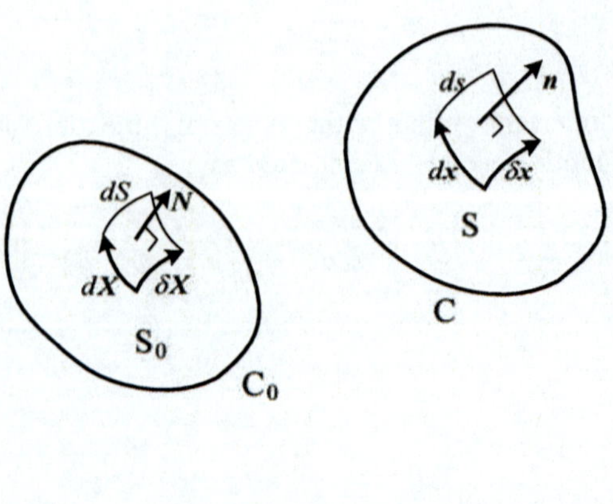

Fig. 2.5 Surface elements in the reference body C_0 and the current body C

Its vector form is given by

$$ds = n\,ds = J\,F^{-T} N\,dS = J\,F^{-T} dS \tag{2.86}$$

where $F^{-T} = (F^{-1})^T$. Equation 2.86 is referred to as *Nanson's formula.*

The time-differentiation of $F\,F^{-1} = I$ together with (2.65) gives

$$\dot{F}^{-1} = -F^{-1} L,$$

therefore the time-derivative of (2.86) can be written as

$$\frac{d}{dt}(ds) = (\dot{F}^{-1})^T J N\,dS + (F^{-1})^T \dot{J} N\,dS = \left[(\mathrm{tr}\,L) I - L^T\right] ds.$$

Thus the *transport theorem for the surface integral* of a scalar-valued function is given by

$$\frac{d}{dt}\int_S \phi\,ds = \int_S \left[\frac{d\phi}{dt} + \phi\,\mathrm{tr}\,L - \phi L^T\right] ds. \tag{2.87}$$

For a vector function q we have $q \cdot L^T ds = L q \cdot ds$ and

$$\frac{dq}{dt} + q\,\mathrm{tr}\,L - Lq = \frac{\partial q}{\partial t} + \nabla \wedge (q \wedge v) + v\,(\nabla \cdot q).$$

Thus the transport theorem (2.87) can be rewritten as

$$\begin{aligned} \frac{d}{dt}\int_S q \cdot ds &= \int_S \left[\frac{\partial q}{\partial t} + \nabla \wedge (q \wedge v) + v\,(\nabla \cdot q)\right] \cdot ds \\ &= \int_S \left[\frac{\partial q}{\partial t} + (\nabla \cdot q)\,v\right] \cdot ds + \int_C (v \wedge q) \cdot dx \end{aligned} \tag{2.88}$$

where C is a line surrounding the surface S.

Next, using (2.65) the time-derivative of the line element dx is given as

$$\frac{d}{dt}(dx) = \dot{F}\,dX = L\,dx.$$

Thus the *transport theorem for the line integral* is derived as

$$\frac{d}{dt}\int_C \phi\,dx = \int_C \left(\frac{d\phi}{dt} + \phi\,L\right) dx. \tag{2.89}$$

For a vector function $\boldsymbol{q}$ this can be written as

$$\frac{d}{dt}\int_C \boldsymbol{q}\cdot d\boldsymbol{x} = \int_C \left(\frac{d\boldsymbol{q}}{dt} + \boldsymbol{L}^T\boldsymbol{q}\right)\cdot d\boldsymbol{x}. \tag{2.90}$$

We next consider transport theorems involving a singular surface Σ that separates a body Ω into Ω^+ and Ω^- (Fig. 2.6). For example, a singular surface corresponds to the front of a shock wave causing a sonic boom or a migrating frozen front during ground freezing. Let the velocity of the singular surface Σ be $\boldsymbol{V}$, $\partial\Omega^+$ be the surface of Ω^+ except for Σ, $\partial\Omega^-$ be the surface of Ω^- except for Σ, and $\boldsymbol{n}$ be a unit normal on Σ directed to Ω^+.

Let ϕ be any function. ϕ^+ implies the value of ϕ on Σ approached from Ω^+, and, similarly, ϕ^- is the value approached from Ω^-. By applying the transport theorem (2.84) at each domain we have

$$\frac{d}{dt}\int_{\Omega^+}\phi\, dv = \int_{\Omega^+}\frac{\partial\phi}{\partial t}dv + \int_{\partial\Omega^+}\phi\boldsymbol{v}\cdot\boldsymbol{n}\, ds - \int_\Sigma \phi^+\boldsymbol{V}\cdot\boldsymbol{n}\, ds$$
$$\frac{d}{dt}\int_{\Omega^-}\phi\, dv = \int_{\Omega^-}\frac{\partial\phi}{\partial t}dv + \int_{\partial\Omega^-}\phi\boldsymbol{v}\cdot\boldsymbol{n}\, ds + \int_\Sigma \phi^-\boldsymbol{V}\cdot\boldsymbol{n}\, ds$$

where $\boldsymbol{v}$ is the velocity of each material point. Adding both equations under $V_n = \boldsymbol{V}\cdot\boldsymbol{n}$ yields the following *Reynolds' transport theorem with a singular surface*:

$$\frac{d}{dt}\int_\Omega \phi\, dv = \int_\Omega \frac{\partial\phi}{\partial t}dv + \int_{\partial\Omega}\phi\boldsymbol{v}\cdot\boldsymbol{n}\, ds - \int_\Sigma [\![\phi]\!] V_n\, ds \tag{2.91}$$

where

$$[\![\phi]\!] = \phi^+ - \phi^- \tag{2.92}$$

implies a *jump* of the function ϕ on Σ.

We will show in Sects. 2.3–2.5 that physical conservation laws can be written in the following form:

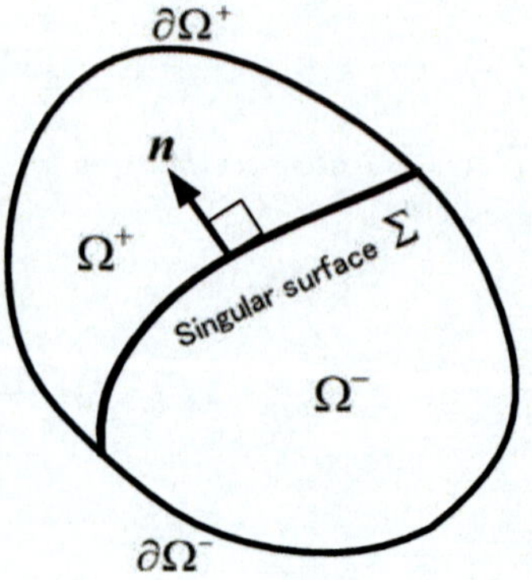

Fig. 2.6 Singular surface

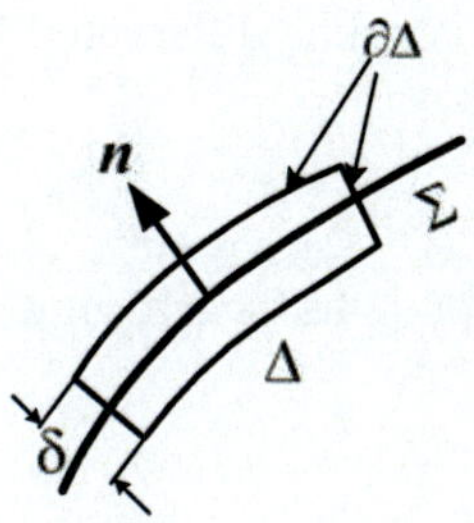

Fig. 2.7 Infinitesimal domain involving a singular surface

$$\frac{d}{dt}\int_{\Omega}\rho\varphi\,dv = \int_{\Omega}\rho s\,dv + \int_{\partial\Omega} q\,ds \tag{2.93}$$

where s is a source per mass of the field variable φ and q is a flux flowing into the body Ω through the surface $\partial\Omega$. We then apply the transport theorem (2.91) for an infinitesimal domain Δ with its boundary $\partial\Delta$ (Fig. 2.7) using the conservation law (2.93):

$$\int_{\Delta}\frac{\partial(\rho\varphi)}{\partial t}dv + \int_{\partial\Delta}\rho\varphi\boldsymbol{v}\cdot\boldsymbol{n}\,ds - \int_{\Sigma}[\![\rho\varphi]\!]V_n\,ds = \int_{\Delta}\rho s\,dv + \int_{\partial\Delta} q\,ds.$$

If the thickness δ of the domain Δ approaches zero, all the volumetric terms vanish:

$$\int_{\Sigma}[\![\rho\varphi\,(\boldsymbol{V}-\boldsymbol{v})\cdot\boldsymbol{n} + q]\!]\,ds = 0.$$

We can conclude that, for the conservation law (2.93) involving the singular surface Σ, we have the following singular surface equation:

$$[\![\rho\varphi\,(\boldsymbol{V}-\boldsymbol{v})\cdot\boldsymbol{n} + q]\!] = 0. \tag{2.94}$$

2.3 Mass Conservation Law

We refer again to Fig. 2.2. If there is no mass flux, the total mass $\mathcal{M}$ of the undeformed body Ω_0 is conserved in the deformed body Ω:

$$\mathcal{M} = \int_{\Omega_0}\rho_0\,dV = \int_{\Omega}\rho\,dv \tag{2.95}$$

where ρ_0 and ρ are the mass densities before and after deformation, respectively. Substituting (2.81) into (2.95) yields

$$\rho_0 - \rho J = 0 \quad\Rightarrow\quad J = \frac{\rho_0}{\rho} \tag{2.96}$$

The time differential form of (2.95) using the Reynolds' transport theorem (2.84) gives

$$\frac{d\mathcal{M}}{dt} = \frac{d}{dt}\int_{\Omega}\rho\, dv = \int_{\Omega}\frac{\partial\rho}{\partial t}dv + \int_{\partial\Omega}\rho\boldsymbol{v}\cdot\boldsymbol{n}\, ds = 0,$$

or in the local form we have the following *mass conservation law*:

$$\frac{d\rho}{dt} + \rho\nabla\cdot\boldsymbol{v} = \frac{\partial\rho}{\partial t} + \nabla\cdot(\rho\boldsymbol{v}) = 0. \tag{2.97}$$

Equation 2.97 is sometimes referred to as the *continuity equation*. Then using (2.81), (2.83) and (2.97) we can see that

$$\dot{dv} = -\frac{\dot{\rho}}{\rho}dv = \operatorname{tr}\boldsymbol{D}\, dV \quad \Rightarrow \quad \nabla\cdot\boldsymbol{v} = \operatorname{tr}\boldsymbol{D} = -\frac{\dot{\rho}}{\rho} = -\frac{d(\ln\rho)}{dt}. \tag{2.98}$$

If the material is incompressible, ρ is constant, and the mass conservation law (2.97) can be written as

$$\nabla\cdot\boldsymbol{v} = \frac{\partial v_i}{\partial x_i} = 0, \tag{2.99}$$

which gives the *incompressibility condition*.

The mass conservation law (2.97) gives an alternative form of the Reynolds' transport theorem (2.84) as

$$\frac{d}{dt}\int_{\Omega}\rho\phi\, dv = \int_{\Omega}\rho\frac{d\phi}{dt}dv. \tag{2.100}$$

2.4 Law of Conservation of Linear Momentum and Stress

Newton's second law states that in an inertial frame the rate of linear momentum is equal to the applied force. Here, by applying the second law to a continuum region, we define the Cauchy stress, and derive the equation of motion.

2.4.1 Eulerian Descriptions

Linear momentum of the deformed body Ω is given by

$$\mathcal{L} = \int_{\Omega}\rho\boldsymbol{v}\, dv.$$

If an external force per unit area $\boldsymbol{t}$, called the *traction* or *stress vector*, acts on a boundary $\partial\Omega$, with the body force per unit volume $\boldsymbol{b}$ acting in the volume, the total force is

$$\mathcal{F} = \int_{\partial\Omega} \boldsymbol{t}\, ds + \int_{\Omega} \rho \boldsymbol{b}\, dv.$$

Since Newton's second law is given by $\dot{\mathcal{L}} = \mathcal{F}$,

$$\frac{d}{dt}\int_{\Omega} \rho \boldsymbol{v}\, dv = \int_{\partial\Omega} \boldsymbol{t}\, ds + \int_{\Omega} \rho \boldsymbol{b}\, dv. \tag{2.101}$$

An example of the body force $\boldsymbol{b} = (b_1, b_2, b_3)$ is the gravitational force. If it acts in the negative z-direction, we have $\rho\boldsymbol{b} = (0, 0, -\gamma)$ where $\gamma = \rho g$ (ρ is the mass density, g the acceleration due to gravity and γ the unit weight).

We consider a surface S within the body Ω. One part of Ω bisected by S is denoted by Ω^+ and the other part is denoted by Ω^- (Fig. 2.8). The outward normal vector $\boldsymbol{n}$ is set on the surface S observed from Ω^+. Let an infinitesimal rectangular parallelepiped be located on the surface S with the thickness δ. A surface of the $\boldsymbol{n}$ side of the parallelepiped is referred to as ΔS^+, the opposite side is ΔS^- and other lateral surfaces are ΔS^δ. The surface area of ΔS^+ and ΔS^- is ΔS, and the total area of the lateral surfaces is ΔS^δ. Tractions, i.e., forces per unit area, acting on ΔS^+, ΔS^- and ΔS^δ are $\boldsymbol{t}^+$, $\boldsymbol{t}^-$ and $\boldsymbol{t}^\delta$, respectively. Applying Newton's second law (2.101) to the parallelepiped gives

$$\rho \frac{d\boldsymbol{v}}{dt} \delta \Delta S = \boldsymbol{t}^+ \Delta S + \boldsymbol{t}^- \Delta S + \boldsymbol{t}^\delta \Delta S^\delta + \rho \boldsymbol{b}\, \delta \Delta S,$$

and $\delta \to 0$ under $\Delta S = \text{constant}$. The terms in the above equation that are dependent on the volume and the lateral surface all vanish, with the result

$$\boldsymbol{t}^- = -\boldsymbol{t}^+. \tag{2.102}$$

This relation is known as *Cauchy's lemma*.

We next consider an infinitesimal tetrahedron OABC in the body Ω (Fig. 2.9). Let the area of ABC be ds, the outward unit vector on ABC be $\boldsymbol{n}$, the traction acting on ABC be $\boldsymbol{t}$, while the distance between ABC and O is h. In addition, let

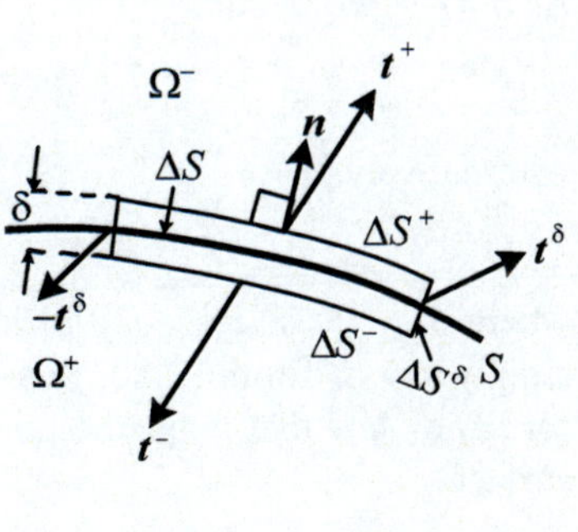

Fig. 2.8 Stress vector defined on an internal surface

Fig. 2.9 Tetrahedron defining the stress

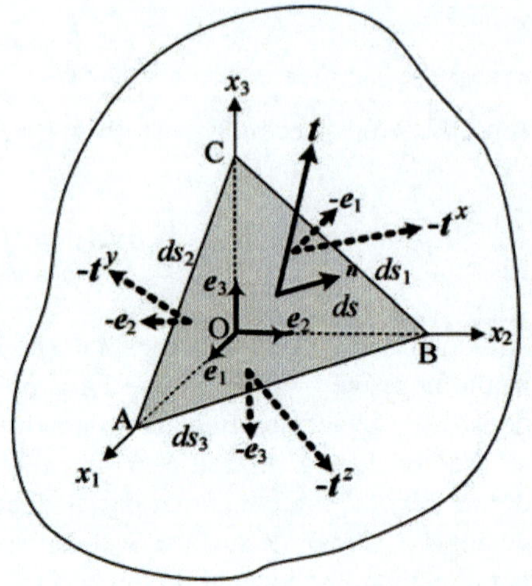

the areas of OBC, OCA and OAB be ds_1, ds_2 and ds_3, respectively, so that we have

$$ds_i = ds\ \cos(\boldsymbol{n}, x_i) = n_i\ ds \quad (i = 1, 2, 3).$$

The volume of the tetrahedron is given as $dv = h\,ds/3$. Let the traction acting on the x^+-surface be $\boldsymbol{t}^x = \boldsymbol{t}^1$. Similarly, $\boldsymbol{t}^y = \boldsymbol{t}^2$ and $\boldsymbol{t}^z = \boldsymbol{t}^3$, which are tractions acting on the y^+- and z^+-surface, respectively. The outward unit normal on OBC is $-\boldsymbol{e}_1$, therefore it is the $\boldsymbol{x}^-$-surface. Then using Cauchy's lemma (2.102) the traction acting on this surface is given by

$$-\boldsymbol{t}^x ds_1 = (-t_1^x, -t_2^x, -t_3^x)^T n_1 ds.$$

Similar results can be obtained for the other surfaces. Now let the body force acting in this tetrahedron be $\boldsymbol{b}$, and the linear momentum be $\rho\boldsymbol{v}$. Then the law of conservation of linear momentum for the tetrahedron states that

$$-\boldsymbol{t}^x n_1 ds - \boldsymbol{t}^y n_2 ds - \boldsymbol{t}^z n_3 ds + \boldsymbol{t}\,ds + \rho\boldsymbol{b}\frac{1}{3}h\,ds = \rho\dot{\boldsymbol{v}}\frac{1}{3}h\,ds.$$

As $h \to 0$, we obtain

$$\boldsymbol{t} = \boldsymbol{t}^x n_1 + \boldsymbol{t}^y n_2 + \boldsymbol{t}^z n_3.$$

Alternatively,

$$\boldsymbol{t} = \boldsymbol{\sigma}^T \boldsymbol{n} = \sigma_{ji}\, n_i\, \boldsymbol{e}_i, \qquad \boldsymbol{\sigma} = \sigma_{ij}\, \boldsymbol{e}_i \otimes \boldsymbol{e}_j \tag{2.103}$$

where $\sigma_{ij} = t_j^i$. The second-order tensor $\boldsymbol{\sigma}$ is referred to as the *Cauchy stress* or simply *stress*. From (2.103) we understand that the stress tensor $\boldsymbol{\sigma}$ gives a transformation law that maps the unit outward normal $\boldsymbol{n}$ to the traction $\boldsymbol{t}$ acting on that surface.

Returning to the law of conservation of linear momentum (2.101) and using the mass conservation law, we have

$$\frac{d}{dt}\int_{\Omega} \rho \boldsymbol{v}\, dv = \int_{\Omega} \rho \frac{d\boldsymbol{v}}{dt}\, dv.$$

Therefore, by substituting (2.103) into the first term of the r.h.s. of (2.101) and applying the divergence theorem, we have

$$\int_{\partial\Omega} \boldsymbol{t}\, ds = \int_{\partial\Omega} \boldsymbol{\sigma}^T \boldsymbol{n}\, ds = \int_{\Omega} \operatorname{div} \boldsymbol{\sigma}^T\, dv$$

Then we can obtain the following *Eulerian form of the equation of motion* defined in the current deformed body:

$$\rho \frac{d\boldsymbol{v}}{dt}(\boldsymbol{x}, t) = \rho \left(\frac{\partial \boldsymbol{v}}{\partial t} + \boldsymbol{v} \cdot \nabla \boldsymbol{v} \right) = \operatorname{div} \boldsymbol{\sigma}^T(\boldsymbol{x}, t) + \rho\, \boldsymbol{b}(\boldsymbol{x}, t) \tag{2.104}$$

where

$$\operatorname{div} \boldsymbol{\sigma}^T = \nabla \cdot \boldsymbol{\sigma}^T = \frac{\partial \sigma_{ji}}{\partial x_j} \boldsymbol{e}_i.$$

The component form of (2.104) is given by

$$\rho \frac{dv_i}{dt} = \frac{\partial \sigma_{ji}}{\partial x_j} + \rho\, b_i \tag{2.105}$$

For a static equilibrium problem the partial differential equation system in Eulerian form together with the boundary conditions is given by

$$\nabla \cdot \boldsymbol{\sigma}^T + \rho \boldsymbol{b} = \boldsymbol{0} \tag{2.106}$$

$$\boldsymbol{u}(\boldsymbol{x}) = \bar{\boldsymbol{u}} \quad \text{on} \quad \partial\Omega_u, \tag{2.107}$$

$$\boldsymbol{\sigma}^T \boldsymbol{n}(\boldsymbol{x}) = \bar{\boldsymbol{t}} \quad \text{on} \quad \partial\Omega_t \tag{2.108}$$

2.4.2 Lagrangian Descriptions ♣

We introduce a relationship between the Cauchy stress $\boldsymbol{\sigma}$ defined in the deformed body with its basis $\{\boldsymbol{e}_i\}$ and the *first Piola-Kirchhoff stress* $\boldsymbol{\Pi}$ defined in the undeformed body with its basis $\{\boldsymbol{E}_I\}$ as follows:

$$\boldsymbol{t}\, ds = \boldsymbol{\sigma}^T \boldsymbol{n}\, ds = \boldsymbol{t}^0\, dS = \boldsymbol{\Pi}^T \boldsymbol{N}\, dS \tag{2.109}$$

where $\boldsymbol{n}$ and N are unit outward normals defined on ds for the deformed body and on dS for the undeformed body. The vector $\boldsymbol{t}$ is a traction defined on ds, and $\boldsymbol{t}^0$ is the shifted vector of $\boldsymbol{t}$ on dS (see Fig. 2.10). By substituting Nanson's relation (2.86), such that $\boldsymbol{n}\, ds = J\, \boldsymbol{F}^{-T} \boldsymbol{N}\, dS$, into (2.109), the first Piola-Kirchhoff stress can be written as

$$\boldsymbol{\Pi} = \Pi_{Ii} \boldsymbol{E}_I \otimes \boldsymbol{e}_i = J \boldsymbol{F}^{-1} \boldsymbol{\sigma}, \qquad \Pi_{Ii} = J \frac{\partial X_I}{\partial x_j} \sigma_{ji} \tag{2.110}$$

The transpose of the first Piola-Kirchhoff stress $\boldsymbol{S} = \boldsymbol{\Pi}^T$ is known as the nominal stress:

$$\boldsymbol{S} = S_{iI} \boldsymbol{e}_i \otimes \boldsymbol{E}_I = J \boldsymbol{\sigma} \boldsymbol{F}^{-T}, \qquad S_{iI} = J \sigma_{ij} \frac{\partial X_I}{\partial x_j} \tag{2.111}$$

Note that in some books, e.g., Kitagawa (1987) pp. 33, the first Piola-Kirchhoff stress and the nominal stress are defined in an opposite sense.

Since the first Piola-Kirchhoff stress $\boldsymbol{\Pi}$ is not symmetric as understood by (2.110), we introduce a symmetrized tensor $\boldsymbol{T}$, called the *second Piola-Kirchhoff stress*, and the Euler stress $\boldsymbol{\tau}$, which is the transformed tensor of $\boldsymbol{T}$, into the deformed body using the rotation tensor $\boldsymbol{R}$:

$$\boldsymbol{T} = J \boldsymbol{F}^{-1} \boldsymbol{\sigma} \boldsymbol{F}^{-T} = \boldsymbol{\Pi}\, \boldsymbol{F}^{-T} = \boldsymbol{F}^{-1} \boldsymbol{S}, \tag{2.112}$$

$$\boldsymbol{\tau} = \boldsymbol{R} \boldsymbol{T} \boldsymbol{R}^T. \tag{2.113}$$

Referring to (2.109), Newton's equation of motion (2.101) can be expressed by the Lagrangian description as

$$\int_{\Omega_0} \rho \frac{d\boldsymbol{v}}{dt}(\boldsymbol{X}, t)\, J\, dV = \int_{\partial\Omega_0} \boldsymbol{\Pi}^T \boldsymbol{N}(\boldsymbol{X}, t)\, dS + \int_{\Omega_0} \rho \boldsymbol{b}(\boldsymbol{X}, t)\, J\, dV.$$

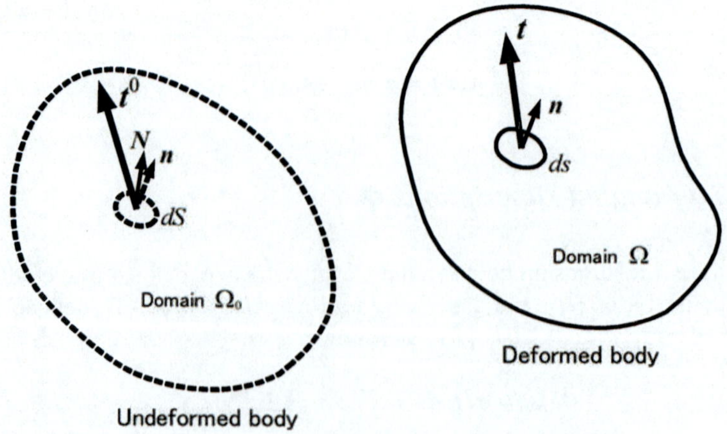

Fig. 2.10 Traction vectors represented for undeformed and deformed bodies

Recalling $J = \rho_0/\rho$ and applying the divergence theorem to the first term of the r.h.s. yields the following *Lagrangian form of the equation of motion*:

$$\rho_0 \frac{d\boldsymbol{v}}{dt}(\boldsymbol{X},t) = \text{Div}\, \boldsymbol{\Pi}^T(\boldsymbol{X},t) + \rho_0 \boldsymbol{b}(\boldsymbol{X},t) \tag{2.114}$$

where we have

$$\text{Div}\, \boldsymbol{\Pi}^T = (\Pi_{Ii} \boldsymbol{e}_i \otimes \boldsymbol{E}_I)\left(\boldsymbol{e}_j \frac{\partial}{\partial X_J}\right) = \frac{\partial \Pi_{Ii}}{\partial X_I} \boldsymbol{e}_i .$$

Then a component form of (2.114) is given by

$$\rho_0 \frac{dv_i}{dt} = \frac{\partial \Pi_{Ii}}{\partial X_I} + \rho_0\, b_i . \tag{2.115}$$

For a static equilibrium problem, the system of partial differential equations in Lagrangian form together with the boundary conditions is given by

$$\text{Div}\, \boldsymbol{\Pi}^T(\boldsymbol{X}) + \rho_0 \boldsymbol{b}(\boldsymbol{X}) = \boldsymbol{0} \tag{2.116}$$

$$\boldsymbol{u}(\boldsymbol{X}) = \bar{\boldsymbol{u}}^0 \qquad \text{on} \quad \partial\Omega_u, \tag{2.117}$$

$$\boldsymbol{\Pi}^T \boldsymbol{N}(\boldsymbol{X}) = \bar{\boldsymbol{t}}^0 \qquad \text{on} \quad \partial\Omega_t \tag{2.118}$$

2.5 Conservation of Moment of Linear Momentum and Symmetry of Stress

Let $\boldsymbol{x}$ be a position vector in the deformed body Ω, then the total moment of linear momentum of the body with respect to the origin O is calculated by

$$\mathcal{H} = \int_\Omega \boldsymbol{x} \wedge \rho \boldsymbol{v}\, dv,$$

and the total torque $\mathcal{T}$ due to an external force $\boldsymbol{t}$ and a body force $\boldsymbol{b}$ is given by

$$\mathcal{T} = \int_{\partial\Omega} \boldsymbol{x} \wedge \boldsymbol{t}\, ds + \int_\Omega \boldsymbol{x} \wedge \rho\, \boldsymbol{b}\, dv.$$

The *conservation law for the moment of linear momentum* states that $\dot{\mathcal{H}} = \mathcal{T}$; therefore we have

$$\frac{d}{dt} \int_\Omega \boldsymbol{x} \wedge \rho \boldsymbol{v}\, dv = \int_{\partial\Omega} \boldsymbol{x} \wedge \boldsymbol{t}\, ds + \int_\Omega \boldsymbol{x} \wedge \rho\, \boldsymbol{b}\, dv. \tag{2.119}$$

Applying the transport theorem (2.84) to the l.h.s. of (2.119) and noting $d\boldsymbol{x}/dt \wedge \boldsymbol{v} = \boldsymbol{v} \wedge \boldsymbol{v} = \boldsymbol{0}$ yields

$$e_{ijk}x_j\,\rho\frac{dv_k}{dt} = e_{ijk}\frac{\partial x_j \sigma_{lk}}{\partial x_l} + e_{ijk}x_j\,\rho b_k$$

$$\Rightarrow e_{ijk}x_j\left(\rho\frac{dv_k}{dt} - \frac{\partial\sigma_{lk}}{\partial x_l} - b_k\right) - e_{ijk}\sigma_{jk} = 0.$$

The terms in ( ) of this equation vanish because of the equation of motion (2.104), and eventually the following result is obtained:

$$e_{ijk}\sigma_{jk} = 0.$$

This implies

$$\sigma_{jk} = \sigma_{kj} \quad \Rightarrow \quad \boldsymbol{\sigma} = \boldsymbol{\sigma}^T. \tag{2.120}$$

That is, if we have the conservation law of moment of linear momentum and assume no point-wise source term of moment, the Cauchy stress is symmetric. A further exposition of the symmetry property of the Cauchy stress is given by Selvadurai (2000b).

2.6 Incremental Forms of the Equation of Equilibrium ♣

For nonlinear problems such as elasto-plastic materials it is necessary to use a formulation based on an incremental form of the equation of equilibrium. We can introduce either the *total Lagrangian form* or the *updated Lagrangian form*. In the former case the incremental form is expressed in Lagrangian terms, while in the latter case the incremental form is given in an Eulerian description.

2.6.1 Total Lagrangian Form

The total Lagrangian form of the equation of equilibrium is obtained by differentiating (2.116) directly. Thus the partial differential equation system together with the boundary conditions is given by

$$\text{Div}\,\dot{\boldsymbol{\Pi}}^T(\boldsymbol{X}) + \rho_0\dot{\boldsymbol{b}}(\boldsymbol{X}) = \boldsymbol{0}, \tag{2.121}$$

$$\boldsymbol{v}(\boldsymbol{X}) = \bar{\boldsymbol{v}}^0 \qquad \text{on} \quad \partial\Omega_u, \tag{2.122}$$

$$\dot{\boldsymbol{\Pi}}^T\boldsymbol{N}(\boldsymbol{X}) = \dot{\bar{\boldsymbol{t}}}^0 \qquad \text{on} \quad \partial\Omega_t. \tag{2.123}$$

The component form can be written as

$$\frac{\partial \dot{\Pi}_{Ii}(X)}{\partial X_I} + \rho_0 \dot{b}_i(X) = 0, \tag{2.124}$$

$$v_i(X) = \bar{v}_i^0 \qquad \text{on} \quad \partial\Omega_u^0, \tag{2.125}$$

$$\dot{\Pi}_{Ii} N_I(X) = \dot{\bar{t}}_i^0 \qquad \text{on} \quad \partial\Omega_t^0. \tag{2.126}$$

2.6.2 Updated Lagrangian Form

By integrating each term of Nanson's relation $\boldsymbol{n}\, ds = J\, \boldsymbol{F}^{-T} \boldsymbol{N}\, dS$ we have

$$\int_{\partial\Omega} \boldsymbol{n}\, ds = \int_{\partial\Omega_0} J\, \boldsymbol{F}^{-T} \boldsymbol{N}\, dS = \int_{\Omega_0} \mathrm{Div}(J\, \boldsymbol{F}^{-T})\, dV = \boldsymbol{0},$$

$$\int_{\partial\Omega_0} \boldsymbol{N}\, dS = \int_{\partial\Omega} J^{-1}\, \boldsymbol{F}^T \boldsymbol{n}\, ds = \int_{\Omega} \mathrm{div}(J^{-1}\, \boldsymbol{F}^T)\, dv = \boldsymbol{0}.$$

These give the following equations:

$$\mathrm{Div}(J\, \boldsymbol{F}^{-T}) = \frac{\partial}{\partial X_I} \left(J\, F_{Ii}^{-1}\right) \boldsymbol{e}_i = \boldsymbol{0}, \quad \mathrm{div}(J^{-1}\, \boldsymbol{F}^T) = \frac{\partial}{\partial x_i} \left(J^{-1}\, F_{iI}\right) \boldsymbol{E}_I = \boldsymbol{0}. \tag{2.127}$$

Next, (2.110) implies that

$$J\boldsymbol{\sigma} = \boldsymbol{F}\boldsymbol{\Pi},$$

which can be time-differentiated to yield the *nominal stress rate* $\overset{\circ}{\boldsymbol{\Pi}}$ defined by

$$\overset{\circ}{\boldsymbol{\Pi}} \equiv J^{-1} \boldsymbol{F} \dot{\boldsymbol{\Pi}} = \dot{\boldsymbol{\sigma}} - \boldsymbol{L}\boldsymbol{\sigma} + \boldsymbol{\sigma}\, \mathrm{tr}\, \boldsymbol{D}. \tag{2.128}$$

Equation 2.128 implies that $\overset{\circ}{\boldsymbol{\Pi}}$ is an image of $\dot{\boldsymbol{\Pi}}$ in the deformed body mapped from the undeformed body by $J^{-1}\boldsymbol{F}$. A component form of (2.128) is given by

$$\overset{\circ}{\Pi}_{ij} = J^{-1} F_{iI} \dot{\Pi}_{Ij} \tag{2.129}$$

We take the divergence of (2.129) using $(2.127)_2$ and obtain

$$\frac{\partial \overset{\circ}{\Pi}_{ji}}{\partial x_j} = J^{-1} F_{jI} \frac{\partial \dot{\Pi}_{Ii}}{\partial x_j} = J^{-1} \frac{\partial x_j}{\partial X_I} \frac{\partial \dot{\Pi}_{Ii}}{\partial x_j} = J^{-1} \frac{\partial \dot{\Pi}_{Ii}}{\partial X_I}.$$

Since $J = \rho_0/\rho$, (2.124) for the deformed body can be given as

$$\frac{\partial \overset{\circ}{\Pi}_{ji}}{\partial x_j} + \rho \dot{b}_i(\boldsymbol{x}) = 0. \tag{2.130}$$

Thus the partial differential equation system of the updated Lagrangian form together with the boundary conditions is given by

$$\operatorname{div} \overset{\circ}{\boldsymbol{\Pi}}^T(\boldsymbol{x}) + \rho \dot{\boldsymbol{b}}(\boldsymbol{x}) = \boldsymbol{0}, \tag{2.131}$$

$$\boldsymbol{v}(\boldsymbol{x}) = \bar{\boldsymbol{v}} \qquad \text{on} \quad \partial\Omega_u, \tag{2.132}$$

$$\overset{\circ}{\boldsymbol{\Pi}}^T \boldsymbol{n}(\boldsymbol{x}) = \bar{\dot{\boldsymbol{t}}} \qquad \text{on} \quad \partial\Omega_t. \tag{2.133}$$

It must be noted that, as understood from (2.131), "the updated Lagrangian form is expressed in Eulerian terms".

2.7 Specific Description of the Equation of Motion ♣

As we shall discuss in Chap. 3, specific descriptions of stress and stress increments are preferable when formulating energy theorems. Hence we will now rewrite the equations of motion in the specific forms.

2.7.1 *Eulerian Equation of Motion*

Let us define the normalized measure of Cauchy stress $\boldsymbol{\sigma}^\ddagger$ by

$$\boldsymbol{\sigma}^\ddagger(\boldsymbol{x}, t) = \frac{1}{\rho} \boldsymbol{\sigma}(\boldsymbol{x}, t). \tag{2.134}$$

Then the Eulerian equation of motion (2.104) can be written as follows:

$$\rho \frac{d\boldsymbol{v}}{dt} = \operatorname{div} (\rho \boldsymbol{\sigma}^\ddagger)^T + \rho \boldsymbol{b}. \tag{2.135}$$

It is interesting to note that the dimension of Cauchy stress $\boldsymbol{\sigma}$ is M/LT^2 (N/m^2 = Pa in MKS) while for $\boldsymbol{\sigma}^\ddagger$ it is L^2/T^2 (J/kg in MKS), which has the dimension of energy per unit mass.

2.7.2 Lagrangian Equation of Motion

The specific first Piola-Kirchhoff stress $\boldsymbol{\Pi}^{\ddagger}$, specific second Piola-Kirchhoff stress $\boldsymbol{T}^{\ddagger}$ and specific Euler stress $\boldsymbol{\tau}^{\ddagger}$ are defined by

$$\boldsymbol{\Pi}^{\ddagger} = \frac{1}{\rho}\boldsymbol{\Pi}, \qquad \boldsymbol{T}^{\ddagger} = \frac{1}{\rho_0}\boldsymbol{T}, \qquad \boldsymbol{\tau}^{\ddagger} = \boldsymbol{R}\boldsymbol{T}^{\ddagger}\boldsymbol{R}^T \tag{2.136}$$

Since $J = \rho_0/\rho$, and using (2.110) and (2.112),

$$\boldsymbol{\Pi}^{\ddagger} = \boldsymbol{F}^{-1}\boldsymbol{\sigma}^{\ddagger}, \tag{2.137}$$

$$\boldsymbol{T}^{\ddagger} = \boldsymbol{F}^{-1}\boldsymbol{\sigma}^{\ddagger}\boldsymbol{F}^{-T}. \tag{2.138}$$

Then the Lagrangian equation of motion (2.114) can be written as

$$\rho_0 \frac{d\boldsymbol{v}}{dt} = \text{Div}\,(\rho_0\boldsymbol{\Pi}^{\ddagger})^T + \rho_0\boldsymbol{b}. \tag{2.139}$$

2.7.3 Incremental Form of the Total Lagrangian Equation of Motion

Differentiating (2.139) with respect to time yields

$$\rho_0 \frac{d\dot{\boldsymbol{v}}(\boldsymbol{X},t)}{dt} = \text{Div}\left(\rho_0\dot{\boldsymbol{\Pi}}^{\ddagger}(\boldsymbol{X},t)\right)^T + \rho_0\dot{\boldsymbol{b}}(\boldsymbol{X},t) \tag{2.140}$$

where $\dot{\boldsymbol{v}} = d\boldsymbol{v}/dt = \boldsymbol{a}$ ($\boldsymbol{a}$ is the acceleration), and the l.h.s. of (2.140) is the rate of acceleration, which can be regarded as an increment of the velocity. Equation 2.140 is equivalent to (2.121) if $\boldsymbol{v} = \boldsymbol{0}$. The component form of (2.140) is given by

$$\rho_0 \frac{d\dot{v}_i}{dt} = \frac{\partial}{\partial X_I}\left(\rho_0\dot{\Pi}^{\ddagger}_{Ii}\right) + \rho_0\dot{b}. \tag{2.141}$$

2.7.4 Incremental Form of the Updated Lagrangian Equation of Motion

From (2.137), $\boldsymbol{\sigma}^{\ddagger} = \boldsymbol{F}\boldsymbol{\Pi}^{\ddagger}$ and differentiating this yields

$$\dot{\boldsymbol{\sigma}}^{\ddagger} = \boldsymbol{L}\boldsymbol{\sigma}^{\ddagger} + \boldsymbol{F}\dot{\boldsymbol{\Pi}}^{\ddagger}.$$

Thus we can define the specific nominal stress rate by

$$\overset{\circ}{\boldsymbol{\Pi}}^{\ddagger} \equiv \boldsymbol{F}\dot{\boldsymbol{\Pi}}^{\ddagger} = \dot{\boldsymbol{\sigma}}^{\ddagger} - \boldsymbol{L}\boldsymbol{\sigma}^{\ddagger} \tag{2.142}$$

On the other hand we have

$$\frac{\partial}{\partial X_I} = F_{jI} \frac{\partial}{\partial x_j}.$$

Then (2.140) can be expressed as

$$\rho_0 \frac{d\dot{\boldsymbol{v}}(\boldsymbol{x},t)}{dt} = \operatorname{div} \left(\rho_0 \overset{\circ}{\boldsymbol{\Pi}}^{\ddagger}(\boldsymbol{x},t) \right)^T + \rho_0 \dot{\boldsymbol{b}}(\boldsymbol{x},t) \tag{2.143}$$

The l.h.s. of (2.143) can be written in the normal Eulerian description as

$$\rho_0 \frac{d\dot{\boldsymbol{v}}(\boldsymbol{x},t)}{dt} = \rho_0 \left(\frac{\partial \dot{\boldsymbol{v}}}{\partial t} + \boldsymbol{v} \cdot \operatorname{div} \dot{\boldsymbol{v}} \right). \tag{2.144}$$

The component form of (2.143) is given as follows:

$$\rho_0 \left(\frac{\partial \dot{v}_i}{\partial t} + v_j \frac{\partial \dot{v}_i}{\partial x_j} \right) = \frac{\partial}{\partial x_j} \left(\rho_0 \overset{\circ}{\Pi}{}^{\ddagger}_{ji} \right) + \rho_0 \dot{b}_i. \tag{2.145}$$

Equation 2.143 is equivalent to (2.131) if $\boldsymbol{v} = \boldsymbol{0}$; however, it is interesting to note that the body force $\dot{\boldsymbol{b}}$ is modified by ρ, while each term of (2.143) is modified by ρ_0.

2.8 Response of Materials: Constitutive Theory

The governing equations that control material responses are given by the mass conservation law (2.97) and the equation of motion (2.104) if no energy conservation is considered. Note that the Cauchy stress is symmetric under the conservation law of moment of linear momentum. Furthermore, if the change of mass density is small (or it may be constant), the equation to be solved is given by (2.104). The unknowns in this equation are the velocity $\boldsymbol{v}$ (or displacement $\boldsymbol{u}$ in the small strain theory) and the stress $\boldsymbol{\sigma}$, i.e. giving a total of nine, that is, three for $\boldsymbol{v}$ (or $\boldsymbol{u}$) and six for $\boldsymbol{\sigma}$. However, the equation of motion (2.104) consists of three components, therefore it cannot be solved, suggesting that we must introduce a relationship between $\boldsymbol{v}$ (or $\boldsymbol{u}$) and $\boldsymbol{\sigma}$. The framework that provides this relationship is referred to as a constitutive theory.

2.8.1 Fundamental Principles of Material Response

The constitutive law is fundamentally determined for each material, and it gives an empirical rule. To establish constitutive laws the following physical conditions are required:

Principle of determinism: The stress is determined by the history of the motion undergone by the body.
Principle of local action: The stress at a point is not influenced by far-field motions.
Principle of frame indifference: The response of a material must be described under the frame indifference (see Sect. 2.2.2).

The principle of frame indifference is sometimes called *objectivity*.

Let a stress $\boldsymbol{\sigma}$ be described as a function of a material point $\boldsymbol{x}$ in the deformed body at time t:

$$\boldsymbol{\sigma} = \boldsymbol{\sigma}(\boldsymbol{x}, t). \tag{2.146}$$

As shown by (2.18), a coordinate transformation of the point $\boldsymbol{x}$ between two different coordinate systems defined in the deformed body Ω is written as

$$\boldsymbol{x}^* = \boldsymbol{x}_0^* + \boldsymbol{Q}(\boldsymbol{x} - \boldsymbol{x}_0), \qquad t^* = t - a, \tag{2.147}$$

and the frame indifference of the stress $\boldsymbol{\sigma}$ is therefore

$$\boldsymbol{\sigma}^*(\boldsymbol{x}^*, t^*) = \boldsymbol{Q}(t)\, \boldsymbol{\sigma}(\boldsymbol{x}, t)\, \boldsymbol{Q}^T(t). \tag{2.148}$$

Because (2.147) implies that $\boldsymbol{F}^* d\boldsymbol{X} = \boldsymbol{Q}\boldsymbol{F}\, d\boldsymbol{X}$, the deformation gradient $\boldsymbol{F}$ is transformed as

$$\boldsymbol{F}^* = \boldsymbol{Q}\, \boldsymbol{F}, \qquad \boldsymbol{Q} = Q_{ij} \boldsymbol{e}_i^* \otimes \boldsymbol{e}_j. \tag{2.149}$$

Thus $\boldsymbol{F}$ is not frame indifferent. Some tensors introduced in Sect. 2.2.4 are verified as follows:

$$\boldsymbol{C}^* = \boldsymbol{F}^{*T} \boldsymbol{F}^* = (\boldsymbol{Q}\boldsymbol{F})^T (\boldsymbol{Q}\boldsymbol{F}) = \boldsymbol{F}^T \boldsymbol{F} = \boldsymbol{C}, \tag{2.150}$$

$$\boldsymbol{B}^* = \boldsymbol{F}^* \boldsymbol{F}^{*T} = (\boldsymbol{Q}\boldsymbol{F})(\boldsymbol{Q}\boldsymbol{F})^T = \boldsymbol{Q}\, \boldsymbol{B}\, \boldsymbol{Q}^T, \tag{2.151}$$

$$\boldsymbol{E}^* = \boldsymbol{F}^* \boldsymbol{F}^{*T} - \boldsymbol{I}^* = (\boldsymbol{Q}\boldsymbol{F})^T (\boldsymbol{Q}\boldsymbol{F}) - \boldsymbol{I} = \boldsymbol{E}, \tag{2.152}$$

$$\boldsymbol{L}^* = \dot{\boldsymbol{F}}^* (\boldsymbol{F}^*)^{-1} = (\boldsymbol{Q}\dot{\boldsymbol{F}} + \dot{\boldsymbol{F}}\boldsymbol{Q})(\boldsymbol{Q}\boldsymbol{F})^{-1} = \boldsymbol{Q}\, \boldsymbol{L}\, \boldsymbol{Q}^T + \boldsymbol{\Omega}, \tag{2.153}$$

$$\boldsymbol{D}^* = \frac{1}{2}\left(\boldsymbol{L}^* + \boldsymbol{L}^{*T}\right) = \boldsymbol{Q}\, \boldsymbol{D}\, \boldsymbol{Q}^T, \tag{2.154}$$

$$\boldsymbol{W}^* = \frac{1}{2}\left(\boldsymbol{L}^* - \boldsymbol{L}^{*T}\right) = \boldsymbol{Q}\, \boldsymbol{W}\, \boldsymbol{Q}^T + \boldsymbol{\Omega}. \tag{2.155}$$

Note that the right Cauchy-Green tensor $\boldsymbol{C}$ and the Green strain $\boldsymbol{E}$ are not frame indifferent, but frame invariant.

2.8.2 *Convected Derivative, Corotational Derivative and Frame Indifference* ♣

If we consider time-differentiation of vectors and tensors, not only the components but also the basis must be differentiated. Thus even if the original vectors and tensors are frame indifferent, their time-derivatives are not in general frame indifferent. In order to avoid this difficulty several time-differential forms are considered.

Embedded coordinates are used (recalling Note 2.5) with the covariant basis $\{\boldsymbol{g}_i\}$ and contravariant basis $\{\boldsymbol{g}^i\}$ in the deformed body (i.e., Eulerian description), so that a vector $\boldsymbol{v}$ can be written as

$$\boldsymbol{v} = v^i\, \boldsymbol{g}_i = v_i\, \boldsymbol{g}^i. \tag{2.156}$$

Its material time derivative is given by

$$\dot{\boldsymbol{v}} = \dot{v}^i\, \boldsymbol{g}_i + v^i\, \dot{\boldsymbol{g}}_i = \dot{v}_i\, \boldsymbol{g}^i + v_i\, \dot{\boldsymbol{g}}^i. \tag{2.157}$$

Recalling (2.78), the time-derivative of the covariant basis is $\dot{\boldsymbol{g}}_i = \boldsymbol{L}\boldsymbol{g}_i$. On the other hand, since $\boldsymbol{g}^i = \boldsymbol{F}^{-T}\boldsymbol{G}^i$ due to (2.74) and $\dot{\overline{\boldsymbol{F}^{-1}}} = -\boldsymbol{F}^{-1}\boldsymbol{L}$ due to (2.68), the time-derivative of the contravariant basis is given by

$$\dot{\boldsymbol{g}}^i = -\boldsymbol{L}^T \boldsymbol{g}^i. \tag{2.158}$$

Now we can define the *convected derivatives* by

$$\frac{\delta^c \boldsymbol{v}}{\delta t} = \dot{v}^i\, \boldsymbol{g}_i, \qquad \frac{\delta_c \boldsymbol{v}}{\delta t} = \dot{v}_i\, \boldsymbol{g}^i. \tag{2.159}$$

Then from (2.158) we have

$$\frac{\delta^c \boldsymbol{v}}{\delta t} = \dot{\boldsymbol{v}} - \boldsymbol{L}\boldsymbol{v} = \boldsymbol{F}\frac{d(\boldsymbol{F}^{-1}\boldsymbol{v})}{dt}, \tag{2.160}$$

$$\frac{\delta_c \boldsymbol{v}}{\delta t} = \dot{\boldsymbol{v}} + \boldsymbol{L}^T\boldsymbol{v} = \boldsymbol{F}^{-T}\frac{d(\boldsymbol{F}^T\boldsymbol{v})}{dt}. \tag{2.161}$$

As observed in (2.159), $\delta^c\boldsymbol{v}/\delta t$, $\delta_c\boldsymbol{v}/\delta t$ denotes a part of $\boldsymbol{v}$ excluding the change of basis, which corresponds to the change of $\boldsymbol{v}$ when the observer is moving along the same coordinate system of the deformed body. $\delta^c\boldsymbol{v}/\delta t$ is known as the *contravariant derivative* or *upper convected rate*, and $\delta_c\boldsymbol{v}/\delta t$ is known as the *covariant derivative* or *lower convected rate*.[5]

---

[5]The convected derivatives of a vector $\boldsymbol{v}$ are sometimes written as $\delta^c\boldsymbol{v}/\delta t = \overset{\triangleleft}{\boldsymbol{v}}$, $\delta_c\boldsymbol{v}/\delta t = \overset{\triangleright}{\boldsymbol{v}}$. For the second-order tensor $\boldsymbol{T}$ these are $\delta^{cc}\boldsymbol{T}/\delta t = \overset{\triangleleft}{\boldsymbol{T}}$, $\delta_{cc}\boldsymbol{T}/\delta t = \overset{\triangleright}{\boldsymbol{T}}$.

For a second-order tensor $\boldsymbol{T}$ four convected derivatives are introduced as follows:

$$\frac{\delta^{cc}\boldsymbol{T}}{\delta t} = \dot{T}^{ij}\boldsymbol{g}_i \otimes \boldsymbol{g}_j, \quad \frac{\delta^{c}_{\cdot c}\boldsymbol{T}}{\delta t} = \dot{T}^{i}_{\cdot j}\boldsymbol{g}_i \otimes \boldsymbol{g}^j,$$
$$\frac{\delta^{\cdot c}_{c}\boldsymbol{T}}{\delta t} = \dot{T}^{\cdot j}_{i}\boldsymbol{g}^i \otimes \boldsymbol{g}_j, \quad \frac{\delta_{cc}\boldsymbol{T}}{\delta t} = \dot{T}_{ij}\boldsymbol{g}^i \otimes \boldsymbol{g}^j. \tag{2.162}$$

Applying (2.78) and (2.158) and noting that $\boldsymbol{L}^T = \boldsymbol{g}^i \otimes \dot{\boldsymbol{g}}_j$ yields

$$\begin{aligned}
T^{ij}\dot{\boldsymbol{g}}_i \otimes \boldsymbol{g}_j &= \boldsymbol{L}\, T^{ij}\boldsymbol{g}_i \otimes \boldsymbol{g}_j = \boldsymbol{L}\boldsymbol{T},\\
T^{ij}\boldsymbol{g}_i \otimes \dot{\boldsymbol{g}}_j &= T^{ij}\boldsymbol{g}_i \otimes \boldsymbol{L}\boldsymbol{g}_j = (T^{ij}\boldsymbol{g}_i \otimes \boldsymbol{g}_j)\boldsymbol{L}^T = \boldsymbol{T}\boldsymbol{L}^T,\\
T^{i}_{\cdot j}\dot{\boldsymbol{g}}_i \otimes \boldsymbol{g}^j &= \boldsymbol{L}\, T^{i}_{\cdot j}\boldsymbol{g}_i \otimes \boldsymbol{g}^j = \boldsymbol{L}\boldsymbol{T},\\
T^{i}_{\cdot j}\boldsymbol{g}_i \otimes \dot{\boldsymbol{g}}^j &= -T^{i}_{\cdot j}\boldsymbol{g}_i \otimes \boldsymbol{L}^T\boldsymbol{g}^j = -\boldsymbol{T}\boldsymbol{L},\\
T^{\cdot j}_{i}\dot{\boldsymbol{g}}^i \otimes \boldsymbol{g}_j &= -\boldsymbol{L}^T T^{\cdot j}_{i}\boldsymbol{g}^i \otimes \boldsymbol{g}_j = -\boldsymbol{L}^T\boldsymbol{T},\\
T^{\cdot j}_{i}\boldsymbol{g}^i \otimes \dot{\boldsymbol{g}}_j &= T^{\cdot j}_{i}\boldsymbol{g}_i \otimes \boldsymbol{L}\boldsymbol{g}^j = \boldsymbol{T}\boldsymbol{L}^T,\\
T_{ij}\dot{\boldsymbol{g}}^i \otimes \boldsymbol{g}^j &= -\boldsymbol{L}^T T_{ij}\boldsymbol{g}^i \otimes \boldsymbol{g}^j = -\boldsymbol{L}^T\boldsymbol{T},\\
T_{ij}\boldsymbol{g}^i \otimes \dot{\boldsymbol{g}}^j &= -T_{ij}\boldsymbol{g}^i \otimes \boldsymbol{L}^T\boldsymbol{g}^j = -\boldsymbol{T}\boldsymbol{L}.
\end{aligned}$$

Thus (2.162) can be written as

$$\frac{\delta^{cc}\boldsymbol{T}}{\delta t} = \dot{\boldsymbol{T}} - \boldsymbol{L}\boldsymbol{T} - \boldsymbol{T}\boldsymbol{L}^T = \boldsymbol{F}\frac{d(\boldsymbol{F}^{-1}\boldsymbol{T}\boldsymbol{F}^{-T})}{dt}\boldsymbol{F}^T, \tag{2.163}$$

$$\frac{\delta^{c}_{\cdot c}\boldsymbol{T}}{\delta t} = \dot{\boldsymbol{T}} - \boldsymbol{L}\boldsymbol{T} + \boldsymbol{T}\boldsymbol{L} = \boldsymbol{F}\frac{d(\boldsymbol{F}^{-1}\boldsymbol{T}\boldsymbol{F})}{dt}\boldsymbol{F}^{-1}, \tag{2.164}$$

$$\frac{\delta^{\cdot c}_{c}\boldsymbol{T}}{\delta t} = \dot{\boldsymbol{T}} + \boldsymbol{L}^T\boldsymbol{T} - \boldsymbol{T}\boldsymbol{L}^T = \boldsymbol{F}^{-T}\frac{d(\boldsymbol{F}^T\boldsymbol{T}\boldsymbol{F}^{-T})}{dt}\boldsymbol{F}^T, \tag{2.165}$$

$$\frac{\delta_{cc}\boldsymbol{T}}{\delta t} = \dot{\boldsymbol{T}} + \boldsymbol{L}^T\boldsymbol{T} + \boldsymbol{T}\boldsymbol{L} = \boldsymbol{F}^{-T}\frac{d(\boldsymbol{F}^T\boldsymbol{T}\boldsymbol{F})}{dt}\boldsymbol{F}^{-1}. \tag{2.166}$$

If an orthonormal coordinate transformation tensor $\boldsymbol{Q}$ ($\boldsymbol{Q}^{-1} = \boldsymbol{Q}^T$) is used instead of the deformation gradient $\boldsymbol{F}$, the concept of the convected derivative can be extended. That is, let $\boldsymbol{\Omega} = \dot{\boldsymbol{Q}}\boldsymbol{Q}^T$ be an antisymmetric rotation tensor generated by $\boldsymbol{Q}$ as shown in (2.27), then the *corotational derivative* of a second-order tensor $\boldsymbol{T}$ due to $\boldsymbol{Q}$ is defined by

$$\frac{\mathfrak{D}_Q\boldsymbol{T}}{\mathfrak{D}t} = \dot{\boldsymbol{T}} + \boldsymbol{T}\boldsymbol{\Omega} - \boldsymbol{\Omega}\boldsymbol{T} = \boldsymbol{Q}\frac{d(\boldsymbol{Q}^T\boldsymbol{T}\boldsymbol{Q})}{dt}\boldsymbol{Q}^T. \tag{2.167}$$

Here $\mathfrak{D}_Q \boldsymbol{T}/\mathfrak{D}t$ represents an objective part of the time derivative of the second-order tensor $\boldsymbol{T}$ (the proof is similar to (2.170) as shown below). For example if the rotation tensor $\boldsymbol{R}$ and the spin tensor $\boldsymbol{W}$ are used instead of $\boldsymbol{Q}$ and $\boldsymbol{\Omega}$, we can introduce the *Zaremba-Jaumann rate* as follows:

$$\frac{\mathfrak{D}\boldsymbol{T}}{\mathfrak{D}t} = \dot{\boldsymbol{T}} + \boldsymbol{T}\boldsymbol{W} - \boldsymbol{W}\boldsymbol{T} = \boldsymbol{R}\frac{d(\boldsymbol{R}^T\boldsymbol{T}\boldsymbol{R})}{dt}\boldsymbol{R}^T. \tag{2.168}$$

The material time derivative of a vector-valued or tensor-valued function is not always objective as described above even if the original function is objective. It can be said that the convected derivative and corotational derivative are introduced to ensure objectivity of the time-derivative. For example we have

$$\frac{\delta_c \boldsymbol{v}^*}{\delta t} = \frac{d\boldsymbol{v}^*}{dt^*} + \boldsymbol{L}^{*T}\boldsymbol{v}^* = \frac{d(\boldsymbol{Q}\boldsymbol{v})}{dt} + (\boldsymbol{Q}\,\boldsymbol{L}^T\,\boldsymbol{Q}^T - \boldsymbol{\Omega})\,(\boldsymbol{Q}\boldsymbol{v}) = \boldsymbol{Q}\frac{\delta_c \boldsymbol{v}}{\delta t} \tag{2.169}$$

$$\begin{aligned}\frac{\delta_{cc}\boldsymbol{T}^*}{\delta t} &= \frac{d\boldsymbol{T}^*}{dt^*} + \boldsymbol{L}^{*T}\boldsymbol{T}^* + \boldsymbol{T}^*\boldsymbol{L}^* \\ &= \frac{d(\boldsymbol{Q}\boldsymbol{T}\boldsymbol{Q}^T)}{dt} + (\boldsymbol{Q}\,\boldsymbol{L}^T\,\boldsymbol{Q}^T - \boldsymbol{\Omega})\,\boldsymbol{Q}\boldsymbol{T}\boldsymbol{Q}^T + \boldsymbol{Q}\boldsymbol{T}\boldsymbol{Q}^T(\boldsymbol{Q}\boldsymbol{L}\boldsymbol{Q}^T + \boldsymbol{\Omega}) \\ &= \boldsymbol{Q}\frac{\delta_{cc}\boldsymbol{T}}{\delta t}\boldsymbol{Q}^T.\end{aligned} \tag{2.170}$$

Note that the time derivative $\dot{\boldsymbol{\sigma}}$ of Cauchy stress $\boldsymbol{\sigma}$ is not objective, but the Zaremba-Jaumann rate $\mathfrak{D}\boldsymbol{\sigma}/\mathfrak{D}t$ is objective.

2.8.2.1 Spin of Eulerian Triads $\boldsymbol{\Omega}^E$

Recall that the Eulerian triads $\{\boldsymbol{n}_i\}$ was introduced in (2.57). Let us define the spin $\boldsymbol{\Omega}^E$ of $\{\boldsymbol{n}_i\}$ by

$$\dot{\boldsymbol{n}}_i = \boldsymbol{\Omega}^E\,\boldsymbol{n}_i \quad \Rightarrow \quad \boldsymbol{\Omega}^E = \dot{\boldsymbol{n}}_i \otimes \boldsymbol{n}_i = \overline{\Omega}^E_{ij}\,\boldsymbol{n}_i \otimes \boldsymbol{n}_i, \quad \overline{\Omega}^E_{ij} = \boldsymbol{n}_i \cdot \dot{\boldsymbol{n}}_i. \tag{2.171}$$

Since $\boldsymbol{n}_i \otimes \boldsymbol{n}_i = \boldsymbol{i}$ ($\boldsymbol{i}$ is the unit tensor for the deformed body), the time-differential yields

$$\boldsymbol{\Omega}^E = \dot{\boldsymbol{n}}_i \otimes \boldsymbol{n}_i = -\boldsymbol{n}_i \otimes \dot{\boldsymbol{n}}_i = -(\boldsymbol{\Omega}^E)^T.$$

This shows that $\boldsymbol{\Omega}^E$ is antisymmetric.

We can define the following Lagrangian tensor $\boldsymbol{\Omega}^{ER}$, which is the pull-back of $\boldsymbol{\Omega}^E$ to the undeformed body by the rotation tensor $\boldsymbol{R} = \boldsymbol{n}_i \otimes \boldsymbol{N}_i$:

$$\boldsymbol{\Omega}^{ER} = \boldsymbol{R}^T\boldsymbol{\Omega}^E\boldsymbol{R} = \overline{\Omega}^E_{ij}\,\boldsymbol{N}_i \otimes \boldsymbol{N}_i \quad \Rightarrow \quad \boldsymbol{R}^T\dot{\boldsymbol{n}}_i = \boldsymbol{\Omega}^{ER}\boldsymbol{N}_i \tag{2.172}$$

2.8.2.2 Spin of Lagrangian Triads $\boldsymbol{\Omega}^L$

Recall that the Lagrangian triads $\{\boldsymbol{N}_i\}$ was introduced in (2.55). Let us also define the spin $\boldsymbol{\Omega}^L$ of $\{\boldsymbol{N}_i\}$ by

$$\dot{\boldsymbol{N}}_i = \boldsymbol{\Omega}^L \boldsymbol{N}_i \quad \Rightarrow \quad \boldsymbol{\Omega}^L = \dot{\boldsymbol{N}}_i \otimes \boldsymbol{N}_i = \overline{\Omega}_{ij}^L \boldsymbol{N}_i \otimes \boldsymbol{N}_i, \quad \overline{\Omega}_{ij}^L = \boldsymbol{N}_i \cdot \dot{\boldsymbol{N}}_i. \tag{2.173}$$

It is obvious that $\boldsymbol{\Omega}^L$ is antisymmetric. We can define the following Eulerian tensor $\boldsymbol{\Omega}^{RL}$, which is the push-forward of $\boldsymbol{\Omega}^L$ to the deformed body by the rotation tensor $\boldsymbol{R} = \boldsymbol{n}_i \otimes \boldsymbol{N}_i$:

$$\boldsymbol{\Omega}^{RL} = \boldsymbol{R}\boldsymbol{\Omega}^L \boldsymbol{R}^T = \overline{\Omega}_{ij}^L \boldsymbol{n}_i \otimes \boldsymbol{n}_i \quad \Rightarrow \quad \boldsymbol{R}\dot{\boldsymbol{N}}_i = \boldsymbol{\Omega}^{RL}\boldsymbol{n}_i. \tag{2.174}$$

2.8.2.3 Eulerian Spin $\boldsymbol{\omega}^R$ and Lagrangian Spin $\boldsymbol{\omega}^{RR}$

Time-differentiating $\boldsymbol{R}\boldsymbol{R}^T = \boldsymbol{i}$ yields $\dot{\boldsymbol{R}}\boldsymbol{R}^T + \boldsymbol{R}\dot{\boldsymbol{R}}^T = \boldsymbol{0}$, therefore the Eulerian spin $\boldsymbol{\omega}^R$ and Lagrangian spin $\boldsymbol{\omega}^{RR}$, which is the pull-back of the Eulerian spin into the undeformed body, can be defined by

$$\boldsymbol{\omega}^R = \dot{\boldsymbol{R}}\boldsymbol{R}^T = \omega_{ij}^R \boldsymbol{n}_i \otimes \boldsymbol{n}_i, \qquad \boldsymbol{\omega}^{RR} = \boldsymbol{R}^T \boldsymbol{\omega}^R \boldsymbol{R} = \boldsymbol{R}^T \dot{\boldsymbol{R}} = \omega_{ij}^R \boldsymbol{N}_i \otimes \boldsymbol{N}_i. \tag{2.175}$$

On the other hand, because of (2.171) and the relation $\boldsymbol{n}_i = \boldsymbol{R}\boldsymbol{N}_i$, we have

$$\begin{aligned}\overline{\Omega}_{ij}^E &= \boldsymbol{n}_i \cdot \dot{\boldsymbol{n}}_j = \boldsymbol{n}_i \cdot (\dot{\boldsymbol{R}}\boldsymbol{N}_i + \boldsymbol{R}\dot{\boldsymbol{N}}_i) = \boldsymbol{n}_i \cdot (\dot{\boldsymbol{R}}\boldsymbol{R}^T)\boldsymbol{n}_j + \boldsymbol{n}_i \cdot \boldsymbol{R}\boldsymbol{\Omega}^L \boldsymbol{N}_j \\ &= \boldsymbol{n}_i \cdot \boldsymbol{\omega}^R \boldsymbol{n}_j + \boldsymbol{n}_i \cdot \boldsymbol{\Omega}^{RL}\boldsymbol{n}_j.\end{aligned}$$

Thus the component form of the Eulerian spin is

$$\omega_{ij}^R = \overline{\Omega}_{ij}^E - \overline{\Omega}_{ij}^L. \tag{2.176}$$

The direct notations are given by

$$\boldsymbol{\omega}^R = \boldsymbol{\Omega}^E - \boldsymbol{\Omega}^{RL}, \quad \boldsymbol{\omega}^{RR} = \boldsymbol{\Omega}^{ER} - \boldsymbol{\Omega}^L. \tag{2.177}$$

2.8.2.4 Corotational Derivatives

Since the deformation gradient $\boldsymbol{F}$ can be written in terms of the polar decomposition as defined in (2.46), its time-differentiation gives

$$\dot{\boldsymbol{F}} = \dot{\boldsymbol{R}}\boldsymbol{U} + \boldsymbol{R}\dot{\boldsymbol{U}}.$$

By using (2.65) we obtain the following corotational derivative $\overset{\circ}{V}$ due to ω^R:

$$L = \dot{F}F^{-1} = W + D = \dot{R}R^T + R\dot{U}R^T(RU^{-1}R^T) = \omega^R + \overset{\circ}{V}V^{-1} \tag{2.178}$$

$$\overset{\circ}{V} = R\dot{U}R^T = \frac{\mathfrak{D}_R V}{\mathfrak{D}t} = \dot{V} + V\omega^R - \omega^R V. \tag{2.179}$$

It should be noted that the term $\overset{\circ}{V}V^{-1}$ includes an antisymmetric part. Both U and $\dot{U}$ are symmetric, therefore we have

$$D = \frac{1}{2}R\left(\dot{U}U^{-1} + U^{-1}\dot{U}\right)R^T, \qquad W = \omega^R + \frac{1}{2}R\left(\dot{U}U^{-1} - U^{-1}\dot{U}\right)R^T. \tag{2.180}$$

On the other hand, by recalling that $V = \sum \lambda_i\, n_i \otimes n_i$, $\Omega^E = \dot{n}_i \otimes n_i$, the time-derivative $\dot{V}$ can be given as

$$\dot{V} = \sum\left(\dot{\lambda}_i\, n_i \otimes n_i + \lambda_i\, \dot{n}_i \otimes n_i + \lambda_i\, n_i \otimes \dot{n}_i\right) = \sum \dot{\lambda}_i\, n_i \otimes n_i + \Omega^E V - V\Omega^E.$$

Thus we can define the following corotational derivative $\overset{\triangledown}{V}$ due to Ω^E:

$$\overset{\triangledown}{V} = \frac{\mathfrak{D}_E V}{\mathfrak{D}t} = \dot{V} + V\Omega^E - \Omega^E V = \sum \dot{\lambda}_i\, n_i \otimes n_i. \tag{2.181}$$

Equation 2.179 gives the relationship between both corotational derivatives as

$$\overset{\circ}{V} = \overset{\triangledown}{V} + \Omega^{RL}V - V\Omega^{RL}. \tag{2.182}$$

It should be emphasized that from (2.179), $\overset{\circ}{V} = \mathfrak{D}_R V/\mathfrak{D}t$ gives the corotational derivative of V due to ω^R, while from (2.181), $\overset{\triangledown}{V} = \mathfrak{D}_E V/\mathfrak{D}t$ gives the corotational derivative of V due to Ω^E.

2.8.3 *Invariants of Stress and Strain and Isotropic Elastic Solids*

2.8.3.1 Invariants and Spherical Decomposition of a Second-Order Real-Valued Symmetric Tensor

A second-order real-valued symmetric tensor T gives three real eigenvalues λ which are determined by the characteristic equation:

$$\det(T - \lambda I) = -\lambda^3 + I_1\lambda^2 - I_2\lambda + I_3 = 0, \tag{2.183}$$

$$I_1 = \mathrm{tr}\, \boldsymbol{T}, \quad I_2 = \frac{1}{2}\left[(\mathrm{tr}\, \boldsymbol{T})^2 - \mathrm{tr}\, \boldsymbol{T}^2\right], \quad I_3 = \det \boldsymbol{T} \tag{2.184}$$

where I_1, I_2, I_3 are the first, second and third principal invariants, respectively.

The mean or volumetric tensor $\overline{\boldsymbol{T}}$ and the deviatoric tensor $\boldsymbol{T}'$ are defined by

$$\overline{\boldsymbol{T}} = \frac{1}{3}(\mathrm{tr}\, \boldsymbol{T})\boldsymbol{I}, \qquad \boldsymbol{T}' = \boldsymbol{T} - \overline{\boldsymbol{T}}. \tag{2.185}$$

Since the first invariant of the deviatoric tensor $\boldsymbol{T}'$ is zero ($J_1 = \mathrm{tr}\, \boldsymbol{T}' \equiv 0$), its second and third invariants are

$$J_2 = \frac{1}{2} T'_{ik} T'_{ki} = \frac{1}{2} \mathrm{tr}\, (\boldsymbol{T}')^2, \qquad J_3 = \det \boldsymbol{T}'. \tag{2.186}$$

Let us define the k-th moment $\bar{I}_k$ of a tensor $\boldsymbol{T}$ by

$$\bar{I}_k = \mathrm{tr}\, \boldsymbol{T}^k. \tag{2.187}$$

Note 2.6 (Cayley-Hamilton Theorem). The well-known Cayley-Hamilton theorem states that

$$C(\boldsymbol{T}) = -\boldsymbol{T}^3 + I_1 \boldsymbol{T}^2 - I_2 \boldsymbol{T} + I_3 \boldsymbol{I} = \boldsymbol{0} \tag{2.188}$$

which is similar to the characteristic equation (2.183).

Proof. Let us introduce an orthonormal basis $\{\boldsymbol{e}_i\}$ $(i = 1, 2, 3)$, and let the coefficient matrix of $\boldsymbol{T}$ be T_{lk} ($\boldsymbol{T} = T_{lk}\, \boldsymbol{e}_l \otimes \boldsymbol{e}_k$), which gives

$$\boldsymbol{T} \boldsymbol{e}_k = T_{lk}\, \boldsymbol{e}_l. \tag{2.189}$$

We define a tensor $\boldsymbol{B}_{lk}$ with tensorial components given by

$$\boldsymbol{B}_{lk} = T_{lk} \boldsymbol{I} - \delta_{lk} \boldsymbol{T}$$

(note that $\boldsymbol{I} = \delta_{ij}\, \boldsymbol{e}_i \otimes \boldsymbol{e}_j$); thus (2.189) is equivalent to

$$\boldsymbol{B}_{lk} \boldsymbol{e}_l = \boldsymbol{0}. \tag{2.190}$$

We should recall that the adjoint A^*_{ij} of a regular matrix A_{ij} is given by $A^*_{km}\, A_{lk} = (\det \boldsymbol{A})\, \delta_{ml}$, and if we multiply the adjoint $\boldsymbol{B}^*_{km}$ (with tensorial components) by (2.190), we obtain

$$\boldsymbol{B}^*_{km} \boldsymbol{B}_{lk} \boldsymbol{e}_l = \det(\boldsymbol{B}_{ij})\, \delta_{ml} \boldsymbol{e}_l = \boldsymbol{0}, \quad \Rightarrow \quad C(\boldsymbol{T})\, \boldsymbol{e}_l = \det(T_{lk} \boldsymbol{I} - \delta_{lk} \boldsymbol{T})\, \boldsymbol{e}_l = \boldsymbol{0} \tag{2.191}$$

$$\delta_{ml} = \begin{cases} \boldsymbol{I} & \text{if } m = l \\ \boldsymbol{0} & \text{if } m \neq l. \end{cases}$$

By multiplying v_l with this result and setting $\boldsymbol{v} = v_l \boldsymbol{e}_l$, we have $\boldsymbol{C}(\boldsymbol{T})\,\boldsymbol{v} = \boldsymbol{0}$ for an arbitrary $\boldsymbol{v}$, which implies that we have (2.188) ■

Operating the trace on (2.188) and using the definition of the k-th moment given by (2.187) yields $I_3 = \bar{I}_3 - I_1\,\bar{I}_2 + I_2\,I_1$. Since $I_1 = \bar{I}_1$, the relationships between the invariants and moments are given by

$$I_1 = \bar{I}_1, \qquad I_2 = \frac{1}{2}\left[(\bar{I}_1)^2 - \bar{I}_2\right], \qquad I_3 = \frac{1}{3}\bar{I}_3 - \frac{1}{2}\bar{I}_1\bar{I}_2 + \frac{1}{6}(\bar{I}_1)^3. \quad (2.192)$$

Thus the third invariant J_3 of the deviatoric tensor $\boldsymbol{T}'$ is given by the third moment $\bar{J}_3$:

$$J_3 = \frac{1}{3}T'_{ik}T'_{kl}T'_{li} = \frac{1}{3}\mathrm{tr}\,(\boldsymbol{T}')^3 = \frac{1}{3}\bar{J}_3. \quad (2.193)$$

Let us introduce the *norm*[6] of a second-order tensor $\boldsymbol{T}$ by

$$||\boldsymbol{T}|| = (\boldsymbol{T} : \boldsymbol{T})^{1/2} = T_{ij}\,T_{ij}.$$

Since $||\boldsymbol{I}|| = \sqrt{3}$, the 'signed magnitude' of the volumetric tensor $\overline{\boldsymbol{T}}$ is calculated as

$$\overline{T} = \frac{1}{\sqrt{3}}T_{kk}. \quad (2.194)$$

If the 'basis tensor' of $\overline{\boldsymbol{T}}$ is introduced by

$$\boldsymbol{n}^{(1)} = n_{ij}^{(1)}\,\boldsymbol{e}_i \otimes \boldsymbol{e}_j, \qquad n_{ij}^{(1)} = \frac{\partial \overline{T}}{\partial T_{ij}} = \frac{T_{ij}}{\overline{T}} = \frac{\delta_{ij}}{\sqrt{3}}, \quad (2.195)$$

we have

$$\overline{\boldsymbol{T}} = \overline{T}\,\boldsymbol{n}^{(1)}. \quad (2.196)$$

For the deviatoric tensor $\boldsymbol{T}'$ we introduce the norm T' and the 'basis tensor' $\boldsymbol{n}^{(2)}$ by

$$T' = (\boldsymbol{T}' : \boldsymbol{T}')^{1/2} = (T_{ij}\,T_{ij} - \overline{T}^2)^{1/2}, \quad (2.197)$$

[6]The inner product of the second-order tensors $\boldsymbol{A}, \boldsymbol{B}$ is introduced by $\boldsymbol{A} : \boldsymbol{B} = \mathrm{tr}\,(\boldsymbol{A}^T\boldsymbol{B}) = A_{ij}\,B_{ij}$. $\boldsymbol{A}$ and $\boldsymbol{B}$ are *orthogonal* if $\boldsymbol{A} : \boldsymbol{B} = 0$.

$$n^{(2)} = n_{ij}^{(2)} \, e_i \otimes e_j, \qquad n_{ij}^{(2)} = \frac{\partial T'}{\partial T'_{ij}} = \frac{T'_{ij}}{T'} \tag{2.198}$$

$$\Rightarrow \quad T' = T' n^{(2)}. \tag{2.199}$$

Lode's angle T_θ of $\boldsymbol{T}$ and the Lode parameter y_T can be introduced by

$$y_T = \cos(3T_\theta) = \frac{3\sqrt{3}J_3}{2(J_2)^{3/2}} = \sqrt{6}\,\mathrm{tr}\,(n^{(2)})^3. \tag{2.200}$$

If the tensor of Lode's angle is defined by

$$\boldsymbol{T}_\theta = T_\theta \, n^{(3)}, \tag{2.201}$$

its 'basis tensor' $n^{(3)} = n_{ij}^{(3)} \, e_i \otimes e_j$ can be calculated as

$$n_{ij}^{(3)} = T' \frac{\partial T_\theta}{\partial T_{ij}} = T' \frac{\partial T_\theta}{\partial y_T}\frac{\partial y_T}{\partial T_{ij}} = \frac{\sqrt{6}}{\sin(3\tau_\theta)} \left[n_{ij}^{(2)} \mathrm{tr}\,(n^{(2)})^3 - n_{ik}^{(2)} n_{kj}^{(2)} + \frac{1}{\sqrt{3}} n_{ij}^{(1)} \right]$$

$$\Rightarrow \quad n^{(3)} = \frac{\sqrt{6}}{\sin(3\tau_\theta)} \left[n^{(2)} \mathrm{tr}\,(n^{(2)})^3 - (n^{(2)})^2 + \frac{1}{\sqrt{3}} n^{(1)} \right] \tag{2.202}$$

where we used the relationship $\mathrm{tr}\,(n^{(2)})^2 = 1$.

The basis tensors $n^{(1)}$, $n^{(2)}$, $n^{(3)}$ are mutually orthogonal in the sense of

$$n^{(\alpha)} : n^{(\beta)} = \delta_{\alpha\beta}. \tag{2.203}$$

Thus the second-order real-valued symmetric tensor $\boldsymbol{T}$ is written in orthogonal components by

$$\boldsymbol{T} = T_{(\alpha)} \, n^{(\alpha)} \qquad (\alpha : \text{summed}) \tag{2.204}$$

where we set

$$T_{(1)} = \overline{T}, \qquad T_{(2)} = T', \qquad T_{(3)} = T_\theta.$$

The result (2.204) is referred to as the *spherical decomposition* of $\boldsymbol{T}$.

2.8.3.2 Geometrical Interpretation of Spherical Decomposition in the Principal Space

Let the eigenvalue representation of the second-order real-valued symmetric tensor $\boldsymbol{T}$ be given by

$$\boldsymbol{T} = \sum_{i=1}^{3} T_i e'_i \otimes e'_i.$$

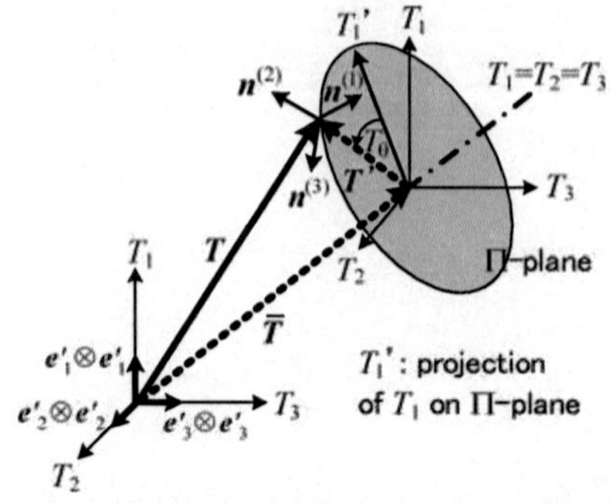

Fig. 2.11 Deviatoric, volumetric and Lode's components in the principal space

Then the volumetric, deviatoric and Lode's components $\overline{\boldsymbol{T}}$, $\boldsymbol{T}'$, T_θ are as shown in Fig. 2.11 together with the base tensors $\boldsymbol{n}^{(\alpha)}$ $(\alpha = 1, 2, 3)$ (since the tensors in the principal space are termed 'vectors', we will use this designation). $\overline{\boldsymbol{T}}$ is the projection of $\boldsymbol{T}$ on the diagonal axis $T_1 = T_2 = T_3$ (which is referred to as the hydrostatic axis for the stress). The difference vector $\boldsymbol{T} - \overline{\boldsymbol{T}}$ gives the deviatoric vector $\boldsymbol{T}'$. The orthogonal plane to $\overline{T}$ including the vector $\boldsymbol{T}'$ is referred to as the Π-*plane*. . If the projected axis of T_1 on the Π-plane is T_1', the angle between T_1' and $\boldsymbol{T}'$ gives the Lode's angle T_θ.

This implies that, by spherical decomposition, cylindrical polar coordinates are introduced in terms of the 'hydrostatic' axis.

2.8.3.3 Spherical Decompositions of Stress and Strain and the Response of an Isotropic Elastic Solid

The stress $\boldsymbol{\sigma}$ is a second-order real-valued symmetric tensor, and the spherical decomposition is given as follows:

$\bar{\boldsymbol{\sigma}} = I_1^\sigma \boldsymbol{I}/3$	Volumetric stress
$I_1^\sigma = \mathrm{tr}\,(\boldsymbol{\sigma})$	First invariant of stress
$\bar{\sigma} = I_1^\sigma/\sqrt{3}$	Magnitude of volumetric stress
$\boldsymbol{\sigma}' = \boldsymbol{\sigma} - \bar{\boldsymbol{\sigma}}$	Deviatoric stress
$\sigma' = \Vert\boldsymbol{\sigma}'\Vert = (\boldsymbol{\sigma}' : \boldsymbol{\sigma}')^{1/2}$	Magnitude of deviatoric stress
$\sigma_\theta = \frac{1}{3}\cos^{-1}\{3\sqrt{3}J_3^\sigma/2(J_2^\sigma)^{3/2}\}$	Lode's angle for stress
$J_2^\sigma = \boldsymbol{\sigma}' : \boldsymbol{\sigma}'/2$	Second invariant of deviatoric stress
$J_3^\sigma = \det\,(\boldsymbol{\sigma}')$	Third invariant of deviatoric stress

For the strain $\boldsymbol{\varepsilon}$ we can introduce the spherical decomposition as follows:

$\bar{\boldsymbol{\varepsilon}} = I_1^\varepsilon \boldsymbol{I}/3$	Volumetric strain
$I_1^\varepsilon = \mathrm{tr}\,(\boldsymbol{\varepsilon})$	First invariant of strain
$\bar{\varepsilon} = I_1^\varepsilon/\sqrt{3}$	Magnitude of volumetric strain
$\boldsymbol{\varepsilon}' = \boldsymbol{\varepsilon} - \bar{\boldsymbol{\varepsilon}}$	Deviatoric strain

$\varepsilon' = \Vert\boldsymbol{\varepsilon}'\Vert = (\boldsymbol{\varepsilon}' : \boldsymbol{\varepsilon}')^{1/2}$	Magnitude of deviatoric strain
$\varepsilon_\theta = \frac{1}{3}\cos^{-1}\{3\sqrt{3}J_3^\varepsilon/2(J_2^\varepsilon)^{3/2}\}$	Lode's angle for strain
$J_2^\varepsilon = \boldsymbol{\varepsilon}' : \boldsymbol{\varepsilon}'/2$	Second invariant of deviatoric strain
$J_3^\varepsilon = \det(\boldsymbol{\varepsilon}')$	Third invariant of deviatoric strain

If a material body is an isotropic solid, the stress and strain are decomposed by using the same basis $\boldsymbol{n}^{(\alpha)}$:

$$\boldsymbol{\sigma} = \bar{\boldsymbol{\sigma}} + \boldsymbol{\sigma}' + \boldsymbol{\sigma}_\theta, \qquad \bar{\boldsymbol{\sigma}} = \bar{\sigma}\,\boldsymbol{n}^{(1)}, \qquad \boldsymbol{\sigma}' = \sigma'\boldsymbol{n}^{(2)}, \qquad \boldsymbol{\sigma}_\theta = \sigma'\sigma_\theta\,\boldsymbol{n}^{(3)}, \tag{2.205}$$

$$\boldsymbol{\varepsilon} = \bar{\boldsymbol{\varepsilon}} + \boldsymbol{\varepsilon}' + \boldsymbol{\varepsilon}_\theta, \qquad \bar{\boldsymbol{\varepsilon}} = \bar{\varepsilon}\,\boldsymbol{n}^{(1)}, \qquad \boldsymbol{\varepsilon}' = \varepsilon'\boldsymbol{n}^{(2)}, \qquad \boldsymbol{\varepsilon}_\theta = \varepsilon'\varepsilon_\theta\,\boldsymbol{n}^{(3)}, \tag{2.206}$$

and the linear elastic response is written in terms of the volumetric and deviatoric components independently (cf. Note 2.7). Thus, referring to Fig. 2.11, the response gives a state that is symmetric about the hydrostatic axis as follows:

$$\bar{\boldsymbol{\sigma}} = 3\lambda\bar{\boldsymbol{\varepsilon}}, \qquad \boldsymbol{\sigma}' = 2\mu\boldsymbol{\varepsilon}'. \tag{2.207}$$

The coefficients λ, μ are called Lamé's constants. Then the response of the linear elastic solid, called the Hookean solid, is written as

$$\sigma_{ij} = \lambda\varepsilon_{kk}\,\delta_{ij} + 2\mu\,\varepsilon_{ij}. \tag{2.208}$$

Young's modulus E and Poisson's ratio ν are related to Lamé's constants λ, μ, the shear modulus G and bulk modulus K as

$$\lambda = \frac{E\nu}{(1+\nu)(1-2\nu)}, \quad \mu = \frac{E}{2(1+\nu)} = G, \quad K = \lambda + \frac{2}{3}\mu = \frac{E}{3(1-2\nu)}. \tag{2.209}$$

Note 2.7 (Lode's angle and the response of isotropic solids). If the elastic response of solids is written using Hooke's law as

$$\boldsymbol{\sigma} = \boldsymbol{D}^e\,\boldsymbol{\varepsilon}, \tag{2.210}$$

the most general form of the fourth order tensor $\boldsymbol{D}^e$ for isotropic materials is given by

$$D^e_{ijkl} = \lambda\delta_{ij}\delta_{kl} + \mu(\delta_{ik}\delta_{jl} + \delta_{il}\delta_{jk}) + \nu(\delta_{ik}\delta_{jl} - \delta_{il}\delta_{jk}) \tag{2.211}$$

(cf. Malvern 1969, pp. 277). Since $\boldsymbol{\sigma}$ and $\boldsymbol{\varepsilon}$ are symmetric ($\sigma_{ij} = \sigma_{ji}$, $\varepsilon_{ij} = \varepsilon_{ji}$), we have the condition $\nu = 0$, which causes no change in the Lode's angle component for isotropic solids, and the axi-symmetric response with respect to the hydrostatic axis. Another result is that there exist two independent elastic constants, although the number of eigenvalues of stress and strain is three.

We also note that a tensor function defined as

$$D^e_{ijkl} = \lambda\delta_{ij}\delta_{kl} + \mu\delta_{ik}\delta_{jl} + \nu\delta_{il}\delta_{jk} \tag{2.212}$$

is also isotropic (Little 1973; Spencer 2004). The constitutive equation (2.210) now becomes

$$\sigma = \lambda\delta_{ij}\varepsilon_{kk} + \mu\varepsilon_{ij} + \nu\varepsilon_{ij}. \tag{2.213}$$

Since $\varepsilon_{ij} = \varepsilon_{ji}$, no generality is lost by setting $\mu = \nu$ such that $\sigma = \lambda\delta_{ij}\varepsilon_{kk} + 2\mu\varepsilon_{ij}$. ■

The inverse relation of (2.208) is

$$\varepsilon_{ij} = -\frac{\nu}{E}\sigma_{kk}\,\delta_{ij} + \frac{1+\nu}{E}\sigma_{ij}. \tag{2.214}$$

For two dimensional problems we can consider two idealized states: the plane strain state where $\varepsilon_{zz} = \varepsilon_{xz} = \varepsilon_{yz} = 0$ and the plane stress state in which $\sigma_{zz} = \sigma_{xz} = \sigma_{yz} = 0$. Under these conditions, Hooke's law is rewritten for the vector forms of stress and strain as

$$\boldsymbol{\sigma} = \boldsymbol{D}^e\,\boldsymbol{\varepsilon}, \qquad \boldsymbol{\sigma} = [\sigma_{xx}\ \sigma_{yy}\ \sigma_{xy}]^T, \quad \boldsymbol{\varepsilon} = [\varepsilon_{xx}\ \varepsilon_{yy}\ \gamma_{xy}]^T \tag{2.215}$$

$$\text{Plane strain :} \quad \boldsymbol{D}^e = \frac{E(1-\nu)}{(1+\nu)(1-2\nu)}\begin{bmatrix} 1 & \dfrac{\nu}{1-\nu} & 0 \\ \dfrac{\nu}{1-\nu} & 1 & 0 \\ 0 & 0 & \dfrac{1-2\nu}{2(1-\nu)} \end{bmatrix} \tag{2.216}$$

$$\text{Plane stress :} \quad \boldsymbol{D}^e = \frac{E}{1-\nu^2}\begin{bmatrix} 1 & \nu & 0 \\ \nu & 1 & 0 \\ 0 & 0 & \dfrac{1-\nu}{2} \end{bmatrix}. \tag{2.217}$$

Here we have used the engineering shear strain $\gamma_{xy} = 2\varepsilon_{xy}$. The representation of stress and strain given by (2.215) is referred to as the contracted form.

If the material body involves an initial stress $\boldsymbol{\sigma}_0$ and/or initial strain $\boldsymbol{\varepsilon}_0$, Hooke's law is transformed to

$$\boldsymbol{\sigma} = \boldsymbol{D}^e(\boldsymbol{\varepsilon} + \boldsymbol{\varepsilon}_0) + \boldsymbol{\sigma}_0 \tag{2.218}$$

If the initial strain is caused by a temperature difference $T - T_0$, we have $\boldsymbol{\varepsilon}_0 = \alpha(T - T_0)\boldsymbol{i}$ for an isotropic material body, therefore the above equation becomes

$$\varepsilon_{ij} = -\frac{\nu}{E}\sigma_{kk}\,\delta_{ij} + \frac{1+\nu}{E}\sigma_{ij} + \alpha(T - T_0)\,\delta_{ij} \tag{2.219}$$

where T_0 is the reference temperature and α is the thermal expansion coefficient.

Substituting Hooke's law (2.208) into the equation of motion (2.104) yields the following *Navier's equation* where the unknown variable is the displacement $\boldsymbol{u}$:

$$\rho \frac{d^2 u_i}{dt^2} = (\lambda + \mu) \frac{\partial^2 u_j}{\partial x_i \partial x_j} + \mu \frac{\partial^2 u_i}{\partial x_j \partial x_j} + \rho b_i. \tag{2.220}$$

Note 2.8 (Solid and fluid). The term "solid" is used for the material body where the response is between the stress $\boldsymbol{\sigma}$ and the strain $\boldsymbol{\varepsilon}$ or between the stress increment $d\boldsymbol{\sigma}$ and the strain increment $d\boldsymbol{\varepsilon}$. The term "fluid" is used for the material body where the response is between the stress $\boldsymbol{\sigma}$ and the strain rate $\dot{\boldsymbol{\varepsilon}}$ (or the stretch tensor $\boldsymbol{D}$). For a fluid we have to introduce a time-integration constant, which is referred to as the pressure p. ∎

2.8.4 Newtonian Fluid

For simplicity, we describe the equations without mass density ρ. Since the specific stress $\boldsymbol{\sigma}^{\ddagger}(\boldsymbol{x}, t)$ is given in terms of an Eulerian description, we treat here the simplest response for that description. To satisfy the principle of determinism and the principle of local action mentioned in the previous section, the stress $\boldsymbol{\sigma}^{\ddagger}(\boldsymbol{x}, t)$ can be written in terms of $\boldsymbol{v}$ and $\nabla \boldsymbol{v}$:

$$\boldsymbol{\sigma}^{\ddagger} = \boldsymbol{\sigma}^{\ddagger}(\boldsymbol{v}, \nabla \boldsymbol{v}). \tag{2.221}$$

The frame indifference of the stress $\boldsymbol{\sigma}^{\ddagger}(\boldsymbol{x}, t)$ is a natural conclusion of Newtonian mechanics in that the force vector is frame indifferent. Since the stretch tensor $\boldsymbol{D}$ is frame invariant by (2.154), we use $\boldsymbol{D}$ instead of $\nabla \boldsymbol{v}$. From (2.27) we have

$$\boldsymbol{v}^* = \frac{d\boldsymbol{x}_0^*}{dt} + \frac{d\boldsymbol{Q}}{dt}(\boldsymbol{x} - \boldsymbol{x}_0) + \boldsymbol{Q}\boldsymbol{v}.$$

Therefore the frame indifference requires the following condition:

$$\boldsymbol{\sigma}^{\ddagger *}(\boldsymbol{v}^*, \boldsymbol{D}^*) = \boldsymbol{Q}\boldsymbol{\sigma}^{\ddagger}\left(\dot{\boldsymbol{x}}_0^* + \dot{\boldsymbol{Q}}(\boldsymbol{x} - \boldsymbol{x}_0) + \boldsymbol{Q}\boldsymbol{v},\ \boldsymbol{Q}\boldsymbol{D}\boldsymbol{Q}^T\right)\boldsymbol{Q}^T$$

We define $\boldsymbol{x}_0^*$ as

$$\dot{\boldsymbol{x}}_0^* = -\dot{\boldsymbol{Q}}(\boldsymbol{x} - \boldsymbol{x}_0) - \boldsymbol{Q}\boldsymbol{v}$$

Then we can see that if we have

$$\boldsymbol{Q}\boldsymbol{\sigma}^{\ddagger}(\boldsymbol{D})\boldsymbol{Q}^T = \boldsymbol{\sigma}^{\ddagger}(\boldsymbol{Q}\boldsymbol{D}\boldsymbol{Q}^T) \quad \Rightarrow \quad \boldsymbol{\sigma}^{\ddagger} = \boldsymbol{\sigma}^{\ddagger}(\boldsymbol{D}) \tag{2.222}$$

the fundamental principles mentioned in the previous section are satisfied. Since $\boldsymbol{D}$ is symmetric and non-negative definite, the most general form (Truesdell and Noll 1965, pp. 32; Malvern 1969, pp. 194) can be given by

$$\boldsymbol{\sigma} = \rho \boldsymbol{\sigma}^{\ddagger} = \phi_0 \boldsymbol{i} + \phi_1 \boldsymbol{D} + \phi_2 \boldsymbol{D}^2 \tag{2.223}$$

where ϕ_i $(i=0,1,2)$ are functions of the invariants I_i^D $(i=1,2,3)$ of $\boldsymbol{D}$:

$$\phi_i = \phi_i(I_1^D, I_2^D, I_3^D). \tag{2.224}$$

The invariants I_i^D are calculated by the following characteristic equation for specifying the eigenvalue ζ:

$$\det(\boldsymbol{D} - \zeta \boldsymbol{i}) = -\zeta^3 + I_1^D \zeta^2 - I_2^D \zeta^1 + I_3^D = 0, \tag{2.225}$$

$$I_1^D = \operatorname{tr} \boldsymbol{D} = \nabla \cdot \boldsymbol{v},$$

$$I_2^D = \frac{1}{2}\left[(I_1^D)^2 - \hat{I}_2^D \right], \qquad \hat{I}_2^D = \operatorname{tr}(\boldsymbol{D}^2),$$

$$I_3^D = \det \boldsymbol{D}$$

We omit the third term of the r.h.s. of (2.223) so as to linearize it:

$$\boldsymbol{\sigma} = \rho \boldsymbol{\sigma}^{\ddagger} = (-p + \lambda \operatorname{tr} \boldsymbol{D}) \boldsymbol{i} + 2\mu \boldsymbol{D} \tag{2.226}$$

where p is the pressure and λ, μ are viscosities (μ is the shearing viscosity, and $\kappa = \lambda + 2\mu/3$ is the bulk viscosity: described below). The pressure p appears in this equation because $\boldsymbol{v}$ (and also $\boldsymbol{D}$) is a material time derivative of the position vector $\boldsymbol{x}$ of a material point in the deformed body, which needs an integration constant; this corresponds to the pressure. Note that usually the "pressure" is set positive for compression, therefore a negative sign of p appears in (2.226). Materials that behave as (2.226) are referred to as *Newtonian fluids*.

Let us resolve the stress $\boldsymbol{\sigma}$ and stretch tensor $\boldsymbol{D}$ into direct sums of volumetric and deviatoric components, respectively:

$$\boldsymbol{\sigma} = \bar{\boldsymbol{\sigma}} + \boldsymbol{\sigma}', \tag{2.227}$$

$$\bar{\boldsymbol{\sigma}} = \frac{1}{3}(\operatorname{tr} \boldsymbol{\sigma})\, \boldsymbol{i}, \qquad \boldsymbol{\sigma}' = \boldsymbol{\sigma} - \bar{\boldsymbol{\sigma}}, \tag{2.228}$$

$$\boldsymbol{D} = \overline{\boldsymbol{D}} + \boldsymbol{D}', \tag{2.229}$$

$$\overline{\boldsymbol{D}} = \frac{1}{3}(\operatorname{tr} \boldsymbol{D})\, \boldsymbol{i}, \qquad \boldsymbol{D}' = \boldsymbol{D} - \overline{\boldsymbol{D}}. \tag{2.230}$$

$\bar{\boldsymbol{\sigma}}$ and $\overline{\boldsymbol{D}}$ are the volumetric components of each tensor, and $\boldsymbol{\sigma}'$ and $\boldsymbol{D}'$ are the deviatoric (or shearing) components. The volumetric component is orthogonal to the deviatoric one in the following sense:

$$\bar{\boldsymbol{\sigma}} : \boldsymbol{\sigma}' = \mathrm{tr}\left(\bar{\boldsymbol{\sigma}}^T \boldsymbol{\sigma}'\right) = 0, \qquad \overline{\boldsymbol{D}} : \boldsymbol{D}' = \mathrm{tr}\left(\overline{\boldsymbol{D}}^T \boldsymbol{D}'\right) = 0. \tag{2.231}$$

Since $(\nabla \cdot \boldsymbol{v})\boldsymbol{I} = 3\overline{\boldsymbol{D}}$, we can rewrite (2.226) as

$$\bar{\boldsymbol{\sigma}} + \boldsymbol{\sigma}' = -p\boldsymbol{i} + 3\left(\lambda + \frac{2}{3}\mu\right)\overline{\boldsymbol{D}} + 2\mu\boldsymbol{D}'$$

Recalling the orthogonality of the volumetric and deviatoric components, each component will give an independent response:

$$\bar{\boldsymbol{\sigma}} = -p\boldsymbol{i} + 3\kappa\overline{\boldsymbol{D}}, \qquad \boldsymbol{\sigma}' = 2\mu\boldsymbol{D}' \tag{2.232}$$

This is a direct result of the response of an isotropic linear fluid. In this equation the constant

$$\kappa = \lambda + \frac{2}{3}\mu \tag{2.233}$$

gives the bulk (i.e., volumetric) viscosity and μ is the shearing viscosity.

Thus the most fundamental constitutive law for a fluid is understood to be given as a Newtonian fluid defining a linear relationship between the stress $\boldsymbol{\sigma}$ and the stretch tensor $\boldsymbol{D}$ (recall that the stretch tensor $\boldsymbol{D}$ is equal to the strain rate for the solid with small strain). The constitutive law is also called *Stokes' law*, and can be rewritten as

$$\sigma_{ij} = -p\,\delta_{ij} + \lambda\, D_{kk}\,\delta_{ij} + 2\mu\, D_{ij} \tag{2.234}$$

Substituting Stokes' law (2.234) into the equation of motion (2.104) under the Eulerian description yields the following equation of motion for the unknown velocity $\boldsymbol{v}$:

$$\rho\left(\frac{\partial v_i}{\partial t} + v_j\frac{\partial v_i}{\partial x_j}\right) = -\frac{\partial p}{\partial x_i} + (\lambda + \mu)\frac{\partial^2 v_j}{\partial x_i \partial x_j} + \mu\frac{\partial^2 v_i}{\partial x_j \partial x_j} + \rho b_i \tag{2.235}$$

These are the *Navier-Stokes equations*. If the body force can be set as $\boldsymbol{b} = -\nabla\phi$ by a potential ϕ, we define

$$p^* = p + \rho\phi, \tag{2.236}$$

and the Navier-Stokes equations can be written as

$$\rho\left(\frac{\partial v_i}{\partial t} + v_j\frac{\partial v_i}{\partial x_j}\right) = -\frac{\partial p^*}{\partial x_i} + (\lambda + \mu)\frac{\partial^2 v_j}{\partial x_i \partial x_j} + \mu\frac{\partial^2 v_i}{\partial x_j \partial x_j}. \tag{2.237}$$

If the fluid is incompressible, the condition (2.99) applies and we have

$$\rho\left(\frac{\partial v_i}{\partial t}+v_j\frac{\partial v_i}{\partial x_j}\right)=-\frac{\partial p}{\partial x_i}+\mu\frac{\partial^2 v_i}{\partial x_j\partial x_j}+\rho b_i\,. \tag{2.238}$$

2.9 Small Strain Viscoelasticity Theory

The one-dimensional viscoelastic response is schematically illustrated in Fig. 2.12. Note that the 'stress relaxation' is a phenomenon that appears under a constant strain condition, while 'creep' is one that appears under a constant stress condition. The response shown is represented by a model based on an excitation-response theory together with a data management procedure. Note that we assume an isotropic material response.

2.9.1 *Boltzmann Integral and Excitation-response Theory*

The viscoelastic response is commonly described by using a form of *Boltzmann's hereditary integral*, referred to as the *excitation-response theory* (Gurtin and Sternberg 1962; Yamamoto 1972; Christensen 2003).

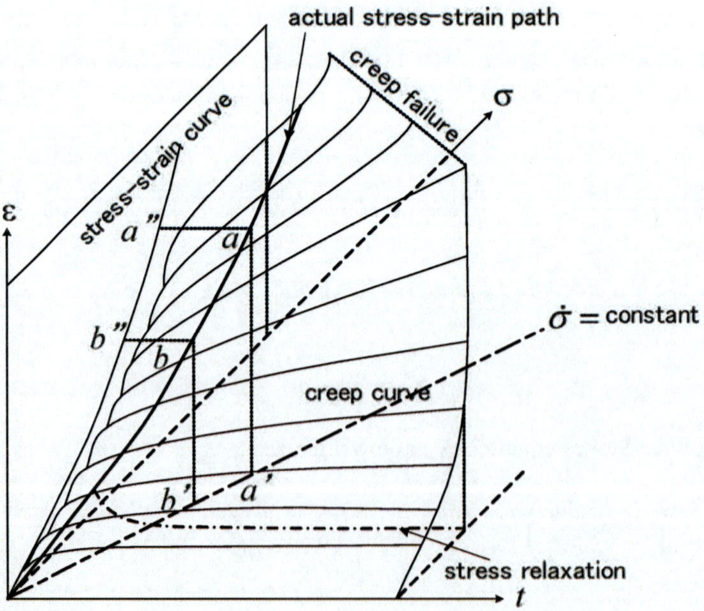

Fig. 2.12 Viscoelastic response

Let us consider a step input function

$$x(t) = \begin{cases} 0 & \text{for } t < 0, \\ x_0 = \text{constant} & \text{for } t > 0. \end{cases} \tag{2.239}$$

The corresponding response to this input can be written as

$$y(t) = \phi(t)\, x_0 \tag{2.240}$$

where $\phi(t)$ is referred to as the *after-effect function* (Fig. 2.13a) which satisfies the condition

$$\phi(t) = 0 \qquad \text{for } t < 0.$$

If the input $x(t)$ is given by a collection of step functions as shown in Fig. 2.13b, the response is written as

$$y(t) = \sum_i \phi(t - t_i)\Delta x_i . \tag{2.241}$$

Then for a general form of the input function $x(t)$, we have

$$y(t) = \int_{-\infty}^{t} \phi(t - s)\frac{dx(s)}{ds} ds. \tag{2.242}$$

This is referred to as *Boltzmann's superposition principle.*

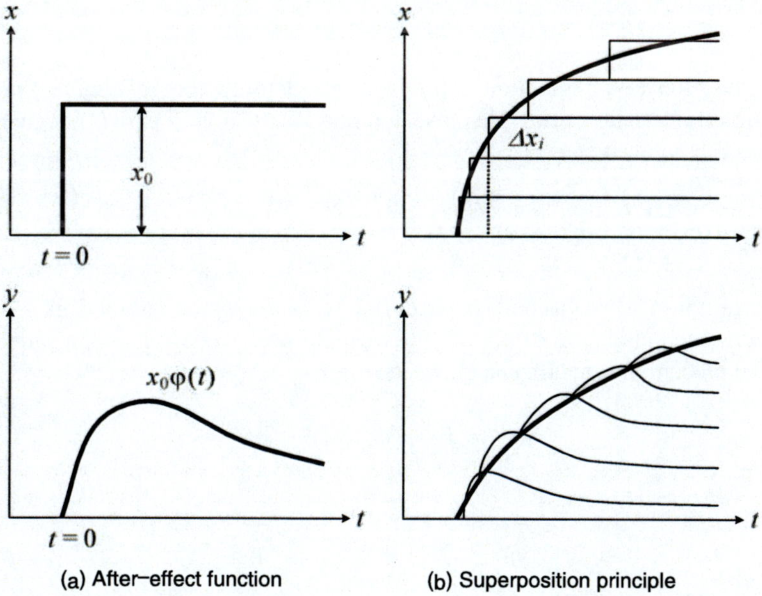

Fig. 2.13 Boltzmann's hereditary integral

Integrating (2.242) by parts yields

$$y(t) = \phi(0+)\,x(t) + \int_{-\infty}^{t} \frac{d\phi(t-s)}{ds} x(s)\,ds \equiv \int_{-\infty}^{t} \mu(t-s)\,x(s)\,ds \quad (2.243)$$

where

$$\phi(0+) = \lim_{t \to +0} \phi(t)$$

and

$$\mu(t-s) = \frac{d\phi(t-s)}{ds} + \delta(s)\,\phi(s) \quad (2.244)$$

is referred to as the *response function*.[7]

2.9.2 Stress Relaxation and the Relaxation Spectra: Generalized Maxwell Model

First we consider a simple uniaxial response. If a strain ε is given by

$$\varepsilon(t) = \begin{cases} 0 & \text{for } t < 0, \\ \varepsilon_0 & \text{for } t > 0, \end{cases} \quad (2.245)$$

we write the relaxation stress as

$$\sigma(t) = E(t)\varepsilon_0, \qquad E(t) = 0 \quad \text{for } t < 0 \quad (2.246)$$

where the after-effect function $E(t)$ is referred to as the *relaxation function*. Following Boltzmann's principle presented previously, if an input $\varepsilon(t)$ is given, the response can be written as

$$\sigma(t) = \int_{-\infty}^{t} E(t-s)\frac{d\varepsilon(s)}{ds} ds = \int_{-\infty}^{t} \Lambda(t-s)\,\varepsilon(s)\,ds. \quad (2.247)$$

The response of a viscoelastic material is generally represented as a model containing a combination of elastic, viscous and pure stress-relaxation properties. Then the relaxation function can be written as

$$E(t) = E_0 + \eta_\infty\,\delta(t) + \overline{E}(t) \quad (2.248)$$

[7] The definition of the δ-function is given by $\int_\Omega dy\,\delta(y-x)f(y) = f(x)$.

where $\overline{E}(t)$ is a smooth, monotonically decreasing function such that

$$\overline{E}(0) = \kappa < +\infty, \qquad \overline{E}(+\infty) = 0,$$

and can be represented by a Laplace transformation:

$$\overline{E}(t) = \int_0^\infty N(s) \exp(-st)\, ds. \tag{2.249}$$

In order to provide a discrete approximation of this function, the variable s is changed into τ by $\tau = 1/s$. Then (2.249) can be written as

$$\overline{E}(t) = \int_0^\infty H(\tau) \exp(-t/\tau)\, d(\ln \tau), \tag{2.250}$$

$$H(\tau) = \frac{1}{\tau} N\left(\frac{1}{\tau}\right), \tag{2.251}$$

and $H(\tau)$ is referred to as the *relaxation spectrum*.

Substituting (2.250) into (2.247) and changing the order of integration, we have

$$\sigma(t) = \sigma_0(t) + \sigma_\infty(t) + \int_0^\infty \sigma(t,\tau)\, d(\ln \tau), \tag{2.252}$$

$$\sigma(t,\tau) = H(\tau) \int_{-\infty}^t \exp\left(-(t-s)/\tau\right) \frac{d\varepsilon(s)}{ds} ds, \tag{2.253}$$

$$\sigma_0(t) = E_0\, \varepsilon(t), \qquad \sigma_\infty(t) = \eta_\infty \frac{d\varepsilon(t)}{dt}. \tag{2.254}$$

Differentiating (2.253)[8] we obtain

$$H(\tau)\frac{d\varepsilon(\tau)}{dt} = \frac{d}{dt}\sigma(t,\tau) + \frac{1}{\tau}\sigma(t,\tau). \tag{2.255}$$

---

[8]Leibnitz rule: If we have an integral of a continuous function f such as

$$\phi(x) = \int_{h_0(x)}^{h_1(x)} f(x,\xi)\, d\xi,$$

and if $h_1(x)$ and $h_0(x)$ are continuous on $R = \{\,(x,\xi) : a \le x \le b,\ c \le \xi \le d\,\}$, then

$$\frac{d\phi(x)}{dx} = f\left(x, h_1(x)\right) \frac{dh_1(x)}{dx} - f\left(x, h_0(x)\right) \frac{dh_0(x)}{dx} + \int_{h_0(x)}^{h_1(x)} \frac{\partial f(x,\xi)}{\partial x}\, d\xi$$

(see, e.g., Protter and Morrey 1977, pp. 284).

A mechanical representation of a simple Maxwell model is shown in Fig. 2.14a and can be indicated in the same form as (2.255): i.e.

$$E\frac{d\varepsilon}{dt} = \frac{d\sigma}{dt} + \frac{1}{\tau}\sigma$$

where E and η are the elastic and viscous constants, respectively, and $\tau = \eta/E$ is referred to as the *relaxation time*. Thus we understand that (2.252) gives a synthesis of each spectral response corresponding to τ, and this suggests that, for a discrete case, the generalized Maxwell model can be represented by Fig. 2.15a.

If the material is isotropic, the response can be written in both deviatoric and volumetric forms:

$$\boldsymbol{s}(t) = 2\int_{-\infty}^{t} G(t-s)\frac{d\boldsymbol{e}(s)}{ds}ds, \qquad \bar{\sigma}(t) = 3\int_{-\infty}^{t} K(t-s)\frac{d\bar{\varepsilon}(s)}{ds}ds \quad (2.256)$$

where

$$G(t) = G_0 + \eta_\infty^s\,\delta(t) + \overline{G}(t), \qquad K(t) = K_0 + \eta_\infty^v\,\delta(t) + \overline{K}(t), \quad (2.257)$$

$$\overline{G}(t) = \int_0^\infty \Phi^s(\tau)\exp(-t/\tau)\,d(\ln\tau), \qquad \overline{K}(t) = \int_0^\infty \Phi^v(\tau)\exp(-t/\tau)\,d(\ln\tau). \quad (2.258)$$

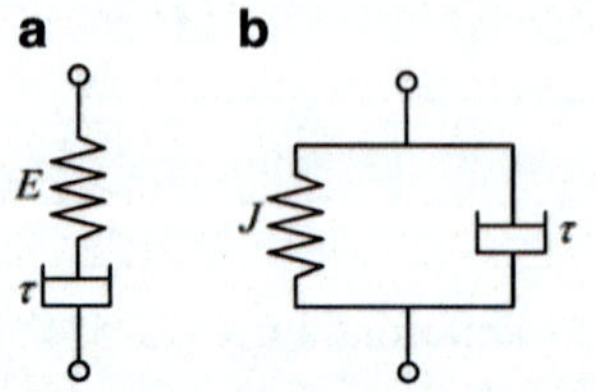

Fig. 2.14 **(a)** Maxwell model, **(b)** Kelvin-Voigt model

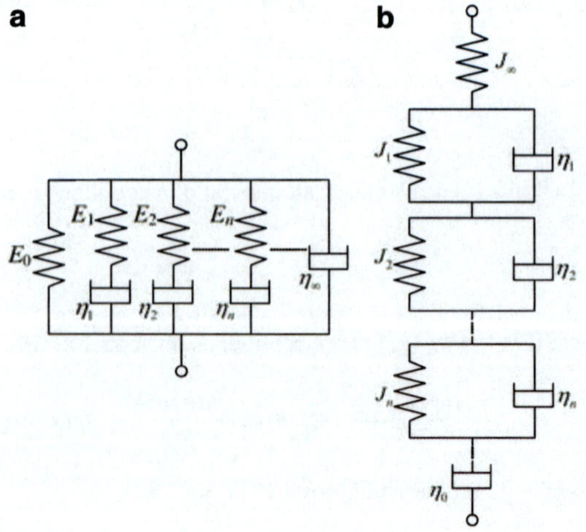

Fig. 2.15 **(a)** Generalized Maxwell model, **(b)** generalized Kelvin-Voigt model

Since $\boldsymbol{\sigma}(t) = \boldsymbol{s}(t) + \bar{\boldsymbol{\sigma}}(t)$, the total response can be represented as

$$\boldsymbol{\sigma}(t) = \int_{-\infty}^{t} \left[2G(t-s)\frac{d\boldsymbol{e}(s)}{ds} + 3K(t-s)\frac{d\bar{\boldsymbol{\varepsilon}}(s)}{ds} \right] ds = \int_{-\infty}^{t} \boldsymbol{D}(t-s)\frac{d\boldsymbol{\varepsilon}(s)}{ds}\, ds. \tag{2.259}$$

2.9.3 *Creep and the Retardation Spectra: Generalized Kelvin-Voigt Model*

Let us consider a simple uniaxial creep response. If a creep stress σ is given by

$$\sigma(t) = \begin{cases} 0 & \text{for } t < 0, \\ \sigma_0 & \text{for } t > 0, \end{cases} \tag{2.260}$$

and the corresponding strain response is written as

$$\varepsilon(t) = J(t)\,\sigma_0, \qquad J(t) = 0 \quad \text{for } t < 0, \tag{2.261}$$

then for a general input $\sigma(t)$ the strain response can be represented as

$$\varepsilon(t) = \int_{-\infty}^{t} J(t-s)\frac{d\sigma(s)}{ds} ds = \int_{-\infty}^{t} \Gamma(t-s)\,\sigma(s)\,ds \tag{2.262}$$

where $J(t)$ is referred to as *creep function*.

In the same way as we did for stress relaxation, we combine the instantaneous elastic, viscous and pure stress-relaxation properties, and write the response as

$$J(t) = J_\infty + \frac{t}{\eta_0} + \bar{J}(t) \tag{2.263}$$

where $J_\infty = 1/E_\infty$. Since $\bar{J}(t)$ is a smooth, monotone decreasing function, we have

$$\bar{J}(0) = 0, \qquad \bar{J}(+\infty) = \rho < +\infty.$$

Therefore it can be represented by a Laplace transformation:

$$\bar{J}(t) = \int_0^\infty M(s)\,(1 - \exp(-st))\,ds = \int_0^\infty L(\tau)\,(1 - \exp(-t/\tau))\,d(\ln\tau), \tag{2.264}$$

$$L(\tau) = \frac{1}{\tau} M\left(\frac{1}{\tau}\right). \tag{2.265}$$

$L(\tau)$ is referred to as the *retardation spectrum*.

Substituting (2.264) into (2.262) and changing the order of integration, we have

$$\varepsilon(t) = \varepsilon_\infty(t) + \varepsilon_0(t) + \int_0^\infty \varepsilon(t,\tau)\, d(\ln\tau) \tag{2.266}$$

$$\varepsilon(t,\tau) = L(\tau)\int_0^\infty (1 - \exp(-(t-s)/\tau))\frac{d\sigma(s)}{ds}ds, \tag{2.267}$$

$$\varepsilon_\infty(t) = J_\infty\,\sigma, \qquad \varepsilon_0(t) = \frac{1}{\eta_0}\int_{-\infty}^t \sigma(s)\,ds. \tag{2.268}$$

Differentiating (2.267), we obtain

$$L(\tau)\sigma(t) = \tau\frac{d}{dt}\varepsilon(t,\tau) + \varepsilon(t,\tau) \tag{2.269}$$

The response of a simple one-unit Kelvin-Voigt model shown by Fig. 2.14b can be written in a form similar to (2.269) as

$$J\sigma = \tau\frac{d\varepsilon}{dt} + \varepsilon$$

where $J = 1/E$ and η are the elastic compliance and viscous constant, respectively, and $\tau = \eta/E$ is referred to as the *retardation time*. Thus we can see that (2.266) gives a synthesis of each spectral response corresponding to τ, and this suggests that for a discrete case the generalized Kelvin-Voigt model can be represented by Fig. 2.15b.

If the material is isotropic, the response can be given separately for the deviatoric and volumetric deformations as follows:

$$\boldsymbol{e}(t) = \frac{1}{2}\int_{-\infty}^t B(t-s)\frac{d\boldsymbol{s}(s)}{ds}ds, \qquad \bar{\boldsymbol{\varepsilon}}(t) = \frac{1}{3}\int_{-\infty}^t C(t-s)\frac{d\bar{\boldsymbol{\sigma}}(s)}{ds}ds \tag{2.270}$$

where

$$B(t) = B_\infty + \frac{t}{\eta_0^s} + \overline{B}(t), \qquad C(t) = C_\infty + \frac{t}{\eta_0^v} + \overline{C}(t) \tag{2.271}$$

$$\overline{B}(t) = \int_0^\infty \Psi^s(\tau)\,(1 - \exp(-t/\tau))\;d(\ln\tau),$$

$$\overline{C}(t) = \int_0^\infty \Psi^v(\tau)\,(1 - \exp(-t/\tau))\;d(\ln\tau). \tag{2.272}$$

Since $\boldsymbol{\varepsilon}(t) = \boldsymbol{e}(t) + \bar{\boldsymbol{\varepsilon}}(t)$, the total response is thus represented as

$$\boldsymbol{\varepsilon}(t) = \int_{-\infty}^t \left[\frac{1}{2}B(t-s)\frac{d\boldsymbol{s}(s)}{ds} + \frac{1}{3}C(t-s)\frac{d\bar{\boldsymbol{\sigma}}(s)}{ds}\right] ds = \int_{-\infty}^t \boldsymbol{C}(t-s)\frac{d\boldsymbol{\sigma}(s)}{ds}\,ds. \tag{2.273}$$

2.9.4 Relaxation and Retardation Spectra and Their Asymptotic Expansion

The k-th derivative of (2.249) yields

$$\overline{E}^{(k)}(t) = \frac{d^k \overline{E}(t)}{dt^k} = (-1)^k \int_0^\infty N(s) s^k \exp(-st)\, ds$$

where $\overline{E}^{(k)}$ implies the k-time differentiation of $\overline{E}$. Since $s^k \exp(-st)$ shows the peak value at $s = k/t$ and the value increases with k, we can replace it by the δ-function so that we have

$$\int_0^\infty s^k \exp(-st)\, ds = \frac{k!}{t^{k+1}}.$$

Therefore we set

$$\overline{E}^{(k)}(t) = (-1)^k \frac{k!}{t^{k+1}} \int_0^\infty N(s)\delta(s - k/t)\, ds = (-1)^k \frac{k!}{t^{k+1}} N(k/t).$$

Now we change the variable as $t = k/s$, and obtain

$$\overline{E}^{(k)}(k/s) = (-1)^k s^{k+1} \frac{k!}{k^{k+1}} N(s).$$

This gives a function $N(s)$, and, when $k \to \infty$, we have the following asymptotic form:

$$N(s) = \lim_{k \to \infty} \frac{(-1)^k}{k!} \left(\frac{k}{s}\right)^{k+1} \overline{E}^{(k)}(k/s).$$

Since $\overline{E}$ is a function of k/s, and using (2.251) we have a function $H(\tau)$ instead of $N(s)$[9]:

$$H(\tau) = \frac{1}{\tau} N\left(\frac{1}{\tau}\right) = \lim_{k \to \infty} \frac{(-1)^k}{(k-1)!} (k\tau)^k \overline{E}^{(k)}(k\tau). \qquad (2.274)$$

Equation 2.274 shows that the relaxation spectrum $H(\tau)$ can be approximated by

$$H_1(\tau) = -\tau \frac{\partial \overline{E}(\tau)}{\partial \tau}, \quad H_2(\tau/2) = \tau^2 \frac{\partial^2 \overline{E}(\tau)}{\partial \tau^2}, \quad H_3(\tau/3) = -\frac{\tau^3}{2} \frac{\partial^3 \overline{E}(\tau)}{\partial \tau^3}, \quad \ldots$$

When using actual experimental data, the differentiation frequently involves some numerical errors, therefore we can use either H_2 or, more often, H_1 for the spectral approximation. The practical procedures will be given in a later section.

---

[9]Note that the differentiation implies $\overline{E}^{(k)}(k\tau) = \dfrac{d^k \overline{E}(k\tau)}{d(k\tau)^k}$. Others are the same.

The most important advantage of this procedure is that we can determine the spectral points that govern the corresponding viscoelastic response, which gives their coefficients, specified by the least squares method.

The retardation spectrum $L(\tau)$ of (2.264) is written as

$$L(\tau) = \lim_{k\to\infty} \frac{(-1)^{k+1}}{(k-1)!} (k\tau)^k \bar{J}^{(k)}(k\tau). \tag{2.275}$$

For an isotropic material the relaxation spectra are given by

$$\Phi^s(\tau) = \lim_{k\to\infty} \frac{(-1)^k}{(k-1)!} (k\tau)^k \overline{G}^{(k)}(k\tau), \quad \Phi^v(\tau) = \lim_{k\to\infty} \frac{(-1)^k}{(k-1)!} (k\tau)^k \overline{K}^{(k)}(k\tau), \tag{2.276}$$

and the retardation spectra are

$$\Psi^s(\tau) = \lim_{k\to\infty} \frac{(-1)^k}{(k-1)!} (k\tau)^k \overline{B}^{(k)}(k\tau), \quad \Psi^v(\tau) = \lim_{k\to\infty} \frac{(-1)^k}{(k-1)!} (k\tau)^k \overline{C}^{(k)}(k\tau). \tag{2.277}$$

2.9.5 *Experiments for Determining Viscous Properties*

Several experiments are used to determine the viscoelastic properties of a material: creep tests (stress = constant), relaxation tests (strain = constant) and a constant stress rate test. Figure 2.12 shows the loading paths of each test for the one-dimensional case.

We first determine the creep function $J(t)$ based on the results of a creep test: If the creep stress is given by (2.260), the response is given by (2.261), therefore $\varepsilon(t) = J(t)\sigma_0$. Let us assume that the relaxation spectrum is represented by the first approximation of (2.275), i.e.,

$$L(\tau) = -\tau \frac{\partial \overline{J}(\tau)}{\partial \tau}.$$

By substituting (2.261) into this result we have

$$L(\tau) = -\frac{1}{\sigma_0} \tau \frac{\partial \varepsilon(\tau)}{\partial \tau}. \tag{2.278}$$

We can then plot the values of (2.278) calculated by the creep curve shown, for example, in Fig. 2.16. In this figure we specify discrete values $\tau_1, \tau_2, \ldots$. Since the discrete representation of (2.264) is given by

$$\overline{J}(t) = \sum_i J_i \left(1 - \exp(-t/\tau_i)\right), \qquad J_i = L(\tau_i)\, \Delta \ln \tau_i, \tag{2.279}$$

Fig. 2.16 Implication of retardation spectrum

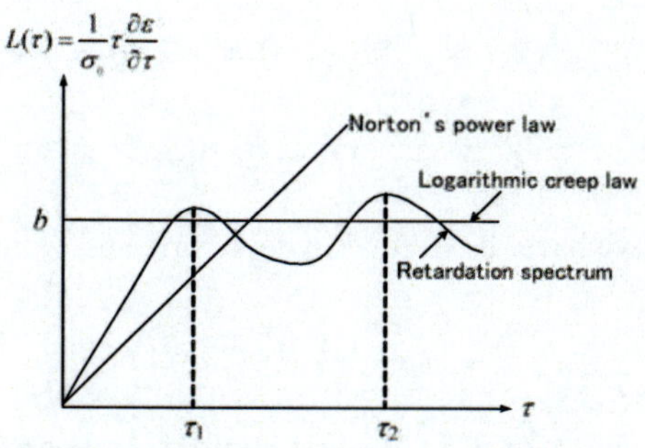

we can substitute the above τ_i's into (2.279), and, using (2.261) and (2.263), we obtain the following:

$$\varepsilon(t) = \left[J_\infty + \frac{t}{\eta_0} + \sum_i J_i \left(1 - \exp(-t/\tau_i)\right) \right] \sigma_0. \tag{2.280}$$

We now need a procedure to determine η_0. From (2.280) the slope at the final elapsed time t^* is given by

$$\left. \frac{1}{\sigma_0} \frac{d\varepsilon}{dt} \right|_{t=t^*} = \sum \frac{J_i}{\tau_i} \exp(-t^*/\tau_i) + \frac{1}{\eta_0}. \tag{2.281}$$

Thus if $\tau_i \ll t^*$, we can specify the slope $1/\eta_0$. If $\tau_i \approx t^*$, the reader is referred to Akagi (1980) for details of the method to specify η_0.

If τ_i $(i = 1, 2, \ldots)$ and η_0 are known, J_∞ and J_i $(i = 1, 2, \ldots)$ can be specified by a linear least squares method. That is, setting the error estimate as

$$\Pi = \frac{1}{2}\sigma_0^2 \left[J_\infty + \frac{t}{\eta_0} + \sum_i J_i \left(1 - \exp(-t/\tau_i)\right) \right]^2, \tag{2.282}$$

J_∞ and J_i can be calculated by solving the following simultaneous equations:

$$\frac{\partial \Pi}{\partial J_\infty} = 0, \qquad \frac{\partial \Pi}{\partial J_i} = 0, \qquad i = 1, 2, \ldots \tag{2.283}$$

For an isotropic material, the creep functions $B(t)$ and $C(t)$ are determined using the same procedure. For an axisymmetric triaxial stress state, let σ_1 be the axial stress, $\sigma_3 (< 0)$ (with $\sigma_2 = \sigma_3$) the confining pressure, ε_1 the measured axial strain and ε_3 the measured lateral strain. Then setting

$$s_{11} = \frac{2}{3}(\sigma_1 - \sigma_3) = -2s_{22} = -2s_{33} \equiv q_0, \qquad s_{23} = s_{31} = s_{13} = 0,$$

$$\bar{\sigma} = \frac{1}{\sqrt{3}}(\sigma_1 + 2\sigma_3) \equiv p_0,$$

$$e_{11} = \frac{2}{3}(\varepsilon_1 - \varepsilon_3) = -2e_{22} = -2e_{33} \equiv \varepsilon^s, \qquad e_{23} = e_{31} = e_{13} = 0,$$
$$\bar{\varepsilon} = \frac{1}{\sqrt{3}}(\varepsilon_1 + 2\varepsilon_3) \equiv \varepsilon^v,$$

we have the following deviatoric and volumetric creep responses:

$$\varepsilon^s(t) = \frac{1}{2}B(t)\,q_0, \qquad \varepsilon^v(t) = \frac{1}{3}C(t)\,p_0. \tag{2.284}$$

Due to (2.277), the first approximations of the relaxation spectra of each component are given by

$$\Psi^s(\tau^s) = -\frac{2}{q_0}\tau^s\frac{\partial\varepsilon^s(\tau^s)}{\partial\tau^s}, \qquad \Psi^v(\tau^v) = -\frac{3}{p_0}\tau^v\frac{\partial\varepsilon^v(\tau^v)}{\partial\tau^v}. \tag{2.285}$$

Therefore the discrete relaxation times $\tau_1^s, \tau_2^s, \ldots,\ \tau_1^v, \tau_2^v, \ldots$ can be obtained by creep curves (similar to Fig. 2.16). This gives the discrete forms of (2.272) as

$$\begin{aligned}\overline{B}(t) &= \sum_i B_i\left(1 - \exp(-t/\tau_i^s)\right), \qquad B_i = \Psi^s(\tau_i^s)\,\Delta\ln\tau_i^s \\ \overline{C}(t) &= \sum_i C_i\left(1 - \exp(-t/\tau_i^v)\right), \qquad C_i = \Psi^v(\tau_i^v)\,\Delta\ln\tau_i^v\end{aligned} \tag{2.286}$$

and we obtain

$$\begin{aligned}\varepsilon^s(t) &= \frac{1}{2}\left[B_\infty + \frac{t}{\eta_0^s} + \sum_i B_i\left(1 - \exp(-t/\tau_i^s)\right)\right]q_0 \\ \varepsilon^v(t) &= \frac{1}{3}\left[C_\infty + \frac{t}{\eta_0^v} + \sum_i C_i\left(1 - \exp(-t/\tau_i^v)\right)\right]p_0.\end{aligned} \tag{2.287}$$

Under the condition $\tau_i^s,\ \tau_i^v \ll t^*$, η_0^s and η_0^v can be specified by using the slopes of the creep curves at the final time stage t^* by

$$\frac{2}{q_0}\frac{d\varepsilon^s}{dt}\bigg|_{t=t^*} \doteq \frac{1}{\eta_0^s}, \qquad \frac{3}{p_0}\frac{d\varepsilon^v}{dt}\bigg|_{t=t^*} \doteq \frac{1}{\eta_0^v}. \tag{2.288}$$

The coefficients B_∞ and B_i $(i=1,2,\ldots)$, C_∞ and C_i $(i=1,2,\ldots)$ can be calculated by the least squares method where the error estimates are given by

$$\Pi^s = \frac{1}{8}q_0^2\left[B_\infty + \frac{t}{\eta_0^s} + \sum_i B_i\left(1 - \exp(-t/\tau_i^s)\right)\right]^2,$$
$$\Pi^v = \frac{1}{18}p_0^2\left[C_\infty + \frac{t}{\eta_0^v} + \sum_i C_i\left(1 - \exp(-t/\tau_i^v)\right)\right]^2,$$

which gives the simultaneous equations

$$\frac{\partial \Pi^s}{\partial B_\infty} = 0, \qquad \frac{\partial \Pi^s}{\partial B_i} = 0, \tag{2.289}$$

$$\frac{\partial \Pi^v}{\partial C_\infty} = 0, \qquad \frac{\partial \Pi^v}{\partial C_i} = 0. \tag{2.290}$$

Finally, we can see the relationship between the response given by the hereditary integral form and that given by conventional creep laws such as the logarithmic form

$$\varepsilon(t) = a + b \ln t, \tag{2.291}$$

and Norton's power law

$$\frac{\varepsilon}{t} = a\,\sigma^n. \tag{2.292}$$

Differentiating (2.291) yields

$$t \frac{d\varepsilon}{dt} = b.$$

Therefore the logarithmic creep law employs an averaged relaxation spectrum for all elapsed time (i.e., only one time-dependent mechanism is assumed) as shown in Fig. 2.16, which may cause difficulty under real, complex situations such as the long term behavior of rock. On the other hand the power law (2.292) gives

$$t \frac{d\varepsilon}{dt} = t\,(a\sigma_0^n)$$

for a creep stress σ_0, which implies a linear distribution of the spectra (Fig. 2.16).

2.10 Small Strain Plasticity: Flow Theory

In this book we are considering porous materials. Therefore, the stress treated here must be an effective stress $\boldsymbol{\sigma}' = \boldsymbol{\sigma} + p\boldsymbol{I}$ where $\boldsymbol{\sigma}$ is the total stress, and p is the pore fluid pressure. Note that in this section we are using the sign convention for stresses adopted in continuum mechanics, therefore the tension stress/strain is considered positive, and the pore fluid pressure is positive, since $\boldsymbol{\sigma}' = \boldsymbol{\sigma} + p\boldsymbol{I}$ (details are described in Chap. 6). In this Section we denote the stress as $\boldsymbol{\sigma}$ instead of the effective stress $\boldsymbol{\sigma}'$ for simplicity. Readers can see that all results in this section also work for the effective stress. It should be noted that in this section the deviatoric stress is denoted as $\boldsymbol{s}$ whereas in other expositions the deviatoric stress is written as $\boldsymbol{\sigma}'$. Similarly, the deviatoric strain is denoted as $\boldsymbol{e}$.

Symbols and notations used in this section are given below:

$\boldsymbol{\sigma}$	(Total) stress tensor (tension: +)
$I_1 = \mathrm{tr}\,(\boldsymbol{\sigma}) = \sigma_{ii}$	First invariant of stress
$I_2 = \{(\mathrm{tr}\,\boldsymbol{\sigma})^2 - \mathrm{tr}\,(\boldsymbol{\sigma}^2)\}/2$	Second invariant of stress
$I_3 = \det(\boldsymbol{\sigma})$	Third invariant of stress
$\bar{\boldsymbol{\sigma}} = \mathrm{tr}\,(\boldsymbol{\sigma})\boldsymbol{I}/3$	Volumetric stress tensor (tension, +)
$\bar{\sigma} = \mathrm{tr}\,(\boldsymbol{\sigma})/\sqrt{3}$	Magnitude of volumetric stress (tension, +)
$\boldsymbol{s} = \boldsymbol{\sigma} - \bar{\boldsymbol{\sigma}}$	Deviatoric stress tensor
$s = \lvert\boldsymbol{s}\rvert = (s_{ij}\,s_{ij})^{1/2}$	Norm of deviatoric stress
$\sigma_\theta = \frac{1}{3}\cos^{-1}[3\sqrt{3}\,J_3^\sigma/\{2(J_2^\sigma)^{3/2}\}]$	Lode's angle for stress
$J_2 = s_{ij}s_{ij}/2 = 2s^2$	Second invariant of deviatoric stress
$J_3 = \det(\boldsymbol{s})$	Third invariant of deviatoric stress
$d\boldsymbol{\varepsilon} = d\boldsymbol{\varepsilon}^e + d\boldsymbol{\varepsilon}^p$	Strain increment tensor
$d\boldsymbol{\varepsilon}^e$	Elastic strain increment tensor
$d\boldsymbol{\varepsilon}^p$	Plastic strain increment tensor
$d\bar{\boldsymbol{\varepsilon}}^p = \mathrm{tr}\,(d\boldsymbol{\varepsilon}^p)\boldsymbol{I}/3$	Volumetric plastic strain increment tensor
$d\bar{\varepsilon}^p = \mathrm{tr}\,(d\boldsymbol{\varepsilon}^p)/\sqrt{3}$	Magnitude of volumetric plastic strain increment
$d\boldsymbol{e}^p = d\boldsymbol{\varepsilon}^p - d\bar{\boldsymbol{\varepsilon}}^p$	Deviatoric plastic strain increment tensor
$de^p = \lvert d\boldsymbol{e}^p\rvert = (de_{ij}^p\,de_{ij}^p)^{1/2}$	Norm of deviatoric plastic strain increment
κ	Hardening parameter

The result of a simple tension experiment for a metal is schematically shown in Fig. 2.17 with axes of axial stress σ_1 and axial strain ε_1 or deviatoric stress s and deviatoric strain e. In metals the volumetric plastic strain can generally be ignored ($\bar{\varepsilon}^p = 0$); therefore we can treat the behavior as a uniaxial response. On the other hand, the shearing behavior of geomaterials is inevitably accompanied by volume changes that are plastic, therefore we have to modify the original flow theory developed for metallic materials (Kachanov 2005; Lubliner 1990). Note that in small strain plasticity we assume that the plastic increment $d\boldsymbol{\varepsilon}$ can be decomposed into incremental elastic and plastic components:

$$d\boldsymbol{\varepsilon} = d\boldsymbol{\varepsilon}^e + d\boldsymbol{\varepsilon}^p. \tag{2.293}$$

2.10.1 Yield Function and Hardening Law

Let us denote the yield condition by

$$f(\boldsymbol{\sigma}, \kappa) = 0 \tag{2.294}$$

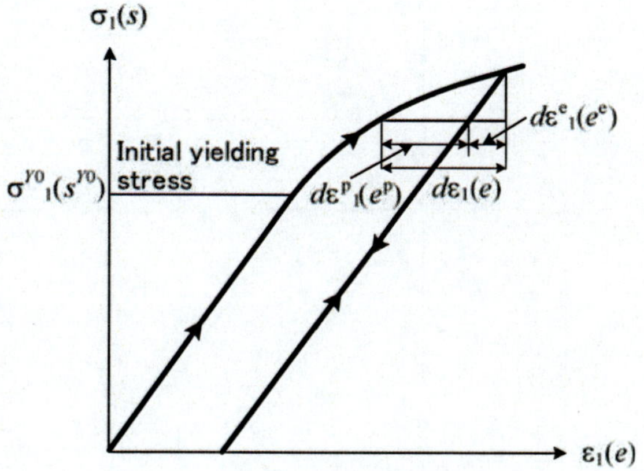

Fig. 2.17 Uniaxial stress-strain relationship

which includes the initial and subsequent yield surfaces (Fig. 2.17). Here κ is the hardening parameter which denotes a history of past stress and strain, and the increment can be denoted by

$$d\kappa = dW^p = \boldsymbol{\sigma} : d\boldsymbol{\varepsilon}^p \quad \text{: work hardening rule} \tag{2.295}$$

$$d\kappa = |\, d\boldsymbol{\varepsilon}^p| \quad \text{: strain hardening rule} \tag{2.296}$$

where dW^p is an increment of plastic work. Note that the initial yielding condition is given by $f(\boldsymbol{\sigma}, \kappa=0) = 0$.

The yield function $f(\boldsymbol{\sigma}, \kappa)$ is classified into several hardening models depending on the history of loading:

$$f(\boldsymbol{\sigma}, \kappa) = f_1(\boldsymbol{\sigma}) - K(\kappa) \quad \text{: isotropic hardening model} \tag{2.297}$$

$$f(\boldsymbol{\sigma}, \kappa) = f_1(\boldsymbol{\sigma} - \boldsymbol{\alpha}(\kappa)) \quad \text{: kinematic hardening model} \tag{2.298}$$

$$d\boldsymbol{\alpha} = c\, d\boldsymbol{\varepsilon}^p \quad \text{Prager model}$$

$$d\boldsymbol{\alpha} = c(\boldsymbol{\sigma} - \boldsymbol{\alpha})|\, d\boldsymbol{\varepsilon}^p| \quad \text{Ziegler model}$$

$$f(\boldsymbol{\sigma}, \kappa) = f_1(\boldsymbol{\sigma} - \boldsymbol{\alpha}(\kappa)) - f_2(\kappa) \quad \text{: anisotropic hardening model} \tag{2.299}$$

We illustrate the isotropic and kinematic hardening models for the one-dimensional problem in Fig. 2.18a, b, and for the two dimensional problem in Fig. 2.19. Note that the behavior shown in Fig. 2.18b is known as the Bauschinger effect. In geotechnical engineering practice the isotropic hardening model is widely used because, except in earthquake situations, it is rare that the loading direction is completely reversed.

The stress-dependent part of the yield function has been widely investigated using experimental methods. We summarize the results for the case of the isotropic

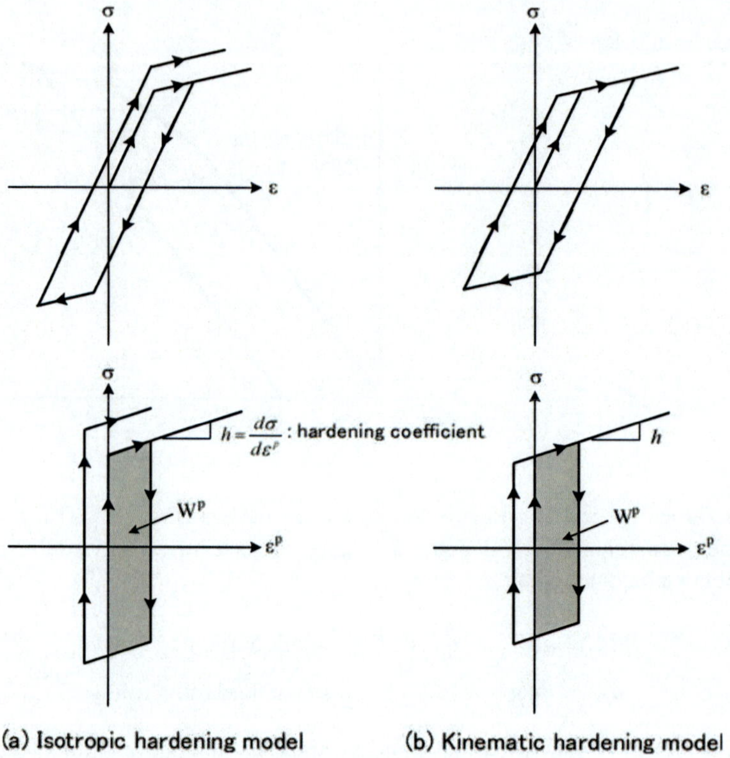

Fig. 2.18 Axial stress-strain relations and hardening models

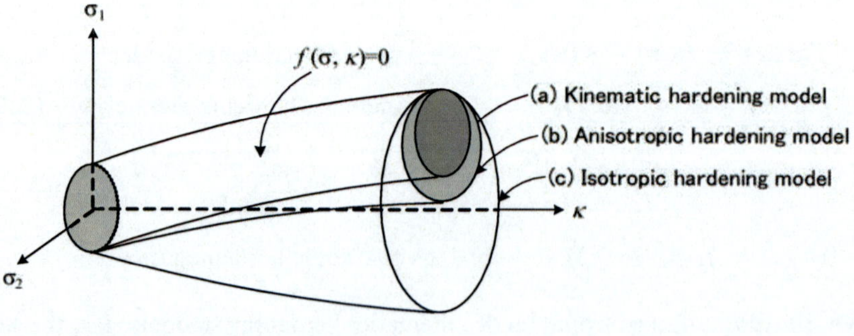

Fig. 2.19 Hardening model for a 2D problem

hardening (2.297) (see also Desai and Siriwardane 1984; Darve 1990; Davis and Selvadurai 2002; Pietruszczak 2010):

$$f_1(\boldsymbol{\sigma}) = \sqrt{J_2} \qquad \text{: von Mises}$$
$$f_1(\boldsymbol{\sigma}) = \sqrt{J_2} + \alpha I_1 \qquad \text{: Drucker-Prager}$$

$f_1(\boldsymbol{\sigma}) = |\sigma_1 - \sigma_3|$: Tresca

$\sigma_1 \geq \sigma_2 \geq \sigma_3$: principal stresses

$f_1(\boldsymbol{\sigma}) = \pm\frac{1}{2}(\sigma_1 - \sigma_3) + \frac{1}{2}(\sigma_1 + \sigma_3)\cdot\sin\phi - C\cos\phi$: Mohr-Coulomb

ϕ: internal friction angle, C: cohesion

$f_1(\boldsymbol{\sigma}) = (I_1)^3/I_3$: Lade-Duncan

$f_1(\boldsymbol{\sigma}) = (I_1 I_2)/I_3$: Matsuoka-Nakai

$f_1(\boldsymbol{\sigma}) = MD\ln\left(p'/p'_0\right) + D\,(q/p')$: Roscoe (Cam clay model)

(For the Cam clay model we use the effective stress $\boldsymbol{\sigma}'$.)

$p' = -\frac{1}{3}\sigma'_{ii}, \quad q = -(\sigma_1 - \sigma_3) = -(\sigma'_1 - \sigma'_3)$ $\quad$ (σ_i, σ'_i: positive for tension)

Under biaxial conditions, $\quad p' = -\frac{1}{3}(\sigma'_1 + 2\sigma'_3)$

p'_0: preload

M: slope of critical state line

$D = (\lambda - \kappa)/M(1+e)$

$e = n/(1-n)$: void ratio

λ: slope of normal consolidation line

κ: swelling index (slope of unloading-reloading line)

The original Cam clay model employs isotropic hardening with strain hardening such as

$$K(\kappa) = \int d\kappa, \qquad d\kappa = |d\varepsilon_v^p| = \sqrt{d\varepsilon_v^p d\varepsilon_v^p}$$

(Schofield and Wroth 1968). Here $e = n/(1-n)$ is the void ratio, n is the porosity and $d\varepsilon_v^p = -\,d\varepsilon_{ii}^p$ is $\sqrt{3}$ times the volumetric plastic strain (positive for compression); it is related to the void ratio e and plastic increment of the void ratio de^p by

$$d\varepsilon_v^p = -\frac{de^p}{1+e}.$$

2.10.2 Prager's Consistency Condition

Subsequent yielding occurs at the stress $\boldsymbol{\sigma} + d\boldsymbol{\sigma}$ and plastic state $\kappa + d\kappa$ after the initial yielding if the yielding condition

$$f(\boldsymbol{\sigma} + d\boldsymbol{\sigma},\, \kappa + d\kappa) = 0$$

is satisfied. Using this with (2.294) results in

$$df = \frac{\partial f}{\partial \boldsymbol{\sigma}} : d\boldsymbol{\sigma} + \frac{\partial f}{\partial \kappa} d\kappa = 0. \tag{2.300}$$

This is *Prager's consistency condition*. Substituting the work hardening law (2.295) into (2.300) gives

$$\frac{\partial f}{\partial \boldsymbol{\sigma}} : d\boldsymbol{\sigma} + \frac{\partial f}{\partial \kappa} \boldsymbol{\sigma} : d\boldsymbol{\varepsilon}^p = 0 \tag{2.301}$$

If we apply the strain hardening law (2.296), the consistency condition (2.300) remains unchanged.

2.10.3 Flow Rule and Incremental Constitutive Law

Let g be a scalar function such that the plastic strain increment $d\boldsymbol{\varepsilon}^p$ can be obtained as follows:

$$d\boldsymbol{\varepsilon}^p = \lambda \frac{\partial g}{\partial \boldsymbol{\sigma}} \quad \text{or} \quad d\varepsilon_{ij}^p = \lambda \frac{\partial g}{\partial \sigma_{ij}} \tag{2.302}$$

This is referred to as the *flow rule of plasticity*. The parameter λ is determined by the hardening law (mentioned later).

The flow rule (2.302) implies that the direction of the plastic strain increment $d\boldsymbol{\varepsilon}^p$ is normal to the surface $g =$ constant, and coincides with the stress $\boldsymbol{\sigma}$. For isotropic materials this can be described as follows. We introduce the unit tensors (see Sect. 2.8.3) as

$$\boldsymbol{n}^{(1)} = \frac{\partial \bar{\sigma}}{\partial \bar{\boldsymbol{\sigma}}} = \frac{\partial \bar{\sigma}}{\partial \boldsymbol{\sigma}} = \frac{\bar{\boldsymbol{\sigma}}}{\bar{\sigma}}, \quad \boldsymbol{n}^{(2)} = \frac{\partial s}{\partial \boldsymbol{s}} = \frac{\partial s}{\partial \boldsymbol{\sigma}} = \frac{\boldsymbol{s}}{s}, \quad \boldsymbol{n}^{(3)} = \frac{1}{\sigma_\theta} \frac{\partial \sigma_\theta}{\partial \boldsymbol{\sigma}} \tag{2.303}$$

These are orthonormal as described in Sect. 2.8.3, which gives

$$d\boldsymbol{\varepsilon}^p = \lambda \frac{\partial g}{\partial \boldsymbol{\sigma}} = \lambda \left(\frac{\partial g}{\partial \bar{\sigma}} \boldsymbol{n}^{(1)} + \frac{\partial g}{\partial s} \boldsymbol{n}^{(2)} + \sigma_\theta \frac{\partial g}{\partial \sigma_\theta} \boldsymbol{n}^{(3)} \right).$$

Therefore the direction of $d\boldsymbol{\varepsilon}^p$ coincides with the direction of the global stress $\boldsymbol{\sigma}$ if the material is isotropic.

Note 2.9 (Drucker's stability postulate). If the response of an elasto-plastic body is stable, the plastic work, W^p, must be non-negative:

$$W^p = \int (\boldsymbol{\sigma} - \boldsymbol{\sigma}_0) : d\boldsymbol{\varepsilon}^p \geq 0. \tag{2.304}$$

This is referred to as *Drucker's stability postulate*. Here $\boldsymbol{\sigma}_0$ is an arbitrary stress which satisfies

$$f(\boldsymbol{\sigma}_0, \kappa) < 0.$$

The condition (2.304) is written in the local form as

$$(\boldsymbol{\sigma} - \boldsymbol{\sigma}_0) \cdot d\boldsymbol{\varepsilon}^p \geq 0, \tag{2.305}$$

or by setting $\boldsymbol{\sigma} = \boldsymbol{\sigma}_0 + d\boldsymbol{\sigma}$ we can write

$$d\boldsymbol{\sigma} \cdot d\boldsymbol{\varepsilon}^p \geq 0. \tag{2.306}$$

Substituting the flow rule (2.302) into the inequality (2.305) gives

$$(\boldsymbol{\sigma} - \boldsymbol{\sigma}_0) \cdot \lambda \frac{\partial g}{\partial \boldsymbol{\sigma}} \geq 0.$$

Since $\boldsymbol{\sigma}_0$ is arbitrary, the above condition is satisfied if $g = f$ (the associated flow rule) and f is convex (Fig. 2.20). This gives a strong restriction for elasto-plastic materials, especially for granular media, since most of the experimental data show that if we apply the associated flow rule with a yield function, such as the Coulomb or Drucker-Prager type, the dilatancy (i.e., the volume change due to shearing) is over-estimated. ■

The incremental constitutive equation for applying the flow rule can be obtained as follows: Substituting the flow rule (2.302) into the consistency condition, we obtain

$$\lambda = \frac{1}{h} \frac{\partial f}{\partial \boldsymbol{\sigma}} : d\boldsymbol{\sigma}, \tag{2.307}$$

$$h = \begin{cases} -\dfrac{\partial f}{\partial W^p} \boldsymbol{\sigma} : \dfrac{\partial g}{\partial \boldsymbol{\sigma}} & : \text{work hardening} \\[2ex] -\dfrac{\partial f}{\partial \kappa} \left| \dfrac{\partial g}{\partial \boldsymbol{\sigma}} \right| & : \text{strain hardening} \end{cases} \tag{2.308}$$

where h is the hardening parameter, which will be determined later. Equation 2.307 is again substituted into the flow rule to give

$$d\boldsymbol{\varepsilon}^p = \frac{1}{h} \frac{\partial g}{\partial \boldsymbol{\sigma}} \left(\frac{\partial f}{\partial \boldsymbol{\sigma}} : d\boldsymbol{\sigma} \right) = \frac{1}{h} \left(\frac{\partial g}{\partial \boldsymbol{\sigma}} \otimes \frac{\partial f}{\partial \boldsymbol{\sigma}} \right) d\boldsymbol{\sigma} \tag{2.309}$$

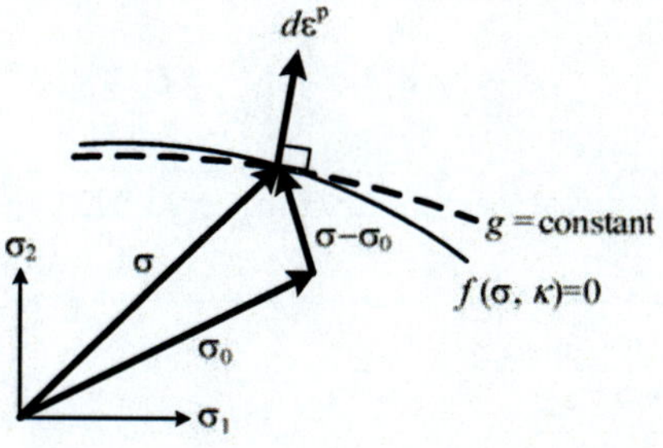

Fig. 2.20 Drucker's stability condition

where $\otimes$ denotes the tensor product, and if $\partial g/\partial \boldsymbol{\sigma}$ and $\partial f/\partial \boldsymbol{\sigma}$ are represented as vectors, it becomes

$$\frac{\partial g}{\partial \boldsymbol{\sigma}} \otimes \frac{\partial f}{\partial \boldsymbol{\sigma}} = \left\{\frac{\partial g}{\partial \boldsymbol{\sigma}}\right\}\left\{\frac{\partial f}{\partial \boldsymbol{\sigma}}\right\}^T . \tag{2.310}$$

Equation 2.309 can be written in the indicial notation:

$$d\varepsilon_{ij}^p = \frac{1}{h}\frac{\partial g}{\partial \sigma_{ij}}\frac{\partial f}{\partial \sigma_{kl}} d\sigma_{kl}. \tag{2.311}$$

Let $\boldsymbol{C}^p$ be the plastic compliance tensor (cf. (2.309)) given by

$$\boldsymbol{C}^p = \frac{1}{h}\frac{\partial g}{\partial \boldsymbol{\sigma}} \otimes \frac{\partial f}{\partial \boldsymbol{\sigma}} \qquad \text{or} \qquad C_{ij}^p = \frac{1}{h}\frac{\partial g}{\partial \sigma_{ij}}\frac{\partial f}{\partial \sigma_{kl}}, \tag{2.312}$$

then we have

$$d\boldsymbol{\varepsilon}^p = \boldsymbol{C}^p d\boldsymbol{\sigma} \qquad \text{or} \qquad d\varepsilon_{ij}^p = C_{ijkl}^p d\sigma_{kl}. \tag{2.313}$$

Equation 2.309 (or 2.313) is referred to as *Melan's formula*.

Since the plastic compliance tensor $\boldsymbol{C}^p$ of (2.312), determined by the flow rule, is represented by a product of two second-order tensors, the determinant is identically zero ($\det \boldsymbol{C}^p = 0$, if we set the second-order tensors as vectors as mentioned in (2.310)). Since it is not possible to obtain the inverse of $\boldsymbol{C}^p$ directly, we use the properties of the elastic compliance $\boldsymbol{C}^e$, which has the inverse, along with the direct sum of the strain increment given by (2.293). That is,

$$d\boldsymbol{\varepsilon} = d\boldsymbol{\varepsilon}^e + d\boldsymbol{\varepsilon}^p = (\boldsymbol{C}^e + \boldsymbol{C}^p)\, d\boldsymbol{\sigma},$$

$$\Rightarrow \quad d\boldsymbol{\sigma} = \boldsymbol{D}^{ep} d\boldsymbol{\varepsilon}, \qquad \boldsymbol{D}^{ep} = (\boldsymbol{C}^e + \boldsymbol{C}^p)^{-1}. \tag{2.314}$$

$\boldsymbol{D}^{ep}$ is determined explicitly as follows. Let

$$d\boldsymbol{\sigma} = \boldsymbol{D}^e d\boldsymbol{\varepsilon}^e = \boldsymbol{D}^e (d\boldsymbol{\varepsilon} - d\boldsymbol{\varepsilon}^p) = \boldsymbol{D}^e d\boldsymbol{\varepsilon} - \boldsymbol{D}^e \frac{1}{h}\frac{\partial g}{\partial \boldsymbol{\sigma}}\left(\frac{\partial f}{\partial \boldsymbol{\sigma}} : d\boldsymbol{\sigma}\right). \tag{2.315}$$

Taking the inner-product with $\partial f/\partial \boldsymbol{\sigma}$, the above gives

$$\frac{\partial f}{\partial \boldsymbol{\sigma}} : d\boldsymbol{\sigma} = \frac{h}{H}\frac{\partial f}{\partial \boldsymbol{\sigma}} : \boldsymbol{D}^e d\boldsymbol{\varepsilon}, \qquad H = h + \frac{\partial f}{\partial \boldsymbol{\sigma}} : \left(\boldsymbol{D}^e \frac{\partial g}{\partial \boldsymbol{\sigma}}\right). \tag{2.316}$$

Using the symmetry property of $\boldsymbol{D}^e$ and substituting (2.316) into (2.315), we obtain

$$d\boldsymbol{\sigma} = \boldsymbol{D}^{ep} d\boldsymbol{\varepsilon}, \tag{2.317}$$

where

$$\boldsymbol{D}^{ep} = \boldsymbol{D}^e - \frac{1}{H}\left(\boldsymbol{D}^e \frac{\partial g}{\partial \boldsymbol{\sigma}}\right) \otimes \left(\boldsymbol{D}^e \frac{\partial f}{\partial \boldsymbol{\sigma}}\right). \tag{2.318}$$

The indicial form of $\boldsymbol{D}^{ep}$ is given by

$$D^{ep}_{ijkl} = D^{e}_{ijkl} - \frac{1}{H}\left(D^{e}_{ijmn}\frac{\partial g}{\partial \sigma_{mn}}\right)\left(D^{e}_{klst}\frac{\partial f}{\partial \sigma_{st}}\right). \tag{2.319}$$

We now need to determine the hardening parameter h. For simplicity we use the isotropic hardening model (2.307), and from (2.308) we have

$$h = \begin{cases} \dfrac{\partial K}{\partial W^p}\boldsymbol{\sigma} : \dfrac{\partial g}{\partial \boldsymbol{\sigma}} & : \text{work-hardening}, \\[2ex] \dfrac{\partial K}{\partial \kappa}\left|\dfrac{\partial g}{\partial \boldsymbol{\sigma}}\right| & : \text{strain-hardening}. \end{cases} \tag{2.320}$$

A function $\phi(\boldsymbol{x})$ is said to be m-th order homogeneous of $\boldsymbol{x}$ if, for any scalar t,

$$\phi(t\boldsymbol{x}) = t^m \phi(\boldsymbol{x}).$$

Then we have the following Euler's theorem (cf. Note 3.8, p. 108):

$$\boldsymbol{x} \cdot \frac{\partial \phi}{\partial \boldsymbol{x}} = m\,\phi(\boldsymbol{x})$$

It should be noted that the von Mises ($f_1 = \sqrt{J_2}$) and Drucker-Prager ($f_1 = \sqrt{J_2} + \alpha I_1$) yield functions are first order homogeneous functions; however

$$f_1 = J_2 + \alpha I_1$$

is not a homogeneous function.

Let us apply the associated flow rule ($g = f$) and assume that f is an m-th order homogeneous function, so that the work-hardening rule (2.320) can be written as

$$h = m\frac{\partial K}{\partial W^p} f_1. \tag{2.321}$$

If we define the equivalent stress by

$$f_1(\boldsymbol{\sigma}) = \sigma_e, \tag{2.322}$$

and determine the equivalent plastic strain increment by

$$dW^p = \boldsymbol{\sigma} : d\boldsymbol{\varepsilon}^p \equiv \sigma_e\, d\varepsilon^p_e, \tag{2.323}$$

the hardening parameter can be specified by

$$h = m\frac{\partial \sigma_e}{\partial \varepsilon^p_e}. \tag{2.324}$$

Let us now consider the physical implications of the equivalent stress and strain. For metallic materials, which show no volumetric plastic straining, we have

$$\boldsymbol{\sigma} : d\boldsymbol{\varepsilon}^p = s\, de^p + \bar{\sigma} d\bar{\varepsilon}^p = s\, de^p.$$

where de^p is the deviatoric plastic strain increment and $d\bar{\varepsilon}^p$ is the volumetric plastic strain increment, and therefore (2.323) is meaningful. However, for porous materials that show a considerable amount of volumetric plastic strain the concept of equivalent stress and strain defined by (2.323) is not appropriate. Note that in the metallic plasticity theory, in order to utilize the result of simple tension we sometimes define

$$\sigma_e = \sqrt{3J_2} \quad (= \sigma_{11}),$$

and introduce

$$d\varepsilon_e^p = \sqrt{\frac{2}{3} d\mathbf{e}^p : d\mathbf{e}^p}.$$

This, of course, causes an adjustment of the coefficients.

Chapter 3
Non-equilibrium Thermodynamics

We first review the principles of Classical Thermodynamics (see also Appendix D), and proceed to give an alternative formulation of Thermodynamics in the context of a true dynamical process.

3.1 Review of Classical Thermodynamics

Classical thermodynamics provides valuable information that can assist in the solution of practical problems in many fields of science and engineering. In particular, when applied to coupled problems that include mechano-thermal and chemo-thermal phenomena, it gives extremely valuable results (see e.g., de Groot and Mazur 1962; Kestin 1979; Kondepudi and Prigogine 1998). An informative account of thermodynamics of solids is given by Ericksen (1991).

Classical thermodynamics involves three central concepts: the principle of conservation of energy (i.e., the First Law of Thermodynamics), the existence of entropy (i.e., the first part of the Second Law of Thermodynamics) and the concept of increasing entropy (i.e., the second part of the Second Law of Thermodynamics). There are, however, some as yet unresolved issues in classical thermodynamics that are open to interpretation.

The outline of the classical theory of thermodynamics for increasing entropy is summarized as follows: First, certain equilibrium states are defined, and then a transition process between equilibrium states is introduced for the purpose of defining both reversible and irreversible processes. The latter process results in the *entropy increasing principle*. Every thermodynamician states that "natural phenomena are essentially irreversible", but this is an axiom or postulate. It should be noted that this *principle* was established as a basis for defining the efficiency of a steam engine in the nineteenth century. The steam engine is a mechano-thermal *system* that always involves some internal loss of energy. If we try to prove whether the process is either reversible or irreversible through experimentation, we would need to gain access to the system using a measuring device. However, it becomes

Y. Ichikawa and A.P.S. Selvadurai, *Transport Phenomena in Porous Media*, DOI 10.1007/978-3-642-25333-1_3, © Springer-Verlag Berlin Heidelberg 2012 impossible to prove theoretically whether the process is reversible or irreversible because the measuring device itself can consume energy.

3.1.1 An Application of Classical Thermodynamics for a Newtonian Fluid

We first consider the following mechano-thermal coupling problem in a fluid, showing how classical thermodynamics may be applied within the framework of the principle of *conservation of energy*.

Let $\boldsymbol{\sigma}$ and p be the Cauchy stress and pressure (compression positive), respectively. We first introduce the Kirchhoff energy stress $\boldsymbol{\sigma}^{\ddagger}(\boldsymbol{x},t)$ and the corresponding pressure $p^{\ddagger}(\boldsymbol{x},t)$ by

$$\boldsymbol{\sigma}^{\ddagger} = \frac{\boldsymbol{\sigma}}{\rho}, \qquad p^{\ddagger} = \frac{p}{\rho} \tag{3.1}$$

where ρ is the mass density.

Recall that the fluid is a material in which the residual stress, reduced by the reversible part of the stress (i.e., the pressure), is dissipative and irreversible; therefore, for an isotropic Newtonian fluid, we have

$$\boldsymbol{\sigma} - (-p\boldsymbol{I}) = \rho\left(\boldsymbol{\sigma}^{\ddagger} + p^{\ddagger}\boldsymbol{I}\right) = \lambda\,(\mathrm{tr}\boldsymbol{D})\boldsymbol{I} + \mu\boldsymbol{D} \tag{3.2}$$

where $\boldsymbol{D}$ is the stretch tensor, μ is the shearing viscosity, $\kappa = \lambda + 2\mu/3$ is the bulk viscosity and $\boldsymbol{I}$ is the unit tensor.

We assume that the internal energy $u(v,s)$ and Helmholtz free energy $f(v,T)$ exist. Here v is the volume, s is the entropy and T is the absolute temperature (details of these variables will be discussed later). The increments are written as

$$\begin{aligned} &\rho\,du = -p\,dv + \rho T\,ds, \qquad \rho\,df = -p\,dv - \rho s\,dT \\ \Rightarrow\quad &du = -p^{\ddagger}\,dv + T\,ds, \qquad df = -p^{\ddagger}\,dv - s\,dT \end{aligned} \tag{3.3}$$

Since $-p\,dv/\rho = -p^{\ddagger}\,dv$ is the elastically stored internal energy due to stress, it can be written as

$$du^{e}_{\sigma} = -p^{\ddagger}\,dv \tag{3.4}$$

Since the entropy, s, can also be considered as a potential, at least in the classical sense, we write $s = s(v,T)$. The increment is calculated as follows:

$$ds = \frac{\partial s}{\partial v}dv + \frac{\partial s}{\partial T}dT = -\frac{\alpha}{T}dv + \frac{1}{T}c_v\,dT \;\Rightarrow\; T\,ds = -\alpha\,dv + c_v\,dT \tag{3.5}$$

$$\alpha = T\frac{\partial}{\partial v}\left(\frac{\partial f}{\partial T}\right), \qquad c_v = \frac{\partial u(s(T))}{\partial T}\bigg|_v = T\frac{\partial s}{\partial T}\bigg|_v \tag{3.6}$$

where α is the thermal pressure coefficient and c_v is the isovolume heat capacity.

The change of internal energy du is produced by the change of the total internal work $\boldsymbol{\sigma}^{\ddagger} : \boldsymbol{D}\,dt$ and the change of heat dq:

$$du = \boldsymbol{\sigma}^{\ddagger} : \boldsymbol{D}\,dt + dq \tag{3.7}$$

where : denotes the inner product of two second-order tensors. Using (3.5) and (3.7) we have

$$c_v\,dT = (\boldsymbol{\sigma}^{\ddagger} : \boldsymbol{D}\,dt - du_\sigma^e) + \alpha\,dv + dq = dq + (r_1 + r_2) \tag{3.8}$$

$$r_1 = \boldsymbol{\sigma}^{\ddagger} : \boldsymbol{D}\,dt - du_\sigma^e = \frac{1}{\rho}\left[\lambda\,(\mathrm{tr}\boldsymbol{D})^2 + \mu\boldsymbol{D} : \boldsymbol{D}\right]dt \tag{3.9}$$

$$r_2 = \alpha\,dv \tag{3.10}$$

The term r_1 is the heat generated by deformation and r_2 is that generated by entropy change. Equation 3.8 results from classical thermodynamics, which represents a coupling of mechano-thermal effects. In addition, we recall that the right-hand-side terms of (3.2) give a dissipative stress; thus the 'dissipation' of this system can be written as

$$d\mathcal{D} = T\,ds - dq = (\boldsymbol{\sigma}^{\ddagger} + p^{\ddagger}\boldsymbol{I}) : \boldsymbol{D}\,dt = \frac{1}{\rho}\left[\lambda\,(\mathrm{tr}\boldsymbol{D})^2 + \mu\boldsymbol{D} : \boldsymbol{D}\right]dt \geq 0 \tag{3.11}$$

The relation (3.11), which is equivalent to the classical Clausius-Duhem inequality, implies that the dissipative part of the internal work due to stress is *converted* into heat. This suggests that the *irreversible process* of classical thermodynamics gives a conversion of energy from one form to another. The result is evident, because its framework is built on a potential theory.

Note that, in the above discussion, we have not yet used the second part of the Second Law of Thermodynamics explicitly. The results are derived solely from the First Law of Thermodynamics (i.e., the conservation of energy law), using the thermodynamic potentials $u(v, s)$ and $f(v, T)$, where the first part of the Second Law was used because in (3.3) we introduced the definition of entropy s.

3.1.2 The Role of the Second Law of Thermodynamics in Classical Theory

As we observed in the preceding discussion, the framework of classical thermodynamics is established on the basis of thermodynamic potentials - such as the internal energy, enthalpy, Helmholtz free energy and Gibbs free energy. The existence of such potentials is assured by the first part of the Second Law of Thermodynamics. This is the essence of classical thermodynamics. Recall that the Second Law of Thermodynamics in the classical theory of thermodynamics consists of two parts (see Kestin 1979; Raniecki 1976). In the first part, the existence of equilibrium entropy and a thermodynamical (absolute) temperature is deduced for a class of reversible processes, using Carathéodory's Theorem. The second part states that entropy increases in irreversible processes of adiabatic systems. We will not discuss the details of the proof of the classical theory here, which can be found in Buchdahl (1966) and Kestin (1979).

3.1.3 The Entropy Inequality in Classical Continuum Mechanics

In theories of continuum thermodynamics, comprehensive procedures that formally introduce the entropy inequality are commonplace. The schemes are classified into two major groups: classical irreversible thermodynamics (CIT) and rational thermodynamics (RT). The form of the entropy inequality is different in both theories. Note that here the molecular-based discussions are disregarded because of the basic assumption of a continuum formulation.

The fundamental hypothesis of CIT is the existence of a local-equilibrium condition. A series of finite volume cells is considered in a material body, in which local variables such as temperature and entropy are uniform and in equilibrium, but time-dependent. The variables can take different values from cell to cell. The majority of textbooks are written using this formulation (see, e.g., Kestin 1979, which refers to this as the principle of local state). The most important result of CIT under the local-equilibrium hypothesis is that, as a natural result of the Second Law of Thermodynamics in the course of a mechano-thermal process, we have the following entropy inequality:

$$\rho \frac{ds}{dt} \geq -\frac{1}{T} \operatorname{div} \boldsymbol{q} + \rho \frac{r}{T} \tag{3.12}$$

where ρ is the mass density, $\boldsymbol{q}$ is the heat flux and r is the heat generation. The local-equilibrium hypothesis, however, has an inevitable drawback for CIT, since non-equilibrium processes, such as chemical reactions that are extremely important in thermodynamics, cannot be treated within a framework that is based on local-equilibrium states.

The basic framework of RT was developed by Truesdell and Toupin (1960) and Coleman (1964), which excludes the local-equilibrium hypothesis (see Note 3.3, p. 95). Let us define $\boldsymbol{q}/T$ as the entropy flux and r/T as the entropy supply. RT introduces the rate of entropy production Γ in a part $\mathcal{P}$ of the body as

$$\Gamma = \frac{d}{dt}\int_{\mathcal{P}} \rho s \, dv + \int_{\partial\mathcal{P}} \frac{\boldsymbol{q}}{T} \cdot \boldsymbol{n} \, ds - \int_{\mathcal{P}} \frac{\rho r}{T} \, dv, \tag{3.13}$$

where $\boldsymbol{n}$ is the unit outward normal to $\partial\mathcal{P}$. The implementation of the Second Law of Thermodynamics requires that

$$\Gamma \geq 0. \tag{3.14}$$

Consequently, the local form of (3.13) is obtained as

$$\rho \frac{ds}{dt} \geq -\mathrm{div}\left(\frac{\boldsymbol{q}}{T}\right) + \rho \frac{r}{T}. \tag{3.15}$$

This procedure is still followed in most of the recent works that utilize the RT framework. However, as discussed in Note 3.3, p. 95, the RT theory contains an inevitable shortcoming in that the understanding of the 'entropy flux' introduced in that framework is difficult to justify physically. Thus no continuum mechanics formulation has been successful in establishing a rational non-equilibrium thermodynamics framework.

3.1.4 Note on the Proposed Framework

We will develop an alternative framework for examining non-equilibrium thermodynamics processes in continuum mechanics. In this approach we invoke a strong constraint of the entropy inequality, i.e., the second part of the Second Law of Thermodynamics. The essence of the alternative framework is that the *lost energy* in the sense of the classical irreversible process does not completely disappear but it is converted into another form (fundamentally into heat, especially in a mechano-thermal problem), and in most cases the lost energy is not measurable.

The entropy inequality is thus considered not as a universal law, but as a working hypothesis for examining phenomena in a given thermodynamic system where non-measurable energy components exist. This non-measurable energy occurs in the entropy inequality, which is satisfied only in a local form (see Sect. 3.3.2). In addition, following the original concept of Clausius (see Yamamoto 1987, p. 390), we show that the second part of the Second Law of Thermodynamics is represented as a positive definite condition for the heat flux. The resulting framework is simple but noteworthy.

It is important to note that an irreversible process is essential in classical theories of thermodynamics, since the entropy inequality is derived for this irreversible process. However, the concept of reversible and irreversible processes plays no role in our framework, because the energy conversion process described by the First Law of Thermodynamics is essential, and the entropy inequality is not formulated as a universal law as mentioned above.

3.2 Conservation of Energy: The First Law of Thermodynamics

Energy can be imparted to a system in several forms, which are mutually convertible (Fig. 3.1). In this Section the conservation of energy law, that is the First Law of Thermodynamics, for mechanical and heat energies is discussed as a typical example. The First Law of Thermodynamics asserts that both mechanical and heat fluxes contribute to the increase of internal energy. Here the energy, power and mass flux received from the surroundings are denoted as positive except for the heat flux, which follows the classical thermodynamics convention.

3.2.1 *Stokes' Power Formulation as Mechanical Conservation of Energy in a Continuum: Eulerian Description*

We derive the conservation law of mechanical energy, referred to as Stokes' power formula, starting with the following equation of motion in an Eulerian framework:

$$\rho \frac{d\boldsymbol{v}}{dt}(\boldsymbol{x},t) = \rho \left(\frac{\partial \boldsymbol{v}}{\partial t} + \boldsymbol{v} \cdot \operatorname{grad} \boldsymbol{v} \right) = \operatorname{div} \boldsymbol{\sigma}^T(\boldsymbol{x},t) + \rho \boldsymbol{b}(\boldsymbol{x},t). \tag{3.16}$$

In an isolated or closed system the above conservation law of linear momentum is satisfied under mass conservation.

The work per unit volume $d\Omega \subset \Omega$ due to the body force $\boldsymbol{b}$ at a position $\boldsymbol{x} \in \Omega$ can be written as

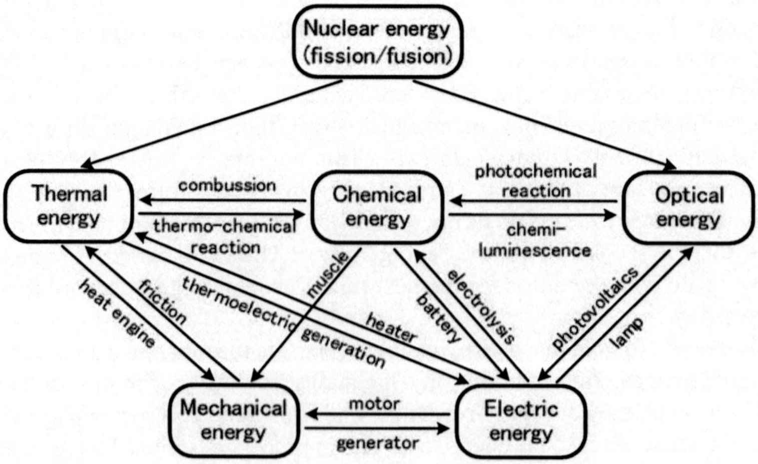

Fig. 3.1 Several forms of energy and their mutual conversion

$$\int_{x\in d\Omega} \rho \boldsymbol{b}\cdot d\boldsymbol{x} = \int_t \rho \boldsymbol{b}\cdot \frac{d\boldsymbol{x}}{dt}\,dt = \int_t \rho \boldsymbol{b}\cdot \boldsymbol{v}\,dt,$$

while the work per unit area $d(\partial\Omega) \subset \partial\Omega$ due to the traction $\boldsymbol{t} \in \partial\Omega$ at a position $\boldsymbol{x} \in \Omega$ can be written as

$$\int_{x\in d(\partial\Omega)} \boldsymbol{t}\cdot d\boldsymbol{x} = \int_t \boldsymbol{t}\cdot \frac{d\boldsymbol{x}}{dt}\,dt = \int_t \boldsymbol{t}\cdot \boldsymbol{v}\,dt.$$

Then the total work $\mathcal{W}$ given for the body Ω with a surface $\partial\Omega$ is denoted as

$$\begin{aligned}\mathcal{W} &= \int_\Omega \left(\int_t \rho \boldsymbol{b}\cdot \boldsymbol{v}dt\right) dv + \int_{\partial\Omega}\left(\int_t \boldsymbol{t}\cdot \boldsymbol{v}\,dt\right) ds \\ &= \int_t \left(\int_\Omega \rho \boldsymbol{b}\cdot \boldsymbol{v}\,dv + \int_{\partial\Omega} \boldsymbol{t}\cdot \boldsymbol{v}\,ds\right) dt.\end{aligned}$$

The power input to the body is given by

$$\frac{d\mathcal{W}}{dt} = \int_{\partial\Omega} \boldsymbol{t}\cdot \boldsymbol{v}\,ds + \int_\Omega \rho \boldsymbol{b}\cdot \boldsymbol{v}\,dv. \tag{3.17}$$

Let us construct an inner product with a velocity $\boldsymbol{v}(\boldsymbol{x}, t)$ in the equation of motion (3.16), and recall that (using the symmetry of $\boldsymbol{\sigma}$)

$$\rho\frac{d\boldsymbol{v}}{dt}\cdot \boldsymbol{v} = \rho\frac{d}{dt}\left(\frac{1}{2}\boldsymbol{v}\cdot \boldsymbol{v}\right), \qquad (\operatorname{div}\boldsymbol{\sigma})\cdot \boldsymbol{v} = \operatorname{div}(\boldsymbol{\sigma}\boldsymbol{v}) - \boldsymbol{\sigma} : \boldsymbol{D},$$

and integrate this to obtain

$$\int_\Omega \rho\frac{d}{dt}\left(\frac{1}{2}\boldsymbol{v}\cdot \boldsymbol{v}\right)(\boldsymbol{x}, t)\,dv = \int_{\partial\Omega} \boldsymbol{t}\cdot \boldsymbol{v}\,ds + \int_\Omega \rho \boldsymbol{b}\cdot \boldsymbol{v}\,dv - \int_\Omega \boldsymbol{\sigma} : \boldsymbol{D}\,dv \tag{3.18}$$

where $\boldsymbol{\sigma} : \boldsymbol{D} = \operatorname{tr}(\boldsymbol{\sigma}^T \boldsymbol{D})$, and $\boldsymbol{D} = (\dot{\boldsymbol{F}}\boldsymbol{F}^{-1})_{sym} = (\operatorname{grad}\boldsymbol{v})_{sym}$ is the stretch tensor defined by (2.70). We applied the divergence theorem to the first term of the r.h.s. of (3.18) under the condition $\boldsymbol{\sigma}\boldsymbol{v}\cdot \boldsymbol{n} = \boldsymbol{v}\cdot \boldsymbol{\sigma}^T\boldsymbol{n} = \boldsymbol{v}\cdot \boldsymbol{t}$. Equation 3.18 is known as *Stokes' power formula* in an Eulerian frame.

We define the *kinetic energy* by

$$\mathcal{K} = \int_\Omega \rho\kappa\,dv; \qquad \kappa = \frac{1}{2}\boldsymbol{v}\cdot \boldsymbol{v} \tag{3.19}$$

where $\kappa(\boldsymbol{x},t)$ is the kinetic energy per unit mass (i.e., specific kinetic energy). The rate of the internal mechanical energy (or internal work) $d\mathcal{U}_\sigma/dt$ due to stress,[1] and the rate of its specific value $du_\sigma(\boldsymbol{x},t)/dt$ are defined by

$$\frac{d\mathcal{U}_\sigma}{dt} = \int_\Omega \rho \frac{du_\sigma}{dt}(\boldsymbol{x},t)\,dv, \tag{3.20}$$

$$\rho \frac{du_\sigma}{dt}(\boldsymbol{x},t) = \boldsymbol{\sigma} : \boldsymbol{D} = \rho\boldsymbol{\sigma}^\ddagger : \boldsymbol{D} \tag{3.21}$$

$$\Rightarrow \quad \frac{du_\sigma}{dt}(\boldsymbol{x},t) = \boldsymbol{\sigma}^\ddagger : \boldsymbol{D}, \quad \boldsymbol{\sigma}^\ddagger(\boldsymbol{x},t) = \frac{\boldsymbol{\sigma}(\boldsymbol{x},t)}{\rho} \tag{3.22}$$

where $\boldsymbol{\sigma}^\ddagger(\boldsymbol{x},t)$ is the Kirchhoff energy stress (i.e., Cauchy stress per unit mass). It is noted that in deriving (3.20) we have used the following mass conservation law:

$$\frac{d\rho}{dt} + \rho\,\mathrm{div}\,\boldsymbol{v} = 0. \tag{3.23}$$

We recognize that $d\mathcal{U}_\sigma$ consists of an *elastic* part $d\mathcal{U}_\sigma^e$ (i.e., a mechanically reversible part) and an *inelastic* part $d\mathcal{U}_\sigma^i$ (i.e., a mechanically irreversible part), and the rate of its specific value $du_\sigma(\boldsymbol{x},t)$ also consists of $du_\sigma^e(\boldsymbol{x},t)$ and $du_\sigma^i(\boldsymbol{x},t)$[2]:

$$\frac{d\mathcal{U}_\sigma}{dt} = \frac{d\mathcal{U}_\sigma^e}{dt} + \frac{d\mathcal{U}_\sigma^i}{dt}, \tag{3.24}$$

$$\frac{d\mathcal{U}_\sigma^e}{dt} = \int_\Omega \rho \frac{du_\sigma^e}{dt}(\boldsymbol{x},t)\,dv, \quad \frac{d\mathcal{U}_\sigma^i}{dt} = \int_\Omega \rho \frac{du_\sigma^i}{dt}(\boldsymbol{x},t)\,dv, \tag{3.25}$$

$$\frac{du_\sigma}{dt} = \frac{du_\sigma^e}{dt} + \frac{du_\sigma^i}{dt}. \tag{3.26}$$

In the above, $\mathcal{U}_\sigma$ is known as the strain energy, and $u_\sigma^e(\boldsymbol{x},t)$ and $u_\sigma^i(\boldsymbol{x},t)$ are its specific values referred to the deformed body. Note that the mechanically irreversible part of the internal energy $d\mathcal{U}_\sigma^i/dt$ is eventually transformed into heat.

With the above definitions, Stokes' power formula (3.18) can be written as

$$\frac{d\mathcal{E}_{\mathit{æ}}}{dt} = \frac{d\mathcal{K}}{dt} + \frac{d\mathcal{U}_\sigma}{dt} = \frac{d\mathcal{W}}{dt} \tag{3.27}$$

---

[1]The rate of the internal mechanical energy $d\mathcal{U}_\sigma/dt$ due to stress is frequently referred to as the rate of internal work and is written as $d\mathcal{W}_\sigma/dt$. The rate of its specific value $du_\sigma(\boldsymbol{x},t)/dt$ is similarly written as $dw_\sigma(\boldsymbol{x},t)/dt$.

[2]Since the stretch tensor $\boldsymbol{D}$ may be decomposed into a reversible (i.e., elastic) part $\boldsymbol{D}^e$ and an irreversible (i.e., inelastic) part $\boldsymbol{D}^i$ ($\boldsymbol{D} = \boldsymbol{D}^e + \boldsymbol{D}^i$), (3.26) is valid. However this decomposition is not unique. We do not discuss the details here (see Raniecki and Nguyen 2005).

where $\mathcal{E}_\sigma (= \mathcal{K} + \mathcal{U}_\sigma)$ is the total mechanical energy in the absence of thermal and chemical effects. Note that (3.27) is equivalent to the equation of motion (2.104). We can write (3.27) in an incremental form:

$$d\mathcal{E}_\sigma = d\mathcal{K} + d\mathcal{U}_\sigma = d\mathcal{W}. \tag{3.28}$$

Stokes' power formula expresses the relationship that *the rate of 'total mechanical energy' is equal to the rate of kinetic energy plus strain energy* (i.e., elastic and inelastic energies), which is equal to *the 'power input'* in the absence of a heat supply; we understand here that the change of energy $d\mathcal{E}_\sigma$ is equivalent to the corresponding energy flux (in the above case, the power input $d\mathcal{W}$).

It is important to characterize the physical law in (3.18): i.e., the mechanical energy must be transferred smoothly. Then if we formulate a constitutive law, a condition is required for the constitutive equation such as positive definiteness

$$\boldsymbol{\sigma} : \boldsymbol{D} \geq 0 \tag{3.29}$$

for a given set of state variables.

Note 3.1 (The small strain theory). Under the assumption of small strains we have

$$\boldsymbol{\sigma} : \boldsymbol{D} = \boldsymbol{\sigma} : \dot{\boldsymbol{\varepsilon}} \qquad \text{or} \qquad \boldsymbol{\sigma}^\ddagger : \boldsymbol{D} = \boldsymbol{\sigma}^\ddagger : \dot{\boldsymbol{\varepsilon}},$$

then (3.20) is given by

$$\frac{du_\sigma}{dt} = \boldsymbol{\sigma}^\ddagger : \boldsymbol{D} = \boldsymbol{\sigma}^\ddagger : \frac{d\boldsymbol{\varepsilon}}{dt} \quad \Rightarrow \quad du_\sigma = \boldsymbol{\sigma}^\ddagger : d\boldsymbol{\varepsilon}. \tag{3.30}$$

It is understood that under the small strain state the variable of u_σ is $\boldsymbol{\varepsilon}$ and du_σ is composed of the increments of the extensive variable $\boldsymbol{\varepsilon}$ and the intensive variable $\boldsymbol{\sigma}^\ddagger$ (see Appendix D for the definition of the extensive and intensive variables).

The rate of strain $d\boldsymbol{\varepsilon}$ is decomposed into the elastic part $d\boldsymbol{\varepsilon}^e$ and the inelastic part $d\boldsymbol{\varepsilon}^i$:

$$d\boldsymbol{\varepsilon} = d\boldsymbol{\varepsilon}^e + d\boldsymbol{\varepsilon}^i, \tag{3.31}$$

$$du_\sigma = du_\sigma^e + du_\sigma^i, \quad du_\sigma^e = \boldsymbol{\sigma}^\ddagger : d\boldsymbol{\varepsilon}^e, \quad du_\sigma^i = \boldsymbol{\sigma}^\ddagger : d\boldsymbol{\varepsilon}^i. \tag{3.32}$$

Using $(3.32)_2$ we can calculate $\boldsymbol{\sigma}^\ddagger$ as a differential of u_σ with respect to $\boldsymbol{\varepsilon}^e$. Details are found in Sect. 3.4. ■

3.2.2 Generalized Strain Measure and its Conjugate Stress in a Continuum ♣

As shown in the small strain theory (3.30), the increment of internal energy du_σ is formed as a pair of stress $\boldsymbol{\sigma}^\ddagger$ and strain increments $d\boldsymbol{\varepsilon}$. This fundamental requirement allows us to introduce a Legendre transformation to construct the energy u under the inner product of the 'duality pair' of stress $\boldsymbol{\sigma}^\ddagger$ and strain $\boldsymbol{\varepsilon}$ (cf. Sect. 3.4.1). Here we consider the duality pair of the general strain measure and its energy-conjugate stress for the case of a finite strain field following Raniecki and Nguyen (2005) (cf. Note 2.4, p. 21).

As shown in (2.46) the deformation gradient is decomposed by a polar decomposition as $\boldsymbol{F} = \boldsymbol{R}\boldsymbol{U} = \boldsymbol{V}\boldsymbol{R}$. Since $\boldsymbol{U}$ is positive definite, the characteristic form is given by

$$\boldsymbol{U} = \sum_i \lambda_i \boldsymbol{N}_i \otimes \boldsymbol{N}_i .$$

The Lagrangian strain measure $\boldsymbol{E}(n)$ is introduced by using $\lambda_i,\ \boldsymbol{N}_i$ as

$$\boldsymbol{E}(n) = \sum_i \frac{(\lambda_i)^{2n} - 1}{2n} \boldsymbol{N}_i \otimes \boldsymbol{N}_i . \tag{3.33}$$

Note that Green's strain corresponds to the case $\boldsymbol{E}(1)$ as mentioned previously ($\boldsymbol{E} = \boldsymbol{E}(1)$).

From (3.20) we have $\rho\, du_\sigma/dt = \boldsymbol{\sigma} : \boldsymbol{D}$, and substituting $\boldsymbol{D} = \boldsymbol{F}^{-T}\dot{\boldsymbol{E}}\boldsymbol{F}^{-1}$ from (2.71), we obtain [3]

$$\rho \frac{du_\sigma}{dt} = \boldsymbol{\sigma} : \boldsymbol{F}^{-T}\dot{\boldsymbol{E}}\boldsymbol{F}^{-1} = \boldsymbol{F}^{-1}\boldsymbol{\sigma}\boldsymbol{F}^{-T} : \dot{\boldsymbol{E}} = \frac{\boldsymbol{T}}{J} : \dot{\boldsymbol{E}}$$

where we have used the definition of the second Piola-Kirchhoff stress $\boldsymbol{T} = J\boldsymbol{F}^{-1}\boldsymbol{\sigma}\boldsymbol{F}^{-T}$ (cf. (2.112)). Recalling that $J = \rho_0/\rho$, we obtain

$$\frac{du_\sigma}{dt} = \boldsymbol{\sigma}^\ddagger : \boldsymbol{D} = \boldsymbol{T}^\ddagger : \dot{\boldsymbol{E}}, \qquad \boldsymbol{\sigma}^\ddagger = \frac{\boldsymbol{\sigma}}{\rho}, \qquad \boldsymbol{T}^\ddagger = \frac{\boldsymbol{T}}{\rho_0} \tag{3.34}$$

---

[3] $\boldsymbol{\sigma} = \sigma_{ij}\,\boldsymbol{e}_i \otimes \boldsymbol{e}_j,\ \dot{\boldsymbol{E}} = \dot{E}_{IJ}\,\boldsymbol{E}_I \otimes \boldsymbol{E}_J$. From (2.39) we obtain

$$\boldsymbol{F}^{-T}\dot{\boldsymbol{E}}\boldsymbol{F}^{-1} = \frac{\partial X_I}{\partial x_i}\dot{E}_{IJ}\frac{\partial X_J}{\partial x_j}\boldsymbol{e}_i \otimes \boldsymbol{e}_j, \qquad \boldsymbol{F}^{-1}\boldsymbol{\sigma}\boldsymbol{F}^{-T} = \frac{\partial X_I}{\partial x_i}\sigma_{ij}\frac{\partial X_J}{\partial x_j}\boldsymbol{E}_I \otimes \boldsymbol{E}_J$$

The inner product of second order tensors $\boldsymbol{A},\ \boldsymbol{B}$ is given by $\boldsymbol{A} : \boldsymbol{B} = \mathrm{tr}\,(\boldsymbol{A}^T\boldsymbol{B})$; thus (3.34) can be proved.

It is understood that the density function of the second Piola-Kirchhoff stress $\boldsymbol{T}^{\ddagger} = T^{\ddagger}_{IJ} \boldsymbol{E}_I \otimes \boldsymbol{E}_J$ is the 'energy-conjugate' or 'dual' stress to $\dot{\boldsymbol{E}}$ under the definition of the increment of internal energy $du_\sigma = \boldsymbol{\sigma}^{\ddagger} : \boldsymbol{D}$. Note that the r.h.s. term of (3.34)$_1$, $\boldsymbol{T}^{\ddagger} : \dot{\boldsymbol{E}}$, is denoted by a Lagrangian description (cf. Sect. 2.7).

The result (3.34) can be extended by introducing the rate of the generalized Lagrangian strain measure $\dot{\boldsymbol{E}}(n)$:

$$\frac{du_\sigma}{dt} = \boldsymbol{\sigma}^{\ddagger} : \boldsymbol{D} = \boldsymbol{T}^{\ddagger}(n) : \dot{\boldsymbol{E}}(n) \tag{3.35}$$

where $\boldsymbol{T}^{\ddagger}(n)$ is the generalized Lagrangian stress which is energy-conjugate to $\dot{\boldsymbol{E}}(n)$.

For the case of the generalized Eulerian strain measure $\boldsymbol{e}(n)$ the procedure adopted is more complicated. Detailed discussions are not given, only the results. If the generalized Eulerian strain measure is given by $\mathfrak{e} = \boldsymbol{R}\, \mathfrak{E}\, \boldsymbol{R}^T$ as (2.63), the corotational rate given by $\boldsymbol{\omega}^R$ is read as $\mathfrak{D}_R \mathfrak{e}/\mathfrak{D}t = \overset{\circ}{\mathfrak{e}}$, and the energy-conjugate corotational Euler stress is calculated as $\boldsymbol{t}^{E\ddagger} = \boldsymbol{R}\boldsymbol{T}^{\ddagger}\boldsymbol{R}^T$:

$$\frac{du_\sigma}{dt} = \boldsymbol{\sigma}^{\ddagger} : \boldsymbol{D} = \boldsymbol{t}^{E\ddagger} : \overset{\circ}{\mathfrak{e}}. \tag{3.36}$$

Furthermore, if the material body is isotropic, (3.36) is satisfied for any corotational rate $\mathfrak{D}_Q \mathfrak{e}/\mathfrak{D}t$ under the arbitrary spin $\boldsymbol{\Omega} = \dot{\boldsymbol{Q}}\boldsymbol{Q}^T$; then we can define the internal energy rate for the material derivative $\dot{\mathfrak{e}}$ and the same corotational Eulerian stress as follows:

$$\frac{du_\sigma}{dt} = \boldsymbol{t}^{E\ddagger} : \frac{\mathfrak{D}_Q \mathfrak{e}}{\mathfrak{D}t} = \boldsymbol{t}^{E\ddagger} : \dot{\mathfrak{e}}. \tag{3.37}$$

For the deformed body (i.e., Eulerian description) we can introduce a Legendre transformation; it should be noted that material isotropy is imposed. In the analysis of finite strain problems the updated Lagrangian equilibrium equation, which is a form of an Eulerian description (cf. Sect. 2.6.2) is usually used, meaning that the body is required to be isotropic.

3.2.3 Stokes' Power Formula in a Continuum: Lagrangian Description ♣

Recalling Nanson's relation $\boldsymbol{n}\, ds = J\, \boldsymbol{F}^{-T} \boldsymbol{N}\, dS$ given by (2.86) and the relation $\rho_0 = \rho J$, we have the following Lagrangian description of power:

$$\frac{d\mathcal{W}}{dt} = \int_{\Omega_0} \rho_0\, \boldsymbol{b} \cdot \boldsymbol{v}(\boldsymbol{X}, t)\, dV + \int_{\partial\Omega_0} \boldsymbol{t}^0 \cdot \boldsymbol{v}(\boldsymbol{X}, t)\, dS \tag{3.38}$$

where $\boldsymbol{t}^0 = \boldsymbol{\Pi}^T \boldsymbol{N}$ as defined by (2.109), and $\boldsymbol{\Pi} = J\boldsymbol{F}^{-1}\boldsymbol{\sigma}$ is the first Piola-Kirchhoff stress. Using (3.35), the Lagrangian form of Stokes' power formula is obtained as follows:

$$\int_{\Omega_0} \rho_0 \frac{d}{dt}\left(\frac{1}{2}\boldsymbol{v}\cdot\boldsymbol{v}\right)(\boldsymbol{X},t)\,dV = \int_{\Omega_0} \rho_0\,\boldsymbol{b}\cdot\boldsymbol{v}\,dV + \int_{\partial\Omega_0} \boldsymbol{t}^0\cdot\boldsymbol{v}\,dS - \int_{\Omega_0} \rho_0\,\boldsymbol{T}^{\ddagger} : \dot{\boldsymbol{E}}\,dV. \tag{3.39}$$

3.2.4 *First Law of Thermodynamics in a Thermo-mechanical Continuum: Conservation of Mechanical and Thermal Energies*

In addition to the mechanical field we can also consider a thermal energy field with a heat supply $\mathcal{Q}$. If heat is supplied at a rate $d\mathcal{Q}/dt$, the rate of reversible internal energy $d\mathcal{U}/dt$ consists of both the mechanical and heat contributions, and the conservation of energy law can be written as

$$\frac{d\mathcal{E}}{dt} = \frac{d\mathcal{K}}{dt} + \frac{d\mathcal{U}}{dt} = \frac{d\mathcal{W}}{dt} + \frac{d\mathcal{Q}}{dt}, \tag{3.40}$$

or alternatively

$$d\mathcal{E} = d\mathcal{K} + d\mathcal{U} = d\mathcal{W} + d\mathcal{Q}. \tag{3.41}$$

This gives the *First Law of Thermodynamics* indicating that there is conservation of mechanical and thermal energies. It should be noted that the rate of heat supply $d\mathcal{Q}/dt$ is the flux of thermal energy.

Stokes' power formula (3.27), which was derived from the equation of motion, is also satisfied as

$$\frac{d\mathcal{E}_\sigma}{dt} = \frac{d\mathcal{K}}{dt} + \frac{d\mathcal{U}_\sigma}{dt} = \frac{d\mathcal{W}}{dt}. \tag{3.42}$$

By subtracting (3.42) from (3.40) we obtain the following:

$$\frac{d\mathcal{U}}{dt} = \frac{d\mathcal{U}_\sigma}{dt} + \frac{d\mathcal{Q}}{dt}, \tag{3.43}$$

which we refer to as the *dissipative energy law*.

As shown in (3.24), the increment of the internal mechanical energy $d\mathcal{U}_\sigma$ of (3.43) can be divided into the mechanically reversible part $d\mathcal{U}_\sigma^e$ and the mechanically irreversible part $d\mathcal{U}_\sigma^i$, i.e., $d\mathcal{U}_\sigma = d\mathcal{U}_\sigma^e + d\mathcal{U}_\sigma^i$. Therefore, by substituting (3.24) into (3.43), we obtain the following incremental form:

$$d\mathcal{U} - d\mathcal{U}_\sigma^e = d\mathcal{U}_\sigma^i + d\mathcal{Q}. \tag{3.44}$$

The l.h.s. of (3.44) gives a term for the change of the internal energy due to heat, which is denoted as $d\mathcal{U}_q$:

$$d\mathcal{U}_q = d\mathcal{U} - d\mathcal{U}_\sigma^e, \tag{3.45}$$

since the change of mechanically reversible energy $d\mathcal{U}_\sigma^e$ is subtracted from the change in the total reversible internal energy $d\mathcal{U}$ (note that $d\mathcal{U}$ involves both mechanical and heat effects). From (3.44) and (3.45), the dissipative energy law (3.43) is written as

$$d\mathcal{U}_q = d\mathcal{U}_\sigma^i + d\mathcal{Q}. \tag{3.46}$$

We can clearly understand the physical implication of (3.46): The change of internal heat energy $d\mathcal{U}_q$ consists of two parts; i.e., the change of irreversible internal mechanical energy $d\mathcal{U}_\sigma^i$ and the heat supply $d\mathcal{Q}$ due to an energy flux.

If chemical and electromagnetic energies are supplied to the body, the conservation of energy law and the dissipative energy law are of the same form but require additional energy fluxes in the r.h.s. terms of (3.40) and (3.43) to account for chemical and electromagnetic phenomena.

3.2.5 First Law of Thermodynamics in a Thermo-mechanical Continuum: Eulerian Description

If the internal energy $\mathcal{U}$ is sufficiently smooth, there exists a corresponding density function u such that

$$\mathcal{U} = \int_\Omega \rho\, u(\boldsymbol{x}, t)\, dv \tag{3.47}$$

The heat flux $d\mathcal{Q}/dt$ is written as

$$\frac{d\mathcal{Q}}{dt} = -\int_{\partial\Omega} \boldsymbol{q}\cdot\boldsymbol{n}\, ds + \int_\Omega \rho\, r\, dv \tag{3.48}$$

where $\boldsymbol{q}$ is the heat flux per unit area flowing outward from the surface (as a convention of thermodynamics, $\boldsymbol{q}$ is plus-signed for an out-flux) and r is the heat source generated in a unit mass. Thus the First Law of Thermodynamics can be written as

$$\frac{d}{dt}\int_\Omega \left(\frac{1}{2}\rho\boldsymbol{v}\cdot\boldsymbol{v} + \rho u\right) dv = \int_{\partial\Omega} \boldsymbol{t}\cdot\boldsymbol{v}\, ds + \int_\Omega \rho\boldsymbol{b}\cdot\boldsymbol{v}\, dv - \int_{\partial\Omega} \boldsymbol{q}\cdot\boldsymbol{n}\, ds + \int_\Omega \rho\, r\, dv. \tag{3.49}$$

Applying the divergence theorem to the third term of the r.h.s. and using Stokes' power formula (3.18), we obtain the following local Eulerian form of the dissipative energy equation:

$$\rho\frac{du}{dt} = \rho\boldsymbol{\sigma}^\ddagger : \boldsymbol{D} - \operatorname{div}\boldsymbol{q} + \rho r = \rho\frac{du_\sigma}{dt} - \operatorname{div}\boldsymbol{q} + \rho r, \tag{3.50}$$

and (3.50) gives the local form of (3.43).

3.2.6 First Law of Thermodynamics in a Thermo-mechanical Continuum: Lagrangian Description ♣

From (3.47), the internal energy in Eulerian and Lagrangian descriptions is

$$\mathcal{U} = \int_{\Omega} \rho\, u(\boldsymbol{x}, t)\, dv = \int_{\Omega_0} \rho_0\, u(\boldsymbol{X}, t)\, dV. \tag{3.51}$$

The heat flux dQ/dt in the Lagrangian description is written by substituting Nanson's relation $\boldsymbol{n}\, ds = J\, \boldsymbol{F}^{-T} \boldsymbol{N}\, dS$ into (3.48); i.e.

$$\frac{dQ}{dt} = -\int_{\partial\Omega_0} \boldsymbol{q}^0 \cdot \boldsymbol{N}\, dS + \int_{\Omega_0} \rho_0\, r\, dV, \qquad \boldsymbol{q}^0 = J\, \boldsymbol{F}^{-1} \boldsymbol{q}. \tag{3.52}$$

The external power in the Lagrangian description is given by (3.38). Then the First Law of Thermodynamics in the Lagrangian description is written as

$$\frac{d}{dt} \int_{\Omega_0} \left(\frac{1}{2} \rho_0 \boldsymbol{v} \cdot \boldsymbol{v} + \rho_0\, u \right) dV = \int_{\partial\Omega_0} \boldsymbol{t}^0 \cdot \boldsymbol{v}\, dS + \int_{\Omega_0} \rho_0 \boldsymbol{b} \cdot \boldsymbol{v}\, dV - \int_{\partial\Omega_0} \boldsymbol{q}^0 \cdot \boldsymbol{N}\, dS + \int_{\Omega_0} \rho_0\, r\, dV. \tag{3.53}$$

The divergence theorem is applied to the third term of the r.h.s. of this equation and substitution of Stokes' power formula gives the following Lagrangian form of the First Law of Thermodynamics (local balance of energy):

$$\rho_0 \frac{du}{dt}(\boldsymbol{X}, t) = \rho_0\, \boldsymbol{T}^{\ddagger} : \dot{\boldsymbol{E}} - \mathrm{Div}\, \boldsymbol{q}^0 + \rho_0\, r \tag{3.54}$$

where $\boldsymbol{T}^{\ddagger} = \boldsymbol{T}/\rho_0$ is the second Piola-Kirchhoff stress per unit mass.

3.3 Second Law of Thermodynamics

In classical thermodynamics, the Second Law consists of two parts (see Kestin 1979). The first part defines the entropy for 'reversible' processes based on Carathéodory's theorem. The latter part concludes that entropy increases in all 'irreversible' processes. Recall that, as discussed in Sect. 3.1, the terminologies of the reversible and irreversible processes are used here in the sense of classical thermodynamics; however, in this Section irreversible processes acquire a different interpretation to that used in the classical sense. That is, we may be able to introduce an apparent irreversible process for the case that considers a non-measurable energy, which will be discussed in more detail in Sect. 3.3.2.

First we need to show the existence of entropy based on the auxiliary conservation of energy law (3.50). We will then indicate that *irreversible* processes, in the sense of the present work, result in an observational problem with any non-measurable data.

3.3.1 Thermal Energy and the Existence of Entropy: The First Part of the Second Law of Thermodynamics

Recall from (3.26) that $du_\sigma = du_\sigma^e + du_\sigma^i$. Thus (3.50) can be written as

$$\rho \frac{d}{dt}(u - u_\sigma^e) = \rho \frac{du_\sigma^i}{dt} - \operatorname{div} \boldsymbol{q} + \rho r \tag{3.55}$$

The r.h.s. of (3.55) describes a heat supply including a heat source r and the heat generated by the inelastic deformation du_σ^i, while the l.h.s. shows the corresponding rate of internal energy due to these sources, which is written as

$$\frac{du_q}{dt} = \frac{d}{dt}(u - u_\sigma^e). \tag{3.56}$$

The incremental form du_q can be represented by a set of functions (T^*, s^*) where T^* is an empirical temperature and s^* is the corresponding entropy. Since T^* and s^* are not independent, we have either the form $du_q = T^*\, ds^*$ or the form $du_q = s^*\, dT^*$. As to which form is preferable depends on how T^* and s^* are defined physically.

Based on experience it is easier to introduce the temperature T^* first, and since du_q is known from the above discussions, the increment of entropy ds^* can be defined by

$$ds^* = \frac{du_q}{T^*}.$$

It is easily observed that the above representation is not unique. [4] We can choose the absolute temperature T instead of the empirical temperature T^*, and the

[4] Suppose that we have a function f such that $s^\dagger = f(s^*)$, and set

$$T^\dagger = \frac{T^*}{df(s^*)/ds^*}.$$

This is equivalent to

$$T^\dagger\, ds^\dagger = T^*\, ds^*.$$

Thus the representation (3.57) is not unique.

corresponding entropy increment is ds, giving

$$ds = \frac{du_q}{T}. \tag{3.57}$$

Here we should note that from (3.57) we cannot determine the absolute value of s, which is specified by virtue of the Nernst Heat Theorem (i.e., the Third Law of Thermodynamics):

$$s \to 0 \qquad \text{as} \qquad T \to 0. \tag{3.58}$$

Then (3.50) can be rewritten as

$$\rho \frac{du_q}{dt} = \rho T \frac{ds}{dt} = \rho \frac{du_\sigma^i}{dt} - \operatorname{div} \boldsymbol{q} + \rho r. \tag{3.59}$$

Another representation for du_q is based on the Boltzmann formula

$$s = k_B \ln W \tag{3.60}$$

where k_B is Boltzmann's constant and W is the most probable configuration of the system. If we first know the entropy from (3.60), we can introduce the temperature increment dT by

$$dT = \frac{du_q}{s}. \tag{3.61}$$

This gives another interpretation of thermodynamics. However we will not discuss it further here.

3.3.2 *Entropy Inequality: Second Part of the Second Law of Thermodynamics*

The classical notation of thermodynamical equilibrium has no role in the proposed framework since all the phenomena are treated as non-equilibrium dynamical processes.

In classical thermodynamics it is important to distinguish whether a process is reversible or irreversible, since the entropy inequality is established for irreversible processes (the second part of the Second Law of Thermodynamics). In Sect. 3.1.1, the classical theory was formulated within the framework of thermodynamic potentials and the total internal energy is conservative but converted from one form to another. In the classical theory this conversion process is defined as an irreversible process.

The definition of either reversible or irreversible processes is not meaningful in the present framework; since the theory is also built on thermodynamic potentials, the energy is fundamentally conservative, and *in the irreversible process there are some energy components that are undetected by the measuring device*. It should be noted that the material body treated here may not be homogeneous and experiences dynamic processes, and thus we have to consider that heat, which flows out/in from the boundary during the 'reversible process'. In this interpretation the entropy inequality is not considered as a universal law, but as a working hypothesis when examining phenomena in a given thermodynamic system where some non-measurable energy components exist. Furthermore, following the original concept of Clausius (see Yamamoto 1987, p. 390), it is concluded that the second part of the Second Law of Thermodynamics gives a condition of positive definiteness of the heat flux term.

The preceding discussion is applicable to all chemical processes. Note that a chemical process consists of a combination of reaction and diffusion of mass and heat; the chemical reaction is caused by a change of electron orbits, which is of course a conservation of energy process. In the diffusion process, masses and energies are transferred to alternative positions and/or some energy is converted into another form.

3.3.2.1 Inequality for a Non-measurable Energy

The question arises as to whether or not irreversible processes exist in the framework proposed. The answer is affirmative for real systems. Every measuring system involves certain errors; even if the measuring system is perfect, it uses a certain amount of energy in the measurement process itself. For example, consider a viscous fluid which is flowing in a channel. It is difficult to measure the temperature distribution at all positions in the fluid, since an inherent energy conversion process occurs from a mechanical form to heat due to viscous dissipation in the fluid (cf. Sect. 3.4.4). Thus a certain amount of heat energy can be omitted even though it exists in the fluid. This results in an *apparent irreversible process*, which can be accounted for in our framework.

We now examine an implication of (3.59). Let us first rewrite it as

$$\rho T \frac{ds}{dt} - \rho \frac{du_\sigma^i}{dt} = -\mathrm{div}\,\boldsymbol{q} + \rho r. \tag{3.62}$$

The problem is formulated as follows: the terms du_σ^i, $\boldsymbol{q}$, r and the temperature T can be determined, whereas the unknown variable is the entropy increment ds. If all the known variables are correctly measured and evaluated, the entropy can then be correctly estimated, and the equation (3.62) is satisfied. However in an actual mechano-thermal system, the measured data are not always perfect. For example, the term $du_\sigma^i/dt \geq 0$ is usually disregarded in classical thermodynamics. If this term is not taken into account, the 'measured' entropy rate ds/dt becomes larger than the true value. Thus we have

$$\rho \frac{ds}{dt} \geq -\frac{1}{T} \operatorname{div} \boldsymbol{q} + \rho \frac{r}{T}. \tag{3.63}$$

In fact, as mentioned in Sect. 3.1.3, the CIT theory introduces this entropy inequality (3.63), which is known as the Clausius-Duhem inequality. A similar situation to that described for the case of a viscous fluid occurs, for example, in heat engines, where the energy loss due to friction between metals is difficult to quantify.

We can state that the entropy inequality (3.63) is applicable if measured data are provided imperfectly. This is the physical background to the second part of the Second Law of Thermodynamics, which is necessary for application to real thermodynamical systems.

If (3.62) is satisfied, the inequality (3.63) simply implies

$$\frac{du_\sigma^i}{dt} \geq 0. \tag{3.64}$$

In conclusion, if a component of energy is expected to be non-measurable in a real system, the entropy inequality needs to be written in the following form:

$$\frac{ds}{dt} \geq \frac{1}{T}\frac{du_\sigma^i}{dt} - \frac{1}{\rho T} \operatorname{div} \boldsymbol{q} + \frac{1}{T} r. \tag{3.65}$$

Here we assume that the inelastic energy du_σ^i/dt is partially measured.

If the inequality (3.65) is satisfied, the apparent irreversible process in the mechano-thermal system could be caused by the fact that some of the energy is released from the system, for example as electro-magnetic radiation, in a non-measurable way. It is also noted that, in the pure mechano-thermal problem, only equality has a sound physical implication; the entropy increasing inequality (3.65) does not correspond to a universal law but suggests that there exist some non-measurable energy components.

Note 3.2 (Classical theories of entropy inequalities). In the CIT theory of continuum mechanics, the entropy inequality is given by (3.63) and thus can be satisfied only if the term du_σ^i/dt is disregarded.

In the RT theory, the entropy inequality is regarded as a fundamental and universal principle. The entropy S of a part $\mathcal{P} \subset \Omega$ is given and the entropy inequality is satisfied:

$$\frac{dS}{dt} \geq -\int_{\partial \mathcal{P}} \frac{\boldsymbol{q}}{T} \cdot \boldsymbol{n}\, ds + \int_{\mathcal{P}} \rho \frac{r}{T}\, dv \tag{3.66}$$

where $\boldsymbol{q}/T$ is the entropy flux and r/T is the entropy source in a thermodynamical system. By using the entropy density, s, we obtain the following local form of the classical Clausius-Duhem inequality:

$$\rho \frac{ds}{dt} \geq -\text{div}\left(\frac{\boldsymbol{q}}{T}\right) + \rho \frac{r}{T} \tag{3.67}$$

(cf. Coleman 1964, p. 15; Leigh 1968, Sect. 7.9). The inequality (3.67) is not physically admissible in the defined mechano-thermal problem since the entropy flow $\boldsymbol{q}/T$ is introduced without consideration of the exact physical significance.

Since we have $\text{div}\,(\boldsymbol{q}/T) = [\text{div}\boldsymbol{q} - \boldsymbol{q}\cdot\text{grad}\,(\ln T)]/T$ and (3.62), the inequality (3.67) is equivalent to

$$\rho \frac{du_\sigma^i}{dt} - \boldsymbol{q}\cdot\text{grad}\,(\ln T) \geq 0. \tag{3.68}$$

The classical thermodynamics theory states that since the inequality (3.70) is satisfied for all state variables including the temperature T, it is equivalent to the following two conditions:

$$\frac{du_\sigma^i}{dt} \geq 0 \quad \text{and} \quad -\boldsymbol{q}\cdot\text{grad}\,(\ln T) \geq 0. \tag{3.69}$$

If we introduce Fourier's law for an isotropic material

$$\boldsymbol{q} = -k\,\text{grad}\,T \tag{3.70}$$

where k is the thermal conductivity, the inequality $(3.69)_2$ gives $k \geq 0$.

However, the conditions given by (3.69) are necessary for (3.68) but not sufficient. We should treat the condition (3.68) as a constraint for the governing equations (3.16) and (3.50), though the classical process resulting in the entropy inequality (3.68) involves some doubt in the context of a physical understanding of entropy flux. ■

Note 3.3 (On the theory of entropy production). Following the theory of entropy production proposed by Prigogine (1967) (cf. Sect. D.4.4), we have the Second Law of Thermodynamics in the form

$$ds = ds^e + ds^i, \qquad ds^i \geq 0. \tag{3.71}$$

This is known as the theory of entropy production. The essential discussion of this procedure is that (3.71) is derived locally since the second part of the Second Law of Thermodynamics is not a universal law but a working hypothesis concerned with the observation of the thermal field.

From (3.56) and (3.57) we have

$$\frac{du}{dt} = \frac{du_\sigma^e}{dt} + \frac{du_q}{dt} = \frac{du_\sigma^e}{dt} + T\frac{ds}{dt} \tag{3.72}$$

because in the theory of entropy production the change of internal energy du should be dependent on ds^e. Substituting (3.71)$_1$ and (3.72) into the inequality (3.71)$_2$, and substituting the auxiliary conservation of energy equation (3.50) into it, we obtain the following Clausius-Duhem inequality

$$\frac{ds}{dt} \geq \frac{1}{T}\frac{du^i_\sigma}{dt} - \frac{1}{\rho T}\operatorname{div}\boldsymbol{q} + \frac{r}{T}. \tag{3.73}$$

This is the Second Law of Thermodynamics (for a continuum) following the theory of entropy production.

It is noted that (3.73) is equal to (3.65), which implies that the entropy production ds^i is caused by a non-measurable part of energy. Until now, no physical interpretation of entropy production has been given, but using the present framework makes it clear. This is the rationale for the development of this particular approach. ■

3.3.2.2 Positive Definiteness of the Heat Flux

The original proposal of Clausius for the Second Law of Thermodynamics was to formulate the fact that heat flows from a position of higher temperature to a lower one (Yamamoto 1987). In this sequence it is natural to introduce the condition of positive definiteness for the heat flux as the second part of the Second Law of Thermodynamics.

Let $\Omega \subset \mathbb{R}^n$ be a material body with a boundary $\partial\Omega_T$ having the temperature $\hat{T}$ and a boundary $\partial\Omega_q$ having the flux $\hat{q}$ ($\mathbb{R}^n$ is the n dimensional real number space). Let $V = \{T \in H^1_T(\Omega);\ T = \hat{T} \text{ on } \partial\Omega_T\}$ and $V_0 = \{\eta \in H^1_0(\Omega);\ \eta = 0 \text{ on } \partial\Omega_T\}$ where $H^1_T(\Omega)$ and $H^1_0(\Omega)$ are the Sobolev subspaces of the Hilbert space $L_2(\Omega)$. The weak form of (3.62) for the temperature T is given by

$$\int_\Omega \left[\rho\left(T\frac{ds}{dt} - \frac{du^i_\sigma}{dt} - r\right)\eta\right] dv + \int_{\partial\Omega_q} \hat{q}\,\eta\, ds$$
$$- \int_\Omega \boldsymbol{q}\cdot \operatorname{grad}\eta\, dv = 0; \qquad \forall \eta \in H^1_0(\Omega). \tag{3.74}$$

We can then introduce the following positive definite condition for the heat flux $\boldsymbol{q}$, which indicates that heat is flowing from the higher temperature location to the lower in the body Ω:

$$- \int_\Omega \boldsymbol{q}\cdot \operatorname{grad} T\, dv \geq 0 \tag{3.75}$$

for all temperatures T satisfying the boundary condition $T = \hat{T}$ on $\partial\Omega_T$, their derivatives $\partial T/\partial x_i, \cdots$ and internal parameters ξ_i, which designate the history of temperature and/or stress. If we introduce Fourier's law $\boldsymbol{q} = -k \operatorname{grad} T$ for an isotropic material (where k is the thermal conductivity), the condition (3.75) implies

$$k \geq 0. \tag{3.76}$$

Note 3.4 (On the constitutive relation and the Second Law of Thermodynamics). The Second Law of Thermodynamics essentially gives a relationship between the heat flux $\boldsymbol{q}$ that is externally supplied and the induced temperature field. Some scientists have stated that constitutive relations that depend on fields other than the temperature can also be derived by the Second Law; however, as shown above, the Second Law does not consider fields other than the heat flux and temperature. All the constitutive relations can be derived from the First Law of Thermodynamics, internal energy and thermodynamic potentials induced by Legendre transformations (see Sect. 3.4). ■

3.3.3 Second Law of Thermodynamics in a Thermo-mechanical Continuum: Lagrangian Description ♣

As we did for the case of the First Law of Thermodynamics we can obtain the Second Law of Thermodynamics (Clausius-Duhem inequality) in a Lagrangian description. The entropy density in a Lagrangian description can be written as

$$\mathcal{S} = \int_{\Omega} \rho\, s(\boldsymbol{x}, t)\, dv = \int_{\Omega_0} \rho_0\, \mathsf{s}(\boldsymbol{X}, t)\, dV \tag{3.77}$$

By using this, (3.65) is converted to a Lagrangian description:

$$\frac{d\mathsf{s}}{dt}(\boldsymbol{X}, t) \geq \frac{1}{T}\, du^i_\sigma(\dot{\boldsymbol{E}}^i, t) - \frac{1}{\rho_0 T}\,\mathrm{Div}\,\boldsymbol{q}^0 + \frac{r}{T}, \qquad \boldsymbol{q}^0 = J\,\boldsymbol{F}^{-1}\boldsymbol{q} \tag{3.78}$$

The term $du^i_\sigma(\dot{\boldsymbol{E}}^i, t) = \boldsymbol{T}^{\ddagger} : \dot{\boldsymbol{E}}^i$ must be discussed in detail; however it is a topic that is best discussed in the context of a theory of finite strain plasticity (see, e.g., Lubarda 2000, 2004; Selvadurai and Yu 2006).

3.4 Thermodynamic Functions

The internal energy is transformed into several energy forms (i.e., thermodynamical potential functions) through the Legendre transformation. Here we introduce the Legendre transformation and consider the various energy forms. In this Chapter it is understood that the entropy inequality is not a physical law and the conservation of energy law is fundamental; therefore we do not divide the entropy increment as $ds = ds^e + ds^i$ but simply we set $ds = ds^e$.

3.4.1 Legendre Transformation and Convex Functions

Let us consider two smooth, differentiable functions[5] $\phi(\boldsymbol{x})$ and $\psi(\boldsymbol{y})$. If they are related by

$$\boldsymbol{y} = \frac{\partial \phi}{\partial \boldsymbol{x}}, \qquad \boldsymbol{x} = \frac{\partial \psi}{\partial \boldsymbol{y}}, \tag{3.79}$$

they are said to be *conjugate*. Obviously the above relation is equivalent to

$$\phi(\boldsymbol{x}) = \int_0^{\boldsymbol{x}} \boldsymbol{y} \cdot d\boldsymbol{x}, \qquad \psi(\boldsymbol{y}) = \int_0^{\boldsymbol{y}} \boldsymbol{x} \cdot d\boldsymbol{y}. \tag{3.80}$$

Then $\phi(\boldsymbol{x})$ and $\psi(\boldsymbol{y})$ are related by the following *Legendre transformation*:

$$\phi(\boldsymbol{x}) = \boldsymbol{x} \cdot \boldsymbol{y} - \psi(\boldsymbol{y}), \qquad \psi(\boldsymbol{y}) = \boldsymbol{x} \cdot \boldsymbol{y} - \phi(\boldsymbol{x}). \tag{3.81}$$

A function $\phi(\boldsymbol{x})$ is convex if, for any $\boldsymbol{x}_1$ and $\boldsymbol{x}_2$, we have

$$\phi((1-\lambda)\boldsymbol{x}_1 + \lambda \boldsymbol{x}_2) \le (1-\lambda)\phi(\boldsymbol{x}_1) + \lambda \phi(\boldsymbol{x}_2) \tag{3.82}$$

where $\lambda \in [0, 1]$ is an arbitrary real number (Fig. 3.2).

A directional derivative of ϕ at $\boldsymbol{x}$ in the direction $(\boldsymbol{y}-\boldsymbol{x})$ is defined by

$$\lim_{\lambda \to 0} \frac{1}{\lambda} [\phi(\boldsymbol{x} + \lambda(\boldsymbol{y} - \boldsymbol{x})) - \phi(\boldsymbol{x})] \equiv \frac{\partial \phi}{\partial \boldsymbol{x}} \cdot (\boldsymbol{y} - \boldsymbol{x}) \tag{3.83}$$

where $\boldsymbol{y}$ is an arbitrary vector. This is also referred to as the Gâteaux differential.

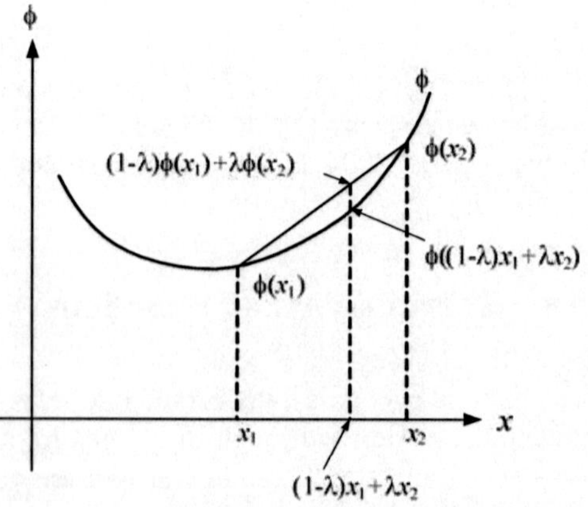

Fig. 3.2 Convex functional

[5]If the domain of a function is (a subset of) one dimensional real number space $\mathbb{R}$, it is said to be a 'functional'. The potential functions, such as the internal energy, are a typical functional.

The inequality (3.82) is changed into

$$\phi(\boldsymbol{x}_2) - \phi(\boldsymbol{x}_1) \geq \frac{1}{\lambda}\left[\phi((1-\lambda)\boldsymbol{x}_1 + \lambda\boldsymbol{x}_2) - \phi(\boldsymbol{x}_1)\right].$$

Then we take the limit $\lambda \to 0$, and substitute the definition (3.83) to obtain

$$\phi(\boldsymbol{x}_2) - \phi(\boldsymbol{x}_1) \geq \frac{\partial \phi}{\partial \boldsymbol{x}_1} \cdot (\boldsymbol{x}_2 - \boldsymbol{x}_1). \tag{3.84}$$

Commuting $\boldsymbol{x}_1$ and $\boldsymbol{x}_2$ in (3.84), we have

$$\phi(\boldsymbol{x}_1) - \phi(\boldsymbol{x}_2) \geq \frac{\partial \phi}{\partial \boldsymbol{x}_2} \cdot (\boldsymbol{x}_1 - \boldsymbol{x}_2). \tag{3.85}$$

Summing (3.84) and (3.85) yields

$$\left(\frac{\partial \phi}{\partial \boldsymbol{x}_2} - \frac{\partial \phi}{\partial \boldsymbol{x}_1}\right) \cdot (\boldsymbol{x}_2 - \boldsymbol{x}_1) \geq 0. \tag{3.86}$$

This property of the gradient $\partial\phi/\partial\boldsymbol{x}$ is known as *monotonicity*.

If a function ϕ is sufficiently smooth, the convexity of ϕ is equivalent to the monotonicity of $\partial\phi/\partial\boldsymbol{x}$, as observed above.

If a function $\phi(\boldsymbol{x})$ is convex, the conjugate function $\psi(\boldsymbol{y}) = \boldsymbol{y} \cdot \boldsymbol{x} - \phi(\boldsymbol{x})$ is also convex. The proof is left up to the reader.

3.4.2 *Thermodynamic Functions in a Thermo-mechanical Field: Solids with Small Strain*

As described in Sect. 3.2, the change of internal energy $du(\boldsymbol{\varepsilon}^e, s)$ under the assumption of a small strain theory is given by

$$du(\boldsymbol{\varepsilon}^e, s) = \boldsymbol{\sigma}^{\ddagger} : d\boldsymbol{\varepsilon}^e + T\, ds, \tag{3.87}$$

$$\boldsymbol{\sigma}^{\ddagger} = \frac{\partial u}{\partial \boldsymbol{\varepsilon}^e}, \qquad T = \frac{\partial u}{\partial s} \tag{3.88}$$

where $\boldsymbol{\sigma}^{\ddagger}$ is the normalized measure of Cauchy stress defined by (3.22) and s is the entropy density exchangeable to the surroundings as mentioned in Sect. 3.2.5.

Note 3.5 (Paffian and existence of potential function). The differential form

$$\boldsymbol{F} \cdot d\boldsymbol{x} = \sum_{\alpha=1}^{n} F_\alpha\, dx_\alpha \tag{3.89}$$

which is given, e.g., as the terms of the r.h.s. of (3.87), is often referred to as the *Paffian*. The condition that the Paffian is written as an exact differential, that is, there exists a scalar function φ such that

$$d\varphi = \frac{\partial \varphi}{\partial x_\alpha} dx_\alpha,$$

is referred to as the integrability condition. The integrability is sometimes referred to as the Born-Carathèodory theorem (see Kestin 1979). The necessary and sufficient condition of the integrability of the Paffian is given by

$$\nabla \wedge \boldsymbol{F} = \boldsymbol{0} \quad \Rightarrow \quad \frac{\partial F_\xi}{\partial x_\eta} = \frac{\partial F_\eta}{\partial x_\xi} \qquad (\xi, \eta = 1, 2, \cdots, n). \tag{3.90}$$

For proof, Stokes' theorem is used. Note that in Sect. 9.10 and Chap. 10 of Kestin (1979) we find detailed descriptions concerning the Born-Carathèodory theorem based on conventional thermodynamics.

If the integrability (3.90) is applied for the case of internal energy $u(\boldsymbol{\varepsilon}^e, s)$, we have

$$\frac{\partial \boldsymbol{\sigma}^\ddagger}{\partial s} = \frac{\partial T}{\partial \boldsymbol{\varepsilon}^e} \quad \Rightarrow \quad \frac{\partial^2 u}{\partial s\, \partial \boldsymbol{\varepsilon}^e} = \frac{\partial^2 u}{\partial \boldsymbol{\varepsilon}^e\, \partial s}, \tag{3.91}$$

since from (3.88) $\boldsymbol{\sigma}^\ddagger = \partial u / \partial \boldsymbol{\varepsilon}^e$, $T = \partial u / \partial s$. Thus we understand that if $u(\boldsymbol{\varepsilon}^e, s)$ is sufficiently smooth (i.e., twice differentiable), $(3.91)_1$ is clearly satisfied.

Mathematically the Paffian is a 1-form in exterior algebra. The theory of exterior algebra and differential geometry are fundamental tools of modern physics (see, e.g., Hassani 1999; Frankel 2004). ■

Note 3.6 (Notation of pressure and volume change for fluid and thermodynamic functions). Most textbooks on thermodynamics (de Groot and Mazur 1962; Kestin 1979; Kondepudi and Prigogine 1998) contain treatments of the perfect gas, therefore we have $\boldsymbol{\sigma} = -p\boldsymbol{i}$ (p is pressure and compression is positive). Let $p^\ddagger = p/\rho$, then (3.87) can be written as

$$du(v^e, s) = -p^\ddagger\, dv^e + T\, ds, \qquad p^\ddagger = -\frac{\partial u}{\partial v^e}, \quad T = \frac{\partial u}{\partial s} \tag{3.92}$$

where $dv^e = d\varepsilon^e_{ii}$ is the change of elastic volumetric strain.

The increment of internal energy du_σ for, e.g., a Newtonian fluid, given by (2.226) is constructed by

$$du_\sigma = \boldsymbol{\sigma}^\ddagger : \boldsymbol{D}\, dt = \left[-p^\ddagger\, \mathrm{tr}\, \boldsymbol{D} + \lambda\, (\mathrm{tr}\, \boldsymbol{D})^2 + 2\mu \boldsymbol{D} : \boldsymbol{D} \right] dt \tag{3.93}$$

It is not possible to directly have a constitutive law from this relation such as $\boldsymbol{\sigma}^\ddagger = \partial u_\sigma / \partial(\cdot)$, i.e., no exact differential form exists. If the second and third terms of the r.h.s. of (3.93) are ignored, i.e., fluid is assumed to be inviscid, (3.93) is reduced to (3.92) and we can consider the potential form. That is, the second and third terms correspond to the internal mechanism of dissipation for the exchange of energies (i.e., kinetic or stress energies) into heat energy. Details are given in Sect. 3.4.4. ■

If the extensive variables are increased k-times, the internal energy u also increases k-times. That is, the internal energy $u(\boldsymbol{\varepsilon}^e, s)$ is a homogeneous function of order one (cf. Note 3.8, p. 108) of the extensive variables:

$$u(k\boldsymbol{\varepsilon}^e, ks) = k\,u(\boldsymbol{\varepsilon}^e, s) \tag{3.94}$$

Let us introduce Taylor's series expansion after setting $k = 1+\epsilon$ ($\epsilon \ll 1$) in this equation to obtain

$$u\left((1+\epsilon)\boldsymbol{\varepsilon}^e,\ (1+\epsilon)s\right) = u + \epsilon\left(\frac{\partial u}{\partial \boldsymbol{\varepsilon}^e} : \boldsymbol{\varepsilon}^e + \frac{\partial u}{\partial s}s\right) = u + \epsilon\left(\boldsymbol{\sigma}^\ddagger : \boldsymbol{\varepsilon}^e + T\,s\right) \tag{3.95}$$

where (3.87) is used. However, since $u(\boldsymbol{\varepsilon}^e, s)$ is a homogeneous function of order one, we have

$$u\left((1+\epsilon)\boldsymbol{\varepsilon}^e,\ (1+\epsilon)s\right) = (1+\epsilon)u = u + \epsilon u. \tag{3.96}$$

Comparing the terms of the r.h.s. of (3.95) and (3.96) yields

$$u(\boldsymbol{\varepsilon}^e, s) = \boldsymbol{\sigma}^\ddagger : \boldsymbol{\varepsilon}^e + T\,s. \tag{3.97}$$

This is referred to as *Euler's equation*. Differentiating this and substituting $(3.87)_1$ we obtain the *Gibbs-Duhem relation* as

$$\boldsymbol{\varepsilon}^e : d\boldsymbol{\sigma}^\ddagger + s\,dT = 0. \tag{3.98}$$

This relation indicates that the conjugate variables $(\boldsymbol{\varepsilon}^e, s)$ and $(\boldsymbol{\sigma}^\ddagger, T)$ are not independent of each other.

Note 3.7 (Euler's equation and the Gibbs-Duhem relation). If the r.h.s. of (3.97) is differentiated in a straightforward way, we have

$$d\boldsymbol{\sigma}^\ddagger : \boldsymbol{\varepsilon}^e + \boldsymbol{\varepsilon}^e : d\boldsymbol{\sigma}^\ddagger + s\,dT + T\,ds.$$

Thus the Gibbs-Duhem relation gives the condition that the variables of internal energy u are $(\boldsymbol{\varepsilon}^e,\ s)$, and the differential is given by (3.87):

$$du(\boldsymbol{\varepsilon}^e, s) = \boldsymbol{\sigma}^\ddagger : d\boldsymbol{\varepsilon}^e + T\,ds. \tag{3.99}$$

Recall that the existence of the energy potential u is proved by the integrability condition (cf. Note 3.5). ■

The *Helmholtz free energy* $f(\boldsymbol{\varepsilon}^e, T)$ is defined by

$$f(\boldsymbol{\varepsilon}^e, T) = u(\boldsymbol{\varepsilon}^e, s) - T\,s \tag{3.100}$$

with its variables $(\boldsymbol{\varepsilon}^e, T)$. Here, because of the convention in thermodynamics, the negative sign of the Legendre transformation is used. Alternatively, the Helmholtz free energy is interpreted as the energy obtained by subtracting the effect of temperature from the internal energy u. Thus from the definition of the Helmholtz free energy (3.100) we have

$$df = du - d(Ts) = \boldsymbol{\sigma}^{\ddagger} : d\boldsymbol{\varepsilon}^e - s\,dT, \qquad \boldsymbol{\sigma}^{\ddagger} = \frac{\partial f}{\partial \boldsymbol{\varepsilon}^e}, \quad s = -\frac{\partial f}{\partial T}. \tag{3.101}$$

The *enthalpy* $h(\boldsymbol{\sigma}^{\ddagger}, s)$ is defined by

$$h(\boldsymbol{\sigma}^{\ddagger}, s) = u(\boldsymbol{\varepsilon}^e, s) - \boldsymbol{\sigma}^{\ddagger} : \boldsymbol{\varepsilon}^e \tag{3.102}$$

with its variables $(\boldsymbol{\sigma}^{\ddagger}, s)$. The enthalpy is interpreted as the energy obtained by subtracting the effect of mechanical work from the internal energy u. Thus from the definition of the enthalpy (3.102) we have

$$dh = du - d(\boldsymbol{\sigma}^{\ddagger} : \boldsymbol{\varepsilon}^e) = T\,ds - \boldsymbol{\varepsilon}^e : d\boldsymbol{\sigma}^{\ddagger}, \qquad \boldsymbol{\varepsilon}^e = -\frac{\partial h}{\partial \boldsymbol{\sigma}^{\ddagger}}, \quad T = \frac{\partial h}{\partial s}. \tag{3.103}$$

The *Gibbs free energy* or *Gibbs energy* $g(\boldsymbol{\sigma}^{\ddagger}, T)$ is defined by

$$g(\boldsymbol{\sigma}^{\ddagger}, T) = h(\boldsymbol{\sigma}^{\ddagger}, s) - T\,s = f(\boldsymbol{\varepsilon}^e, T) - \boldsymbol{\sigma}^{\ddagger} : \boldsymbol{\varepsilon}^e = u(\boldsymbol{\varepsilon}^e, s) - \boldsymbol{\sigma}^{\ddagger} : \boldsymbol{\varepsilon}^e - T\,s \tag{3.104}$$

with its variables $(\boldsymbol{\sigma}^{\ddagger}, T)$. The Gibbs free energy is interpreted as the energy obtained by subtracting the thermal and mechanical effects from the internal energy u. Thus from the definition of the Gibbs free energy (3.104) we have

$$dg = du - d(\boldsymbol{\sigma}^{\ddagger} : \boldsymbol{\varepsilon}^e) - d(T\,s) = -\boldsymbol{\varepsilon}^e : d\boldsymbol{\sigma}^{\ddagger} - s\,dT, \quad \boldsymbol{\varepsilon}^e = -\frac{\partial g}{\partial \boldsymbol{\sigma}^{\ddagger}}, \quad s = -\frac{\partial g}{\partial T}. \tag{3.105}$$

Example 3.1 (Elastic material under small strain condition). Under small strain theory the variables of internal energy u in the course of 'mechanical and thermal fields' are the strain $\boldsymbol{\varepsilon}^e$ and the entropy s which are extensive variables; thus the increment of internal energy density $du(\boldsymbol{\varepsilon}^e, s)$ can be written as

$$du(\boldsymbol{\varepsilon}^e, s) = \boldsymbol{\sigma}^{\ddagger} : d\boldsymbol{\varepsilon}^e + T\,ds, \tag{3.106}$$

$$\boldsymbol{\sigma}^{\ddagger} = \frac{\partial u}{\partial \boldsymbol{\varepsilon}^e}, \qquad T = \frac{\partial u}{\partial s} \tag{3.107}$$

where ds is the change of entropy exchangeable with its surroundings.

As understood from (3.107), the internal energy density u of an elastic material is a function of the strain $\boldsymbol{\varepsilon}^e$ if no temperature change is involved. In this case $u(\boldsymbol{\varepsilon}^e)$ is the strain energy, and we have

$$u(\boldsymbol{\varepsilon}^e) = \int_0^{\boldsymbol{\varepsilon}^e} \boldsymbol{\sigma}^\ddagger(\boldsymbol{\varepsilon}^e) : d\boldsymbol{\varepsilon}^e \quad \Leftrightarrow \quad du = \boldsymbol{\sigma}^\ddagger : d\boldsymbol{\varepsilon}^e. \tag{3.108}$$

Let us introduce the Legendre transformation for $u(\boldsymbol{\varepsilon}^e)$ such that

$$w(\boldsymbol{\sigma}^\ddagger) = \boldsymbol{\sigma}^\ddagger : \boldsymbol{\varepsilon}^e - u(\boldsymbol{\varepsilon}^e)$$

then we obtain the following complementary energy density:

$$w(\boldsymbol{\sigma}^\ddagger) = \int_0^{\boldsymbol{\sigma}^\ddagger} \boldsymbol{\varepsilon}^e(\boldsymbol{\sigma}^\ddagger) : d\boldsymbol{\sigma}^\ddagger \quad \Leftrightarrow \quad dw = \boldsymbol{\varepsilon}^e : d\boldsymbol{\sigma}^\ddagger. \tag{3.109}$$

These functions are schematically shown in Fig. 3.3. It is easy to see that (3.108) and (3.109) are equivalent to

$$\boldsymbol{\sigma}^\ddagger = \frac{\partial u}{\partial \boldsymbol{\varepsilon}^e}, \qquad \boldsymbol{\varepsilon}^e = \frac{\partial w}{\partial \boldsymbol{\sigma}^\ddagger}. \tag{3.110}$$

The material is said to be elastic if

$$\sigma^\ddagger_{ij} = A_{ijkl}\, \varepsilon_{kl}, \qquad A_{ijkl} = \frac{\partial^2 u}{\partial \varepsilon^e_{ij} \partial \varepsilon^e_{kl}} \tag{3.111}$$

$$\varepsilon^e_{ij} = B_{ijkl}\, \sigma^\ddagger_{kl}, \qquad B_{ijkl} = \frac{\partial^2 w}{\partial \sigma^\ddagger_{ij} \partial \sigma^\ddagger_{kl}} \tag{3.112}$$

where the elastic tensor $\boldsymbol{A}$ is the inverse of the compliance tensor $\boldsymbol{B}$ ($B_{ijkl} = A^{-1}_{ijkl}$, $A_{ijkl} = B^{-1}_{ijkl}$). Then we have

$$u = \frac{1}{2} A_{ijkl}\, \varepsilon^e_{ij} \varepsilon^e_{kl}, \qquad w = \frac{1}{2} B_{ijkl}\, \sigma^\ddagger_{ij} \sigma^\ddagger_{kl}.$$

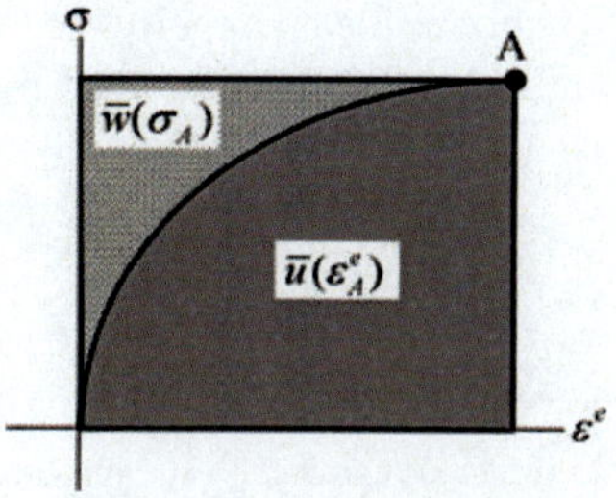

Fig. 3.3 Strain energy and complimentary energy

Since the differentiation of (3.111) and (3.112) is commutative, the elastic tensor $\boldsymbol{A}$ and compliance tensor $\boldsymbol{B}$ show symmetry:

$$A_{ijkl} = A_{klij}, \qquad B_{ijkl} = B_{klij}.$$

For an isotropic linear elastic material we have

$$A_{ijkl} = \lambda\delta_{ij}\delta_{kl} + \mu(\delta_{ik}\delta_{jl} + \delta_{il}\delta_{jk}) \tag{3.113}$$

where λ, μ are Lamè's constants (see Note 2.7, p. 51; this form is obtained by setting $\mu = \nu$ in D^e_{ijkl}). The strain energy density u and the complimentary energy density w are

$$u = \frac{1}{2}\left(\lambda\varepsilon^e_{ii}\varepsilon^e_{jj} + 2\mu\varepsilon^e_{ij}\varepsilon^e_{ij}\right), \qquad w = \frac{1}{2}\left(\frac{1+\nu}{E}\sigma^\ddagger_{ij}\sigma^\ddagger_{ij} - \frac{\nu}{E}\sigma^\ddagger_{ii}\sigma^\ddagger_{jj}\right) \tag{3.114}$$

where E and ν are Young's modulus and Poisson's ratio, respectively, and are related to the Lamè constants μ (=G, the shear modulus) and λ as follows:

$$\mu = G = \frac{E}{2(1+\nu)}, \qquad \lambda = \frac{\nu E}{(1+\nu)(1-2\nu)}. \qquad \blacksquare \tag{3.115}$$

Example 3.2 (Heat conduction problem, Fourier's law and heat capacity). The dissipative energy equation (3.50) defines the governing equation of heat conduction. If force effects are not accounted for, we can ignore the first term of the r.h.s. of (3.50) to obtain

$$C_v \frac{dT}{dt} = \operatorname{div} \boldsymbol{k} \operatorname{grad} T + \rho r. \tag{3.116}$$

Here, we employ Fourier's law which relates the temperature gradient to the heat flux. It should be noted that the temperature plays a role of potential, and C_v is the heat capacity under constant volume, calculated as follows:

$$C_v = \rho \frac{\partial s}{\partial T}\frac{\partial u}{\partial s}\bigg|_{v^e} \tag{3.117}$$

(because u is the function of s).[6] The result (3.116) is the classical *heat conduction equation*, which is a partial differential equation of the parabolic-type.

Another heat capacity, i.e., the *heat capacity under constant pressure* C_p, can be defined, which is calculated by using enthalpy h, as

$$C_p = \rho \frac{\partial h}{\partial T}\bigg|_p. \tag{3.118}$$

[6]In thermodynamics $\left.\dfrac{\partial A}{\partial x}\right|_{a,b}$ implies differentiation of A with respect to x under a, b constant.

If the relationship giving the rate of internal energy $\rho\, du = -p\, dv^e + \rho\, T\, ds$ (cf. (3.92)) is applied, we obtain the relationship of both capacities. Suppose the heat supply is dQ, then we have

$$\rho\, du = dQ - p\, dv^e = \left.\frac{\partial s}{\partial T}\frac{\partial u}{\partial s}\right|_{v^e} dT + \rho \left.\frac{\partial u}{\partial v^e}\right|_T dv^e. \tag{3.119}$$

This implies that

$$dQ = \left.\frac{\partial s}{\partial T}\frac{\partial u}{\partial s}\right|_{v^e} dT + \left(p + \rho \left.\frac{\partial u}{\partial v^e}\right|_T\right) dv^e \tag{3.120}$$

Since the change of internal energy is equal to the heat supply under constant volume, we have

$$C_v = \left.\frac{dQ}{dT}\right|_{v^e} = \left.\frac{\partial s}{\partial T}\frac{\partial u}{\partial s}\right|_{v^e}. \tag{3.121}$$

On the other hand, under constant pressure (3.120) is given as

$$C_p = \left.\frac{dQ}{dT}\right|_p = \left.\frac{\partial s}{\partial T}\frac{\partial u}{\partial s}\right|_{v^e} + \left(p + \rho \left.\frac{\partial u}{\partial v^e}\right|_T\right) \left.\frac{\partial v^e}{\partial T}\right|_p. \tag{3.122}$$

Thus the relationship between both capacities is of the form

$$C_p - C_v = \left(p + \rho \left.\frac{\partial u}{\partial v^e}\right|_T\right) \rho \left.\frac{\partial s}{\partial T}\frac{\partial u}{\partial s}\right|_p. \tag{3.123}$$

This result suggests that C_p is always greater than C_v, and a part of the adsorbed heat under constant pressure results in volume expansion. ■

3.4.3 Thermodynamic Functions in a Thermo-mechanical Field: A Finitely Strained Solid ♣

Problems in solid mechanics mostly employ a Lagrangian description. Here we rewrite the thermodynamic functions treated in the previous Subsection in a Lagrangian description.

As understood by (3.77), the change of internal energy $du(\boldsymbol{E}^e, \mathsf{s})$ is

$$du(\boldsymbol{E}^e, \mathsf{s}) = \boldsymbol{T}^{\ddagger} : d\boldsymbol{E}^e + \mathsf{T}\, d\mathsf{s}, \qquad \boldsymbol{T}^{\ddagger} = \frac{\partial u}{\partial \boldsymbol{E}^e}, \quad \mathsf{T} = \frac{\partial u}{\partial \mathsf{s}} \tag{3.124}$$

where $\boldsymbol{T}^{\ddagger} = \boldsymbol{T}/\rho_0$ is the second Piola-Kirchhoff stress per unit mass defined by (3.34), and T is a temperature corresponding to the Lagrange entropy s.

Euler's equation and the Gibbs-Duhem relation are given as follows:

$$u(\boldsymbol{E}^e, \mathsf{s}) = \boldsymbol{T}^{\ddagger} : \boldsymbol{E}^e + \mathsf{T}\,\mathsf{s}, \tag{3.125}$$

$$\boldsymbol{E}^e : d\boldsymbol{T}^{\ddagger} + \mathsf{s}\, d\mathsf{T} = 0. \tag{3.126}$$

The Helmholtz free energy $f(\boldsymbol{E}^e, \mathsf{T})$ is defined by

$$f(\boldsymbol{E}^e, \mathsf{T}) = u(\boldsymbol{E}^e, \mathsf{s}) - \mathsf{T}\,\mathsf{s} \tag{3.127}$$

with its variables $(\boldsymbol{E}^e, \mathsf{T})$ where $\boldsymbol{E}^e$ is the Green's strain. Thus from the definition of the above Helmholtz free energy we have

$$df = du - d(\mathsf{T}\mathsf{s}) = \boldsymbol{T}^{\ddagger} : d\boldsymbol{E}^e - \mathsf{s}\, d\mathsf{T}, \qquad \boldsymbol{T}^{\ddagger} = \frac{\partial f}{\partial \boldsymbol{E}^e}, \quad \mathsf{s} = -\frac{\partial f}{\partial \mathsf{T}}. \tag{3.128}$$

The enthalpy $h(\boldsymbol{T}^{\ddagger}, \mathsf{s})$ is defined by

$$h(\boldsymbol{T}^{\ddagger}, \mathsf{s}) = u(\boldsymbol{E}^e, \mathsf{s}) - \boldsymbol{T}^{\ddagger} : \boldsymbol{E}^e \tag{3.129}$$

with its variables $(\boldsymbol{T}^{\ddagger}, \mathsf{s})$. Thus from the above equation we have

$$dh = du - d(\boldsymbol{T}^{\ddagger} : \boldsymbol{E}^e) = -\boldsymbol{E}^e : d\boldsymbol{T}^{\ddagger} + \mathsf{T}\, d\mathsf{s}, \qquad \boldsymbol{E}^e = -\frac{\partial h}{\partial \boldsymbol{T}^{\ddagger}}, \quad \mathsf{T} = \frac{\partial h}{\partial \mathsf{s}} \tag{3.130}$$

The Gibbs free energy $g(\boldsymbol{T}^{\ddagger}, \mathsf{T})$ is defined by

$$g(\boldsymbol{T}^{\ddagger}, \mathsf{T}) = h(\boldsymbol{T}^{\ddagger}, \mathsf{s}) - \mathsf{T}\,\mathsf{s} = f(\boldsymbol{E}^e, \mathsf{T}) - \boldsymbol{T}^{\ddagger} : \boldsymbol{E}^e = u(\boldsymbol{E}^e, \mathsf{s}) - \boldsymbol{T}^{\ddagger} : \boldsymbol{E}^e - \mathsf{T}\,\mathsf{s} \tag{3.131}$$

with its variables $(\boldsymbol{T}^{\ddagger}, \mathsf{T})$. Thus the above definition of Gibbs free energy gives

$$dg = du - d(\boldsymbol{T}^{\ddagger} : \boldsymbol{E}^e) - d(\mathsf{T}\,\mathsf{s}) = -\boldsymbol{E}^e : d\boldsymbol{T}^{\ddagger} - \mathsf{s}\, d\mathsf{T},$$
$$\boldsymbol{E}^e = -\frac{\partial g}{\partial \boldsymbol{T}^{\ddagger}}, \quad \mathsf{s} = -\frac{\partial g}{\partial \mathsf{T}}. \tag{3.132}$$

As described in Sect. 3.2.2 the procedure becomes complicated for the case of a generalized Eulerian strain measure $\boldsymbol{e}(n)$. Only an isotropic material body can form the rate of internal energy $du_\sigma(\mathfrak{e}^e, s)$ together with the rate $\dot{\mathfrak{e}}^e$ of the generalized Eulerian strain measure $\mathfrak{e}^e = \boldsymbol{R}\,\mathfrak{E}^e\,\boldsymbol{R}^T$ and its energy-conjugate corotational Euler stress $\boldsymbol{t}^{E\ddagger} = \boldsymbol{R}\boldsymbol{T}^{\ddagger}\boldsymbol{R}^T$ (see (3.37)). That is, for the isotropic body we have

$$du(\mathfrak{e}^e, s) = \boldsymbol{t}^{E\ddagger} : d\mathfrak{e}^e + T\, ds, \qquad \boldsymbol{t}^{E\ddagger} = \frac{\partial u}{\partial \mathfrak{e}^e}, \quad T = \frac{\partial u}{\partial s}. \tag{3.133}$$

Euler's equation and the Gibbs-Duhem relation are given as follows:

$$u(\mathbf{e}^e, s) = \boldsymbol{t}^{E\ddagger} : \mathbf{e}^e + T\, s, \tag{3.134}$$

$$\mathbf{e}^e : d\boldsymbol{t}^{E\ddagger} + s\, dT = 0. \tag{3.135}$$

The Helmholtz free energy $f(\mathbf{e}^e, T)$, enthalpy $h(\boldsymbol{t}^{E\ddagger}, s)$, Gibbs free energy $g(\boldsymbol{t}^{E\ddagger}, T)$ and their changes are given as follows:

$$f(\mathbf{e}^e, T) = u(\mathbf{e}^e, s) - T\, s, \tag{3.136}$$

$$df = du - d(T s) = \boldsymbol{T}^{\ddagger} : d\mathbf{e}^e - s\, dT,$$

$$\boldsymbol{T}^{\ddagger} = \frac{\partial f}{\partial \mathbf{e}^e}, \quad s = -\frac{\partial f}{\partial T}, \tag{3.137}$$

$$h(\boldsymbol{t}^{E\ddagger}, s) = u(\mathbf{e}^e, s) - \boldsymbol{t}^{E\ddagger} : \mathbf{e}^e, \tag{3.138}$$

$$dh = du - d(\boldsymbol{t}^{E\ddagger} : \mathbf{e}^e) = T\, ds - \mathbf{e}^e : d\boldsymbol{t}^{E\ddagger},$$

$$\mathbf{e}^e = -\frac{\partial h}{\partial \boldsymbol{t}^{E\ddagger}}, \quad T = \frac{\partial h}{\partial s}, \tag{3.139}$$

$$g(\boldsymbol{t}^{E\ddagger}, T) = h(\boldsymbol{t}^{E\ddagger}, s) - T\, s = f(\mathbf{e}^e, T) - \boldsymbol{t}^{E\ddagger} : \mathbf{e}^e = u(\mathbf{e}^e, s) - \boldsymbol{t}^{E\ddagger} : \mathbf{e}^e - T\, s, \tag{3.140}$$

$$dg = du - d(\boldsymbol{t}^{E\ddagger} : \mathbf{e}^e) - d(T\, s) = -\mathbf{e}^e : d\boldsymbol{t}^{E\ddagger} - s\, dT,$$

$$\mathbf{e}^e = -\frac{\partial g}{\partial \boldsymbol{t}^{E\ddagger}}, \quad s = -\frac{\partial g}{\partial T}. \tag{3.141}$$

Example 3.3 (Finite strain hyperelastic theory). The hyperelastic theory with assumptions of small strain, as shown in Example 3.1 (p. 102), can be applied to the hyperelastic material undergoing finite strain in a similar manner. In addition, the temperature field is also included.

Let $u(\boldsymbol{E}^e, \mathsf{s})$ be the internal energy. The enthalpy $h(\boldsymbol{T}^{\ddagger}, \mathsf{s})$ is given by (3.129), which is the negative-valued complementary strain energy. Using (3.124) and (3.130) we have

$$\boldsymbol{T}^{\ddagger} = \frac{\partial u}{\partial \boldsymbol{E}^e}, \quad \boldsymbol{E}^e = -\frac{\partial h}{\partial \boldsymbol{T}^{\ddagger}}, \quad \mathsf{T} = \frac{\partial u}{\partial \mathsf{s}} = \frac{\partial h}{\partial \mathsf{s}}. \tag{3.142}$$

The body is said to be hyperelastic if

$$T^{\ddagger}_{IJ} = A_{IJKL} E^{e}_{KL}, \tag{3.143}$$

$$A_{IJKL} = \frac{\partial^2 u}{\partial E^{e}_{IJ}\, \partial E^{e}_{KL}} = \frac{\partial T^{\ddagger}_{IJ}}{\partial E^{e}_{KL}} = \frac{\partial T^{\ddagger}_{KL}}{\partial E^{e}_{IJ}} = A_{KLIJ} \tag{3.144}$$

$$E^{e}_{IJ} = B_{IJKL} T^{\ddagger}_{KL}, \tag{3.145}$$

$$B_{IJKL} = -\frac{\partial^2 h}{\partial T^{\ddagger}_{IJ}\, \partial T^{\ddagger}_{KL}} = \frac{\partial E^{e}_{IJ}}{\partial T^{\ddagger}_{KL}} = \frac{\partial E^{e}_{KL}}{\partial T^{\ddagger}_{IJ}}. = B_{KLIJ} \tag{3.146}$$

Substituting (3.144) into (3.143), we have

$$T^{\ddagger}_{IJ} = \frac{\partial T^{\ddagger}_{IJ}}{\partial E^{e}_{KL}} E^{e}_{KL}.$$

If we recall Euler's theorem (Note 3.8), this shows that $\boldsymbol{T}^{\ddagger}$ is the homogeneous function of order 1. ■

Note 3.8 (Homogeneous function and Euler's theorem). A function $\phi(\boldsymbol{x})$ is said to be the homogeneous function of degree n if, for any scalar k, we have

$$\phi(k\boldsymbol{x}) = k^n \phi(\boldsymbol{x}). \tag{3.147}$$

If $\phi(\boldsymbol{x})$ is the homogeneous function of order n, we have the following Euler's theorem:

$$\boldsymbol{x} \cdot \frac{\partial \phi}{\partial \boldsymbol{x}} = n\, \phi(\boldsymbol{x}). \tag{3.148}$$

Proof. Differentiate (3.147) with respect to k to obtain

$$\frac{\partial \phi}{\partial k} = \frac{\partial \phi}{\partial (k\boldsymbol{x})} \cdot \frac{\partial (k\boldsymbol{x})}{\partial k} = \frac{\partial \phi}{\partial (k\boldsymbol{x})} \cdot \boldsymbol{x} = n\, k^{n-1} \phi(\boldsymbol{x}),$$

and set $k = 1$, which gives (3.148). ■

3.4.4 Thermodynamic Functions in a Thermo-mechanical Field: A Fluid

Let us consider a Newtonian fluid defined by (2.226):

$$\boldsymbol{\sigma}^{\ddagger} = (-p^{\ddagger} + \lambda \operatorname{tr} \boldsymbol{D})\, \boldsymbol{i} + 2\mu \boldsymbol{D} \tag{3.149}$$

where $p^{\ddagger}$ is the normalized measure of pressure defined by (2.226), and $\lambda^{\ddagger} = \lambda/\rho$, $\mu^{\ddagger} = \mu/\rho$ are the kinematic viscosities. Taking the inner product for (3.149) with respect to the stretch tensor $\boldsymbol{D}$, we obtain

$$\boldsymbol{\sigma}^{\ddagger} : \boldsymbol{D} = -p^{\ddagger}\mathrm{tr}\,\boldsymbol{D} + \kappa^{\ddagger}\,(\mathrm{tr}\,\boldsymbol{D})^2 + 2\mu^{\ddagger}\boldsymbol{D}' : \boldsymbol{D}' \tag{3.150}$$

where $\boldsymbol{D}'$ is the deviatoric part of $\boldsymbol{D}$ and $\kappa^{\ddagger} = \lambda^{\ddagger} + 2\mu^{\ddagger}/3$. Recall that from (2.98) we have

$$\mathrm{div}\,\boldsymbol{v} = \mathrm{tr}\,\boldsymbol{D} = -\frac{\dot{\rho}}{\rho} = -\frac{d}{dt}\left(\ln\frac{\rho}{\rho_0}\right) \tag{3.151}$$

where ρ_0 is the mass density of the undeformed reference state, which, in chemistry, is commonly selected as the standard state ($P^{\ominus} = 0.1\,\mathrm{MPa} = 1\,\mathrm{bar}$, $T^{\ominus} = 25\,°\mathrm{C} = 298.15\,\mathrm{K}$). Substituting (3.151) into (3.150) gives

$$\boldsymbol{\sigma}^{\ddagger} : \boldsymbol{D} = p^{\ddagger}\frac{d}{dt}\left(\ln\frac{\rho}{\rho_0}\right) + \kappa^{\ddagger}\,(\mathrm{tr}\,\boldsymbol{D})^2 + 2\mu^{\ddagger}\boldsymbol{D}' : \boldsymbol{D}'$$

Only the first term of the r.h.s. of this equation has an exact differential form. Thus we can introduce the following:

$$\boldsymbol{\sigma}^{\ddagger} : \boldsymbol{D} = \boldsymbol{\sigma}^{\ddagger} : \boldsymbol{D}^e + \boldsymbol{\sigma}^{\ddagger} : \boldsymbol{D}^i = p^{\ddagger}\frac{d\rho^e}{dt} + \Phi(\boldsymbol{D}), \tag{3.152}$$

$$\rho^e = \ln\frac{\rho}{\rho_0}, \qquad \Phi(\boldsymbol{D}) = \kappa^{\ddagger}\,(\mathrm{tr}\,\boldsymbol{D})^2 + 2\mu^{\ddagger}\boldsymbol{D}' : \boldsymbol{D}' \tag{3.153}$$

Here ρ^e gives an elastic strain and $\Phi(\boldsymbol{D})$, referred to as the dissipation function, represents an internal dissipation due to viscosity. It should be noted that in a fluid only the volumetric change (i.e., the density change) forms the elastic deformation. The dissipation function implies that the rate of change of the mechanical energy of fluid into heat is molecular kinetic energy, and it corresponds to the part $\boldsymbol{\sigma}^{\ddagger} : \boldsymbol{D}^i$ of the dissipative energy equation (3.50) or (3.294).

Based on the above discussions the change of internal energy for a compressible fluid is introduced by

$$du(\rho^e, s) = p^{\ddagger}\,d\rho^e + T\,ds, \qquad p^{\ddagger} = \frac{\partial u}{\partial \rho^e}, \quad T = \frac{\partial u}{\partial s}. \tag{3.154}$$

Euler's equation and the Gibbs-Duhem relation are given as

$$u(\rho^e, s) = p^{\ddagger}\,\rho^e + T\,s, \tag{3.155}$$

$$\rho^e\,dp^{\ddagger} + s\,dT = 0. \tag{3.156}$$

The Helmholtz free energy $f(\rho^e, T)$, enthalpy $h(p^{\ddagger}, s)$, Gibbs free energy $g(p^{\ddagger}, T)$ and their changes are given as follows:

$$f(\rho^e, T) = u(\rho^e, s) - T\,s, \tag{3.157}$$

$$df = du - d(Ts) = p^{\ddagger}\,d\rho^e - s\,dT, \qquad p^{\ddagger} = \frac{\partial f}{\partial \rho^e}, \quad s = -\frac{\partial f}{\partial T}, \tag{3.158}$$

$$h(p^{\ddagger}, s) = u(\rho^e, s) - p^{\ddagger}\,\rho^e, \tag{3.159}$$

$$dh = du - d(p^{\ddagger}\,\rho^e) = T\,ds - \rho^e\,dp^{\ddagger}, \qquad \rho^e = -\frac{\partial h}{\partial p^{\ddagger}}, \quad T = \frac{\partial h}{\partial s}, \tag{3.160}$$

$$g(p^{\ddagger}, T) = h(p^{\ddagger}, s) - T\,s = f(\rho^e, T) - p^{\ddagger}\,\rho^e = u(\rho^e, s) - p^{\ddagger}\,\rho^e - T\,s, \tag{3.161}$$

$$dg = du - d(p^{\ddagger}\,\rho^e) - d(T\,s) = -\,\rho^e\,dp^{\ddagger} - s\,dT, \quad \rho^e = -\frac{\partial g}{\partial p^{\ddagger}}, \quad s = -\frac{\partial g}{\partial T}. \tag{3.162}$$

Note 3.9 (On compressibility of a fluid and an incompressible fluid). Let V be an infinitesimal volume element of a fluid at the current state and V_0 be the corresponding volume at the reference state. From mass conservation we have

$$\rho\,V = \rho_0\,V_0 \tag{3.163}$$

Using this and (3.151), tr $\boldsymbol{D}$ is written as

$$\mathrm{tr}\,\boldsymbol{D} = -\frac{d}{dt}\left(\ln\frac{\rho}{\rho_0}\right) = \frac{d}{dt}\left(\ln\frac{V}{V_0}\right) \tag{3.164}$$

As described in Appendix E.1, in classical thermodynamics the change of internal energy due to pressure is treated as $-P\,d(V/V_0)$ (following the above notation for the volume change).

Let $V = V_0 + \Delta V$, $\varepsilon_v = \Delta V/V_0$, and estimate $\ln(V/V_0)$ by Taylor series:

$$\ln\frac{V}{V_0} = \varepsilon_v - \frac{1}{2}(\varepsilon_v)^2 + \cdots. \tag{3.165}$$

This suggests that if we estimate the change of internal energy as $du = -p^{\ddagger}d\varepsilon_v + T\,ds$ under the volumetric strain ε_v, it is acceptable as a first-order approximation. However, if we treat a compressible gas under high pressure, we need to include second-order effects of (3.165).

The assumption of incompressibility is commonly employed for many cases in fluid mechanics. However, as understood from the above discussions, it is impossible to introduce the thermodynamic functions. In fact, the pressure of an incompressible fluid cannot be determined locally, and it is specified by the velocity field globally. In this sense the incompressibility condition should be used under careful consideration. ■

3.5 Chemical Process and Thermodynamics

Suppose that we add n mole of the same species to a thermodynamic system. Since all the molecules added must be in the same motion as the others, the following amount of internal energy is added to the system:

$$d\mathcal{U}_c = \mu \, dn. \tag{3.166}$$

Here, $d\mathcal{U}_c$ represents the change of internal energy due to the addition of mass during the chemical process, and μ is the chemical potential that provides a resistance when one mole of substance is added to the system. Note that the chemical potential μ is an intensive variable and the amount of substance n is the extensive variable. Here a molar mass description is used. In most of the textbooks in chemistry and chemical physics, molar mass descriptions are employed; however, in continuum mechanics we need to use a concentration-based notation. Note that the chemical potential μ corresponds to the Gibbs energy for concentration $c = 1$ (shown later).

We discuss here the fundamental variables and thermodynamic functions due to a chemical process in terms of continuum mechanics.

3.5.1 *Thermodynamic Variables in a Thermo-mechanochemical Field*

We apply the concept of the representative elementary volume (REV) within a volume ΔV for averaging variables, which will be discussed in Sect. 5.1. Suppose that for a mixture solution with N-components of species, an amount of substance of species α is given as n_α mole. Then the volume molar fraction ω_α and the mass density ρ_α are defined as follows:

$$n_\alpha = \omega_\alpha \, \Delta V = \frac{\rho_\alpha}{m_\alpha} \, \Delta V \qquad (!\alpha) \tag{3.167}$$

where m_α is the molecular weight of the αth component. Suppose that the total amount of substance in this mixture is n^*, and the sum of the volume molar fraction ω_α is ω, then we have

$$n^* = \sum_{\alpha=1}^{N} n_\alpha, \qquad \omega = \sum_{\alpha=1}^{N} \omega_\alpha = \frac{n^*}{\Delta V}. \tag{3.168}$$

Table 3.1 Field variables for a mixture

n_α ($\alpha = 1, 2, \ldots, N$)	Amount of substance of species α [mol]
$n^* = \sum_{\alpha=1}^N n_\alpha$	Total amount of substance in the system [mol]
m_α	Molecular weight of species α [dalton]
$m = \sum_{\alpha=1}^N x_\alpha m_\alpha$	Average molecular weight[dalton]
$\omega_\alpha = \rho_\alpha / m_\alpha = n_\alpha / \Delta V$	Volumetric molar concentration of species α [molar = mol/L^3]
$\omega = \sum_{\alpha=1}^N \omega_\alpha = n^* / \Delta V$	Volumetric molar concentration of system [molar = mol/L^3]
$\omega_\alpha^\ddagger = \omega_\alpha / \rho = x_\alpha \omega^\ddagger$	Mass molar concentration of species α [molal = mol/M]
$\omega^\ddagger = \sum_{\alpha=1}^N \omega_\alpha^\ddagger = 1/m$	Mass molar concentration of system [molal = mol/M]
$x_\alpha = \omega_\alpha / \omega = n_\alpha / n^*$	Molar fraction of species α
$\rho_\alpha = \omega_\alpha m_\alpha = n_\alpha m_\alpha / \Delta V$	Mass density of species α [M/L^3]
$\rho = \sum_{\alpha=1}^N \rho_\alpha$	Averaged mass density of system [M/L^3]
$c_\alpha = \rho_\alpha / \rho$	Mass concentration of species α
μ_α	Chemical potential per unit mass of species α [J/M]

Units: L is the space length, T is the time, M is the mass, and J is the energy (e.g., joule)
The summation convention is not used for α (!α)

Let the total mass ρ be given by

$$\rho = \sum_{\alpha=1}^{N} \rho_\alpha, \tag{3.169}$$

and the *mass concentration* (henceforth simply referred to as the concentration) is defined by

$$c_\alpha = \frac{\rho_\alpha}{\rho}. \tag{3.170}$$

Since we have the relationship

$$\sum_{\alpha=1}^{N} c_\alpha = 1, \tag{3.171}$$

the concentration c_α is not totally independent, in that one of them is dependent on the other.

The molar fraction is defined by

$$x_\alpha(\boldsymbol{x}) = \frac{\omega_\alpha}{\omega} = \frac{n_\alpha}{n^*}, \tag{3.172}$$

and the averaged molecular weight m is given by

$$m = \sum_{\alpha=1}^{N} m_\alpha x_\alpha. \tag{3.173}$$

The variables are shown in Table 3.1.

Note 3.10 (Molar description, concentration, units and dimension). Let us consider the relationship between the internal energy per unit mass u_α, the chemical potential per unit mass μ_α and the mass concentration c_α. The dimension of du_α/dt is [J/(M T)], and this dimension is the same as that for $\mu_\alpha\, dc_\alpha/dt$ $(!\alpha)$. Since the dimension of μ_α is [J/M], c_α is dimensionless. Therefore the concentration which is conjugate to the chemical potential μ_α is simply the mass concentration c_α, and the change of internal energy due to a chemical process can be written as $d(u_c)_\alpha = \mu_\alpha\, dc_\alpha$ $(!\alpha)$. ■

3.5.2 Thermodynamic Functions in a Thermo-mechanochemical Field: A Small Strain Solid

Let the change of internal energy due to mass flow be $du_c = \sum_\alpha \mu_\alpha dc_\alpha$ (cf. Note 3.10), then the change of total internal energy per unit mass is given by[7]

$$du(\boldsymbol{\varepsilon}^e, s, c_\alpha) = \boldsymbol{\sigma}^\ddagger : d\boldsymbol{\varepsilon}^e + T\, ds + \sum_\alpha \mu_\alpha\, dc_\alpha, \tag{3.174}$$

$$\boldsymbol{\sigma}^\ddagger = \left.\frac{\partial u}{\partial \boldsymbol{\varepsilon}^e}\right|_{s,\,c_\alpha}, \quad T = \left.\frac{\partial u}{\partial s}\right|_{\boldsymbol{\varepsilon}^e,\,c_\alpha}, \quad \mu_\alpha = \left.\frac{\partial u}{\partial c_\alpha}\right|_{\boldsymbol{\varepsilon}^e,\,s}. \tag{3.175}$$

As mentioned in Sect. 3.4.2 the internal energy $u(\boldsymbol{\varepsilon}^e, s, c_\alpha)$ is a homogeneous function of order one of the extensive variables, and Euler's equation and the Gibbs-Duhem relation are given as follows:

$$u = \boldsymbol{\sigma}^\ddagger : \boldsymbol{\varepsilon}^e + T\, s + \sum_\alpha \mu_\alpha\, c_\alpha, \tag{3.176}$$

$$\boldsymbol{\varepsilon}^e : d\boldsymbol{\sigma}^\ddagger + s\, dT + \sum_\alpha c_\alpha\, d\mu_\alpha = 0. \tag{3.177}$$

The Helmholtz free energy per unit mass $f(\boldsymbol{\varepsilon}^e, T, c_\alpha)$ is calculated using (3.100) and (3.176) as

$$f(\boldsymbol{\varepsilon}^e, T, c_\alpha) = u(\boldsymbol{\varepsilon}^e, s, c_\alpha) - T\, s = \boldsymbol{\sigma}^\ddagger : \boldsymbol{\varepsilon}^e + \sum_\alpha \mu_\alpha\, c_\alpha, \tag{3.178}$$

and the increment is

[7] The differential appeared in (3.174)

$$\left.\frac{\partial u}{\partial c_\alpha}\right|_{\boldsymbol{\varepsilon}^e,\,s}$$

and is sometimes referred to as the partial mass value (this case, partial mass internal energy) (cf. Slattery 1999, p. 447).

$$df = \boldsymbol{\sigma}^{\ddagger} : d\boldsymbol{\varepsilon}^{e} - s\,dT + \sum_{\alpha} \mu_{\alpha}\,dc_{\alpha}, \tag{3.179}$$

$$\boldsymbol{\sigma}^{\ddagger} = \left.\frac{\partial f}{\partial \boldsymbol{\varepsilon}^{e}}\right|_{T,\,c_{\alpha}}, \quad s = -\left.\frac{\partial f}{\partial T}\right|_{c_{\alpha},\,\boldsymbol{\varepsilon}^{e}}, \quad \mu_{\alpha} = \left.\frac{\partial f}{\partial c_{\alpha}}\right|_{\boldsymbol{\varepsilon}^{e},\,T}. \tag{3.180}$$

The enthalpy per unit mass $h(\boldsymbol{\sigma}^{\ddagger}, s, c_{\alpha})$ is calculated from (3.102) and (3.176) as

$$h(\boldsymbol{\sigma}^{\ddagger}, s, c_{\alpha}) = u(\boldsymbol{\varepsilon}^{e}, s, c_{\alpha}) - \boldsymbol{\sigma}^{\ddagger} : \boldsymbol{\varepsilon}^{e} = T\,s + \sum_{\alpha} \mu_{\alpha}\,c_{\alpha}, \tag{3.181}$$

and the increment is

$$dh = -\boldsymbol{\varepsilon}^{e} : d\boldsymbol{\sigma}^{\ddagger} + T\,ds + \sum_{\alpha} \mu_{\alpha}\,dc_{\alpha}, \tag{3.182}$$

$$\boldsymbol{\varepsilon}^{e} = -\left.\frac{\partial h}{\partial \boldsymbol{\sigma}^{\ddagger}}\right|_{s,\,c_{\alpha}}, \quad T = \left.\frac{\partial h}{\partial s}\right|_{c_{\alpha},\,\boldsymbol{\sigma}^{\ddagger}}, \quad \mu_{\alpha} = \left.\frac{\partial h}{\partial c_{\alpha}}\right|_{\boldsymbol{\sigma}^{\ddagger},\,s}. \tag{3.183}$$

The Gibbs free energy per unit mass $g(\boldsymbol{\sigma}^{\ddagger}, T, c_{\alpha})$, calculated using (3.104) and (3.176), is

$$g(\boldsymbol{\sigma}^{\ddagger}, T, c_{\alpha}) = h(\boldsymbol{\sigma}^{\ddagger}, s, c_{\alpha}) - T\,s = u(\boldsymbol{\varepsilon}^{e}, s, c_{\alpha}) - T\,s - \boldsymbol{\sigma}^{\ddagger} : \boldsymbol{\varepsilon}^{e} = \sum_{\alpha} \mu_{\alpha}\,c_{\alpha}, \tag{3.184}$$

and the increment is

$$dg = -\boldsymbol{\varepsilon}^{e} : d\boldsymbol{\sigma}^{\ddagger} - s\,dT + \sum_{\alpha} \mu_{\alpha}\,dc_{\alpha}, \tag{3.185}$$

$$\boldsymbol{\varepsilon}^{e} = -\left.\frac{\partial g}{\partial \boldsymbol{\sigma}^{\ddagger}}\right|_{T,\,c_{\alpha}}, \quad s = -\left.\frac{\partial g}{\partial T}\right|_{c_{\alpha},\,\boldsymbol{\sigma}^{\ddagger}}, \quad \mu_{\alpha} = \left.\frac{\partial g}{\partial c_{\alpha}}\right|_{\boldsymbol{\sigma}^{\ddagger},\,T}. \tag{3.186}$$

The grand potential per unit mass $\phi(\boldsymbol{\varepsilon}^{e}, T, \mu_{\alpha})$ is given by

$$\phi(\boldsymbol{\varepsilon}^{e}, T, \mu_{\alpha}) = u(\boldsymbol{\varepsilon}^{e}, s, c_{\alpha}) - T\,s - \sum_{\alpha} \mu_{\alpha}\,c_{\alpha} = \boldsymbol{\sigma}^{\ddagger} : \boldsymbol{\varepsilon}^{e}, \tag{3.187}$$

and the increment is

$$d\phi = \boldsymbol{\sigma}^{\ddagger} : d\boldsymbol{\varepsilon}^{e} - s\,dT - \sum_{\alpha} c_{\alpha}\,d\mu_{\alpha}, \tag{3.188}$$

$$\boldsymbol{\sigma}^{\ddagger} = \left.\frac{\partial \phi}{\partial \boldsymbol{\varepsilon}^{e}}\right|_{T,\,\mu_{\alpha}}, \quad s = -\left.\frac{\partial \phi}{\partial T}\right|_{\mu_{\alpha},\,\boldsymbol{\varepsilon}^{e}}, \quad c_{\alpha} = -\left.\frac{\partial \phi}{\partial \mu_{\alpha}}\right|_{\boldsymbol{\varepsilon}^{e},\,T}. \tag{3.189}$$

The transformation, changing all the variables of internal energy $u(\boldsymbol{\varepsilon}^{e}, s, c_{\alpha})$ into $(\boldsymbol{\sigma}^{\ddagger}, T, \mu_{\alpha})$, is

$$\psi = u - T\,s - \boldsymbol{\sigma}^{\ddagger} : \boldsymbol{\varepsilon}^{e} - \sum_{\alpha} \mu_{\alpha}\, c_{\alpha},$$

which is meaningless, since $\psi \equiv 0$ by applying Euler's equation (3.176), and this conclusion is also suggested by the Gibbs-Duhem relation (3.177).

Note 3.11 (Interpretation of thermodynamic functions). Based on (3.176), (3.178), (3.181), (3.184) and (3.187) we understand the following implication of thermodynamic functions:

Internal energy $u(\boldsymbol{\varepsilon}^{e}, s, c_{\alpha}) = \boldsymbol{\sigma}^{\ddagger} : \boldsymbol{\varepsilon}^{e} + T\,s + \sum_{\alpha} \mu_{\alpha}\, c_{\alpha}$:
Sum of mechanical, heat and chemical energies
Helmholtz free energy $f(\boldsymbol{\varepsilon}^{e}, T, c_{\alpha}) = \boldsymbol{\sigma}^{\ddagger} : \boldsymbol{\varepsilon}^{e} + \sum_{\alpha} \mu_{\alpha}\, c_{\alpha}$:
Sum of mechanical and chemical energies
Enthalpy $h(\boldsymbol{\sigma}^{\ddagger}, s, c_{\alpha}) = T\,s + \sum_{\alpha} \mu_{\alpha}\, c_{\alpha}$:
Sum of heat and chemical energies
Gibbs free energy $g(\boldsymbol{\sigma}^{\ddagger}, T, c_{\alpha}) = \sum_{\alpha} \mu_{\alpha}\, c_{\alpha}$:
Chemical energy
Grand potential $\phi(\boldsymbol{\varepsilon}^{e}, T, \mu_{\alpha}) = \boldsymbol{\sigma}^{\ddagger} : \boldsymbol{\varepsilon}^{e}$:
Mechanical energy

The result (3.184) shows that the sum of chemical potentials $\mu = \sum \mu_{\alpha}$ for each species is regarded as the Gibbs free energy when the concentration $c_{\alpha} = 1$ per unit volume and per unit mole. ■

As shown by (3.174) the change of internal energy is given by

$$\begin{aligned} du &= \boldsymbol{\sigma}^{\ddagger} : d\boldsymbol{\varepsilon}^{e} + T\,ds + \sum_{\alpha} \mu_{\alpha}\, dc_{\alpha} \\ &= \left.\frac{\partial u}{\partial \boldsymbol{\varepsilon}^{e}}\right|_{s, c_{\alpha}} : d\boldsymbol{\varepsilon}^{e} + \left.\frac{\partial u}{\partial s}\right|_{c_{\alpha}, \boldsymbol{\varepsilon}^{e}} ds + \sum_{\alpha} \left.\frac{\partial u}{\partial c_{\alpha}}\right|_{\boldsymbol{\varepsilon}^{e}, s} dc_{\alpha}. \end{aligned}$$

Since the differentiations commute

$$\frac{\partial}{\partial s}\left(\left.\frac{\partial u}{\partial \boldsymbol{\varepsilon}^{e}}\right|_{s, c_{\alpha}}\right)_{c_{\alpha}, \boldsymbol{\varepsilon}^{e}} = \frac{\partial}{\partial \boldsymbol{\varepsilon}^{e}}\left(\left.\frac{\partial u}{\partial s}\right|_{c_{\alpha}, \boldsymbol{\varepsilon}^{e}}\right)_{s, c_{\alpha}},$$

we have the relationships to determine $(\boldsymbol{\sigma}^{\ddagger}, T, \mu_{\alpha})$:

$$\left.\frac{\partial \boldsymbol{\sigma}^{\ddagger}}{\partial s}\right|_{c_{\alpha}, \boldsymbol{\varepsilon}^{e}} = \left.\frac{\partial T}{\partial \boldsymbol{\varepsilon}^{e}}\right|_{s, c_{\alpha}}, \quad \left.\frac{\partial T}{\partial c_{\alpha}}\right|_{\boldsymbol{\varepsilon}^{e}, s} = \left.\frac{\partial \mu_{\alpha}}{\partial s}\right|_{c_{\alpha}, \boldsymbol{\varepsilon}^{e}}, \quad \left.\frac{\partial \mu_{\alpha}}{\partial \boldsymbol{\varepsilon}^{e}}\right|_{s, c_{\alpha}} = \left.\frac{\partial \boldsymbol{\sigma}^{\ddagger}}{\partial c_{\alpha}}\right|_{\boldsymbol{\varepsilon}^{e}, s}. \tag{3.190}$$

Using a procedure similar to that outlined above for the internal energy $u(\boldsymbol{\varepsilon}^e, s, c_\alpha)$, we obtain the following relationships from (3.179) for the Helmholtz free energy $f(\boldsymbol{\varepsilon}^e, T, c_\alpha)$:

$$\left.\frac{\partial \boldsymbol{\sigma}^\ddagger}{\partial T}\right|_{c_\alpha, \boldsymbol{\varepsilon}^e} = -\left.\frac{\partial s}{\partial \boldsymbol{\varepsilon}^e}\right|_{T, c_\alpha}, \quad -\left.\frac{\partial s}{\partial c_\alpha}\right|_{\boldsymbol{\varepsilon}^e, T} = \left.\frac{\partial \mu_\alpha}{\partial T}\right|_{c_\alpha, \boldsymbol{\varepsilon}^e}, \quad \left.\frac{\partial \mu_\alpha}{\partial \boldsymbol{\varepsilon}^e}\right|_{T, c_\alpha} = \left.\frac{\partial \boldsymbol{\sigma}^\ddagger}{\partial c_\alpha}\right|_{\boldsymbol{\varepsilon}^e, T}. \tag{3.191}$$

From (3.182) for the enthalpy $h(\boldsymbol{\sigma}^\ddagger, s, c_\alpha)$ we have

$$-\left.\frac{\partial \boldsymbol{\varepsilon}^e}{\partial s}\right|_{c_\alpha, \boldsymbol{\sigma}^\ddagger} = \left.\frac{\partial T}{\partial \boldsymbol{\sigma}^\ddagger}\right|_{s, c_\alpha}, \quad \left.\frac{\partial T}{\partial c_\alpha}\right|_{\boldsymbol{\sigma}^\ddagger, s} = \left.\frac{\partial \mu_\alpha}{\partial s}\right|_{c_\alpha, \boldsymbol{\sigma}^\ddagger}, \quad \left.\frac{\partial \mu_\alpha}{\partial \boldsymbol{\sigma}^\ddagger}\right|_{s, c_\alpha} = -\left.\frac{\partial \boldsymbol{\varepsilon}^e}{\partial c_\alpha}\right|_{\boldsymbol{\sigma}^\ddagger, s}. \tag{3.192}$$

Using (3.185) for the Gibbs free energy $g(\boldsymbol{\sigma}^\ddagger, T, c_\alpha)$ gives

$$-\left.\frac{\partial \boldsymbol{\varepsilon}^e}{\partial T}\right|_{c_\alpha, \boldsymbol{\sigma}^\ddagger} = -\left.\frac{\partial s}{\partial \boldsymbol{\sigma}^\ddagger}\right|_{T, c_\alpha}, \quad -\left.\frac{\partial s}{\partial c_\alpha}\right|_{\boldsymbol{\sigma}^\ddagger, T} = \left.\frac{\partial \mu_\alpha}{\partial T}\right|_{c_\alpha, \boldsymbol{\sigma}^\ddagger}, \quad \left.\frac{\partial \mu_\alpha}{\partial \boldsymbol{\sigma}^\ddagger}\right|_{T, c_\alpha} = -\left.\frac{\partial \boldsymbol{\varepsilon}^e}{\partial c_\alpha}\right|_{\boldsymbol{\sigma}^\ddagger, T}. \tag{3.193}$$

Equation 3.188 for the grand potential $\phi(\boldsymbol{\varepsilon}^e, T, \mu_\alpha)$ is used to give

$$\left.\frac{\partial \boldsymbol{\sigma}^\ddagger}{\partial T}\right|_{\mu_\alpha, \boldsymbol{\varepsilon}^e} = -\left.\frac{\partial s}{\partial \boldsymbol{\varepsilon}^e}\right|_{T, \mu_\alpha}, \quad -\left.\frac{\partial s}{\partial \mu_\alpha}\right|_{\boldsymbol{\varepsilon}^e, T} = -\left.\frac{\partial c_\alpha}{\partial T}\right|_{\mu_\alpha, \boldsymbol{\varepsilon}^e}, \quad -\left.\frac{\partial c_\alpha}{\partial \boldsymbol{\varepsilon}^e}\right|_{T, \mu_\alpha} = \left.\frac{\partial \boldsymbol{\sigma}^\ddagger}{\partial \mu_\alpha}\right|_{\boldsymbol{\varepsilon}^e, T}. \tag{3.194}$$

The relations (3.190)–(3.194) are referred to as *Maxwell's relations*.

The closed system (cf. Appendix D) without chemical processes implies that $dc_\alpha = 0$, and using (3.174) we have

$$ds = \frac{1}{T} du - \frac{1}{T} \boldsymbol{\sigma}^\ddagger : d\boldsymbol{\varepsilon}^e = \frac{1}{T} \left(\left.\frac{\partial u}{\partial \boldsymbol{\varepsilon}^e}\right|_T - \boldsymbol{\sigma}^\ddagger \right) : d\boldsymbol{\varepsilon}^e + \frac{1}{T} \left.\frac{\partial u}{\partial T}\right|_{\boldsymbol{\varepsilon}^e} dT$$

where the entropy is assumed to be a function of the temperature T, and u is treated as a function of T. Alternatively, the total differential of entropy is

$$ds = \left.\frac{\partial s}{\partial \boldsymbol{\varepsilon}^e}\right|_T : d\boldsymbol{\varepsilon}^e + \left.\frac{\partial s}{\partial T}\right|_{\boldsymbol{\varepsilon}^e} dT,$$

and, by comparing both equations, we have

$$\left.\frac{\partial s}{\partial \boldsymbol{\varepsilon}^e}\right|_T = \frac{1}{T} \left(\left.\frac{\partial u}{\partial \boldsymbol{\varepsilon}^e}\right|_T - \boldsymbol{\sigma}^\ddagger \right), \qquad \left.\frac{\partial s}{\partial T}\right|_{\boldsymbol{\varepsilon}^e} = \frac{1}{T} \left.\frac{\partial u}{\partial T}\right|_{\boldsymbol{\varepsilon}^e}. \tag{3.195}$$

Again, the differentiations commute:

$$\left.\frac{\partial^2 s}{\partial T\,\partial \boldsymbol{\varepsilon}^e}\right|_{c_\alpha} = \left.\frac{\partial^2 s}{\partial \boldsymbol{\varepsilon}^e\,\partial T}\right|_{c_\alpha}.$$

Differentiating (3.195) yields

$$\left[\frac{\partial}{\partial T}\left\{\frac{1}{T}\left(\left.\frac{\partial u}{\partial \boldsymbol{\varepsilon}^e}\right|_T - \boldsymbol{\sigma}^\ddagger\right)\right\}\right]_{\boldsymbol{\varepsilon}^e} = \left[\frac{\partial}{\partial \boldsymbol{\varepsilon}^e}\left(\frac{1}{T}\frac{\partial u}{\partial T}\right)_{\boldsymbol{\varepsilon}^e}\right]_T = \left[\frac{\partial}{\partial \boldsymbol{\varepsilon}^e}\left(\frac{s}{T}\right)_{\boldsymbol{\varepsilon}^e}\right]_T = 0.$$

Thus for the closed system we have the following *Helmholtz relation*:

$$\left.\frac{\partial u}{\partial \boldsymbol{\varepsilon}^e}\right|_T = -T^2 \left.\frac{\partial(\boldsymbol{\sigma}^\ddagger/T)}{\partial T}\right|_{\boldsymbol{\varepsilon}^e}. \tag{3.196}$$

From (3.184) and (3.185), the Gibbs free energy $g(\boldsymbol{\sigma}^\ddagger, T, c_\alpha)$ is expressed as

$$g(\boldsymbol{\sigma}^\ddagger, T, c_\alpha) = h(\boldsymbol{\varepsilon}^e, s, c_\alpha) + \left.\frac{\partial g}{\partial T}\right|_{c_\alpha, \boldsymbol{\sigma}^\ddagger} T.$$

This results in the following *Gibbs-Helmholtz relation*[8]:

$$\frac{\partial}{\partial T}\left(\frac{g}{T}\right) = -\frac{h}{T^2}. \tag{3.197}$$

Note 3.12 (Gibbs' phase rule). As described in Note 3.11, the chemical potential μ_α is taken as the Gibbs free energy $g(\boldsymbol{\sigma}^\ddagger, T, c_\alpha)$ of a species α when the concentration $c_\alpha = 1$. Thus the unknown parameters required to determine the chemical potential are temperature T, stress $\boldsymbol{\sigma}^\ddagger$ and concentration c_α. The number of unknown variables are as follows: one for temperature T, six for stress $\boldsymbol{\sigma}^\ddagger$ and $N-1$ for concentration c_α $N-1$ (cf. (3.171), p. 112). It should be noted that in most of the chemical textbooks the 'stress' represented is only the pressure, therefore the number of unknowns is one. Let us assume that each chemical species forms p-phases, and the chemical potential at phase a is $\mu_\alpha^{(a)}$. Then the number of unknowns are $p(N-1)$. If all the phases are in equilibrium, the chemical potentials of all phases are equivalent:

[8]In some textbooks (3.197) is rewritten by using the changes of enthalpy Δh and the Gibbs free energy Δg before and after the reaction, and the Gibbs-Helmholtz relation is given by

$$\frac{\partial}{\partial T}\left(\frac{\Delta g}{T}\right) = -\frac{\Delta h}{T^2}$$

Time-differentiation of (3.197) suggests that the above form is not accurate.

$$\begin{cases} \mu_1^{\ddagger(1)} = \mu_1^{\ddagger(2)} = \cdots = \mu_1^{\ddagger(p)} \\ \vdots \\ \mu_N^{\ddagger(1)} = \mu_N^{\ddagger(2)} = \cdots = \mu_N^{\ddagger(p)}. \end{cases}$$

The number of unknowns for this phase equilibrium is $N(p-1)$. Hence the number of degrees of freedom of this system is (under six components of stress)

$$f = 1 + 6 + p(N-1) - N(p-1) = 7 + N - p. \tag{3.198}$$

If only the pressure is considered, we have

$$f = 1 + 1 + p(N-1) - N(p-1) = 2 + N - p. \tag{3.199}$$

The condition for the degrees of freedom given by (3.198) or (3.199) is referred to as the *Gibbs' phase rule.*

The systems with degrees of freedom of zero, one, two and three are referred to as the non-variant, monovariant, divariant and trivariant systems, respectively. For example, if we have a system with one species and treat only the pressure, $f = 3-p$, therefore the condition for coexistence of gas, liquid and solid phases is $f = 0$, which is referred to as the triple point. ■

3.5.3 *Thermodynamic Functions in a Thermo-mechanochemical Field: A Finitely Strained Solid* ♠

Thermodynamic functions with chemical processes are detailed here for a solid undergoing finite strain. Following the outlines of Sects. 3.4.3 and 3.5.2, the Lagrangian forms are presented first.

The change of internal energy $du(\boldsymbol{E}^e, \mathsf{s}, \mathsf{c}_\alpha)$ in the Lagrangian description is

$$du(\boldsymbol{E}^e, \mathsf{s}, c_\alpha) = \boldsymbol{T}^\ddagger : d\boldsymbol{E}^e + \mathsf{T}\, d\mathsf{s} + \sum_\alpha \mu_\alpha^* \, d\mathsf{c}_\alpha, \tag{3.200}$$

$$\boldsymbol{T}^\ddagger = \frac{\partial u}{\partial \boldsymbol{E}^e}, \quad \mathsf{T} = \frac{\partial u}{\partial \mathsf{s}}, \quad \mu_\alpha^* = \frac{\partial u}{\partial \mathsf{c}_\alpha} \tag{3.201}$$

where c_α is the Lagrangian volume molar concentration defined by

$$\mathsf{c}_\alpha = c_\alpha J \tag{3.202}$$

and μ_α^* is the corresponding chemical potential.

Referring back to (3.176) the internal energy $u(\boldsymbol{E}^e, \mathsf{s}, \mathsf{c}_\alpha)$ is given by

$$u(\boldsymbol{E}^e, \mathsf{s}, \mathsf{c}_\alpha) = \boldsymbol{T}^\ddagger : \boldsymbol{E}^e + \mathsf{T}\mathsf{s} + \sum_\alpha \mu_\alpha^* \mathsf{c}_\alpha. \tag{3.203}$$

This gives Euler's equation. The Gibbs-Duhem relation is obtained by

$$\boldsymbol{E}^e : d\boldsymbol{T}^\ddagger + \mathsf{s}\, d\mathsf{T} + \sum_\alpha \mathsf{c}_\alpha\, d\mu_\alpha^* = 0. \tag{3.204}$$

The Helmholtz free energy $f(\boldsymbol{E}^e, \mathsf{T}, \mathsf{c}_\alpha)$ is calculated as

$$f(\boldsymbol{E}^e, \mathsf{T}, \mathsf{c}_\alpha) = u(\boldsymbol{E}^e, \mathsf{s}, \mathsf{c}_\alpha) - T\, s = \boldsymbol{T}^\ddagger : \boldsymbol{E}^e + \sum_\alpha \mu_\alpha^* \mathsf{c}_\alpha \tag{3.205}$$

where its increment is

$$df = \boldsymbol{T}^\ddagger : d\boldsymbol{E}^e - \mathsf{s}\, d\mathsf{T} + \sum_\alpha \mu_\alpha^*\, d\mathsf{c}_\alpha, \tag{3.206}$$

$$\boldsymbol{T}^\ddagger = \frac{\partial f}{\partial \boldsymbol{E}^e}, \quad \mathsf{s} = -\frac{\partial f}{\partial \mathsf{T}}, \quad \mu_\alpha^* = \frac{\partial f}{\partial \mathsf{c}_\alpha}. \tag{3.207}$$

The enthalpy $h(\boldsymbol{T}^\ddagger, \mathsf{s}, \mathsf{c}_\alpha)$ is given as

$$h(\boldsymbol{T}^\ddagger, \mathsf{s}, \mathsf{c}_\alpha) = u(\boldsymbol{E}^e, \mathsf{s}, \mathsf{c}_\alpha) - \boldsymbol{T}^\ddagger : \boldsymbol{E}^e = \mathsf{T}\mathsf{s} + \sum_\alpha \mu_\alpha^* \mathsf{c}_\alpha \tag{3.208}$$

and its increment is

$$dh = -\boldsymbol{E}^e : d\boldsymbol{T}^\ddagger + T\, ds + \sum_\alpha \mu_\alpha^*\, d\mathsf{c}_\alpha, \tag{3.209}$$

$$\boldsymbol{E}^e = -\frac{\partial h}{\partial \boldsymbol{T}^\ddagger}, \quad \mathsf{T} = \frac{\partial h}{\partial \mathsf{s}}, \quad \mu_\alpha^* = \frac{\partial h}{\partial \mathsf{c}_\alpha}. \tag{3.210}$$

The Gibbs free energy $g(\boldsymbol{T}^\ddagger, \mathsf{T}, \mathsf{c}_\alpha)$ can be written as

$$g(\boldsymbol{T}^\ddagger, \mathsf{T}, \mathsf{c}_\alpha) = h(\boldsymbol{T}^\ddagger, \mathsf{s}, \mathsf{c}_\alpha) - \mathsf{T}\mathsf{s} = u(\boldsymbol{E}^e, \mathsf{s}, \mathsf{c}_\alpha) - \boldsymbol{T}^\ddagger : \boldsymbol{E}^e - \mathsf{T}\mathsf{s} = \sum_\alpha \mu_\alpha^* \mathsf{c}_\alpha \tag{3.211}$$

and its increment is

$$dg = -\boldsymbol{E}^e : d\boldsymbol{T}^\ddagger - \mathsf{s}\, d\mathsf{T} + \sum_\alpha \mu_\alpha^*\, d\mathsf{c}_\alpha, \tag{3.212}$$

$$\boldsymbol{E}^e = -\frac{\partial g}{\partial \boldsymbol{T}^\ddagger}, \quad \mathsf{s} = -\frac{\partial g}{\partial \mathsf{T}}, \quad \mu_\alpha^* = \frac{\partial g}{\partial \mathsf{c}_\alpha}. \tag{3.213}$$

The grand potential $\phi(\boldsymbol{E}^e, \mathsf{T}, \mu_\alpha^*)$ is calculated as

$$\phi(\boldsymbol{E}^e, \mathsf{T}, \mu_\alpha^*) = u(\boldsymbol{E}^e, \mathsf{s}, \mathsf{c}_\alpha) - \mathsf{T}\,\mathsf{s} - \sum_\alpha \mu_\alpha^* \,\mathsf{c}_\alpha = \boldsymbol{T}^\ddagger : \boldsymbol{E}^e \tag{3.214}$$

and its increment is

$$d\phi = \boldsymbol{T}^\ddagger : d\boldsymbol{E}^e - \mathsf{s}\, d\mathsf{T} - \sum_\alpha \mathsf{c}_\alpha \, d\mu_\alpha^*, \tag{3.215}$$

$$\boldsymbol{T}^\ddagger = \frac{\partial \phi}{\partial \boldsymbol{E}^e}, \quad \mathsf{s} = -\frac{\partial \phi}{\partial \mathsf{T}}, \quad \mathsf{c}_\alpha = -\frac{\partial \phi}{\partial \mu_\alpha^*}. \tag{3.216}$$

As mentioned in Sect. 3.4.3, only the isotropic material body can form the rate of internal energy $d u_\sigma(\mathbf{e}^e, s)$ together with the rate $\dot{\mathbf{e}}^e$ of the generalized Eulerian strain measure $\mathbf{e}^e = \boldsymbol{R}\, \mathfrak{E}^e\, \boldsymbol{R}^T$ and its energy-conjugate corotational Euler stress $\boldsymbol{t}^{E\ddagger} = \boldsymbol{R}\boldsymbol{T}^\ddagger \boldsymbol{R}^T$. Then Euler's equation and the change of internal energy are given by

$$u(\mathbf{e}^e, s, c_\alpha) = \boldsymbol{t}^{E\ddagger} : \mathbf{e}^e + T\, s + \sum_\alpha \mu_\alpha \, c_\alpha, \tag{3.217}$$

$$du = \boldsymbol{t}^{E\ddagger} : d\mathbf{e}^e + T\, ds + \sum_\alpha \mu_\alpha \, dc_\alpha,$$

$$\boldsymbol{t}^{E\ddagger} = \frac{\partial u}{\partial \mathbf{e}^e}, \quad T = \frac{\partial u}{\partial s}, \quad \mu_\alpha = \frac{\partial u}{\partial c_\alpha}. \tag{3.218}$$

The Gibbs-Duhem relation is written as

$$\mathbf{e}^e : d\boldsymbol{t}^{E\ddagger} + s\, dT + \sum_\alpha c_\alpha \, d\mu_\alpha = 0. \tag{3.219}$$

The Helmholtz free energy $f(\mathbf{e}^e, T, c_\alpha)$, the enthalpy $h(\boldsymbol{t}^{E\ddagger}, s, c_\alpha)$, the Gibbs free energy $g(\boldsymbol{t}^{E\ddagger}, T, c_\alpha)$, the grand potential $\phi(\mathbf{e}^e, T, \mu_\alpha)$ and their increments are given as follows:

$$f(\mathbf{e}^e, T, c_\alpha) = u(\mathbf{e}^e, s, c_\alpha) - T\, s = \boldsymbol{t}^{E\ddagger} : \mathbf{e}^e + \sum_\alpha \mu_\alpha \, c_\alpha, \tag{3.220}$$

$$df = \boldsymbol{t}^{E\ddagger} : d\mathbf{e}^e - s\, dT + \sum_\alpha \mu_\alpha \, dc_\alpha,$$

$$\boldsymbol{t}^{E\ddagger} = \frac{\partial f}{\partial \mathbf{e}^e}, \quad s = -\frac{\partial f}{\partial T}, \quad \mu_\alpha = \frac{\partial f}{\partial c_\alpha} \tag{3.221}$$

$$h(\boldsymbol{t}^{E\ddagger}, s, c_\alpha) = u(\mathbf{e}^e, s, c_\alpha) - \boldsymbol{t}^{E\ddagger} : \mathbf{e}^e = T\, s + \sum_\alpha \mu_\alpha \, c_\alpha, \tag{3.222}$$

$$dh = T\,ds - \mathbf{e}^e : d\mathbf{t}^{E\ddagger} + \sum_\alpha \mu_\alpha\,dc_\alpha,$$

$$\mathbf{e}^e = -\frac{\partial h}{\partial \mathbf{t}^{E\ddagger}}, \quad T = \frac{\partial h}{\partial s}, \quad \mu_\alpha = \frac{\partial h}{\partial c_\alpha}, \tag{3.223}$$

$$g(\mathbf{t}^{E\ddagger}, T, c_\alpha) = h(\mathbf{t}^{E\ddagger}, s, c_\alpha) - T\,s = u(\mathbf{e}^e, s, c_\alpha) - \mathbf{t}^{E\ddagger} : \mathbf{e}^e - T\,s = \sum_\alpha \mu_\alpha\,c_\alpha, \tag{3.224}$$

$$dg = -\mathbf{e}^e : d\mathbf{t}^{E\ddagger} - s\,dT + \sum_\alpha \mu_\alpha\,dc_\alpha,$$

$$\mathbf{e}^e = -\frac{\partial g}{\partial \mathbf{t}^{E\ddagger}}, \quad s = -\frac{\partial g}{\partial T}, \quad \mu_\alpha = \frac{\partial g}{\partial c_\alpha}, \tag{3.225}$$

$$\phi(\mathbf{e}^e, T, \mu_\alpha) = u(\mathbf{e}^e, s, c_\alpha) - T\,s - \sum_\alpha \mu_\alpha\,c_\alpha = \mathbf{t}^{E\ddagger} : \mathbf{e}^e, \tag{3.226}$$

$$d\phi = \mathbf{t}^{E\ddagger} : d\mathbf{e}^e - s\,dT - \sum_\alpha c_\alpha\,d\mu_\alpha,$$

$$\mathbf{t}^{E\ddagger} = \frac{\partial \phi}{\partial \mathbf{e}^e}, \quad s = -\frac{\partial \phi}{\partial T}, \quad c_\alpha = -\frac{\partial \phi}{\partial \mu_\alpha}. \tag{3.227}$$

3.5.4 Thermodynamic Functions in a Thermo-mechanochemical Field: A Fluid

Referring to Sect. 3.4.4 the change of internal energy $u(\rho^e, s, c_\alpha)$ is given by

$$du = p^\ddagger\,d\rho^e + T\,ds + \sum_\alpha \mu_\alpha\,dc_\alpha,$$

$$p^\ddagger = \frac{\partial u}{\partial \rho^e}, \quad T = \frac{\partial u}{\partial s}, \quad \mu_\alpha = \frac{\partial u}{\partial c_\alpha}. \tag{3.228}$$

Euler's equation and the Gibbs-Duhem relation are

$$u(\rho^e, s, c_\alpha) = p^\ddagger\,\rho^e + T\,s + \sum_\alpha \mu_\alpha\,c_\alpha, \tag{3.229}$$

$$\rho^e\,dp^\ddagger + s\,dT + \sum_\alpha c_\alpha\,d\mu_\alpha = 0. \tag{3.230}$$

The Helmholtz free energy $f(\rho^e, T, c_\alpha)$, the enthalpy $h(p^\ddagger, s, c_\alpha)$, the Gibbs free energy $g(p^\ddagger, T, c_\alpha)$, the grand potential $\phi(\rho^e, T, \mu_\alpha)$ and their changes are given by

$$f(\rho^e, T, c_\alpha) = u(\rho^e, s, c_\alpha) - T\,s = p^\ddagger\,\rho^e + \sum_\alpha \mu_\alpha\,c_\alpha, \tag{3.231}$$

$$df = p^\ddagger\,d\rho^e - s\,dT + \sum_\alpha \mu_\alpha\,dc_\alpha,$$

$$p^\ddagger = \frac{\partial f}{\partial \rho^e}, \quad s = -\frac{\partial f}{\partial T}, \quad \mu_\alpha = \frac{\partial f}{\partial c_\alpha} \tag{3.232}$$

$$h(p^\ddagger, s, c_\alpha) = u(\rho^e, s, c_\alpha) - p^\ddagger\,\rho^e = T\,s + \sum_\alpha \mu_\alpha\,c_\alpha, \tag{3.233}$$

$$dh = T\,ds - \rho^e\,dp^\ddagger + \sum_\alpha \mu_\alpha\,dc_\alpha,$$

$$\rho^e = -\frac{\partial h}{\partial p^\ddagger}, \quad T = \frac{\partial h}{\partial s}, \quad \mu_\alpha = \frac{\partial h}{\partial c_\alpha}, \tag{3.234}$$

$$g(p^\ddagger, T, c_\alpha) = h(p^\ddagger, s, c_\alpha) - T\,s = u(\rho^e, s, c_\alpha) - p^\ddagger\,\rho^e - T\,s = \sum_\alpha \mu_\alpha\,c_\alpha, \tag{3.235}$$

$$dg = -\rho^e\,dp^\ddagger - s\,dT + \sum_\alpha \mu_\alpha\,dc_\alpha,$$

$$\rho^e = -\frac{\partial g}{\partial p^\ddagger}, \quad s = -\frac{\partial g}{\partial T}, \quad \mu_\alpha = \frac{\partial g}{\partial c_\alpha}, \tag{3.236}$$

$$\phi(\rho^e, T, \mu_\alpha) = u(\rho^e, s, c_\alpha) - T\,s - \sum_\alpha \mu_\alpha\,c_\alpha = p^\ddagger\,\rho^e, \tag{3.237}$$

$$d\phi = p^\ddagger\,d\rho^e - s\,dT - \sum_\alpha c_\alpha\,d\mu_\alpha,$$

$$p^\ddagger = \frac{\partial \phi}{\partial \rho^e}, \quad s = -\frac{\partial \phi}{\partial T}, \quad c_\alpha = -\frac{\partial \phi}{\partial \mu_\alpha}. \tag{3.238}$$

3.6 Mixture Theory for a Multi-component Solution

For a multi-component mixture with 1, 2, 3, $\cdots$ N-species in a fundamentally miscible solution, which involves a process of chemical reactions, we establish a framework of continuum thermodynamics theory. The mixture might be a liquid or gas; however, since we mainly use a mass-averaged velocity (defined later), the liquid solution might be more appropriate. The description is Eulerian because we are considering a fluid.

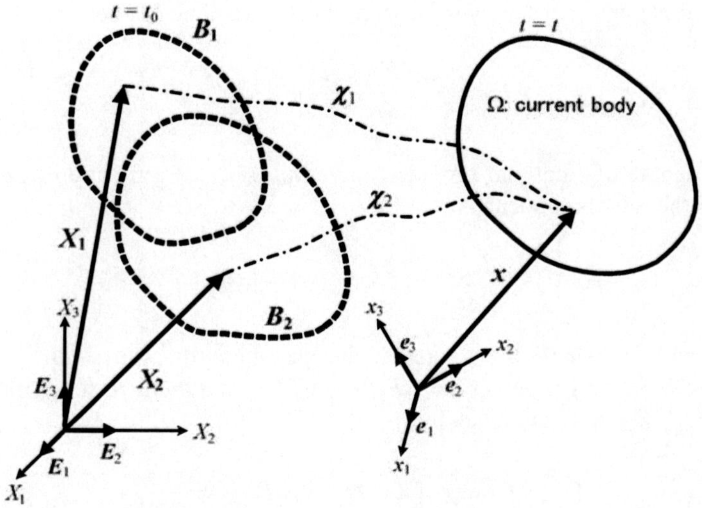

Fig. 3.4 Reference and current bodies for the mixture theory

For any solution we distinguish between solute and solvent. Here we can state that the first species is the solvent and the rest of the (2, 3, $\cdots$ N)-species are solutes. We shall introduce a non-equilibrium thermodynamics theory for this solution that will treat the coupled problem of stress, diffusion with reactions and heat transfer. The field variables for the chemical field are listed in Table 3.1.

We extend the theory of mixture given by Bowen (1976) (see also Drew and Passman 1998; Coussy 1995; Lewis and Schrefler 2000), though the notation is different. A continuum body with N-components is represented as $\mathcal{B}_\alpha$ ($\alpha = 1, 2, \cdots, N$). It should be noted that due to the mixture theory the body $\mathcal{B}_\alpha$ of the αth component defines the independent reference configuration for each component. The case of two components are schematically shown in Fig. 3.4. The motion of a spatial point $\boldsymbol{x}$ in the current body $\Omega \subset \mathbb{R}^3$ can be described by

$$\boldsymbol{x} = \boldsymbol{\chi}_\alpha(\boldsymbol{X}_\alpha, t) = x_i\, \boldsymbol{e}_i \tag{3.239}$$

where $\boldsymbol{e}_i$ ($i = 1, 2, 3$) are the orthogonal Eulerian basis of the spatial coordinates, $\boldsymbol{\chi}_\alpha$ is the deformation function for the αth component, which is sufficiently smooth, $\boldsymbol{X}_\alpha$ is the material point of the αth component in the reference configuration, and t is a time. Let $\partial\Omega$ be the boundary of Ω. Since the function $\boldsymbol{\chi}_\alpha$ is smooth, the reference position $\boldsymbol{X}_\alpha$ can be given by

$$\boldsymbol{X}_\alpha = \boldsymbol{\chi}_\alpha^{-1}(\boldsymbol{x}, t) = X_{\alpha K}\, \boldsymbol{E}_K \tag{3.240}$$

where $\boldsymbol{\chi}_\alpha^{-1}$ is the inverse function of the deformation function, and $\boldsymbol{E}_K$ ($K = 1, 2, 3$) are the orthogonal Lagrangian basis of material coordinates.

The particle velocity $\boldsymbol{v}_\alpha$ of a material point $\boldsymbol{X}_\alpha$ at a time t is obtained as

$$\boldsymbol{v}_\alpha = \left.\frac{d_{(\alpha)}\boldsymbol{\chi}_\alpha(\boldsymbol{X}_\alpha, t)}{dt}\right|_{\boldsymbol{X}_\alpha=\text{constant}}. \tag{3.241}$$

We here denote the material time-derivative $d_{(\alpha)}\phi/dt$ of a function $\phi(\boldsymbol{x},t)$ with respect to the αth component as

$$\frac{d_{(\alpha)}\phi}{dt} = \frac{\partial\phi}{\partial t} + \boldsymbol{v}_\alpha \cdot \operatorname{grad}\phi. \tag{3.242}$$

The symbol 'grad' shows the gradient in the spatial coordinate system.

The deformation gradient $\boldsymbol{F}_\alpha = F_{\alpha iK}\,\boldsymbol{e}_i \otimes \boldsymbol{E}_K$ and the inverse transformation $\boldsymbol{F}_\alpha^{-1} = F_{\alpha Ki}^{-1}\,\boldsymbol{E}_K \otimes \boldsymbol{e}_i$ are given by

$$\boldsymbol{F}_\alpha = \operatorname{Grad}\boldsymbol{\chi}_\alpha(\boldsymbol{X}_\alpha, t), \qquad F_{\alpha iK} = \frac{\partial x_i}{\partial X_{\alpha K}} \tag{3.243}$$

$$\boldsymbol{F}_\alpha^{-1} = \operatorname{grad}\boldsymbol{\chi}_\alpha^{-1}(\boldsymbol{x}, t), \qquad F_{\alpha Ki}^{-1} = \frac{\partial X_{\alpha K}}{\partial x_i} \tag{3.244}$$

where the symbol 'Grad' shows the gradient in the material coordinate system. The time derivative of the Jacobian J_α is calculated as

$$\frac{d_{(\alpha)}J_\alpha}{dt} = J_\alpha \operatorname{div}\boldsymbol{v}_\alpha, \qquad J_\alpha = |\det\boldsymbol{F}_\alpha| \tag{3.245}$$

where the symbol 'div' shows the divergence in the spatial coordinate system.

3.6.1 Mass Conservation Law

Let the total mass of species α be $\mathcal{M}_\alpha$, the mass change due to chemical reaction be $\mathcal{R}_\alpha$. The mass conservation law for the species α can be written as

$$\frac{d_{(\alpha)}\mathcal{M}_\alpha}{dt} = \mathcal{R}_\alpha, \tag{3.246}$$

$$\mathcal{M}_\alpha = \int_\Omega \rho_\alpha\,dv, \quad \mathcal{R}_\alpha = \int_\Omega \rho_\alpha\gamma_\alpha\,dv \qquad (!\alpha) \tag{3.247}$$

(cf. Sect. 3.5). Here γ_α is the volumetric density function of $\mathcal{R}_\alpha$ (i.e., the mass source term per unit volume). Applying Reynolds' transport theorem and the divergence theorem we obtain a local form of the differential equation for mass conservation as

$$\frac{d_{(\alpha)}\rho_\alpha}{dt} + \rho_\alpha \operatorname{div} \boldsymbol{v}_\alpha = \frac{\partial \rho_\alpha}{\partial t} + \operatorname{div}(\rho_\alpha \boldsymbol{v}_\alpha) = \rho_\alpha \gamma_\alpha \qquad (!\alpha) \tag{3.248}$$

where $\boldsymbol{v}_\alpha$ is the particle velocity of the αth species.

By summing up (3.248) for all the species, the reaction terms cancel each other:

$$\sum_\alpha \rho_\alpha \gamma_\alpha = 0. \tag{3.249}$$

Thus we have the total mass conservation law as follows:

$$\frac{d\mathcal{M}}{dt} = 0, \qquad \mathcal{M} = \sum_\alpha \mathcal{M}_\alpha.$$

Its local form is obtained as

$$\frac{d\rho}{dt} + \rho \operatorname{div} \boldsymbol{v} = \frac{\partial \rho}{\partial t} + \operatorname{div}\,(\rho \boldsymbol{v}) = 0 \tag{3.250}$$

where ρ is the total mass density defined by (3.169), and $c_\alpha = \rho_\alpha/\rho$ is the mass concentration (i.e., the mass fraction) shown in Table 3.1. We have defined the mass-averaged velocity $\boldsymbol{v}$ as

$$\boldsymbol{v} = \frac{1}{\rho}\sum_{\alpha=1}^{N} \rho_\alpha \boldsymbol{v}_\alpha = \sum_{\alpha=1}^{N} c_\alpha \boldsymbol{v}_\alpha. \tag{3.251}$$

Note that $\boldsymbol{q}^m = \rho \boldsymbol{v}$ represents the total mass flux. In (3.250) we have introduced the material time-differentiation $d\ \ /dt$ with respect to the mass-averaged velocity $\boldsymbol{v}$ as follows:

$$\frac{d}{dt} = \frac{\partial}{\partial t} + \boldsymbol{v} \cdot \operatorname{grad}(\ ). \tag{3.252}$$

The velocity gradient $\boldsymbol{L}$, the stretch tensor $\boldsymbol{D}$ and the spin tensor $\boldsymbol{W}$ are defined by

$$\boldsymbol{L} = \operatorname{grad} \boldsymbol{v} = \boldsymbol{D} + \boldsymbol{W}, \tag{3.253}$$

$$\boldsymbol{D} = \frac{1}{2}(\boldsymbol{L} + \boldsymbol{L}^T), \quad \boldsymbol{W} = \frac{1}{2}(\boldsymbol{L} - \boldsymbol{L}^T). \tag{3.254}$$

If we use the mass-averaged velocity $\boldsymbol{v}$, (3.248) can be written as

$$\frac{d\rho_\alpha}{dt} + \rho_\alpha \operatorname{div} \boldsymbol{v} = -\operatorname{div}\,(\rho_\alpha \bar{\boldsymbol{v}}_\alpha) + \rho_\alpha \gamma_\alpha \qquad (!\alpha) \tag{3.255}$$

where we have defined the diffusion velocity $\bar{\boldsymbol{v}}_\alpha$ as follows:

$$\bar{\boldsymbol{v}}_\alpha = \boldsymbol{v}_\alpha - \boldsymbol{v}. \tag{3.256}$$

By eliminating $\boldsymbol{v}$ from (3.250) and (3.255), we have

$$\frac{d\rho_\alpha}{dt} - c_\alpha \frac{d\rho}{dt} = -\mathrm{div}\,(\rho_\alpha \overline{\boldsymbol{v}}_\alpha) + \rho_\alpha \gamma_\alpha \qquad (!\alpha). \tag{3.257}$$

We take the material time-differentiation for the mass concentration $c_\alpha = \rho_\alpha/\rho$ with respect to the mass-averaged velocity $\boldsymbol{v}$, which gives

$$\rho \frac{dc_\alpha}{dt} = -\frac{d\rho}{dt} c_\alpha + \frac{d\rho_\alpha}{dt}. \tag{3.258}$$

By substituting (3.258) into (3.257), we finally obtain the following *diffusion equation* for a material in motion:

$$\rho \frac{dc_\alpha}{dt} = -\mathrm{div}\,(\rho_\alpha \overline{\boldsymbol{v}}_\alpha) + \rho_\alpha \gamma_\alpha \quad (!\alpha). \tag{3.259}$$

where

$$\boldsymbol{q}_\alpha^m = \rho_\alpha \overline{\boldsymbol{v}}_\alpha \qquad (!\alpha) \tag{3.260}$$

represents a mass flux of the species α.

Note 3.13 (Evaluation of the source term due to reaction). Assume that in the current state the species α is involved in the r-th stage of the chemical reaction, and the reaction rate can be written in the form $w_r = d\xi_r/dt$ as introduced by (E.39) in Appendix E. If the stoichiometric coefficient of this reaction stage is given by $\nu_{\alpha r}$ for the species α, then the amount of mass produced during all the reaction processes is given by

$$\rho_\alpha \gamma_\alpha \Delta V = \sum_r m_\alpha\, \nu_{\alpha r}\, w_r \qquad (!\alpha)$$

where V is the volume of the reaction system (for simplicity, a homogeneous and isotropic system is assumed), and m_α is the molecular weight of the species α. Since we have the relation $n_\alpha\, m_\alpha = \rho_\alpha \Delta V$, the term γ_α is evaluated by

$$\gamma_\alpha = \sum_r \nu_{\alpha r}^\dagger\, w_r, \qquad \nu_{\alpha r}^\dagger = \frac{\nu_{\alpha r}}{n_\alpha} \qquad (!\alpha) \tag{3.261}$$

where $\nu_{\alpha r}^\dagger$ is the stoichiometric coefficient per unit mole.

As implied by (E.144) in Appendix E.9 concerning the chemical reaction, the reaction rate w is commonly represented in a power form depending on the reaction order, and the term on the r.h.s. of (3.261) is given by

$$\gamma_\alpha^\ddagger = \sum_r \nu_{\alpha r}^\dagger\, w_r = \sum_r \nu_{\alpha r}^\dagger\, k_r \prod_\alpha (c_\alpha^r)^{p_\alpha^r} \qquad (!\alpha) \tag{3.262}$$

where the suffix r implies the reaction stage r, k_r is its reaction rate constant, c_α^r is the concentration of species α and p_α^r is the reaction order of species α. ■

Note 3.14 (On the average velocities). In this section we use the mass-averaged velocity (3.251); however in some cases we may use molar-averaged velocity:

$$v^\diamond = \sum_{\alpha=1}^{N} x_\alpha v_\alpha. \tag{3.263}$$

Let us consider an example shown in Fig. 3.5a of a gas diffusion problem enclosed in two glass bulbs under constant temperature/pressure. The l.h.s. bulb contains gaseous nitrogen (N_2, molecular weight 28.02), and the r.h.s. bulb contains gaseous hydrogen (H_2, molecular weight 2.016). Initially the valve at the center is closed; the amount of both gases are the same, therefore the mass center is in the nitrogen gas. After opening the valve the gases are mixed homogeneously, and the mass center moves to the center. Thus the mass-averaged velocity shows a finite value, whereas the molar center is not transferred, which implies the molar-averaged velocity is null in this case.

For the diffusion problem of 50 wt% glycerin solution as shown in Fig. 3.5b, since the mass density of this solution is 1.1 g/cm^3 which is almost the same as the density of pure water (1.0 g/cm^3), the mass-averaged velocity is almost zero (glycerin: $HOCH_2$ $CH(OH)CH_2OH$, molecular weight 92.09, specific gravity 1.26362 at 20°C), while the molar concentration of this solution is 33.21 mol/l

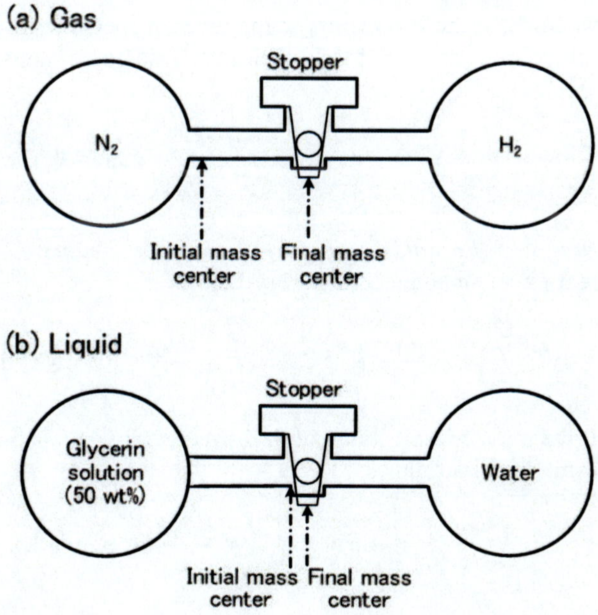

Fig. 3.5 Diffusion experiments for evaluating mean velocities (Cussler 1997, Figs. 3.1–3.2)

which is less than the molar concentration of water 55.56 mol/l, therefore the molar-averaged velocity is of finite value (Cussler 1997; Bird et al. 2002).

In this work we mainly treat solids and liquids, therefore we employ the mass-averaged velocity. For a detailed treatment that uses the molar-averaged velocity, we can consult Chap. 8 of Slattery (1999). Some books treat the volume-averaged velocity; however for a mixture it is almost impossible to measure the volume of each species, therefore descriptions that employ the volume-averaged velocity are not feasible. ■

3.6.2 *Conservations of Linear Momentum and Moment of Momentum*

Suppose an external traction $\boldsymbol{t}_\alpha$ commonly acts on the αth species, then the conservation equation of linear momentum for species α is given by

$$\frac{d_{(\alpha)}\mathcal{L}_\alpha}{dt} = \mathcal{F}_\alpha, \tag{3.264}$$

$$\mathcal{L}_\alpha = \int_\Omega \rho_\alpha \boldsymbol{v}_\alpha \, dv, \quad \mathcal{F}_\alpha = \int_{\partial\Omega} \boldsymbol{t}_\alpha \, ds + \int_\Omega \rho_\alpha \, \boldsymbol{b}_\alpha \, dv + \int_\Omega \boldsymbol{\psi}_\alpha \, dv \qquad (!\alpha), \tag{3.265}$$

$$\boldsymbol{\psi}_\alpha = \sum_{\beta\,(\beta \neq \alpha)} \boldsymbol{\psi}_{\alpha\beta} \tag{3.266}$$

where $\boldsymbol{b}_\alpha$ is the body force acting on each species and $\boldsymbol{\psi}_{\alpha\beta}$ is the interactive force acting from species β to species α. This force vanishes if summed up for all species α:

$$\sum_\alpha \sum_{\beta\,(\beta \neq \alpha)} \boldsymbol{\psi}_{\alpha\beta} = \boldsymbol{0}. \tag{3.267}$$

Note 3.15 (Reynolds' transport theorem for a mixture). Reynolds' transport theorem for a component α can be written as

$$\frac{d_{(\alpha)}}{dt} \int_\Omega \phi \, dv = \int_\Omega \left(\frac{d_{(\alpha)}\phi}{dt} + \phi \operatorname{div} \boldsymbol{v}_\alpha \right) dv \qquad (!\alpha) \tag{3.268}$$

If the mass conservation law (3.248) including a reaction term is satisfied, (3.268) is modified as follows:

$$\frac{d_{(\alpha)}}{dt} \int_\Omega \rho_\alpha \phi \, dv = \int_\Omega \left(\rho_\alpha \frac{d_{(\alpha)}\phi}{dt} + \rho_\alpha \gamma_\alpha \phi \right) dv \qquad (!\alpha) \tag{3.269}$$

where γ_α is a source term of mass per unit mass. ■

By applying Reynolds' transport theorem (3.269) to (3.264), we have the following local form:

$$\rho_\alpha \frac{d_{(\alpha)} \boldsymbol{v}_\alpha}{dt} = \text{div } (\boldsymbol{\sigma}_\alpha)^T + \rho_\alpha \, (\boldsymbol{b}_\alpha - \gamma_\alpha \boldsymbol{v}_\alpha) + \boldsymbol{\psi}_\alpha \qquad (!\alpha) \tag{3.270}$$

where we have used the relation $\boldsymbol{t}_\alpha = \boldsymbol{\sigma}_\alpha^T \boldsymbol{n}$. Note again that we are considering a miscible solution.

Note 3.16 (On the Fundamental Identity (Truesdell and Toupin 1960, pp. 469–474)). For a function φ_α we have the following material time-differential with respect to the velocity $\boldsymbol{v}_\alpha$:

$$\begin{aligned} \rho_\alpha \frac{d_{(\alpha)} \varphi_\alpha}{dt} = {} & \frac{\partial (\rho_\alpha \varphi_\alpha)}{\partial t} + \frac{\partial}{\partial x_k} (\rho_\alpha \varphi_\alpha v_k) + \frac{\partial}{\partial x_k} (\rho_\alpha \varphi_\alpha \bar{v}_{\alpha k}) \\ & - \varphi_\alpha \left[\frac{\partial \rho_\alpha}{\partial t} + \frac{\partial}{\partial x_k} (\rho_\alpha v_{\alpha k}) \right] \qquad (!\alpha) \end{aligned} \tag{3.271}$$

where the mass-averaged velocity and the diffusion velocity are given as $v_k = \sum_\alpha c_\alpha v_{\alpha k}$ $(!\alpha)$ and $\bar{v}_{\alpha k} = v_{\alpha k} - v_k$, respectively.

A mass-average of φ_α is given as

$$\varphi = \sum_\alpha c_\alpha \varphi_\alpha, \tag{3.272}$$

and by adding up (3.271), we obtain the following *Fundamental Identity*:

$$\begin{aligned} \sum_\alpha c_\alpha \frac{d_{(\alpha)} \varphi_\alpha}{dt} = {} & \frac{d\varphi}{dt} + \frac{\varphi}{\rho} \left[\frac{\partial \rho}{\partial t} + \frac{\partial}{\partial x_k} (\rho v_k) \right] + \sum_\alpha \frac{1}{\rho} \frac{\partial}{\partial x_k} (\rho_\alpha \varphi_\alpha \bar{v}_{\alpha k}) \\ & - \sum_\alpha \frac{\varphi_\alpha}{\rho} \left[\frac{\partial \rho_\alpha}{\partial t} + \frac{\partial}{\partial x_k} (\rho_\alpha v_{\alpha k}) \right] \qquad (!\alpha). \end{aligned} \tag{3.273}$$

If we have the mass conservation law (3.248) and (3.250), then (3.273) is modified as follows:

$$\sum_\alpha c_\alpha \frac{d_{(\alpha)} \varphi_\alpha}{dt} = \frac{d\varphi}{dt} + \sum_\alpha \frac{1}{\rho} \frac{\partial}{\partial x_k} (\rho_\alpha \varphi_\alpha \bar{v}_{\alpha k}) - \sum_\alpha \varphi_\alpha \gamma_\alpha \qquad (!\alpha). \quad \blacksquare \tag{3.274}$$

By dividing (3.270) by ρ and applying the Fundamental Identity (3.274), we have

$$\sum_\alpha c_\alpha \frac{d_{(\alpha)} \boldsymbol{v}_\alpha}{dt} = \frac{d\boldsymbol{v}}{dt} + \sum_\alpha \frac{1}{\rho} \left[\text{div } (\rho_\alpha \bar{\boldsymbol{v}}_\alpha \otimes \bar{\boldsymbol{v}}_\alpha) + \text{div } (\rho_\alpha \boldsymbol{v} \otimes \bar{\boldsymbol{v}}_\alpha) - \rho \gamma_\alpha \boldsymbol{v}_\alpha \right]. \tag{3.275}$$

If we apply the diffusion equation (3.259) and the relation

$$\sum_\alpha \rho_\alpha \gamma_\alpha = 0, \qquad \sum_\alpha c_\alpha = 1 \;\Rightarrow\; \sum_\alpha \frac{dc_\alpha}{dt} = 0, \qquad \sum_\alpha \rho_\alpha \overline{\boldsymbol{v}}_\alpha = 0, \quad (3.276)$$

to the third term of the r.h.s. of (3.275), this term becomes null. Since the third term of the r.h.s. of (3.275) is null, the sum of (3.270) results in the following averaged equation of motion:

$$\rho \frac{d\boldsymbol{v}}{dt} = \operatorname{div} \overline{\boldsymbol{\sigma}} + \rho \overline{\boldsymbol{b}} \qquad (!\alpha) \qquad (3.277)$$

where we have introduced the diffusive stress $\overline{\boldsymbol{\sigma}}$ and the diffusive body force $\overline{\boldsymbol{b}}$ as follows:

$$\overline{\boldsymbol{\sigma}} = \sum_\alpha \left[(\boldsymbol{\sigma}_\alpha)^T - \rho_\alpha \overline{\boldsymbol{v}}_\alpha \otimes \overline{\boldsymbol{v}}_\alpha \right], \qquad \overline{\boldsymbol{b}} = \sum_\alpha (\boldsymbol{b}_\alpha + \gamma_\alpha \overline{\boldsymbol{v}}_\alpha). \qquad (3.278)$$

The conservation of moment of linear momentum for a mixture is introduced in the same way as in Sect. 2.5, and we conclude that the averaged stress $\boldsymbol{\sigma}$ for the mixture is also symmetric:

$$\overline{\boldsymbol{\sigma}} = (\overline{\boldsymbol{\sigma}})^T \quad \Rightarrow \quad \boldsymbol{\sigma} = \boldsymbol{\sigma}^T. \qquad (3.279)$$

3.6.3 *First Law of Thermodynamics in a Thermo-mechanochemical Continuum: Conservation of Energy*

For each component we apply Stokes' power formula (3.18):

$$\frac{d_{(\alpha)}\mathcal{E}_{\sigma\alpha}}{dt} = \frac{d_{(\alpha)}\mathcal{K}_\alpha}{dt} + \frac{d_{(\alpha)}\mathcal{U}_{\sigma\alpha}}{dt} = \frac{d_{(\alpha)}\mathcal{W}_\alpha}{dt}, \qquad (3.280)$$

$$\mathcal{K}_\alpha = \int_\Omega \rho_\alpha \kappa_\alpha \, dv, \qquad \kappa_\alpha = \frac{1}{2} \boldsymbol{v}_\alpha \cdot \boldsymbol{v}_\alpha \qquad (!\alpha), \qquad (3.281)$$

$$\frac{d_{(\alpha)}\mathcal{W}_\alpha}{dt} = \int_{\partial\Omega} \boldsymbol{t}_\alpha \cdot \boldsymbol{v}_\alpha \, ds + \int_\Omega \rho_\alpha \boldsymbol{b}_\alpha \cdot \boldsymbol{v}_\alpha \, dv + \int_\Omega \boldsymbol{\psi}_\alpha \cdot \boldsymbol{v}_\alpha \, dv \qquad (!\alpha), \qquad (3.282)$$

$$\frac{d_{(\alpha)}\mathcal{U}_{\sigma\alpha}}{dt} = \int_\Omega \rho_\alpha \frac{d_{(\alpha)} u_{\sigma\alpha}}{dt} \, dv, \qquad \frac{d_{(\alpha)} u_{\sigma\alpha}}{dt} = \boldsymbol{\sigma}_\alpha^\ddagger : \boldsymbol{D}_\alpha \qquad (!\alpha), \qquad (3.283)$$

$$\boldsymbol{\sigma}_\alpha^\ddagger = \frac{1}{\rho_\alpha} \boldsymbol{\sigma}_\alpha, \quad \boldsymbol{D}_\alpha = \frac{1}{2} \left(\operatorname{grad} \boldsymbol{v}_\alpha + \operatorname{grad} \boldsymbol{v}_\alpha^T \right) \qquad (!\alpha). \qquad (3.284)$$

Applying Reynolds' transport theorem (3.269) to Stokes' power formula (3.280) gives the following local form:

$$\rho_\alpha \left(\frac{d_{(\alpha)}\kappa_\alpha}{dt} + \frac{d_{(\alpha)}u_{\sigma\alpha}}{dt} \right) = \operatorname{div}(\boldsymbol{\sigma}_\alpha \boldsymbol{v}_\alpha) + \rho_\alpha \boldsymbol{b}_\alpha \cdot \boldsymbol{v}_\alpha + \boldsymbol{\psi}_\alpha \cdot \boldsymbol{v}_\alpha - \rho_\alpha \gamma_\alpha \kappa_\alpha \qquad (!\alpha). \tag{3.285}$$

The First Law of Thermodynamics is given by (3.40). That is, if we add a flux to the r.h.s. of (3.280), the reversible (i.e., elastic) internal energy is increased. If this procedure is introduced for the multi-component solution of species α, we have

$$\frac{d_{(\alpha)}\mathcal{E}_\alpha}{dt} = \frac{d_{(\alpha)}\mathcal{K}_\alpha}{dt} + \frac{d_{(\alpha)}\mathcal{U}_\alpha}{dt} = \frac{d_{(\alpha)}\mathcal{W}_\alpha}{dt} + \frac{d_{(\alpha)}\mathcal{Q}_\alpha}{dt} + \frac{d_{(\alpha)}\mathcal{C}_\alpha}{dt}. \tag{3.286}$$

where $\mathcal{U}_\alpha$ gives the reversible internal energy, including mechanical, thermal and chemical effects. The mass density function u_α may be written as

$$\frac{d_{(\alpha)}\mathcal{U}_\alpha}{dt} = \int_\Omega \rho_\alpha \frac{d_{(\alpha)}u_\alpha}{dt}\, dv \qquad (!\alpha). \tag{3.287}$$

The heat flux $d_{(\alpha)}\mathcal{Q}_\alpha/dt$ and mass flux $d_{(\alpha)}\mathcal{C}_\alpha/dt$ are given by

$$\frac{d_{(\alpha)}\mathcal{Q}_\alpha}{dt} = -\int_{\partial\Omega} \boldsymbol{q}_\alpha \cdot \boldsymbol{n}\, ds + \int_\Omega \rho_\alpha r_\alpha\, dv \qquad (!\alpha), \tag{3.288}$$

$$\frac{d_{(\alpha)}\mathcal{C}_\alpha}{dt} = \int_\Omega \mu_\alpha \rho_\alpha \gamma_\alpha\, dv \qquad (!\alpha) \tag{3.289}$$

where $\boldsymbol{q}_\alpha$, r_α, γ_α are the heat flux, the heat source per unit mass and the mass source per unit mass (e.g., by a chemical reaction), respectively. The local form of the First Law of Thermodynamics (3.289) can be obtained as

$$\rho_\alpha \left(\frac{d_{(\alpha)}\kappa_\alpha}{dt} + \frac{d_{(\alpha)}u_\alpha}{dt} \right) = \operatorname{div}(\boldsymbol{\sigma}_\alpha \boldsymbol{v}_\alpha) + \rho_\alpha \boldsymbol{b}_\alpha \cdot \boldsymbol{v}_\alpha + \boldsymbol{\psi}_\alpha \cdot \boldsymbol{v}_\alpha - \operatorname{div} \boldsymbol{q}_\alpha + \rho_\alpha r_\alpha + \rho_\alpha \gamma_\alpha \mu_\alpha - \rho_\alpha \gamma_\alpha \kappa_\alpha \quad (!\alpha). \tag{3.290}$$

By subtracting the Stokes' power formula (3.285) from the First Law of Thermodynamics (3.290), we have the following local form of the dissipative energy equation:

$$\rho_\alpha \frac{d_{(\alpha)}u_\alpha}{dt} = \rho_\alpha \frac{d_{(\alpha)}u_{\sigma\alpha}}{dt} - \operatorname{div} \boldsymbol{q}_\alpha + \rho_\alpha r_\alpha + \rho_\alpha \gamma_\alpha \mu_\alpha \qquad (!\alpha). \tag{3.291}$$

If it can be assumed that in the first term $d_{(\alpha)}u_{\sigma\alpha}/dt = \boldsymbol{\sigma}_\alpha^\ddagger : \boldsymbol{D}_\alpha$ of the r.h.s. of (3.291), the stress tensor $\boldsymbol{\sigma}_\alpha^\ddagger$ is coaxial with the stretch tensor $\boldsymbol{D}_\alpha$, we have

$$\sum_\alpha c_\alpha \boldsymbol{\sigma}_\alpha^\ddagger : \boldsymbol{D}_\alpha = \boldsymbol{\sigma}^\ddagger : \boldsymbol{D}, \tag{3.292}$$

$$\boldsymbol{\sigma}^\ddagger = \sum_\alpha c_\alpha \boldsymbol{\sigma}_\alpha^\ddagger, \quad \boldsymbol{D} = \sum_\alpha \boldsymbol{D}_\alpha \tag{3.293}$$

where $c_\alpha = \rho_\alpha/\rho$. Therefore by summing up the component-wise dissipative energy equation (3.291) with α after dividing ρ, we obtain the following dissipative energy equation for the total components:

$$\rho \frac{du}{dt} = \rho \frac{du_\sigma}{dt} - \operatorname{div} \overline{\boldsymbol{q}} + \rho \overline{r} - \sum_r A_r w_r \tag{3.294}$$

where we have

$$\frac{du_\sigma}{dt} = \boldsymbol{\sigma}^\ddagger : \boldsymbol{D} \tag{3.295}$$

$$\overline{\boldsymbol{q}} = \boldsymbol{q} - \boldsymbol{q}^{mv}; \qquad \boldsymbol{q} = \sum_\alpha \boldsymbol{q}_\alpha, \quad \boldsymbol{q}^{mv} = \sum_\alpha \rho_\alpha u_\alpha \overline{\boldsymbol{v}}_\alpha \tag{3.296}$$

$$\overline{r} = r + r^u + r^\mu; \quad r = \sum_\alpha c_\alpha r_\alpha, \quad r^u = \sum_\alpha u_\alpha \gamma_\alpha, \quad r^\mu = \sum_\alpha c_\alpha \mu_\alpha \gamma_\alpha, \tag{3.297}$$

$$A_r = -\sum_\alpha c_\alpha \nu_{\alpha r}^\dagger \mu_\alpha. \tag{3.298}$$

Note that in obtaining the term du/dt, we have used the Fundamental Identity (3.274). In fact, the terms $\boldsymbol{q}^{me}$ and r^u are due to the Fundamental Identity. The term A_r is an affinity at the stage r of the reaction, which is calculated by multiplying the chemical potential μ_α with (3.261), dividing by the molecular weight m_α and summing up with respect to α. A discussion of affinity is given in Appendix E.3(b).

3.6.4 Entropy Inequality in a Thermo-mechanochemical Continuum

In a multi-component mixture the entropy inequality may not be satisfied for each component. Therefore we introduce the inequality for the total mixture. Once again, as discussed in Sect. 3.3.2, we emphasize that the entropy inequality is introduced for a system in which non-measurable energy components exist.

As described in Sect. 3.2.1, the rate of internal energy due to stress $du_\sigma/dt = \boldsymbol{\sigma}^\ddagger : \boldsymbol{D}$ is a combination of the elastic (i.e., reversible) part du^e_σ/dt and the inelastic (i.e., irreversible) part du^i_σ/dt ($du_\sigma = du^e_\sigma + du^i_\sigma$). Thus the rate of internal energy including the chemical process is written as

$$\frac{du}{dt} = \frac{du^e_\sigma}{dt} + T\frac{ds}{dt} + \sum_\alpha \mu_\alpha \frac{dc_\alpha}{dt} \qquad (!\alpha) \tag{3.299}$$

where we have assumed that the temperature T is common to all components, and the term $ds = \sum_\alpha ds_\alpha$ represents the change of entropy of all components.

By substituting the dissipative energy equation (3.294) into (3.299) and applying the diffusion equations (3.259) and (3.276), we obtain

$$T\frac{ds}{dt} = \frac{du^i_\sigma}{dt} - \frac{1}{\rho}\mathrm{div}\bar{\boldsymbol{q}} + r^c + \zeta^c \tag{3.300}$$

$$r^c = r + r^u, \tag{3.301}$$

$$\zeta^c = \frac{1}{\rho}\sum_\alpha c_\alpha \mu_\alpha \bar{\boldsymbol{v}}_\alpha \cdot \mathrm{grad}\,\rho - \frac{1}{\rho}\sum_r A_r w_r. \tag{3.302}$$

As mentioned in Sect. 3.3, if incomplete data is provided due to insufficient measurement, instead of (3.300) we have the following entropy inequality:

$$T\frac{ds}{dt} \geq \frac{du^i_\sigma}{dt} - \frac{1}{\rho}\mathrm{div}\bar{\boldsymbol{q}} + r^c + \zeta^c \tag{3.303}$$

where du^i_σ/dt represents the internal dissipation due to mechanical force, which corresponds to the dissipation function $\Phi(\boldsymbol{D})$ of $(3.153)_2$ for a fluid, and $r^u + \zeta^c$ represents an internal energy dissipation during the chemical process.

3.7 Thermodynamics Laws and Constitutive Theory ♣

The essentials of the constitutive theory are that the experimental results are reproduced after satisfying the requirements of thermodynamics such as the Clausius-Duhem inequality and the Gibbs-Duhem relation. The constitutive theory that includes chemical processes is complex because all the conservation laws for mass, linear momentum/moment of momentum and energy are involved, and experiments to determine the parameters are extremely difficult to conduct.

3.7.1 Constitutive Theory of a Solid with Chemical Processes in the Small Strain Field

In this subsection we show an example of the thermodynamical restrictions required to derive the constitutive equation for a multi-component solid under small strain conditions with chemical processes included.

First let us satisfy the Gibbs-Duhem relation:

$$\boldsymbol{\varepsilon}^e : d\boldsymbol{\sigma}^\ddagger + s\,dT + \sum_{\alpha=1}^{N} c_\alpha\, d\mu_\alpha = 0. \tag{3.304}$$

The Gibbs free energy per unit mass $g(\boldsymbol{\sigma}^\ddagger, T, c_\alpha)$ is defined by (3.184):

$$g(\boldsymbol{\sigma}^\ddagger, T, c_1, \ldots, c_\alpha, \ldots, c_N) = \sum_\alpha \mu_\alpha\, c_\alpha. \tag{3.305}$$

From this definition it is understood that if we retain the α-th concentration c_α and make the other concentrations null, we have $\mu_\alpha\, c_\alpha$:

$$g(\boldsymbol{\sigma}^\ddagger, T, 0, \ldots, c_\alpha, \ldots, 0) = \mu_\alpha\, c_\alpha. \tag{3.306}$$

Using (3.185) we have

$$\boldsymbol{\varepsilon}^e = -\left.\frac{\partial g}{\partial \boldsymbol{\sigma}^\ddagger}\right|_{T, c_\alpha}, \qquad s = -\left.\frac{\partial g}{\partial T}\right|_{c_\alpha, \boldsymbol{\sigma}^\ddagger} \tag{3.307}$$

and we can define the *partial molar strain* $\boldsymbol{\varepsilon}^e_\alpha$ and *partial molar entropy* s_α as

$$\begin{aligned}
\boldsymbol{\varepsilon}^e_\alpha &= -\frac{\partial}{\partial \boldsymbol{\sigma}^\ddagger} g(\boldsymbol{\sigma}^\ddagger, T, 0, \ldots, c_\alpha, \ldots, 0)\bigg|_{T, c_\alpha} = -\frac{\partial \mu_\alpha}{\partial \boldsymbol{\sigma}^\ddagger} c_\alpha, \\
s_\alpha &= -\frac{\partial}{\partial T} g(\boldsymbol{\sigma}^\ddagger, T, 0, \ldots, c_\alpha, \ldots, 0)\bigg|_{c_\alpha, \boldsymbol{\sigma}^\ddagger} = -\frac{\partial \mu_\alpha}{\partial T} c_\alpha
\end{aligned} \tag{3.308}$$

As implied from (3.305) to (3.308), the overall strain and entropy can be written as

$$\boldsymbol{\varepsilon}^e = \sum_\alpha \boldsymbol{\varepsilon}^e_\alpha = \sum_\alpha -\frac{\partial \mu_\alpha}{\partial \boldsymbol{\sigma}^\ddagger} c_\alpha, \qquad s = \sum_\alpha s_\alpha = -\sum_\alpha \frac{\partial \mu_\alpha}{\partial T} c_\alpha \tag{3.309}$$

On the other hand, Maxwell's relation (3.193) gives

$$-\left.\frac{\partial s}{\partial c_\alpha}\right|_{\boldsymbol{\sigma}^\ddagger, T} = \left.\frac{\partial \mu_\alpha}{\partial T}\right|_{c_\alpha, \boldsymbol{\sigma}^\ddagger}, \qquad -\left.\frac{\partial \boldsymbol{\varepsilon}^e}{\partial c_\alpha}\right|_{\boldsymbol{\sigma}^\ddagger, T} = \left.\frac{\partial \mu_\alpha}{\partial \boldsymbol{\sigma}^\ddagger}\right|_{T, c_\alpha} \tag{3.310}$$

Then if we introduce $\boldsymbol{d}_\alpha$ by

$$\begin{aligned}\boldsymbol{d}_\alpha &= c_\alpha \left(\nabla \mu_\alpha - \frac{\partial \mu_\alpha}{\partial \boldsymbol{\sigma}^\ddagger} : \nabla \boldsymbol{\sigma}^\ddagger - \frac{\partial \mu_\alpha}{\partial T} \nabla T \right) \\ &= c_\alpha \left(\nabla \mu_\alpha + \frac{\partial \boldsymbol{\varepsilon}^e}{\partial c_\alpha} : \nabla \boldsymbol{\sigma}^\ddagger + \frac{\partial s}{\partial c_\alpha} \nabla T \right) \end{aligned} \tag{3.311}$$

the Gibbs-Duhem relation (3.304) is satisfied if

$$\sum_\alpha \boldsymbol{d}_\alpha = \mathbf{0}. \tag{3.312}$$

We can now introduce the *extended Fick's law* in terms of diffusion as follows:

$$\boldsymbol{q}_\alpha^m = -\sum_\beta \boldsymbol{D}_{\alpha\beta} \boldsymbol{d}_\beta \tag{3.313}$$

where $\boldsymbol{q}_\alpha^m$ is the mass flux of a species α, which is introduced by (3.260), and $\boldsymbol{D}_{\alpha\beta}$ are the material parameters/functions that are determined by experiments. Thus, under the constraint (3.312), we may be able to determine the constitutive relation from experiments. Note that from (3.260) and since $\sum_\alpha \rho_\alpha \bar{\boldsymbol{v}}_\alpha = 0$, the following constraint must be satisfied:

$$\sum_\alpha \boldsymbol{q}_\alpha^m = -\sum_\alpha \sum_\beta D_{\alpha\beta} \nabla c_\beta = 0. \tag{3.314}$$

Following the same procedure we can introduce the extended Fourier's law for heat conduction and the extended Hooke's law for deformation, which will not be described here.

3.8 Summary of the Framework of Non-equilibrium Thermodynamics ♣

Carnot and Clausius founded the theoretical basis of classical 'equilibrium' thermodynamics, and it has long been applied to many engineering problems, although there are some inconsistencies. The basis of classical thermodynamics lies in the empirical *fact* that the efficiency of practical heat engines never exceeds unity, that is, any heat engine traces the irreversible process, and eventually the entropy of the heat engine system necessarily increases. Such results, especially the Second Law, have been widely applied even though the true implications are beyond complete comprehension, and sometimes it is suggested that the entropy of the *universe* is ever-increasing. Is this true?

If we enclose a rare gas in a box that is closed, adiabatic and of constant volume (i.e., an isolated system of a rare gas) and keep the steady state at a given temperature, the internal energy does not change, therefore the entropy never increases. This system has no autonomous internal mechanism to increase the entropy.

Let us consider a thermodynamical system. If we can correctly measure and estimate all the fluxes (i.e., power, heat flux, mass flux, etc.) and field variables such as temperature and stress, is the process reversible or irreversible? Actually it is impossible to measure and specify all the flux and field variables.

Considering the above discussions, the **Law of Non-equilibrium Thermodynamics** presented here, in Newtonian mechanics, can be summarized as follows:

1. All the systems of thermodynamics are treated under Newtonian mechanics.
2. *Stokes' power law* is satisfied:

$$d\mathcal{E}_\sigma = d\mathcal{K} + d\mathcal{U}_\sigma = d\mathcal{W} \tag{3.315}$$

where $\mathcal{E}_\sigma$ is the total energy in the absence of the effects of heat, chemical action, etc., $\mathcal{K}$ is the kinetic energy, $\mathcal{U}_\sigma$ is the total internal energy due to the stress, including a reversible (i.e., recoverable) part and an irreversible (i.e., dissipative) part and $\mathcal{W}$ is the external work. The increment $d\mathcal{U}_\sigma$ consists of a reversible part $d\mathcal{U}_\sigma^e$ and an irreversible part $d\mathcal{U}_\sigma^i$:

$$d\mathcal{U}_\sigma = d\mathcal{U}_\sigma^e + d\mathcal{U}_\sigma^i. \tag{3.316}$$

3. If all the flux $d\Sigma$ and power $d\mathcal{W}$ that are exchanged with the surroundings are measurable, then the *conservation of energy law*

$$d\mathcal{E} = d\mathcal{K} + d\mathcal{U} = d\mathcal{W} + d\Sigma \tag{3.317}$$

is satisfied, where $\mathcal{E}$ is the total recoverable energy and $\mathcal{U}$ is the 'reversible internal energy'. The flux $d\Sigma$ includes the heat flux $d\mathcal{Q}$, the mass-induced flux $d\mathcal{C}$ in chemical processes and others except for the power $d\mathcal{W}$.
Using the power law and the conservation of energy law the *dissipative energy equation* is derived as

$$d\mathcal{U} = d\mathcal{U}_\sigma + d\Sigma. \tag{3.318}$$

By substituting (3.316) into (3.318), we obtain

$$d\mathcal{U}_q = d\mathcal{U}_\sigma^i + d\Sigma \tag{3.319}$$

where we give $d\mathcal{U}_q = d\mathcal{U} - d\mathcal{U}_\sigma^e$. The meaning of (3.319) is clearly understandable: The change of internal heat energy $d\mathcal{U}_q$ consists of two parts; i.e., the change of irreversible internal mechanical energy $d\mathcal{U}_\sigma^i$ and the heat supply.

We should clearly understand the fact that Stokes' power law (3.315) involves the change of the irreversible internal mechanical energy $d\mathcal{U}_\sigma^i$, while the energy conservation law is given for the change of the reversible internal energy $d\mathcal{U}$.

4. The change of internal energy $d\mathcal{U}_q$ due to the heat flux $d\mathcal{Q}$, a part of $d\mathcal{U}$, is formed by the temperature T and the change of entropy $d\mathcal{S}$:

$$d\mathcal{U}_q = T\,d\mathcal{S}. \tag{3.320}$$

Note that (3.320) is approved if every heat flux and temperature field are correctly identified. For the system where an amount of non-measurable heat flux $\Delta(d\mathcal{Q})$ is expected, the entropy corresponding to $\Delta(d\mathcal{Q}) = T\,d\mathcal{S}^i$ can be introduced. $d\mathcal{S}^i$ is referred to as the entropy production. In this case, the change of total entropy $d\mathcal{S}$ consists of the changes of entropy production $d\mathcal{S}^i$ and the exchangeable entropy $d\mathcal{S}^e$, which corresponds to the measurable flux $d\mathcal{Q}^M = T\,d\mathcal{S}^e$. Because of diffusivity of heat, $d\mathcal{S}^i$ is nonnegative: i.e.

$$d\mathcal{S} - d\mathcal{S}^i = \frac{d\mathcal{Q}}{T}, \qquad d\mathcal{S}^i \geq 0. \tag{3.321}$$

5. In a homogeneous body at absolute zero temperature the entropy $\mathcal{S}$ is null (*Nernst-Planck theorem*).
6. The internal energy $\mathcal{U}$, the total internal energy $\mathcal{U}_\sigma$ induced by stress and the entropy $\mathcal{S}$ are sufficiently smooth and differentiable, and their density functions exist. The derivatives of $\mathcal{U}$ and $\mathcal{U}_\sigma$ are positive definite.

Discussion 1: Diffusive properties in temperature and mass-diffusion problems. It is commonly accepted that the Second Law supports the observation that there is heat and mass transfer taking place in the presence of a gradient. However, if we have the positiveness definite property of the derivatives of internal energy $\mathcal{U}$ with respect to entropy and concentration, respectively (i.e., resulting in Fourier's law and Fick's law), the same results can be obtained. It is surprising that the diffusive processes and properties in temperature and mass-diffusion problems are discussed on the basis of the Second Law, whereas the positive definiteness property of the internal energy $\mathcal{U}$ under Hooke's law is proved without appeal to the Second Law.

Discussion 2: On viscous fluids. In the motion of viscous fluid, we always observe some dissipation of internal energy. This, however, does not cause an increase of entropy, but contributes to a rise of internal temperature. This is because, at the molecular level, the increase of kinetic energy of the molecules causes a temperature rise. Thus for the fluid motion problem it is fundamentally necessary to consider the temperature field; however, we must discuss whether or not the temperature field can be correctly measured in a fluid, non-invasively.

If we can account for every mechanical and thermal effect, the entropy inequality (3.321) must be regarded as an equality. That is, only the first part of the Second Law of Thermodynamics is needed to prove the existence of entropy.

Chapter 4
Virtual Work Equation, Variational Methods and Energy Principles

In this Chapter we consider the virtual work equation of a static problem and its relationship to the variational method and energy principle.

4.1 Variational Method for a One-dimensional Elastic Problem

First, we consider a simple one-dimensional elastic problem.

4.1.1 Strong Form

The system for the one-dimensional elastic problem can be given by a governing equation (4.1a), the Dirichlet (i.e., displacement) boundary condition (4.1b) and the Neumann (i.e., traction) boundary condition (4.1c). This system is referred to as the *strong form*: [SF].

$$\text{[SF]:}\quad \frac{d\sigma}{dx} + \gamma = 0, \qquad x \in \Omega = (x_1, x_2) \tag{4.1a}$$

$$u(x) = \bar{u} \qquad \text{at } x = x_u \tag{4.1b}$$

$$\sigma n = \bar{t} \qquad \text{at } x = x_t \tag{4.1c}$$

where σ is the stress, $\gamma = \rho b_x$ is the body force per unit volume, u is the displacement, n is the unit outward normal ($n = -1$ at $x = x_1, n = +1$ at $x = x_2$), and x_u and x_t are respectively, the boundaries at which displacements and tractions are specified. The strain

$$\varepsilon = \frac{du}{dx} \tag{4.2}$$

Y. Ichikawa and A.P.S. Selvadurai, *Transport Phenomena in Porous Media*,
DOI 10.1007/978-3-642-25333-1_4, © Springer-Verlag Berlin Heidelberg 2012

is related to the stress σ by Hooke's law:

$$\sigma = E\varepsilon \tag{4.3}$$

4.1.2 Weak Form and the Virtual Work Equation

Let v be an arbitrary function, which is null on the displacement boundary:

$$v = 0 \qquad \text{on } x = x_u \tag{4.4}$$

We multiply (4.1a) by $v(x)$, integrate it and introduce an integration by parts to the first term, then we can use the condition (4.4) and the Neumann boundary condition (4.1c), and we finally obtain the following *weak form* [WF], which is 'equivalent' (see Note 4.1) to the strong form (4.1):

$$\text{[WF]}: \ \bar{t}\, v(x_t) - \int_{x_1}^{x_2} \Big(\sigma(u)\, \zeta(v) - \gamma\, v\Big)\, dx = 0 \qquad \forall v \ \ (v = 0 \ \text{ at } \ x = x_u) \tag{4.5}$$

where $\zeta(v) = dv/dx$.

We can regard the arbitrary function v as a virtual displacement δu, then (4.4) corresponds to a null condition of the virtual displacement δu on x_u. We write a virtual strain as

$$\delta\varepsilon = \frac{d(\delta u)}{dx} = \delta\left(\frac{du}{dx}\right) \tag{4.6}$$

In this notation the weak form (4.5) is equivalent to the following *virtual work equation* [VW]:

$$\text{[VW]}: \ \bar{t}\, \delta u(b) - \int_{x_1}^{x_2} \Big(\sigma(u)\, \delta\varepsilon - \gamma\, \delta u\Big)\, dx = 0 \qquad \forall \delta u \ \ (\delta u = 0 \ \text{ at } \ x = x_u) \tag{4.7}$$

Note 4.1 (Equivalence between the strong form and weak form: Fundamental Lemma of the Variational Problem). Since the weak form (4.5) is derived from the strong form (4.1), the solution of the strong form is exactly the solution of the weak form. The converse is not always true. If the solution of the weak form can be regarded as sufficiently smooth, the converse is true, which is proved by the *Fundamental Lemma of the Variational Problem.* Note that the first term defined on the boundary of (4.5) can be considered separately from the rest of the terms defined in the domain, since v is arbitrary both on the boundary and in the domain $\Omega = (x_1, x_2)$. Therefore we have the following *Fundamental Lemma of the Variational Problem* for a domain Ω:

$$\int_{\Omega} \big(\phi(x)\, v(x)\big)\, dx = 0 \quad \forall v \quad \Rightarrow \quad \phi(x) \equiv 0 \quad \forall x \in \Omega \tag{4.8}$$

where the function ϕ is continuous.[1] We use a proof by contradiction (or *reduction ad absurdum*).

Proof. Assume that the given condition (i.e., the integral $= 0\ \forall v$) is satisfied; however since $\phi(\xi) > 0$ at a point $x = \xi$ in $\Omega = (x_1, x_2)$ (note that if $\phi(\xi) < 0$, we can also apply the procedure for the case where $-\phi(\xi) > 0$). Then there exists a constant $\delta > 0$ such that $\phi(x) > 0\ \forall x \in \Omega_\delta = [\xi - \delta, \xi + \delta]$, since ϕ is continuous. We choose the arbitrary function v as $v(x) > 0, \ \ x \in \Omega_\delta$, thus we have

$$\int_{\Omega_\delta} \big(\phi(x)\, v(x)\big)\, dx > 0$$

This contradicts the given condition, which implies that $\phi(x) \equiv 0$ at every point in Ω. ■

4.1.3 Principle of the Energy Minimization and a Variational Method

The weak form (4.5) can be rewritten in the following alternative form:

$$[\mathrm{WF}]: \ \ l(v) - a(u, v) = 0 \qquad \forall v \ \ (v = 0 \ \text{at} \ x = x_u) \tag{4.9a}$$

$$a(u, v) = \int_{x_1}^{x_2} E \frac{du}{dx}\frac{dv}{dx}\, dx, \quad l(v) = \bar{t}\, v(x_t) - \int_{x_1}^{x_2} \gamma\, v\, dx \tag{4.9b}$$

where $a(\cdot, \cdot)$ is referred to as a *bilinear form*, which is positive definite such that

$$a(v, v) \geq 0 \qquad \forall v \tag{4.10}$$

since $E > 0$.

[1] The continuity of ϕ gives an essential condition for ensuring the equivalence between the strong form and the weak form. We must be particularly careful of this requirement if we apply a numerical method such as the finite element method to a problem in which the continuity condition is not completely satisfied, such as a stress concentration problem at a crack tip.

We define the *energy functional* $\Pi(w)$ as follows:

$$\Pi(w) = l(w) - \frac{1}{2}a(w, w) \qquad \text{where } w = \bar{u} \text{ at } x = x_u \tag{4.11}$$

The *problem of the energy minimization* [EM] is then given by the variational problem

$$\text{[EM]: Find } u \text{ such that } \Pi(w) \geq \Pi(u) \qquad \forall w\,(w(x_u) = \bar{u}) \tag{4.12}$$

From Hooke's law (4.3) the problem of the energy minimization (4.12) is equivalent to the weak form (4.9a). A proof is given as follows: We set the arbitrary function w as

$$w = u \pm h\,v \qquad (h > 0 : \text{constant}) \tag{4.13}$$

Thus v is an arbitrary function which takes a fixed value $v = 0$ only at $x = x_u$. The inequality (4.12) is written as

$$\pm h\left[\bar{t}\,v(x_t) - \int_{x_1}^{x_2}\left(E\frac{du}{dx}\frac{dv}{dx} - \gamma\,v\right)dx\right] - \frac{1}{2}h^2\int_{x_1}^{x_2} E\left(\frac{dv}{dx}\right)^2 dx \geq 0$$

Dividing by h (> 0) and taking the limit $h \to 0$, we have

$$\pm\left[\bar{t}\,v(x_t) - \int_{x_1}^{x_2}\left(E\frac{du}{dx}\frac{dv}{dx} - \gamma\,v\right)dx\right] \geq 0.$$

Since this inequality must be satisfied for both negative and positive cases, we have only one possibility of equality:

$$\bar{t}\,v(x_t) - \int_{x_1}^{x_2}\left(E\frac{du}{dx}\frac{dv}{dx} - \gamma\,v\right)dx = 0 \tag{4.14}$$

Thus the problem (4.12) is equivalent to (4.9a).

Let us retrace the process of the energy minimization (4.12)–(4.14). We define a derivative of the functional Π in a direction v (i.e., a directional derivative toward v) as

$$\langle d\Pi(u), v\rangle = \lim_{h\to 0}\frac{1}{h}\left[\Pi(u + h\,v) - \Pi(u)\right] \tag{4.15}$$

Then the problem of the energy minimization can be symbolically denoted as follows:

$$\langle d\Pi(u), v\rangle = 0 \quad \Rightarrow \quad l(v) - a(u, v) = 0 \tag{4.16}$$

If we follow the style of the variational theory, the above discussions are equivalent to a process to obtain Euler's equation by taking a variation of the functional (4.11), i.e., from the null condition of the first variation of Π, we have

$$\delta\Pi(u) = \bar{t}\,\delta u(x_t) - \int_{x_1}^{x_2} \left(E\frac{du}{dx}\frac{d(\delta u)}{dx} - \gamma\,\delta u \right) dx = 0 \tag{4.17}$$

It is clear that (4.17) is equivalent to the virtual work equation and the weak form. If we recall (4.16) and set $\delta u = v$, we have

$$\delta\Pi(u) \;=\; \langle d\Pi(u), \delta u \rangle = l(\delta u) - a(u, \delta u) = 0 \tag{4.18}$$

Therefore the procedure of taking a variation is equivalent to that of calculating a directional derivative.

4.2 Variational Method for a Three-dimensional Elasticity Problem

We consider a variational method for the classical three-dimensional (3D) elasticity problem for bodies that satisfy static equilibrium.

4.2.1 *Strong Form*

The system of the equations (i.e., the strong form [SF]) is given by a governing equation (4.19a), the displacement boundary condition (4.19b) and the traction boundary condition (4.19c):

$$\text{[SF]:} \quad \nabla \cdot \boldsymbol{\sigma} + \rho \boldsymbol{b} = \boldsymbol{0} \qquad \text{in } \Omega \tag{4.19a}$$

$$\boldsymbol{u} = \bar{\boldsymbol{u}} \qquad \text{on } \partial\Omega_u \tag{4.19b}$$

$$\boldsymbol{\sigma}\boldsymbol{n} = \bar{\boldsymbol{t}} \qquad \text{on } \partial\Omega_t \tag{4.19c}$$

Additionally we have the following relations:
(Displacement-strain relationship)

$$\boldsymbol{\varepsilon} = \frac{1}{2}\left[\nabla\boldsymbol{u} + (\nabla\boldsymbol{u})^T\right] \tag{4.20}$$

(Constitutive law: Hooke's law)

$$\boldsymbol{\sigma} = \boldsymbol{D}\boldsymbol{\varepsilon} \tag{4.21}$$

4.2.2 Weak Form and the Virtual Work Equation

Let $\boldsymbol{v}$ be an arbitrary vector-valued function, which is null on $\partial\Omega_u$:

$$\boldsymbol{v} = \boldsymbol{0} \qquad \text{on } \partial\Omega_u \tag{4.22}$$

By taking an inner product between $\boldsymbol{v}$ and (4.19a), applying Gauss-Green's theorem to the first term, and then applying the conditions (4.19c) and (4.22), we have

$$[\text{WF}]: \quad \int_{\partial\Omega_t} \bar{\boldsymbol{t}} \cdot \boldsymbol{v}\, ds - \int_{\Omega} \left(\boldsymbol{\sigma} : \boldsymbol{\zeta} - \rho \boldsymbol{b} \cdot \boldsymbol{v} \right) dv = 0 \qquad \forall \boldsymbol{v} \;\; (\boldsymbol{v} = \boldsymbol{0} \text{ on } \partial\Omega_u) \tag{4.23}$$

where

$$\boldsymbol{\zeta} = \frac{1}{2}\left[\nabla \boldsymbol{v} + (\nabla \boldsymbol{v})^T\right] \quad \Rightarrow \quad \boldsymbol{\sigma} : \nabla \boldsymbol{v} = \boldsymbol{\sigma} : \frac{1}{2}\left[\nabla \boldsymbol{v} + (\nabla \boldsymbol{v})^T\right] = \boldsymbol{\sigma} : \boldsymbol{\zeta} \tag{4.24}$$

The result (4.23) gives a weak form of the strong form (4.19).

In a manner similar to the one-dimensional problem (4.9), the weak form (4.23) is formally represented as follows:

$$[\text{WF}]: \quad a(\boldsymbol{u}, \boldsymbol{v}) = l(\boldsymbol{v}) \qquad \forall \boldsymbol{v} \;\; (\boldsymbol{v} = \boldsymbol{0} \text{ on } \partial\Omega_u) \tag{4.25a}$$

$$a(\boldsymbol{u}, \boldsymbol{v}) = \int_{\Omega} \boldsymbol{D}\boldsymbol{\varepsilon}(\boldsymbol{u}) : \boldsymbol{\zeta}(\boldsymbol{v})\, dv, \tag{4.25b}$$

$$l(\boldsymbol{v}) = \int_{\partial\Omega_t} \bar{\boldsymbol{t}} \cdot \boldsymbol{v}\, ds + \int_{\Omega} \rho \boldsymbol{b} \cdot \boldsymbol{v}\, dv. \tag{4.25c}$$

The bilinear form is also positive definite:

$$a(\boldsymbol{v}, \boldsymbol{v}) \geq 0 \qquad \forall \boldsymbol{v}. \tag{4.26}$$

If we regard the arbitrary function $\boldsymbol{v}$ as a virtual displacement $\delta\boldsymbol{u}$ (i.e., $\boldsymbol{v} = \delta\boldsymbol{u}$), the condition (4.22) corresponds to

$$\delta\boldsymbol{u} = \boldsymbol{0} \qquad \text{on } \partial\Omega_u \tag{4.27}$$

Then the term $\boldsymbol{\zeta} = \left[\delta(\nabla\boldsymbol{u}) + \delta(\nabla\boldsymbol{u})^T\right]/2 = \delta\boldsymbol{\varepsilon}$ is regarded as a virtual strain. Therefore (4.23) can be rewritten as

$$[\text{VW}]: \quad \int_{\partial\Omega_t} \bar{\boldsymbol{t}} \cdot \delta\boldsymbol{u}\, ds + \int_{\Omega} \rho \boldsymbol{b} \cdot \delta\boldsymbol{u}\, dv = \int_{\Omega} \boldsymbol{\sigma} : \delta\boldsymbol{\varepsilon}\, dv \tag{4.28}$$

The result (4.28) is known as the *virtual work equation*.

4.2.3 Principle of the Energy Minimization and a Variational Method

An energy functional is given by

$$\Pi(\boldsymbol{w}) = \int_{\partial\Omega_t} \bar{\boldsymbol{t}} \cdot \boldsymbol{w} \, ds - \int_{\Omega} \left(\frac{1}{2} \boldsymbol{D}\boldsymbol{\xi} : \boldsymbol{\xi} - \rho \boldsymbol{b} \cdot \boldsymbol{w} \right) dv \qquad \text{where } \boldsymbol{w} = \bar{\boldsymbol{u}} \text{ on } \partial\Omega_u \tag{4.29}$$

where we should note that $\boldsymbol{D} = \boldsymbol{D}^T$, and

$$\boldsymbol{\xi} = \frac{1}{2}\Big[\nabla\boldsymbol{w} + (\nabla\boldsymbol{w})^T\Big]$$

The problem of the energy minimization [EM] is defined as follows:

$$\text{[EM]: Find } \boldsymbol{u} \text{ such that } \Pi(\boldsymbol{w}) \geq \Pi(\boldsymbol{u}) \qquad \forall \boldsymbol{w}\,(\boldsymbol{w} = \bar{\boldsymbol{u}} \text{ on } \partial\Omega_u) \tag{4.30}$$

It is not difficult to show that the problem of the energy minimization is ‘equivalent’ to the weak form if we recall the procedure used in the one-dimensional problem.

The directional derivative along $\boldsymbol{v}$ is given by

$$\langle d\Pi(\boldsymbol{u}), \boldsymbol{v}\rangle = \lim_{h\to 0} \frac{1}{h}\Big[\Pi(\boldsymbol{u} + h\boldsymbol{v}) - \Pi(\boldsymbol{u})\Big]. \tag{4.31}$$

Then the energy minimization problem becomes

$$\langle d\Pi(\boldsymbol{u}), \boldsymbol{v}\rangle = \int_{\partial\Omega_t} \bar{\boldsymbol{t}}\cdot\boldsymbol{v}\, ds - \int_{\Omega} \Big(\boldsymbol{D}\boldsymbol{\varepsilon}(\boldsymbol{u}) : \boldsymbol{\zeta}(\boldsymbol{v}) - \rho\boldsymbol{b}\cdot\boldsymbol{v}\Big)\, dv = 0 \ \ \forall \boldsymbol{v}\,(\boldsymbol{v} = \boldsymbol{0} \text{ on } \partial\Omega_u) \tag{4.32}$$

where

$$\boldsymbol{\varepsilon}(\boldsymbol{u}) = \frac{1}{2}\Big[\nabla\boldsymbol{u} + (\nabla\boldsymbol{u})^T\Big], \qquad \boldsymbol{\zeta}(\boldsymbol{v}) = \frac{1}{2}\Big[\nabla\boldsymbol{v} + (\nabla\boldsymbol{v})^T\Big]$$

The proof is similar to the one-dimensional problem given by (4.12)–(4.14).

If we set $\delta\boldsymbol{u} = \boldsymbol{v}$, $\delta\boldsymbol{\varepsilon} = \boldsymbol{\zeta}$, the variational form can be obtained as

$$\delta\Pi(\boldsymbol{u}) = \langle d\Pi(\boldsymbol{u}), \delta\boldsymbol{u}\rangle = \int_{\partial\Omega_t} \bar{\boldsymbol{t}} \cdot \delta\boldsymbol{u}\, ds - \int_{\Omega} \big(\boldsymbol{D}\boldsymbol{\varepsilon} : \delta\boldsymbol{\varepsilon} - \rho\boldsymbol{b}\cdot\delta\boldsymbol{u}\big)\, dv = 0. \tag{4.33}$$

4.3 The Penalty Method and the Lagrangian Multiplier Method

Difficulties can be encountered when applying the weak form and the variational method described in the previous sections and in such cases the solution must be sought in the functions that are constrained on the displacement boundary (i.e., $\boldsymbol{u} = \bar{\boldsymbol{u}}$ on $\partial\Omega_u$). We can remove this constraint by using a *penalty method* and a *Lagrangian multiplier method*.

First, we illustrate a *penalty variational method*, that is, a method of the penalized energy minimization [PEM]:

$$\text{[PEM]: Find } \boldsymbol{u} \text{ such that } \Pi_p(\boldsymbol{w}) \geq \Pi_p(\boldsymbol{u}) \quad \forall \boldsymbol{w} \tag{4.34a}$$

$$\Pi_p(\boldsymbol{w}) = \int_{\partial\Omega_t} \bar{\boldsymbol{t}} \cdot \boldsymbol{w}\, ds - \int_{\Omega} \left(\frac{1}{2} \boldsymbol{D} \boldsymbol{\xi} : \boldsymbol{\xi} - \rho \boldsymbol{b} \cdot \boldsymbol{w} \right) dv + \frac{1}{2\delta} \int_{\partial\Omega_u} (\boldsymbol{w} - \bar{\boldsymbol{u}}) \cdot (\boldsymbol{w} - \bar{\boldsymbol{u}})\, ds \tag{4.34b}$$

where δ is a positive constant such that $0 < \delta \ll 1$.

Note that in the penalized variational method we seek the solution $\boldsymbol{u}$ from arbitrary functions $\boldsymbol{w}$, which are unconstrained. It is observed that if we do not have the condition $\boldsymbol{u} - \bar{\boldsymbol{u}} = \boldsymbol{0}$ on $\partial\Omega_u$, the final term of (4.34b) gives a large value, therefore the energy functional Π_p cannot approach a minimum, which is why this is designated as the penalty method.

By calculating the directional derivative with $\boldsymbol{w} = \boldsymbol{u} \pm h\boldsymbol{v}$, the *penalized weak form* [PWF] is obtained as follows:

$$\text{[PWF]}: \ \langle d\Pi(\boldsymbol{u}), \boldsymbol{v} \rangle = \int_{\partial\Omega_t} \bar{\boldsymbol{t}} \cdot \boldsymbol{v}\, ds + \frac{1}{\delta} \int_{\partial\Omega_u} (\boldsymbol{u} - \bar{\boldsymbol{u}}) \cdot \boldsymbol{v}\, ds - \int_{\Omega} (\boldsymbol{D}\boldsymbol{\varepsilon} : \boldsymbol{\zeta} - \rho \boldsymbol{b} \cdot \boldsymbol{v})\, dv = 0 \quad \forall \boldsymbol{v} \tag{4.35}$$

By applying the divergence theorem to (4.35), the *penalized strong form* [PSF], which is equivalent to (4.35), is given by the following governing equation (4.36a), displacement boundary condition (4.36b) and traction boundary condition (4.36c):

$$\text{[PSF]}: \ \nabla \cdot \boldsymbol{\sigma} + \rho \boldsymbol{b} = \boldsymbol{0} \quad \text{in } \Omega \tag{4.36a}$$

$$\boldsymbol{t} = \frac{1}{\delta} (\boldsymbol{u} - \bar{\boldsymbol{u}}) \quad \text{on } \partial\Omega_u \tag{4.36b}$$

$$\boldsymbol{\sigma}^T \boldsymbol{n} = \bar{\boldsymbol{t}} \quad \text{on } \partial\Omega_t \tag{4.36c}$$

The condition (4.36b) on $\partial\Omega_u$ gives $\boldsymbol{u} = \bar{\boldsymbol{u}}$ when $\delta \to 0$, therefore the displacement boundary condition is satisfied.

An *energy minimization method using the Lagrangian multiplier* [LEM] can be introduced as follows:

$$\text{[LEM]: Find } \boldsymbol{u} \text{ and } \boldsymbol{\lambda} \text{ such that } \Pi_l(\boldsymbol{w}, \boldsymbol{\mu}) \geq \Pi_l(\boldsymbol{u}, \boldsymbol{\lambda}) \quad \forall \boldsymbol{w}, \boldsymbol{\mu} \tag{4.37a}$$

$$\Pi_l(\boldsymbol{w}, \boldsymbol{\mu}) = \int_{\partial\Omega_t} \bar{\boldsymbol{t}} \cdot \boldsymbol{w}\, ds - \int_{\Omega} \left(\frac{1}{2} \boldsymbol{D} \boldsymbol{\xi} : \boldsymbol{\xi} - \rho \boldsymbol{b} \cdot \boldsymbol{w} \right) dv + \int_{\partial\Omega_u} \boldsymbol{\mu} \cdot (\boldsymbol{w} - \bar{\boldsymbol{u}})\, ds \tag{4.37b}$$

Note that δ in the penalty method was constant, whereas $\boldsymbol{\mu}$ is a (vector-valued) function, which is introduced as the dual of the displacement that corresponds to a reactive force on the displacement boundary. We see that, as in the penalty method, the solution is sought from all functions $\boldsymbol{w}$ without constraint.

By calculating the directional derivative as $\boldsymbol{w} = \boldsymbol{u} \pm h\boldsymbol{v}$, $\boldsymbol{\mu} = \boldsymbol{\lambda} \pm h\boldsymbol{\upsilon}$, we obtain the following *Lagrangian weak form* [LWF]:

$$
\begin{aligned}
\text{[LWF]}: \quad \langle d\Pi(\boldsymbol{u}), \boldsymbol{v}\rangle &= \int_{\partial\Omega_t} \bar{\boldsymbol{t}} \cdot \boldsymbol{v}\, ds + \int_{\partial\Omega_u} (\boldsymbol{u} - \bar{\boldsymbol{u}}) \cdot \boldsymbol{\upsilon}\, ds \\
&+ \int_{\partial\Omega_u} \boldsymbol{\lambda} \cdot \boldsymbol{v}\, ds - \int_{\Omega} \left(\boldsymbol{D}\boldsymbol{\varepsilon} : \boldsymbol{\zeta} - \rho \boldsymbol{b} \cdot \boldsymbol{v}\right) dv = 0 \quad \forall \boldsymbol{v},\ \boldsymbol{\upsilon}
\end{aligned}
\tag{4.38}
$$

Applying the divergence theorem to the fourth term, we obtain

$$
\int_{\partial\Omega_t} (\bar{\boldsymbol{t}} - \boldsymbol{t}) \cdot \boldsymbol{v}\, ds + \int_{\partial\Omega_u} \left[(\boldsymbol{u} - \bar{\boldsymbol{u}}) \cdot \boldsymbol{\upsilon} - (\boldsymbol{t} - \boldsymbol{\lambda}) \cdot \boldsymbol{v}\right] \cdot \boldsymbol{v}\, ds + \int_{\Omega} (\nabla \cdot \sigma - \rho \boldsymbol{b}) \cdot \boldsymbol{v}\, dv = 0.
$$

Thus the *Lagrangian strong form* [LSF] which is equivalent to the weak form (4.38) is given by

$$
\text{[LSF]:} \quad \nabla \cdot \boldsymbol{\sigma} + \rho \boldsymbol{b} = \boldsymbol{0} \qquad \text{in } \Omega \tag{4.39a}
$$

$$
\boldsymbol{u} = \bar{\boldsymbol{u}}, \quad \boldsymbol{t} = \boldsymbol{\lambda} \qquad \text{on } \partial\Omega_u \tag{4.39b}
$$

$$
\boldsymbol{\sigma}^T \boldsymbol{n} = \bar{\boldsymbol{t}} \qquad \text{on } \partial\Omega_t \tag{4.39c}
$$

The condition (4.39b) shows that $\boldsymbol{\lambda}$ corresponds to the traction on the displacement boundary $\partial\Omega_u$.

4.4 Convolution Integral and Energy Forms for Parabolic and Hyperbolic PDEs

In a conventional variational scheme, we cannot introduce an energy functional for a parabolic type of partial differential equation (PDE), since the first order partial differential term of time is involved. In order to overcome this problem (formally), we introduce a method based on the convolution integral.

The *convolution integral* of functions $\varphi(t)$ and $\psi(t)$ is defined by

$$
(\varphi * \psi)(t) = \int_0^t \varphi(t - \tau)\, \psi(\tau)\, d\tau = \int_0^t \psi(t - \tau)\, \varphi(\tau)\, d\tau. \tag{4.40}
$$

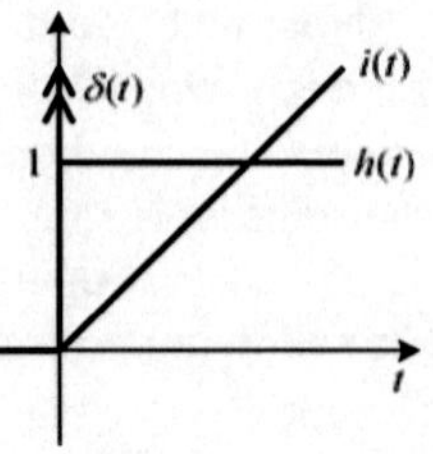

Fig. 4.1 Functions $i(t)$ and $h(t)$, and their generalized derivatives

Let the function $i(t)$ be defined by

$$i(t) = \begin{cases} 0 & \text{if} \quad t < 0, \\ t & \text{if} \quad t \geq 0. \end{cases} \tag{4.41}$$

A generalized derivative[2] of $i(t)$ is the Heaviside step function $h(t)$, and the derivative of the step function is the delta function (cf. Fig. 4.1). The proof is given as follows:

$$\langle i', \phi \rangle = \int_{-\infty}^{\infty} \frac{di}{dt}\phi(t)\,dt = -\int_{0}^{\infty} t\frac{d\phi}{dt}\,dt = \int_{0}^{\infty} \phi(t)\,dt = \langle h, \phi \rangle\,, \tag{4.42}$$

$$\langle h', \phi \rangle = \int_{-\infty}^{\infty} \frac{dh}{dt}\phi(t)\,dt = -\int_{0}^{\infty} \frac{d\phi}{dt}\,dt = \phi(0) = \int_{0}^{\infty} \delta(t)\,\phi(t-0)\,dt = \langle \delta, \phi \rangle\,. \tag{4.43}$$

We can calculate convolutions between a function $\varphi(t)$ and $i(t)$, $h(t)$:

$$h * \frac{d\varphi}{dt} = \int_0^t 1\frac{d\varphi(\tau)}{d\tau}\,d\tau = \varphi(t) - \varphi(0), \tag{4.44}$$

$$\begin{aligned} i * \frac{d\varphi}{dt} &= \int_0^t (t-\tau)\frac{d\varphi(\tau)}{d\tau}\,d\tau = (t-\tau)\varphi(\tau)\Big|_0^t \\ &+ \int_0^t 1\cdot\varphi(\tau)\,d\tau = h * \varphi(t) - t\,\varphi(0), \end{aligned} \tag{4.45}$$

---

[2]Let f be a general function that is bounded ($||f|| < \infty$) but may not be smooth, whereas ϕ is a sufficiently smooth function where $\phi(t) \to 0$ if $t \to \pm\infty$. Then a *generalized derivative* f' can be defined by

$$\langle f', \phi \rangle = \int_{-\infty}^{\infty} f'(t)\,\phi(t)\,dt = f(t)\,\phi(t)\Big|_{-\infty}^{\infty} - \int_{-\infty}^{\infty} f(t)\,\phi(t)'(t)\,dt = -\langle f, \phi' \rangle\,.$$

$$i * \frac{d^2\varphi}{dt^2} = \int_0^t (t-\tau)\frac{d^2\varphi(\tau)}{d\tau^2}\,d\tau = (t-\tau)\frac{d\varphi(\tau)}{d\tau}\Big|_0^t$$
$$+\int_0^t 1\cdot\frac{d\varphi(\tau)}{d\tau}\,d\tau = \varphi(t) - t\,\dot{\varphi}(0) - \varphi(0). \tag{4.46}$$

4.4.1 Energy Form of a Parabolic PDE

The strong form of a parabolic PDE, that is, the governing equation (for example, the classical heat conduction equation with constant heat c, variable thermal conductivity k and heat production f; see, e.g., Carslaw and Jaeger 1959; Selvadurai 2000a), the boundary conditions (BC) and the initial conditions (IC) are given as follows:

(Strong form: [SF])

$$c\frac{\partial u}{\partial t} - \nabla\cdot(k\nabla u) - f = 0 \qquad \text{in } \Omega \tag{4.47a}$$

$$\text{BC:}\quad u(\boldsymbol{x},t) = \bar{u}(t) \qquad \text{on } \partial\Omega_u, \tag{4.47b}$$

$$-k\nabla u\cdot\boldsymbol{n} = \bar{q}(t) \qquad \text{on } \partial\Omega_q, \tag{4.47c}$$

$$\text{IC:}\quad u(\boldsymbol{x},0) = u_0(\boldsymbol{x}) \qquad \text{in } \Omega. \tag{4.47d}$$

A convolution integral between (4.47a) and the step function $h(t)$ is given by

$$h * \left[c\frac{\partial u}{\partial t} - \nabla\cdot(k\nabla u) - f\right] = c\,u(\boldsymbol{x},t) - h * [\nabla\cdot(k\nabla u)] - \bar{f} = 0, \tag{4.48a}$$

$$\bar{f} = h * (f + c\,u_0). \tag{4.48b}$$

Therefore the weak form and the variational principle (i.e., the principle of the energy minimization) of the parabolic PDE are obtained as follows:

(Weak form: [WF])

$$\int_{\partial\Omega_q} h * (\bar{q}\,v)\,ds + \int_\Omega \left[c\,u\,v + h * (k\nabla u\cdot\nabla v) - \bar{f}\,v\right] dv = 0 \quad \forall v\ (v=0 \text{ on } \partial\Omega_u). \tag{4.49}$$

(Variational principle: [EM])

$$\text{Find } \boldsymbol{u} \text{ such that } \Pi(w) \geq \Pi(u) \qquad \forall w \; (w = \bar{u} \text{ on } \partial\Omega_u) \tag{4.50a}$$

$$\Pi(w) = \int_{\partial\Omega_q} h * (\bar{q}\, w)\, ds + \int_{\Omega} \left[\frac{1}{2} c\, w^2 + \frac{1}{2} h * (k \nabla w \cdot \nabla w) - \bar{f}\, w \right] dv \tag{4.50b}$$

4.4.2 Energy Form of a Hyperbolic PDE

The strong form of a hyperbolic PDE, that is, the governing equation, the boundary conditions (BC), the displacement-strain relation (DS) and the initial conditions (IC) are given as follows (for example, the equation of classical elasto-dynamics for a Hookean solid with mass density ρ and elasticity constant tensor $\boldsymbol{D}$; see, e.g., Gurtin 1972; Selvadurai 2000b):

(Strong form: [SF])

$$\rho \frac{d^2 \boldsymbol{u}}{dt^2} - \nabla \cdot \boldsymbol{\sigma} - \boldsymbol{b} = \boldsymbol{0} \qquad \text{in } \Omega \tag{4.51a}$$

$$\text{BC:} \quad \boldsymbol{u}(\boldsymbol{x}, t) = \bar{\boldsymbol{u}}(t) \qquad \text{on } \partial\Omega_u, \tag{4.51b}$$

$$\boldsymbol{\sigma n} = \bar{\boldsymbol{t}}(t) \qquad \text{on } \partial\Omega_t, \tag{4.51c}$$

$$\text{IC:} \quad \boldsymbol{u}(\boldsymbol{x}, 0) = \boldsymbol{u}_0(\boldsymbol{x}) \qquad \text{in } \Omega, \tag{4.51d}$$

$$\dot{\boldsymbol{u}} = \bar{\boldsymbol{v}}_0 \qquad \text{in } \Omega, \tag{4.51e}$$

$$\text{DS:} \quad \boldsymbol{\varepsilon} = \left[\nabla \boldsymbol{u} + \nabla \boldsymbol{u}^T \right] / 2, \tag{4.51f}$$

$$\text{HK:} \quad \boldsymbol{\sigma} = \boldsymbol{D\varepsilon}. \tag{4.51g}$$

A convolution integral between (4.51a) and the step function $i(t)$ is given by

$$i * \left[\rho \frac{d^2 \boldsymbol{u}}{dt^2} - \nabla \cdot \boldsymbol{\sigma} - \boldsymbol{b} \right] = \rho \boldsymbol{u}(\boldsymbol{x}, t) - i * (\nabla \cdot \boldsymbol{\sigma}) - \bar{\boldsymbol{b}} = \boldsymbol{0}, \tag{4.52a}$$

$$\bar{\boldsymbol{b}} = i * \boldsymbol{b} + \rho\, (t\, \boldsymbol{v}_0 + \boldsymbol{u}_0) \tag{4.52b}$$

Therefore the weak form and the variational principle (i.e., the principle of the energy minimization) of the hyperbolic PDE are obtained as follows:

(Weak form: [WF])

$$\int_{\partial\Omega_t} i * (\bar{\boldsymbol{t}} \cdot \boldsymbol{v})\, ds + \int_{\Omega} \left[\rho \boldsymbol{u} \cdot \boldsymbol{v} + i * (\boldsymbol{\sigma} : \nabla \boldsymbol{v}) - \bar{\boldsymbol{b}} \cdot \boldsymbol{v} \right] dv = 0 \quad \forall \boldsymbol{v} \; (\boldsymbol{v} = \boldsymbol{0} \text{ on } \partial\Omega_u). \tag{4.53}$$

(Variational principle: [EM])

$$\text{Find } \boldsymbol{u} \text{ such that } \Pi(\boldsymbol{w}) \geq \Pi(\boldsymbol{u}) \qquad \forall \boldsymbol{w}\,(\boldsymbol{w} = \bar{\boldsymbol{u}} \text{ on } \partial\Omega_u) \tag{4.54a}$$

$$\Pi(\boldsymbol{w}) = \int_{\partial\Omega_t} i * (\bar{\boldsymbol{t}} \cdot \boldsymbol{w})\, ds + \int_{\Omega} \left[\frac{1}{2} \rho \boldsymbol{w} \cdot \boldsymbol{w} + \frac{1}{2} i * \{ \boldsymbol{D}\boldsymbol{\varepsilon}(\boldsymbol{w}) : \boldsymbol{\varepsilon}(\boldsymbol{w}) \} - \bar{\boldsymbol{b}} \cdot \boldsymbol{w} \right] dv. \tag{4.54b}$$

4.5 Interpolation, Approximation and Galerkin Method

An interpolation of a function ϕ is introduced when a sequence of functions is selected, and the function ϕ is approximately represented by the sequence. For example, let us consider a function ϕ such that

$$\phi(x) = \begin{cases} 0, & \text{for } -1 \leq x < 0 \\ \sin \pi x, & \text{for } 0 \leq x \leq 1 \end{cases} \tag{4.55}$$

We interpolate ϕ by the following sequence of polynomials:

$$e_1 = \frac{1}{\sqrt{2}}, \quad e_2 = \sqrt{\frac{3}{2}} x, \quad e_3 = \sqrt{\frac{8}{5}} (3x^2 - 1), \quad e_4 = \sqrt{\frac{7}{8}} (5x^3 - 3x), \quad \cdots \tag{4.56}$$

Here $\{e_i\}$ $(i = 1, 2, 3, \cdots)$ give a sequence of interpolation functions. Note that the sequence of functions (4.56) is referred to as the Legendre polynomials, which are orthonormal such that

$$\langle e_i, e_j \rangle = \int_{-1}^{1} e_i\, e_j\, dx = \begin{cases} 0, & \text{for } i \neq j \\ 1, & \text{for } i = j \end{cases} \tag{4.57}$$

where we have introduced an inner product of functions f and g by

$$\langle f, g \rangle = \int_{-1}^{1} f(x)\, g(x)\, dx \tag{4.58}$$

Then the norm of a function f can be calculated by

$$\| f \| = \langle f, f \rangle^{1/2} \tag{4.59}$$

It is not necessary to use the Legendre polynomials for the approximation of the function (4.55); the following simple polynomials can also be employed:

$$e_1 = 1, \quad e_2 = x, \quad e_3 = x^2, \quad e_4 = x^3, \quad e_5 = x^4, \quad \cdots \tag{4.60}$$

However we should note that the polynomials provided by (4.60) are not orthonormal ($\langle e_i, e_j \rangle \neq \delta_{ij}$). We can use other sequences of functions provided they have a Fourier basis (i.e., sequence of trigonometric functions).[3]

By using the provided sequence of interpolation functions $\{e_i\}$ $(i = 1, 2, 3, \cdots)$, a function $\phi(x)$ can be interpolated by a linear combination:

$$\phi(x) \simeq \phi_h(x) = \phi_1 e_1 + \phi_2 e_2 + \phi_3 e_3 + \cdots = \sum_i \phi_i e_i. \tag{4.61}$$

The coefficients $\phi_1, \phi_2, \phi_3, \cdots$ can then be determined by an appropriate approximation method, such as the least squares method. The procedure for the least squares method is summarized as follows: A square error between a given function ϕ and the approximation ϕ_h based on the interpolation functions $\{e_i\}$ is defined by

$$E(\phi_i) = \frac{1}{2} \langle \phi - \phi_h, \phi - \phi_h \rangle = \frac{1}{2} \int_{-1}^{1} \left[\phi(x) - \phi_h\right]^2(x)\, dx \tag{4.62}$$

The minimum of this error can be attained if

$$\frac{\partial E}{\partial \phi_1} = 0, \qquad i = 1, 2, 3, \cdots \tag{4.63}$$

The approximation (4.61) is then substituted into the square error (4.62), giving

$$\begin{aligned} E(\phi_i) &= \frac{1}{2} \left\langle \phi - \sum_i \phi_i e_i, \phi - \sum_j \phi_j e_j \right\rangle \\ &= \frac{1}{2} \langle \phi, \phi \rangle - \sum_i \phi_i \langle \phi, e_i \rangle + \frac{1}{2} \sum_i \sum_j \phi_i \phi_j \langle e_i, e_j \rangle \end{aligned} \tag{4.64}$$

where we have used

$$\sum_i \phi_i \langle e_i, \phi \rangle = \sum_j \phi_j \langle \phi, e_j \rangle = \sum_i \phi_i \langle \phi, e_i \rangle .$$

If $\{e_i\}$ are orthonormal, the square error (4.64) is reduced to

$$E(\phi_i) = \frac{1}{2} \langle \phi, \phi \rangle - \sum_i \phi_i \langle \phi, e_i \rangle + \frac{1}{2} \sum_i \phi_i^2, \tag{4.65}$$

[3]Sequences such as (4.56) form a basis of the function space which consists of integrable functions whose inner product is introduced by (4.58), whereas the integral is defined in the sense of Lebesgue but not of Riemann.

and from (4.63) the coefficients ϕ_i can be easily obtained as

$$\phi_i = \langle \phi, e_i \rangle . \tag{4.66}$$

On the other hand if $\{e_i\}$ are non-orthogonal, the equation of error minimization (4.63) can be written as

$$\frac{\partial E}{\partial \phi_k} = -\langle \phi, e_k \rangle + \sum_i \phi_i \langle e_i, e_k \rangle = 0 \qquad k = 1, 2, 3, \cdots$$

Thus in order to determine the coefficients $\{\phi_i\}$ $(i = 1, 2, 3, \cdots)$, we must solve the following simultaneous equations:

$$\begin{bmatrix} \langle e_1, e_1 \rangle & \langle e_1, e_2 \rangle & \cdots & \langle e_1, e_n \rangle \\ \langle e_2, e_1 \rangle & \langle e_2, e_2 \rangle & \cdots & \langle e_2, e_n \rangle \\ \vdots & \vdots & & \vdots \\ \langle e_n, e_1 \rangle & \langle e_n, e_2 \rangle & \cdots & \langle e_n, e_n \rangle \end{bmatrix} \begin{bmatrix} \phi_1 \\ \phi_2 \\ \vdots \\ \phi_n \end{bmatrix} = \begin{bmatrix} \langle \phi, e_1 \rangle \\ \langle \phi, e_2 \rangle \\ \vdots \\ \langle \phi, e_n \rangle \end{bmatrix} . \tag{4.67}$$

In fact, we consider the function (4.55) and determine the coefficients ϕ_i based on the Legendre polynomials (4.56) using the least squares method:

$$\begin{aligned} \phi_i &= \int_0^1 \sin \pi x \cdot e_i \, dx \\ &= \left[\frac{\sqrt{2}}{\pi}, \sqrt{\frac{3}{2}}\frac{1}{\pi}, \frac{\sqrt{5}}{\pi} - \frac{12\sqrt{5}}{\pi^3}, \sqrt{\frac{7}{2}}\left(\frac{1}{\pi} - \frac{15}{\pi^3}\right), \sqrt{\frac{9}{92}}\left(\frac{11}{\pi} - \frac{300}{\pi^3} + \frac{1680}{\pi^5}\right), \cdots \right] \end{aligned}$$

In Fig. 4.2 we show the approximation functions $\phi_h(x)$ and the original function $\phi(x)$.

For the system involving a differential equation, we can apply the same procedure. For example, let us consider the following system:

$$-L(\phi) - f = -k\frac{d^2\phi}{dx^2} - f = 0, \tag{4.68a}$$

$$\text{Dirichlet BC:} \quad \phi = \bar{\phi} \quad \text{at } x = x_1 \tag{4.68b}$$

$$\text{Neumann BC:} \quad -k\frac{d\phi}{dx} = \bar{q} \quad \text{at } x = x_2 \tag{4.68c}$$

Assuming that the following sequence of interpolation functions satisfy the boundary conditions (4.68b) and (4.68c):

$$\phi \simeq \phi_h = \sum_\alpha \phi_\alpha N_\alpha(x) \tag{4.69}$$

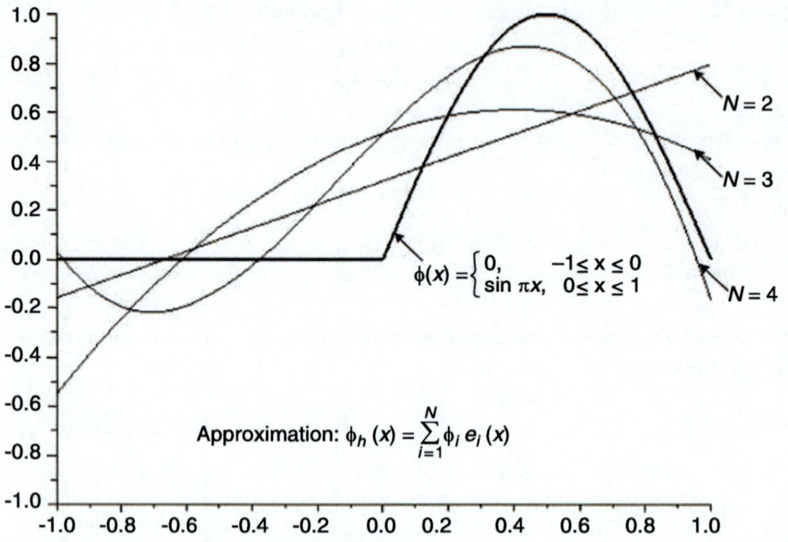

Fig. 4.2 Approximation of $\phi(x)$ by the Legendre polynomials

and substituting (4.69) into the governing equation (4.68a), the residual error is given by

$$\varepsilon = -L(\phi_h) - f = -k\frac{d^2\phi_h}{dx^2} - f \neq 0.$$

Then the square error can be calculated as

$$E(\phi_i) = \frac{1}{2}\langle \varepsilon, \varepsilon \rangle = \frac{1}{2}\int_\Omega \varepsilon^2 dx = \frac{1}{2}\int_\Omega \left[-L(\phi_h) - f\right]^2 dx. \tag{4.70}$$

In order to determine the coefficients ϕ_i, we minimize the square error (4.70)

$$\frac{\partial E}{\partial \phi_i} = 0, \qquad i = 1, 2, 3, \cdots \tag{4.71}$$

We can then substitute (4.69) into (4.70) and use the constraint to (4.71) to obtain the following set of simultaneous equations:

$$\sum_j \big\langle L(N_i),\, L(N_j)\big\rangle \phi_j + \big\langle L(N_j),\, f\big\rangle = 0, \qquad i = 1, 2, 3, \cdots \tag{4.72}$$

or alternatively

$$\begin{bmatrix} \langle L(N_1), L(N_1)\rangle & \langle L(N_1), L(N_2)\rangle & \cdots & \langle L(N_1), L(N_n)\rangle \\ \langle L(N_2), L(N_1)\rangle & \langle L(N_2), L(N_2)\rangle & \cdots & \langle L(N_2), L(N_n)\rangle \\ \vdots & \vdots & & \vdots \\ \langle L(N_n), L(N_1)\rangle & \langle L(N_n), L(N_2)\rangle & \cdots & \langle L(N_n), L(N_n)\rangle \end{bmatrix} \begin{bmatrix} \phi_1 \\ \phi_2 \\ \vdots \\ \phi_n \end{bmatrix} = \begin{bmatrix} -\langle L(N_1), f\rangle \\ -\langle L(N_2), f\rangle \\ \vdots \\ -\langle L(N_n), f\rangle \end{bmatrix} \tag{4.73}$$

We note that (4.73) corresponds to (4.67).

In the least squares method shown above, we have directly defined the square error (4.70), and minimized it; however we can also introduce an energy functional such as

$$\Pi(\psi) = \int_\Omega \left[\frac{1}{2} k \left(\frac{\partial \psi}{\partial x} \right)^2 - f\psi \right] dx + \bar{q}\,\psi(x_2) \equiv \frac{1}{2} a(\psi, \psi) - l(\psi), \tag{4.74}$$

$$a(\phi, \psi) = \int_\Omega k \frac{d\phi}{dx}\frac{d\psi}{dx} dx, \qquad l(\psi) = \int_\Omega f\psi\, dx - \bar{q}\,\psi(x_2), \tag{4.75}$$

and introduce an approximation

$$\phi \simeq \phi_h = \sum_\alpha \phi_\alpha N_\alpha(x) \tag{4.76}$$

where the interpolation functions $N_\alpha(x)$ satisfy only the Dirichlet boundary condition (4.68b). By substituting (4.76) into the energy functional (4.74), we obtain

$$\Pi(\psi_h) = \frac{1}{2} a(\phi_h, \phi_h) - l(\phi_h) = \frac{1}{2} \sum_i \sum_j \phi_i \phi_j\, a(N_i, N_j) - \sum_i \phi_i\, l(N_i), \tag{4.77}$$

and minimize the error by

$$\frac{\partial \Pi}{\partial \phi_i} = 0, \qquad i = 1, 2, 3, \cdots \tag{4.78}$$

This gives the equations that determine the coefficients ϕ_i. In fact, the following simultaneous equation is obtained:

$$\sum_j a(N_i, N_j)\phi_j - l(N_i) = 0, \qquad i = 1, 2, 3, \cdots \tag{4.79}$$

That is,

$$\begin{bmatrix} a(N_1, N_1) & a(N_1, N_2) & \cdots & a(N_1, N_n) \\ a(N_2, N_1) & a(N_2, N_2) & \cdots & a(N_2, N_n) \\ \vdots & \vdots & & \vdots \\ a(N_n, N_1) & a(N_n, N_2) & \cdots & a(N_n, N_n) \end{bmatrix} \begin{bmatrix} \phi_1 \\ \phi_2 \\ \vdots \\ \phi_n \end{bmatrix} = \begin{bmatrix} l(N_1) \\ l(N_2) \\ \vdots \\ l(N_n) \end{bmatrix} \tag{4.80}$$

This procedure is referred to as the Rayleigh-Ritz method, which is equivalent to the Galerkin method.

By using the notation of (4.75) and (4.76), the weak form is given by

$$a(\phi, \eta) - l(\eta) = 0 \quad \forall \eta \ (\eta = 0 \ \text{on} \ x_1) \tag{4.81}$$

We substitute the approximations

$$\phi \simeq \phi_h = \sum_{\alpha} \phi_\alpha N_\alpha(x), \qquad \eta \simeq \eta_h = \sum_{\beta} \eta_\beta N_\beta(x)$$

into (4.81), and obtain the following Galerkin approximation:

$$a(\phi_h, \eta_h) - l(\eta_h) = 0 \quad \forall \eta_h \ (\eta_h = 0 \ \text{on} \ x_1). \tag{4.82}$$

Here $\eta_h = 0$ on the Dirichlet boundary x_1. For the exact solution ϕ, we also have

$$a(\phi, \eta_h) - l(\eta_h) = 0. \tag{4.83}$$

By subtracting (4.83) from (4.82), we obtain

$$a(\phi - \phi_h, \eta_h) = 0 \tag{4.84}$$

where

$$e = \phi - \phi_h \tag{4.85}$$

represents the error between the solution ϕ and the approximation ϕ_h. Therefore, in the Galerkin method the error $e = \phi - \phi_h$ is orthogonal to the approximated function η_h in the sense of (4.84). In addition, since we can choose $\phi_h - \eta_h$ as η_h of (4.84), we have

$$\begin{aligned} a(\phi - \eta_h, \phi - \eta_h) &= a(e + \phi_h - \eta_h, e + \phi_h - \eta_h) \\ &= a(e, e) + a(\phi_h - \eta_h, \phi_h - \eta_h) \geq 0 \end{aligned} \tag{4.86}$$

where the condition $a(e, \phi_h - \eta_h) = 0$ (see (4.84)) and the following property of positive definiteness of $a(\cdot, \cdot)$ are used:

$$a(\zeta, \zeta) \geq 0 \qquad \forall \zeta \tag{4.87}$$

where $k > 0$. The result (4.86) shows that even if the objective function ϕ is not known explicitly, the Galerkin method, which uses $\eta_h = \phi_h$, gives the best approximation since $a(\phi_h - \eta_h, \phi_h - \eta_h) = 0$.

Chapter 5
Classical Theory of Diffusion and Seepage Problems in Porous Media

In this chapter we develop the coupled diffusion and seepage problem using the theory of mixtures. It is clearly understood that the diffusion problem is strongly linked to the seepage problem through the mass conservation law. Adsorption on the solid surface is treated using the concept of an 'adsorption isotherm'.

5.1 Representative Elementary Volume and Averaging

Let us consider a volume ΔV, referred to as the *representative elementary volume* (REV), and with Δl being the size of the porous body for averaging any independent variable within this volume. Figure 5.1 schematically shows how to decide on the REV. If ΔV_v is the volume of voids in the REV, the porosity n is defined by

$$n = \lim_{\Delta V \to 0} \frac{\Delta V_v}{\Delta V}. \tag{5.1}$$

Assume that the center of the REV is located at $\boldsymbol{x}$ and $\boldsymbol{x}'$ is any arbitrary point within the REV. The mean value of a function for this REV is defined by

$$\overline{f}(\boldsymbol{x}, t) = \frac{1}{\Delta V} \int_{\Delta V} f(\boldsymbol{x}', t, \boldsymbol{x})\, dv(\boldsymbol{x}') \tag{5.2}$$

As shown in Fig. 5.2 the mean value $\overline{f}(\boldsymbol{x}, t)$ depends on the REV; i.e., if the representative size l is too small, the mean value $\overline{f}(\boldsymbol{x}, t)$ represents only the material at the center, whereas if the size l is too large (e.g., the case of an inhomogeneous medium) the mean value converges to another limit from its original mean value $\bar{f}(\boldsymbol{x}, t)$. We note that there are upper and lower limits for the representative size, depending on the microscale geometrical properties of the material.

Y. Ichikawa and A.P.S. Selvadurai, *Transport Phenomena in Porous Media*,
DOI 10.1007/978-3-642-25333-1_5, © Springer-Verlag Berlin Heidelberg 2012

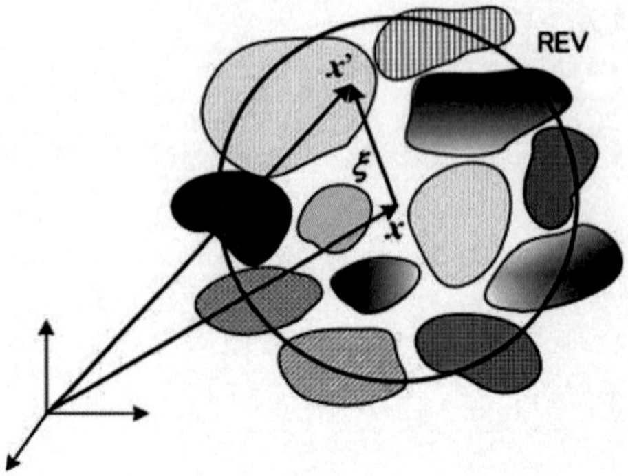

Fig. 5.1 Representative elementary volume (REV)

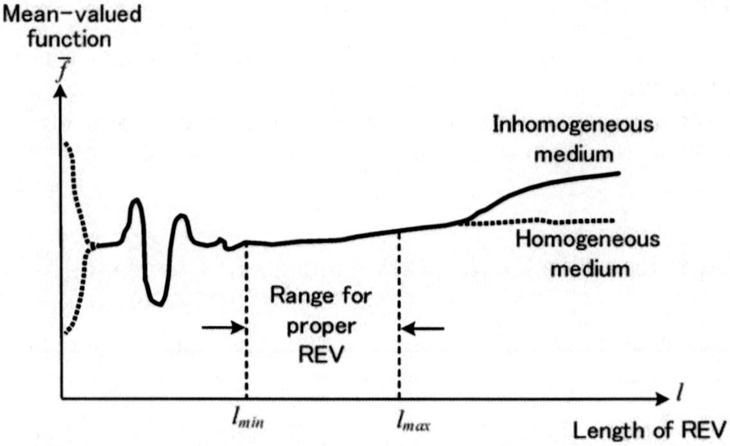

Fig. 5.2 REV and the mean-valued function

The deviation of a function $f(\boldsymbol{x}', t, \boldsymbol{x})$ from its mean value $\overline{f}(\boldsymbol{x}, t)$ is given by

$$\mathring{f}(\boldsymbol{x}', t, \boldsymbol{x}) = f(\boldsymbol{x}', t, \boldsymbol{x}) - \overline{f}(\boldsymbol{x}, t). \tag{5.3}$$

Let $g(\boldsymbol{x}', t, \boldsymbol{x})$ be another dependent variable specified in dv. Since we have

$$\overline{\overline{f}\,\mathring{g}} = \frac{\overline{f}}{\Delta V}\int_{\Delta V} \mathring{g}(\boldsymbol{x}', t, \boldsymbol{x})\, dv(\boldsymbol{x}') \equiv 0, \quad \overline{\mathring{f}\,\overline{g}} = \frac{\overline{g}}{\Delta V}\int_{\Delta V} \mathring{f}(\boldsymbol{x}', t, \boldsymbol{x})\, dv(\boldsymbol{x}') \equiv 0,$$

the mean value of $f\,g$ is calculated by

$$\overline{f\,g} = \overline{(\overline{f} + \mathring{f})(\overline{g} + \mathring{g})} = \overline{f}\,\overline{g} + \overline{\mathring{f}\,\mathring{g}}. \tag{5.4}$$

Note that the discussions in the following sections are based on this mean value.

5.2 Diffusion and Seepage Problem for a Multi-component Solution in Saturated Porous Media

We treat the problem of coupled diffusion and seepage of a multi-component solution in a saturated porous medium with a deformable porous skeleton. The porosity is assumed to be n as defined by (5.1).

The fundamental concept of the mixture theory is given in Sects. 3.5 and 3.6 where the notation for a fluid phase are found. The important notation is reworded to include a notation for a solid phase.

We represent a continuum body of the αth component of the fluid phase with N-components as $\mathcal{B}_\alpha$ $(\alpha = 1, 2, \cdots, N)$. The motion of a spatial point $\boldsymbol{x}$ in the current body $\Omega \subset \mathbb{R}^3$ is given by $\boldsymbol{x} = \chi_\alpha(\boldsymbol{X}_\alpha, t) = x_i\, \boldsymbol{e}_i$, and the reference position $\boldsymbol{X}_\alpha$ is $\boldsymbol{X}_\alpha = \chi_\alpha^{-1}(\boldsymbol{x}, t) = X_{\alpha K}\, \boldsymbol{E}_K$.

The particle velocity $\boldsymbol{v}_\alpha$ of a material point $\boldsymbol{X}_\alpha$ at time t is obtained as $\boldsymbol{v}_\alpha = d_{(\alpha)}\chi_\alpha(\boldsymbol{X}_\alpha, t)/dt$. We denote the material time-derivative $d_{(\alpha)}\phi/dt$ of a function $\phi(\boldsymbol{x}, t)$ with respect to the αth component as

$$\frac{d_{(\alpha)}\phi}{dt} = \frac{\partial \phi}{\partial t} + \boldsymbol{v}_\alpha \cdot \operatorname{grad} \phi. \tag{5.5}$$

The deformation gradient $\boldsymbol{F}_\alpha = F_{\alpha i K}\, \boldsymbol{e}_i \otimes \boldsymbol{E}_K$ and the inverse transformation $\boldsymbol{F}_\alpha^{-1} = F_{\alpha K i}^{-1}\, \boldsymbol{E}_K \otimes \boldsymbol{e}_i$ are given by

$$\boldsymbol{F}_\alpha = \operatorname{Grad} \chi_\alpha(\boldsymbol{X}_\alpha, t), \qquad \boldsymbol{F}_\alpha^{-1} = \operatorname{grad} \chi_\alpha^{-1}(\boldsymbol{x}, t). \tag{5.6}$$

The time derivative of the Jacobian J_α is calculated as

$$\frac{d_{(\alpha)} J_\alpha}{dt} = J_\alpha \operatorname{div} \boldsymbol{v}_\alpha, \qquad J_\alpha = |\det \boldsymbol{F}_\alpha| \tag{5.7}$$

Matric diffusion in the solid phase (see Sect. 5.2.2) can be developed using an approach similar to that adopted to describe transport within the fluid phase. Let $\mathcal{B}_\alpha^*$ $(\alpha = 1, 2, \cdots, N)$ be a continuum region of the αth component of the solid phase, and the motion of a spatial point $\boldsymbol{x}$ can be given by

$$\boldsymbol{x} = \chi_\alpha^*(\boldsymbol{X}_\alpha^*, t) = x_i^*\, \boldsymbol{e}_i \tag{5.8}$$

$$\boldsymbol{X}_\alpha^* = (\chi_\alpha^*)^{-1}(\boldsymbol{x}, t) = X_{\alpha K}^*\, \boldsymbol{E}_K. \tag{5.9}$$

As with the fluid phase, several variables such as the velocity are given by

$$\boldsymbol{v}_\alpha^* = \left.\frac{d_{(*\alpha)}\chi_\alpha^*(\boldsymbol{X}_\alpha^*, t)}{dt}\right|_{\boldsymbol{X}_\alpha^* = \text{constant}} \tag{5.10}$$

$$\frac{d_{(*\alpha)}\phi}{dt} = \frac{\partial \phi}{\partial t} + \boldsymbol{v}_\alpha^* \cdot \mathrm{grad}\,\phi \tag{5.11}$$

$$\boldsymbol{F}_\alpha^* = \mathrm{Grad}\,\boldsymbol{\chi}_\alpha^*(\boldsymbol{X}_\alpha^*, t), \quad F_{\alpha iK}^* = \frac{\partial x_i}{\partial X_{\alpha K}^*} \tag{5.12}$$

$$(\boldsymbol{F}_\alpha^*)^{-1} = \mathrm{grad}\,(\boldsymbol{\chi}_\alpha^*)^{-1}(\boldsymbol{x}, t), \quad (F_{\alpha K i}^*)^{-1} = \frac{\partial X_{\alpha K}^*}{\partial x_i} \tag{5.13}$$

$$\frac{d_{(*\alpha)} J_\alpha^*}{dt} = J_\alpha^* \,\mathrm{div}\, \boldsymbol{v}_\alpha^*, \qquad J_\alpha^* = |\,\mathrm{det}\boldsymbol{F}_\alpha^*\,|. \tag{5.14}$$

5.2.1 *Mass Conservation for the Fluid Phase*

If we include the porosity n, the discussions of Sect. 3.6 can be directly used. Assume that within the REV of volume ΔV all variables are homogeneous. If the mass of the αth component of the fluid phase is denoted as n_α, the volume fraction (occasionally referred to as the volume molar concentration) ω_α and the component mass density ρ_α are defined by

$$n_\alpha = n\,\omega_\alpha\,\Delta V = n\,\frac{\rho_\alpha}{m_\alpha}\,\Delta V \qquad (!\alpha) \tag{5.15}$$

where m_α is the molecular weight of the αth component.[1]

Let γ_α be the mass supply of the αth component due to, for example, a chemical reaction, and let ζ_α be the mass that is absorbed per unit area on the surface of the porous fabric that composes the solid phase. Then the mass conservation law of the αth component can be written as

$$\frac{d_{(\alpha)}}{dt}\int_\Omega n\,\rho_\alpha\,dv = \int_\Omega n\,\gamma_\alpha\,dv - \sum_{i=1}^{M}\int_{\Gamma_{fs}^i} \zeta_\alpha\,ds \tag{5.16}$$

where M is the total number of solid particles that contribute to absorption.

If we apply Reynolds' transport theorem to the l.h.s. of (5.16) under (5.7), the local form of the mass conservation law can be obtained as

$$\frac{d_{(\alpha)} n\rho_\alpha}{dt} + n\rho_\alpha \mathrm{div}\,\boldsymbol{v}_\alpha = \frac{\partial(n\rho_\alpha)}{\partial t} + \mathrm{div}\left(n\rho_\alpha \boldsymbol{v}_\alpha\right) = n\gamma_\alpha - \zeta_\alpha^* \qquad (!\alpha) \tag{5.17}$$

---

[1]The volume molar concentration ω_α given by (5.15) is frequently used in the field of gas chemistry, whereas in continuum mechanics the concentration c_α based on a mass-fraction is commonly used, as defined by (5.27). See e.g., Bowen 1976.

where ζ_α^* is the absorbed mass per unit volume, which is calculated from the absorbed mass per unit area ζ_α. Details of the procedure are given in Note 5.1.

By summing up the total components, we have

$$\frac{d(n\rho)}{dt} + n\rho \,\mathrm{div}\, \boldsymbol{v} = \frac{\partial(n\rho)}{\partial t} + \mathrm{div}\,(n\rho\boldsymbol{v}) = 0. \tag{5.18}$$

Here we have assumed that a chemical species exchange can take place between the fluid phase and the solid phase only through adsorption. Therefore the total mass is balanced, giving

$$\sum_\alpha \left(n\gamma_\alpha - \zeta_\alpha^*\right) = 0 \quad \Rightarrow \quad \sum_\alpha n\gamma_\alpha = \zeta^*, \tag{5.19}$$

and we have used (5.18) when deriving (5.19). Note that the mean velocity $\boldsymbol{v}$, the total mass density ρ and the total absorbed mass ζ^* are defined by

$$\boldsymbol{v} = \frac{1}{\rho}\sum_\alpha \rho_\alpha \boldsymbol{v}_\alpha, \quad \rho = \sum_\alpha \rho_\alpha, \quad \zeta^* = \sum_\alpha \zeta_\alpha^*. \tag{5.20}$$

The material time derivative of a function ϕ at a space-time $(\boldsymbol{x}, t)$ with reference to the mean velocity $\boldsymbol{v}$ is given by

$$\frac{d\phi}{dt} = \frac{\partial \phi}{\partial t} + \boldsymbol{v} \cdot \mathrm{grad}\, \phi. \tag{5.21}$$

The velocity gradient $\boldsymbol{L}$, the stretch tensor $\boldsymbol{D}$ and the spin tensor $\boldsymbol{W}$ are defined as

$$\boldsymbol{L} = \mathrm{grad}\, \boldsymbol{v} = \boldsymbol{D} + \boldsymbol{W}, \tag{5.22}$$

$$\boldsymbol{D} = \frac{1}{2}(\boldsymbol{L} + \boldsymbol{L}^T), \quad \boldsymbol{W} = \frac{1}{2}(\boldsymbol{L} - \boldsymbol{L}^T). \tag{5.23}$$

If the mean velocity $\boldsymbol{v}$ is used, (5.17) can be rewritten as

$$\frac{d(n\rho_\alpha)}{dt} + n\rho_\alpha \mathrm{div}\, \boldsymbol{v} = -\mathrm{div}\left(n\rho_\alpha \bar{\boldsymbol{v}}_\alpha\right) + n\gamma_\alpha - \zeta_\alpha^* \qquad (!\alpha) \tag{5.24}$$

where we have defined the diffusion velocity $\bar{\boldsymbol{v}}_\alpha$ as

$$\bar{\boldsymbol{v}}_\alpha = \boldsymbol{v}_\alpha - \boldsymbol{v}. \tag{5.25}$$

By eliminating $\boldsymbol{v}$ from (5.18) and (5.24), we obtain

$$\frac{d(n\rho_\alpha)}{dt} - c_\alpha \frac{d(n\rho)}{dt} = -\mathrm{div}\left(n\rho_\alpha \bar{\boldsymbol{v}}_\alpha\right) + n\gamma_\alpha - \zeta_\alpha^* \qquad (!\alpha) \tag{5.26}$$

where c_α is the concentration based on the mass fraction, defined by

$$c_\alpha = \frac{n\rho_\alpha}{n\rho} = \frac{\rho_\alpha}{\rho}. \tag{5.27}$$

The material time derivative of c_α with reference to $\boldsymbol{v}$ gives the following:

$$n\rho\frac{dc_\alpha}{dt} = -\frac{d(n\rho)}{dt}c_\alpha + \frac{d(n\rho_\alpha)}{dt}. \tag{5.28}$$

Substituting (5.28) into (5.26) gives the diffusion equation as

$$n\rho\frac{dc_\alpha}{dt} = -\mathrm{div}\left(n\rho_\alpha\bar{\boldsymbol{v}}_\alpha\right) + n\gamma_\alpha - \zeta_\alpha^* \quad (!\alpha). \tag{5.29}$$

From (5.27), we have the following constraint on the concentration c_α:

$$\sum_\alpha c_\alpha = 1. \tag{5.30}$$

Note 5.1 (Estimation of the volume adsorption ζ_α^*). Let us assume that the porous medium is microscopically periodic as shown in Fig. 5.3, and N identical micro-cells are connected in the complete macro-domain Ω_0 with the internal solid/fluid interfaces Γ^i_{fs}. Referring to (5.17), we can rewrite (5.16) as

$$\int_{\Omega_0} \Phi(n\rho_\alpha)\,dv = -\sum_{i=1}^{N}\int_{\Gamma^i_{fs}} \zeta_\alpha\,ds, \tag{5.31}$$

$$\Phi(n\rho_\alpha) = \frac{\partial(n\rho_\alpha)}{\partial t} + \mathrm{div}\left(n\rho_\alpha\boldsymbol{v}_\alpha\right) - n\gamma_\alpha \qquad (!\alpha). \tag{5.32}$$

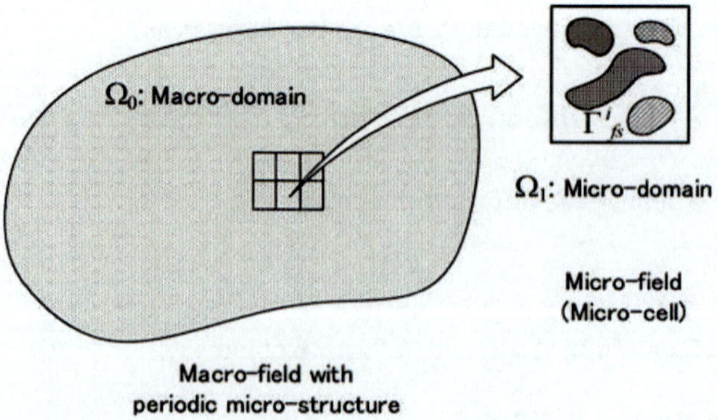

Fig. 5.3 Micro/macro problem and periodic boundary condition

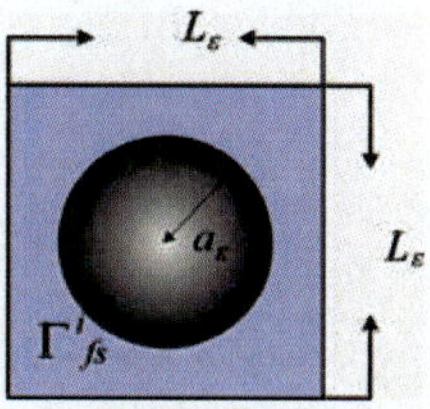

Fig. 5.4 Spherical absorbent body in the micro-cell

We have assumed that in (5.32) the terms on the boundaries cancel out due to periodicity. Let the size of the micro-cell be L_ε, and a sphere of radius a_ε is immersed in the solution (Fig. 5.4). This causes adsorption of the amount ζ^a_α per unit area on the surface of this spherical absorbent body. The volume of the macro-domain is given by $V = N(L_\varepsilon)^3$,and the surface area of one sphere is $4\pi a_\varepsilon^2$. If the adsorption field is uniform, (5.31) gives

$$\Phi(n\rho_\alpha)\, V = -N 4\pi a_\varepsilon^2\, \zeta^a_\alpha$$

This gives the following estimate for ζ^*_α as defined by (5.17):

$$\zeta^*_\alpha = 4\pi \frac{a_\varepsilon^2}{(L_\varepsilon)^3} \zeta^a_\alpha. \tag{5.33}$$

■

5.2.2 *Mass Conservation in the Solid Phase*

We here introduce a process of diffusion and reaction in the solid phase similar to that in the fluid phase. By drawing an analogy to the analysis of the fluid phase, we have the following equation of mass conservation for the αth component:

$$\frac{d_{(*\alpha)}}{dt} \int_\Omega (1-n)\, \rho^*_\alpha\, dv = \int_\Omega (1-n)\, \gamma^*_\alpha\, dv + \sum_{i=1}^{M} \int_{\Gamma^i_{fs}} \boldsymbol{\zeta}_\alpha \cdot \boldsymbol{n}\, ds \tag{5.34}$$

where γ^*_α is the mass supply of the αth component in the solid phase, and ρ^*_α is the component mass density of the species α in the solid phase which is defined in the REV as

$$n^*_\alpha = (1-n)\, \omega^*_\alpha\, \Delta V = (1-n)\, \frac{\rho^*_\alpha}{m_\alpha}\, \Delta V \quad (!\alpha) \tag{5.35}$$

where n^*_α and ω^*_α are the amount of mass and the volume fraction of the αth component, respectively. Note that ζ_α is the mass per unit area absorbed on the solid particles as described in (5.16).

The local form of (5.34) can be obtained as

$$\frac{d_{(*\alpha)}\left((1-n)\,\rho_\alpha^*\right)}{dt} + (1-n)\,\rho_\alpha^* \operatorname{div} \boldsymbol{v}_\alpha^* = (1-n)\,\gamma_\alpha^* + \zeta_\alpha^* \quad (!\alpha) \tag{5.36}$$

By summing up the total components, we have

$$\frac{d_*\left((1-n)\,\rho^*\right)}{dt} + (1-n)\,\rho^* \operatorname{div} \boldsymbol{v}^* = \frac{\partial\left((1-n)\,\rho^*\right)}{\partial t} + \operatorname{div}\left((1-n)\,\rho^* \boldsymbol{v}^*\right) = 0 \tag{5.37}$$

where $\boldsymbol{v}^*$ and ρ^* are the mean velocity and the total mass density of the solid phase which are defined by

$$\boldsymbol{v}^* = \frac{1}{\rho^*}\sum_\alpha \rho_\alpha^* \boldsymbol{v}_\alpha^*, \qquad \rho^* = \sum_\alpha \rho_\alpha^*. \tag{5.38}$$

As in (5.19), we have a constraint

$$\sum_\alpha (1-n)\,\gamma_\alpha^* = \zeta^*. \tag{5.39}$$

The material time derivative of a function ϕ with respect to the mean velocity $\boldsymbol{v}^*(\boldsymbol{x},t)$ is given by

$$\frac{d_*\phi}{dt} \equiv \frac{\partial \phi}{\partial t} + \boldsymbol{v}^* \cdot \operatorname{grad} \phi. \tag{5.40}$$

The velocity gradient $\boldsymbol{L}^*$, the stretch tensor $\boldsymbol{D}^*$ and the spin tensor are defined by

$$\boldsymbol{L}^* = \operatorname{grad} \boldsymbol{v}^* = \boldsymbol{D}^* + \boldsymbol{W}^*; \;\; \boldsymbol{D}^* = \frac{1}{2}\left(\boldsymbol{L}^* + (\boldsymbol{L}^*)^T\right), \;\; \boldsymbol{W}^* = \frac{1}{2}\left(\boldsymbol{L}^* - (\boldsymbol{L}^*)^T\right). \tag{5.41}$$

Adopting a procedure similar to that used in (5.29), the diffusion equation of the αth component in the solid phase can be derived as follows

$$(1-n)\,\rho^* \frac{d_* c_\alpha^*}{dt} = -\operatorname{div}\left((1-n)\,\rho_\alpha^* \overline{\boldsymbol{v}}_\alpha^*\right) + (1-n)\,\gamma_\alpha^* + \zeta_\alpha^* \quad (!\alpha), \tag{5.42}$$

where

$$\overline{\boldsymbol{v}}_\alpha^* = \boldsymbol{v}_\alpha^* - \boldsymbol{v}^* \tag{5.43}$$

is the diffusion velocity of the αth component, and as given in (5.30), we have a constraint on the concentration c_α^*:

$$\sum_\alpha c_\alpha^* = 1. \tag{5.44}$$

The diffusion in the solid phase is referred to as the matric diffusion. Matric diffusion for a fractured rock is presented by Rasilainen (1997); however, there are always small scale fractures even in the 'intact' part. Pure matric diffusion is rarely observed under normal temperature/stress/chemical states and therefore the diffusion velocity in the solid phase can be ignored ($|\bar{v}_\alpha^*| \ll 1$).

Using (5.25) and (5.43) we introduce the diffusion mass flux of the αth component as

$$\boldsymbol{q}_\alpha^m = n\rho_\alpha \bar{\boldsymbol{v}}_\alpha + (1-n)\rho_\alpha^* \bar{\boldsymbol{v}}_\alpha^* \qquad (!\alpha). \tag{5.45}$$

Example 5.1 (A constituent description of bentonite clay). Bentonite consists mainly of smectitic clay minerals such as montmorillonite and beidelite, with macro-grains that are mainly quartz and chalcedony. The chemical formulae of water and quartz are H_2O and SiO_2, respectively, and we assume that the chemical formula of a smectitic clay mineral is $Na_{1/3}Al_2[Si_{11/3}Al_{1/3}]O_{10}(OH)_2$. Note that the sodium ions Na^+ exist in an interlayer space of the clay minerals. For simplicity, the porosity is assumed to be $n = 0.5$, and the composition (i.e., mass fraction) of the solid phase is given as consisting of 50% clay minerals and 50% quartz. The pores are assumed to be filled with pure water. The REV is a 1 cm cube.

Atomic weight of the constituent elements are as follows: Na; 22.989770, Al; 26.981538, Si; 28.05855, O; 15.994, H; 1.00794. The molecular weights of water, the smectitic clay mineral and quartz are thus $m_{\mathrm{w}} = 18.015$, $m_{\mathrm{Sm}} = 367.510$, $m_{\mathrm{Qz}} = 60.057$, respectively.

The mass density of water is 1.00 [g/cm^3], and we assume that the intrinsic mass densities of the smectitic clay mineral and quartz are 2.70 and 2.70 [g/cm^3], respectively. Therefore, the amount of chemical substances of the water, smectitic clay mineral and quartz in the REV are $n_{\mathrm{w}} \simeq 1/36$, $n_{\mathrm{Sm}} \simeq 9/800$, $n_{\mathrm{Qz}} \simeq 81/44,080$ [mol], respectively.

Based on the above data, the component densities of the water, clay mineral and quartz are calculated as $\rho_{\mathrm{w}} \simeq 1.00$, $\rho_{\mathrm{Sm}}^* \simeq 1.35$, $\rho_{\mathrm{Qz}}^* \simeq 1.35$ [g/cm^3], respectively. ■

Note 5.2 (Distribution coefficient K_d and the classical diffusion equation). If the mixture of a fluid (i.e., solution) is dilute, which is what is assumed for an ideal solution, the adsorption can be treated as being the distribution in the solid phase, where a distribution coefficient K_d is introduced as follows:

$$c_\alpha^* = K_d\, c_\alpha. \tag{5.46}$$

For the ideal solution K_d is assumed to be constant, and both $\boldsymbol{v}_\alpha^*$ and $\boldsymbol{v}^*$ can be ignored. Using these assumptions and adding (5.29) and (5.42), we obtain the classical diffusion equation as follows:

$$\begin{aligned} R_d \frac{\partial c_\alpha}{\partial t} &+ n\rho \boldsymbol{v} \cdot \operatorname{grad} c_\alpha \\ &= -\operatorname{div}\left(n\rho_\alpha \bar{\boldsymbol{v}}_\alpha\right) + n\gamma_\alpha + (1-n)\gamma_\alpha^* \quad (!\alpha) \end{aligned} \tag{5.47}$$

where R_d is the retardation coefficient which is given by

$$R_d = n\rho + (1-n)\,\rho^* K_d. \tag{5.48}$$

If we evaluate the adsorption by considering the distribution coefficient K_d, the third term on the r.h.s. of both (5.29) and (5.42) can be neglected. ■

Note 5.3 (Fick's law and the constraint). In the classical diffusion theory we assume that the matric diffusion can be ignored, and we introduce Fick's law for the diffusive flux $\boldsymbol{q}_\alpha^m$ as

$$\boldsymbol{q}_\alpha^m = -\sum_\beta D_{\alpha\beta}\,\mathrm{grad}\,c_\beta \tag{5.49}$$

where $D_{\alpha\beta}$ is the diffusion coefficient, which is positive definite (i.e., $\eta_\alpha\, D_{\alpha\beta}\,\eta_\beta > 0 \;\; \forall \eta_\alpha$). It is noted that (5.49) is an empirical formula. As implied by (5.45), we have $\sum_\alpha \rho_\alpha \bar{\boldsymbol{v}}_\alpha = \sum_\alpha \rho_\alpha^* \bar{\boldsymbol{v}}_\alpha^* = 0$, therefore the following constraint must be satisfied:

$$\sum_\alpha \boldsymbol{q}_\alpha^m = -\sum_\alpha\sum_\beta D_{\alpha\beta}\,\mathrm{grad}\,c_\beta = 0. \tag{5.50}$$

■

5.2.3 *Seepage Theory for the Incompressible Fluid*

When we solve the diffusion problem which is given by, e.g., (5.29), we encounter a problem that relates to the evaluation of the mean velocity $\boldsymbol{v}$ (note that $dc_\alpha/dt = \partial c_\alpha/\partial t + \boldsymbol{v}\cdot\mathrm{grad}\,c_\alpha$). It may be possible to solve a microscale problem based on the Navier-Stokes equation; however, in classical soil mechanics we commonly use the seepage equation to determine $\boldsymbol{v}$. Using the assumption of incompressibility of a fluid, we can derive the seepage equation from (5.18) and (5.37).

Let us assume that both the mixture fluid and the intrinsic part of solid are incompressible (ρ = constant, ρ^* = constant). Under these incompressible conditions and applying (5.41), we obtain alternative forms of (5.18) and (5.37) as follows:

$$\frac{dn}{dt} + \mathrm{div}\,(n\boldsymbol{v}) - \boldsymbol{v}\cdot\mathrm{grad}\,n = 0 \tag{5.51}$$

$$-\frac{d_*n}{dt} + (1-n)\,\mathrm{tr}\,\boldsymbol{D}^* = 0 \tag{5.52}$$

We substitute the relation

$$\frac{d_*n}{dt} = \frac{dn}{dt} - (\boldsymbol{v}-\boldsymbol{v}^*)\cdot\mathrm{grad}\,n = \frac{dn}{dt} - \mathrm{div}\big[n\,(\boldsymbol{v}-\boldsymbol{v}^*)\big] + n\,\mathrm{div}(\boldsymbol{v}-\boldsymbol{v}^*) \tag{5.53}$$

into (5.52), and obtain

$$\frac{dn}{dt} = (1-n)\,\mathrm{tr}\,\boldsymbol{D}^* + (\boldsymbol{v}-\boldsymbol{v}^*)\cdot\mathrm{grad}\,n = \mathrm{div}\big[n\,(\boldsymbol{v}-\boldsymbol{v}^*)\big] - n\,\mathrm{div}\,\boldsymbol{v} + \mathrm{div}\,\boldsymbol{v}^*. \tag{5.54}$$

By substituting (5.54) into (5.51), we finally obtain a seepage equation that includes the volumetric deformation of the solid phase $\mathrm{tr}\,\boldsymbol{D}^*$ as follows:

$$\mathrm{tr}\,\boldsymbol{D}^* + \mathrm{div}\,\tilde{\boldsymbol{v}} = 0 \tag{5.55}$$

$$\frac{dn}{dt} = -n\,\mathrm{div}\,\boldsymbol{v} \tag{5.56}$$

where $(\boldsymbol{v}-\boldsymbol{v}^*)$ is the intrinsic fluid velocity, and

$$\tilde{\boldsymbol{v}} = n\,(\boldsymbol{v}-\boldsymbol{v}^*) \tag{5.57}$$

is referred to as the seepage velocity or Darcy's velocity of a pore fluid, to which we apply Darcy's law given by

$$\tilde{\boldsymbol{v}} = -\boldsymbol{k}\,\mathrm{grad}\,\phi, \qquad \phi = \frac{p}{\rho g} + \zeta \tag{5.58}$$

Using an equation of equilibrium or motion, which determines the deformation of the solid skeleton, and (5.58), a system of differential equations for specifying the mean velocity $\boldsymbol{v}$ (i.e., the conventional consolidation problem) is achieved. Note that in (5.58) ϕ is the reduced Bernoulli potential (i.e., the total head excluding the velocity potential), $\boldsymbol{k}$ is the hydraulic conductivity tensor, p is the pore pressure of the fluid, g is the gravity constant, and ζ is the datum potential. Thus by starting with the mass conservation laws for both fluid and solid phases, we can simultaneously obtain the diffusion equation and the seepage equation which includes a term that accounts for the volumetric deformation of the porous skeleton.

Note 5.4 (On the permeability and flow in a porous medium). The seepage equation can be obtained by substituting Darcy's law into the mass conservation equations of fluid and solid phases, as described above. The effects of the micro-structure and microscale material property are put into the hydraulic conductivity $\boldsymbol{k}$, which is fundamentally specified through experiments. It is not possible to specify the true velocity field by this theory, whereas by applying a homogenization technique, we can determine the velocity field that will be affected by the microscale characteristics. In Chap. 8 we will outline the homogenization theory, which is applied to the problem of water flow in a porous medium, where the microscale flow field is specified.

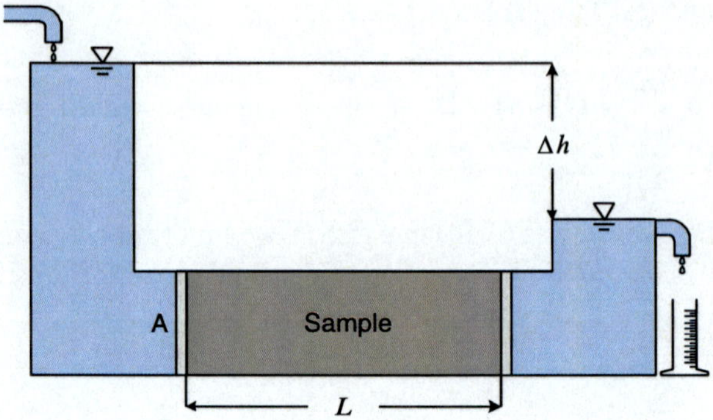

Fig. 5.5 Schematic diagram of the constant head permeability test

Let us assume that a porous medium is isotropic ($k_{ij} = k\,\delta_{ij}$). We briefly describe a method to determine the hydraulic conductivity k through an experiment. Figure 5.5 shows the schematic diagram of a constant head permeability test, in which the flow rate Q is measured as a volume per unit time. If the differential head is fixed as Δh, and the cross section and length of the specimen are A and L, respectively, the hydraulic conductivity k is calculated as

$$Q = k\,A\frac{\Delta h}{L} \tag{5.59}$$

Note that the hydraulic gradient is $i = \Delta h/L$.

The hydraulic conductivity k is strongly affected by the viscosity μ of the fluid, therefore we can introduce the following permeability coefficient k^*:

$$k = k^*\frac{\rho g}{\mu} = k^*\frac{g}{\mu^\ddagger} \tag{5.60}$$

where $\mu^\ddagger = \mu/\rho$ is the kinematic viscosity.

Let M, L and T be the scales of mass, length and time. We denote the dimensions of the above variables together with the pore pressure p as follows:

$$\begin{array}{lll} [Q] = \mathrm{L}^3/\mathrm{T} & [h] = \mathrm{L} & [k] = \mathrm{L}/\mathrm{T} \\ [k^*] = \mathrm{L}^2 & [\mu] = \mathrm{M}/\mathrm{LT} & [\mu^\ddagger] = \mathrm{L}^2/\mathrm{T}^2 \\ [\rho] = \mathrm{M}/\mathrm{L}^3 & [g] = \mathrm{L}/\mathrm{T}^2 & [p] = \mathrm{M}/\mathrm{LT}^2 \end{array} \tag{5.61}$$

The dimensions are given in the MKS units as $[Q]$ =m^3/s, $[h]$ =m, $[k]$ =m/s, $[k^*]$=m^2, $[\mu]$=Pa s, $[\mu^\ddagger]$=m^2/s^2, $[\rho]$=Mg/m^3, $[g]$=m/s^2, $[p]$=Pa ■

5.2.4 Seepage Theory for the Compressible Fluid

We assume that the fluid mixture is compressible ($\rho \neq$ constant) and the intrinsic part of the solid is incompressible ($\rho^*=$ constant). Then (5.18) reduces to

$$\frac{d(n\rho)}{dt} + \operatorname{div}\left(n\rho\boldsymbol{v}\right) - \boldsymbol{v}\cdot\operatorname{grad}(n\rho) = 0. \tag{5.62}$$

By substituting (5.53) into (5.52), we obtain (5.54), which, when substituted into (5.62), gives the following seepage equations for a compressible fluid:

$$n\frac{d(\ln\rho)}{dt} + \operatorname{tr}\boldsymbol{D}^* + \operatorname{div}\tilde{\boldsymbol{v}} = 0 \tag{5.63}$$

$$\frac{dn}{dt} = -n\left[\frac{d(\ln\rho)}{dt} + \operatorname{div}\boldsymbol{v}\right] \tag{5.64}$$

where $\tilde{\boldsymbol{v}} = n(\boldsymbol{v} - \boldsymbol{v}^*)$. Let ρ_0 and K be a reference density and the volumetric elastic constant, respectively, and a constitutive law for the compressible fluid can be written as

$$p = K\ln\left(\frac{\rho}{\rho_0}\right) \quad \Rightarrow \quad \frac{dp}{dt} = K\frac{d(\ln\rho)}{dt}. \tag{5.65}$$

Then (5.63) and (5.64) are rewritten as follows:

$$\frac{n}{K}\frac{dp}{dt} + \operatorname{tr}\boldsymbol{D}^* + \operatorname{div}\tilde{\boldsymbol{v}} = 0 \tag{5.66}$$

$$\frac{dn}{dt} = -n\left(\frac{1}{K}\frac{dp}{dt} + \operatorname{div}\boldsymbol{v}\right) \tag{5.67}$$

It is clear that Darcy's law (5.58) can be applied to (5.66).

5.3 Navier-Stokes Equation and the Classical Permeability Theory

The true velocity $\boldsymbol{v}$ of a fluid mixture flowing in a porous medium is a function not only of the porosity n but also of several other factors, including the particle size and surface conditions due to the viscous properties of the fluid. In this Section we assume that the problem has a simplified geometry in order to solve the problems of viscous flow and discuss the permeability characteristics.

5.3.1 Pipe Flow: Hagen-Poiseulle Flow

Assume that a fluid of mass density ρ flows through a pipe of diameter $d = 2a$ (a is the radius) as shown in Fig. 5.6 (Hagen-Poiseulle flow). When the velocity field is one-dimensional, the differential equation governing the pipe flow problem along with the boundary conditions (BC) is given in cylindrical polar coordinates (r, z) as follows:

$$\frac{1}{r}\frac{d}{dr}\left(r\frac{dv_z}{dr}\right) = \frac{1}{\mu}\frac{dp}{dz}, \tag{5.68}$$

$$\text{BC}: \quad v_z = 0 \quad \text{at} \quad r = a, \tag{5.69}$$

$$v_z = V_{max} < \infty \quad \text{at} \quad r = 0 \tag{5.70}$$

where v_z is the velocity in the z-direction, p is the pressure (dp/dz is the pressure gradient), and μ is the dynamic shear viscosity.

The solution of the ordinary differential equation (5.68) is

$$v_z = \frac{1}{4\mu}\frac{dp}{dz}\left(a^2 - r^2\right) \tag{5.71}$$

and the mass flux q through the pipe is given by

$$q = \int_0^a \rho v_z \cdot 2\pi r\, dr = -\frac{dp}{dz}\frac{\pi a^4}{8\mu^\ddagger} \tag{5.72}$$

where $\mu^\ddagger = \mu/\rho$ is the kinematic viscosity.[2] The mean velocity $\bar{v}$ and maximum velocity V_{max} are calculated as

$$\bar{v} = \frac{1}{\pi a^2}\int_0^a v_z \cdot 2\pi r\, dr = -\frac{dp}{dz}\frac{a^2}{8\mu}, \qquad V_{max} = v_z(r = 0) = 2\,\bar{v}. \tag{5.73}$$

Let L be the length of the pipe, and the pressure loss be Δp. Then we have $-dp/dz \simeq \Delta p/L$, and using (5.73)$_1$

$$\Delta p = -\frac{32\mu\bar{v}L}{d^2}. \tag{5.74}$$

Since the Reynolds' number (cf. Sect. 5.6) for the flow through the pipe is

---

[2]In most textbooks, the shearing kinematic viscosity is denoted as ν, whereas we employ the notation $\mu^\ddagger$ and $\nu^\ddagger$ for shearing and volumetric kinematic viscosities, respectively, since we consider the fluid to be compressible.

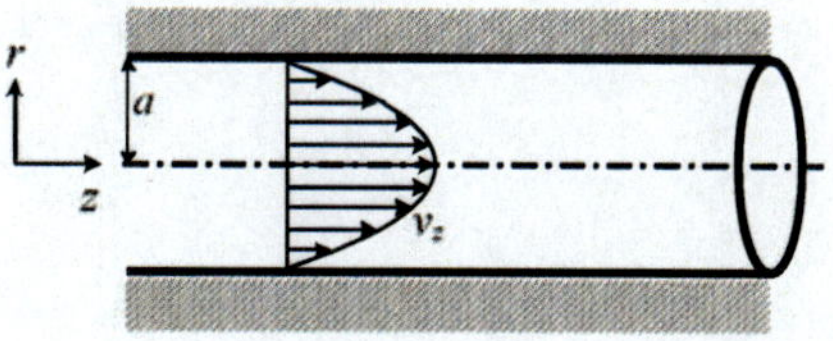

Fig. 5.6 Hagen-Poiseulle flow

$$R_e = \frac{\bar{v} d}{\mu^{\ddagger}}, \tag{5.75}$$

the drag coefficient C_D for the flow in this pipe, with an inner diameter $d = 2a$, is given by

$$C_D = \frac{\text{Pressure loss}}{\text{Kinetic energy of the inflow}} = \frac{-d \cdot dp/dz}{\rho \bar{v}^2/2} = \frac{64}{R_e} \tag{5.76}$$

where (5.73) and (5.75) are used.

The shear stress τ_w on the pipe wall is calculated by considering the equilibrium for pressure loss as

$$\pi a^2 \Delta p = 2\pi a L \tau_w \quad \Rightarrow \quad \tau_w = \frac{d}{4} \frac{\Delta p}{L}, \tag{5.77}$$

and the friction factor f caused by the shear stress is

$$f = \frac{\text{Shear stress at the wall}}{\text{Kinetic energy of the inflow}} = \frac{\tau_w}{\rho \bar{v}^2/2} = \frac{d}{2\rho \bar{v}^2} \frac{\Delta p}{L}. \tag{5.78}$$

This f is known as Fanning's friction factor. Using this f, the frictional energy loss F_f for flow in a pipe of length L is given by

$$F_f = 4f\left(\frac{L}{d}\right)\left(\frac{\bar{v}^2}{2}\right) = \frac{2fL\bar{v}^2}{d}. \tag{5.79}$$

This is referred to as Fanning's formula. The unit of measurement for F_f is [J/kg]. The pressure loss due to friction ΔP_f is then given by

$$\Delta P_f = \rho F_f = \frac{2\rho f L \bar{v}^2}{d} \tag{5.80}$$

where ΔP_f is given in [Pa]. Using (5.76)–(5.78) we can see that the drag coefficient C_D is related to the friction factor f by

$$C_D = 4f. \tag{5.81}$$

In terms of the frictional energy loss we define Darcy's friction factor λ by

$$\lambda = 4f. \tag{5.82}$$

5.3.2 *Flow Through an Assembly of Pipes and Its 'Permeability'*

Consider the viscous flow through an assembly of pipes with an overall square cross section of side length D and an overall longitudinal length L (Fig. 5.7). The diameter of a single pipe is d, and for simplicity we ignore the thickness of the pipe wall. The boundary surface of each pipe is impervious. It should be noted that the porosity $n = 1 - \pi/4$ of the section is the same for any size of pipes.

If the number of pipes is $N = (D/d)^2$, and a flux through one pipe is given by (5.72), the total flux through this pipe assembly is

$$Q = q \cdot N = -\frac{dp}{dz}\frac{\pi a^4}{8\mu^{\ddagger}} N = -\frac{dp}{dz}\frac{\pi}{128\mu^{\ddagger}}(dD)^2. \tag{5.83}$$

From the above equation a 'hydraulic conductivity' k of this system can be defined as

$$k = \frac{\pi D^2}{128\mu^{\ddagger}}\left(\frac{d}{D}\right)^2 \tag{5.84}$$

As we can see, the 'hydraulic conductivity' is proportional to the square of the ratio d/D. That is, if the pipe diameter becomes one tenth under the same D, the 'hydraulic conductivity' becomes one hundredth, since, as understood from (5.71) or (5.73), the velocity is restricted by the viscosity and is proportional to d^2. Figure 5.8 schematically shows the velocity profiles for the case when the pipe diameter becomes one half, then the maximum velocity becomes one fourth.

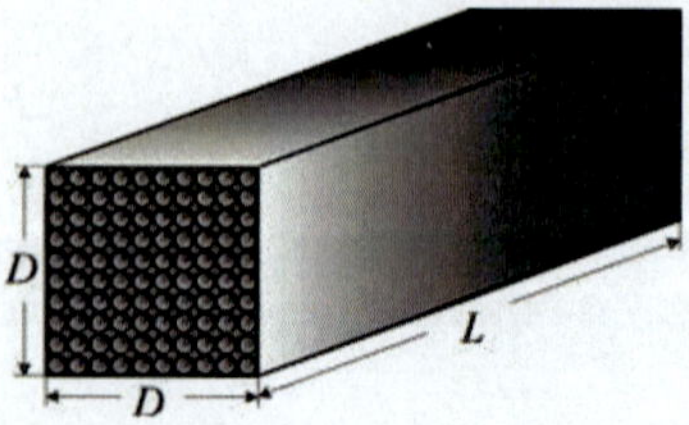

Fig. 5.7 Assembly of pipes

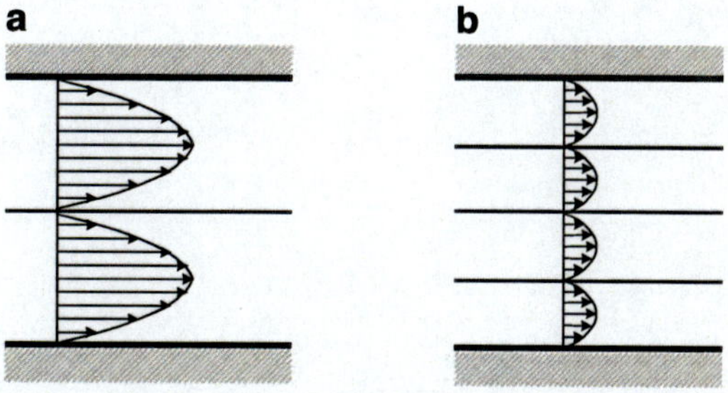

Fig. 5.8 Relationship between the pipe radius and the flow velocity

5.3.3 Flow in a Tank Filled with Solid Particles and Its Permeability

Consider the 'seepage flow' problem in a tank that is filled with solid particles as shown in Fig. 5.9. The total volume of this tank is V and the volume of the voids is V_v, so that the porosity is $n = V_v/V$. The mass flow velocity $\overline{V} = Q/\rho$ is given by the flux Q. The total surface area of solid particles is S_p, and the surface area per unit volume of one particle is s_v (if the particle is a sphere of diameter $d_p = 2r_p$, $s_v = 4\pi r_p^2/(4\pi r_p^3/3) = 6/d_p$). Then the surface area per unit volume S_v of the total solid particles is calculated as $S_v = s_v(1-n)$.

Since there is no known analytical solution to this problem, we simplify the problem by introducing a virtual pathway in the porous medium (Fig. 5.9). The length of the virtual path is L_e, and its longitudinal projection is L_p. The fluid flow passes around each particle, so the wet path length is $l_p = 2\pi r_p$ and the sectional area of flow is $s_p = \pi r_p^2$. Thus a diameter d_e of the virtual pathway can be determined by

$$\frac{s_p}{l_p} = \frac{d_e}{4}. \tag{5.85}$$

This d_e can also be formulated as

$$d_e = 4\frac{s_p}{l_p} = 4\frac{V_v}{S_p} = 4\frac{V\,n}{V\,s_v} = \frac{4n}{S_v(1-n)}. \tag{5.86}$$

The velocity $\bar{v}_e$ of fluid in the virtual pathway is

$$\frac{L_e}{\bar{v}_e} = \frac{L_p}{\overline{V}/n} \quad \Rightarrow \quad \bar{v}_e = \frac{\overline{V}}{n}\frac{L_e}{L_p}. \tag{5.87}$$

Applying the Hagen-Poiseulle flow (5.74) yields

$$\frac{\Delta p}{L_e} = \frac{32\mu\bar{v}_e}{d_e^2} = \kappa\frac{(1-n)^2}{n^3}\mu S_v^2\overline{V}, \qquad \kappa = 32\frac{L_e}{L_p^2} \tag{5.88}$$

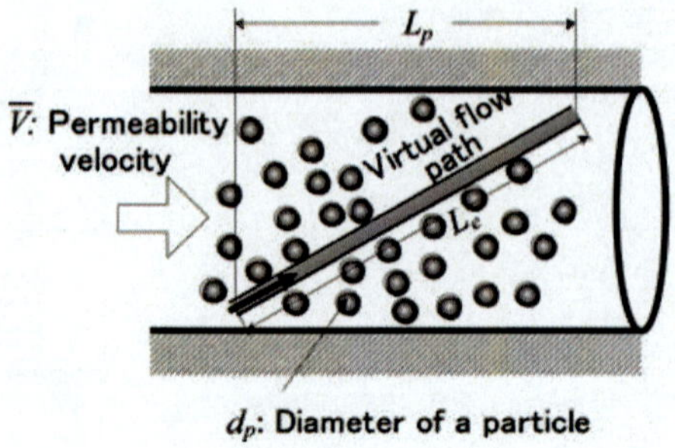

Fig. 5.9 Flow in a tank filled with solid particles

This is known as the Kozeny-Carman formula, and the Kozeny constant κ has been experimentally estimated as $\kappa \simeq 5$. The Kozeny-Carman formula can be simplified to

$$\overline{V} = -\frac{k^*}{\mu}\frac{\Delta p}{L_e} \tag{5.89}$$

which is known as Darcy's formula in which there is no explicitly observed effect of the particle diameter and the permeability k^* is just an experimental constant. It should be noted that in the Kozeny-Carman formula the effect of the particle size is given by $S_v = 6(1-n)/d_p$, so its square effect is explicitly estimated.

5.4 Fick's Law and Evaluation of the Diffusion Coefficient

We suppose that the driving force for the mass flux $\boldsymbol{q}_\alpha^m$, which is introduced by (5.45), is a gradient of the chemical potential. Thus the constitutive law for $\boldsymbol{q}_\alpha^m$ is given by the gradient of the chemical potential (cf. Appendix E.6 and E.7). For example, for a mixture of two elements without intermolecular forces, given by (E.85) and (E.86), we have

$$\boldsymbol{q}_1^m = -D\,\mathrm{grad}\,c_1, \qquad D = D_0\left(1 + \frac{\partial \ln \gamma_1}{\partial \ln c_1}\right) \tag{5.90}$$

where D_0 is a constant, and γ_1 is the activity coefficient of species 1. In the electrolyte solution given by (E.119) and (E.120) we have

$$\boldsymbol{q}_A^m = -D_A \mathrm{grad}\,c_A, \qquad D_A = \frac{u_+ u_-}{u_+ z_+ - u_- z_-} c_A \frac{\partial \mu_A}{\partial c_A} \quad (!A). \tag{5.91}$$

In the following discussion in this Section we treat the phenomenological Fick's law in terms of practical engineering aspects.

In classical diffusion theory, the diffusing mass flux $\boldsymbol{q}^m_\alpha$ is treated as the sum of molecular diffusion, pressure diffusion, thermal diffusion by the Soret effect, and so on, whereas the physical background of these effects are not completely discussed.

Let n be the 'effective' porosity, which directly affects the diffusion. Since isotropic diffusion is considered ($D^{\alpha\beta}_{ij} = D_{\alpha\beta}\delta_{ij}$), the flux of molecular diffusion for the αth component can be given according to the following form of Fick's law:

$$q^m_{\alpha i} = -\rho n \sum_{\beta=1}^{n^*} D^e_{\alpha\beta} \frac{\partial c_\beta}{\partial x_i} \tag{5.92}$$

where ρ is the mean density of the solution, and $D^e_{\alpha\beta}$ is the effective diffusion coefficient, which is related to the molecular diffusion coefficient $D_{\alpha\beta}$ by

$$D^e_{\alpha\beta} = \frac{\delta}{\tau} D_{\alpha\beta}, \tag{5.93}$$

and τ (> 1) is the tortuosity; this is related to the minimum diffusion path of length l and the effective path length l_e by

$$\tau = \left(\frac{l_e}{l}\right)^2 > 1 \tag{5.94}$$

and δ (≤ 1) is the constrictivity, which accounts for the effect of fine pores with a mean diameter d_p, comparable to the diameter of the solute molecules d_m. Then, let λ_p be

$$\lambda_p = \frac{d_m}{d_p} < 1 \tag{5.95}$$

and the empirical constrictivity can be experimentally derived as

$$\delta = (1 - \lambda_p)^2 (1 - 2.104\lambda_p + 2.09\lambda_p^3 - 0.95\lambda_p^5)$$
$$\delta = (1 - \lambda_p)^4$$
$$\delta = \exp(-4.6\lambda_p)$$
$$\delta = 1.03 \exp(-4.5\lambda_p)$$

By combining both coefficients of (5.92) and (5.93), we obtain the formation factor as

$$F = n\frac{\delta}{\tau}$$

This formation factor is related to the microscale geometry of the REV.

Let D_0^α be the diffusion coefficient of the αth component in pure water, and assume that the interactions between the αth and βth components are equivalent; therefore we have

$$D_{\alpha\beta} = \alpha^* D_0^\alpha \delta_{\alpha\beta} \qquad (!\alpha)$$

where α^* is the solute coefficient. Then Fick's law (5.92) is given as

$$q_{\alpha i}^m = -\rho D^e(c^\alpha)\frac{\partial c_\alpha}{\partial x_i} = -\rho n D^{e*}(c^\alpha)\frac{\partial c_\alpha}{\partial x_i} \tag{5.96}$$

$$D^e(c^\alpha) = nD^{e*}(c^\alpha) = \alpha^* F D_0^\alpha = \alpha^* n \frac{\delta}{\tau} D_0^\alpha \qquad (!\alpha). \tag{5.97}$$

Next, suppose that the effect of mechanical dispersion, as shown in Fig. 5.10, accounts for the locally dispersing phenomenon due to the solid particles. We consider the mean values and deviations of the velocity $\boldsymbol{v}$ and the concentration c_α

$$\overline{c_\alpha \boldsymbol{v}} = \overline{c_\alpha}\,\overline{\boldsymbol{v}} + \overline{\mathring{c}_\alpha \mathring{\boldsymbol{v}}} \tag{5.98}$$

($\overline{c_\alpha}\,\overline{\mathring{\boldsymbol{v}}} = \overline{\boldsymbol{v}}\,\overline{\mathring{c}_\alpha} = 0$). From the second term of the r.h.s. we can define the dispersive mass flux as

$$\boldsymbol{q}_\alpha^M = \overline{\mathring{c}_\alpha \mathring{\boldsymbol{v}}} \tag{5.99}$$

Using this dispersive mass flux $\boldsymbol{q}_\alpha^M$, a working hypothesis of Fick's law is introduced:

$$q_{\alpha i}^M = -\rho D_{ij}^{Me}(c^\alpha)\frac{\partial c_\alpha}{\partial x_j} = -\rho n D_{ij}^{Me*}(c^\alpha)\frac{\partial c_\alpha}{\partial x_j} \tag{5.100}$$

$$D_{ij}^{Me}(c^\alpha) = nD_{ij}^{Me*}(c^\alpha) \tag{5.101}$$

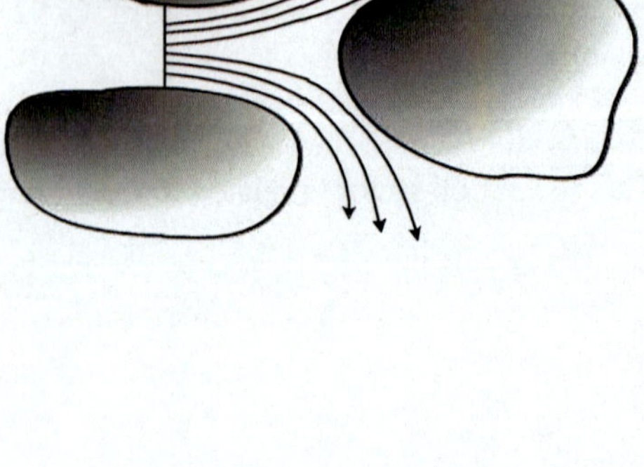

Fig. 5.10 Mechanical dispersion

where the anisotropic property is invoked because of the characteristics of dispersive flux. Based on experimental observations, $D_{ij}^{Me}(c^{\alpha})$ is considered as

$$D_{ij}^{Me}(c^{\alpha}) = \left\{a_T\, v\, \delta_{ij} + (a_L - a_T)\frac{v_i v_j}{v}\right\} f(\text{Pe}, \delta, c^{\alpha}) \tag{5.102}$$

where a_T and a_L are, respectively, the lateral and transverse dispersion coefficients, $v = |\, \boldsymbol{v}\,|$, Pe is the Peclèt number and δ is the constrictivity (Bear and Verruijt 1987). The function $f(\text{Pe}, \delta, c^{\alpha})$ is proposed as

$$f(\text{Pe}, \delta, c^{\alpha}) = \frac{\text{Pe}}{(\text{Pe} + 2 + 4\delta^2)}$$

(Bear and Bachmat 1990). However, usually this is set conveniently as

$$f(\text{Pe}, \delta, c^{\alpha}) \simeq 1.$$

The flux $\boldsymbol{q}_{\alpha}$ that combines both the molecular diffusion and mechanical dispersion can be written as

$$q_{\alpha\, i} \equiv q_{\alpha\, i}^{m} + q_{\alpha\, i}^{M} = -\rho \widetilde{D}_{ij}^{e}(c^{\alpha})\frac{\partial c_{\alpha}}{\partial x_j} = -\rho n \widetilde{D}_{ij}^{e*}(c^{\alpha})\frac{\partial c_{\alpha}}{\partial x_j} \tag{5.103}$$

$$\widetilde{D}_{ij}^{e}(c^{\alpha}) = D^{e}(c^{\alpha})\delta_{ij} + D_{ij}^{Me}(c^{\alpha}), \quad \widetilde{D}_{ij}^{e*}(c^{\alpha}) = D^{e*}\delta_{ij}(c^{\alpha}) + D_{ij}^{Me*}(c^{\alpha}). \tag{5.104}$$

5.5 Adsorption Isotherm and the Distribution Coefficient

Adsorption in a porous medium is processed by the segregation of some species (i.e., adsorbate) from the solution and subsequent precipitation on the surface of the solid (i.e., adsorbent). Adsorption is classified by *physical adsorption* or *physisorption* caused by an electrostatic force (i.e., Coulomb force) and a van der Waals force, while *chemisorption* is caused by surface reactions (Moore 1972). Adsorption of N_2 gas by SiO_2 is a physisorption. Adsorption of O_2 by activated charcoal, and adsorption of H_2 gas for metallic Ni are chemisorption. We need to heat the adsorbent under high pressure in order to remove the chemisorbed material; however, frequently, this resultant material is different from the originally adsorbed one. In physisorption several layers of adsorbed molecules may be formed, while in chemisorption only a monolayer is formed. Occasionally after chemisorption other physisorption processes may occur. The energy of physisorption ranges from 300 to 3,000 J/mol, while the energy required for chemisorption ranges from 40 to 400 KJ/mol. The conditions of temperature, hydrogen-ion exponent pH and redox potential Eh strongly affect the adsorption phenomena.

Note 5.5 (Adsorption, absorption and sorption). The process where a concentration of molecules or ions at an interface between a solid and a fluid becomes incorporated within the solid is referred to as *adsorption*. Note that at the interface of an electrolyte aqueous solution and a solid, the concentration may become dilute. We refer to this as a negative adsorption. When the adsorbed molecules or ions on the surface can be detached due to changes in the chemical conditions, *desorption* occurs.

On a solid-gas interface, we may observe that the gas molecules diffuse into the inner part of the solid. This is referred to as *absorption*. If both adsorption and absorption take place simultaneously, this process is referred to as *sorption*. ■

Note 5.6 (Van der Waals force). Even for nonpolar molecules, the polar character is generated due to instantaneous deviations in the electron orbit. Because of this electric field, the neighboring molecules become polarized, and the energy level of the total system becomes lower if the force is attractive rather than repulsive. Frozen carbon dioxide and crystals of iodine I_2 are examples of crystals formed by van der Waals forces, which are known as molecular crystals. Since the van der Waals forces have no orientation, the molecular crystals occur in a closely-packed structure. The van der Waals force $V(r)$ is inversely proportional to the sixth power of the intermolecular distance r: $V(r) = -C/r^6$. The van der Waals force is extremely small compared to chemical forces such as ionic bonds, covalent bonds and metallic bonds (i.e., less than 1/100). ■

Adsorption can be classified into two types; (1) the *equilibrium adsorption isotherm* in which the reaction process is in equilibrium, and (2) the *nonequilibrium adsorption isotherm* in which the reaction rate is important. Since classical thermodynamics treats only the equilibrium state, the nonequilibrium adsorption isotherm is commonly described in a phenomenological sense (cf. Appendices E.8 and E.9).

Here we list some examples of the equilibrium and nonequilibrium adsorption isotherms. Let the current area of an adsorbed layer be S, and let the final area of the layer after completion of the reaction be S_∞. Then the fractional coverage θ is expressed as

$$\theta = \frac{S}{S_\infty}. \tag{5.105}$$

5.5.1 *Langmuir's Equilibrium Adsorption Isotherm*

Gas adsorption for a solid was first formulated quantitatively by Langmuir (1916, 1918). The model assumes the following conditions:

1. Adsorption proceeds with monolayer coverage.
2. On a perfectly flat surface there are N-sites for adsorption; these are uniformly distributed on the surface, and are mutually equivalent.
3. A molecule adsorbed at a site does not affect the adsorption properties of neighboring sites.

Let k_a and k_d be, respectively, the adsorption and desorption rates of the species α in a gas under the partial pressure P_α. Following the reaction theory given in Appendices E.8 and E.9, the equilibrium reaction of adsorption and desorption can be written as

$$-\frac{d(1-\theta)}{dt} = \frac{d\theta}{dt} = k_a P_\alpha N(1-\theta) - k_d N\theta = 0. \tag{5.106}$$

Then we have

$$\theta = \frac{k_a P_\alpha}{k_d + k_a P_\alpha} = \frac{b_\alpha\, P_\alpha}{1 + b_\alpha\, P_\alpha} \qquad (!\alpha) \tag{5.107}$$

where

$$b_\alpha = \frac{k_a}{k_d} \tag{5.108}$$

is the adsorption coefficient for the species α.

If we plot the fractional coverage θ against the partial pressure P_α, we obtain a hyperbolae. Equation 5.107 can be written in the alternative form as

$$\frac{1}{\theta} = 1 + \frac{1}{b_\alpha\, P_\alpha} \qquad (!\alpha), \tag{5.109}$$

where a plot of $1/\theta$ v.s. $1/P_\alpha$ is linear and referred to as the Langmuir plot, which can be used to determine b_α.

In (5.107) if $b_\alpha\, P_\alpha \ll 1$; $(!\alpha)$, i.e., the partial pressure is small enough or the fractional coverage is small enough, we have

$$\theta = b_\alpha\, P_\alpha \qquad (!\alpha). \tag{5.110}$$

This is known as the linear adsorption isotherm. On the other hand if $b_\alpha\, P_\alpha \gg 1$; $(!\alpha)$, i.e., under high pressure or highly adsorbable, we have

$$1 - \theta = \frac{1}{b_\alpha\, P_\alpha} \qquad (!\alpha). \tag{5.111}$$

In a perfect gas and in a dilute solution, the concentration c_α is proportional to the partial pressure P_α, so that, for the case of a dilute solution that flows in a porous medium, (5.107) becomes

$$c_\alpha^* = \frac{b_\alpha\, c_\alpha}{1 + b_\alpha\, c_\alpha} \qquad (!\alpha) \tag{5.112}$$

where c_α^* and c_α are, respectively, the concentrations of a species α in the solid and in the solution. For the case of $b_\alpha\, c_\alpha \ll 1$; $(!\alpha)$, the equilibrium adsorption isotherm (5.112) is given by

$$c_\alpha^* = K_\alpha^d\, c_\alpha \qquad (!\alpha) \tag{5.113}$$

where $K_\alpha^d\,(=b_\alpha)$ is referred to as the distribution coefficient as described in (5.46).

We sometimes use an alternative form of (5.113)

$$c^*_\alpha = k_1\, c_\alpha + k_2 \tag{5.114}$$

where k_1 and k_2 are constants. This gives a nonlinear adsorption isotherm.

For practical solutions the activity a_α defined by (E.74) is used instead of the concentration c_α. Then (5.112) and (5.113) can be rewritten as

$$c^*_\alpha = \frac{b_\alpha\, a_\alpha}{1 + b_\alpha\, , a_\alpha} \qquad (!\alpha), \tag{5.115}$$

$$c^*_\alpha = K^d_\alpha\, a_\alpha \qquad (!\alpha). \tag{5.116}$$

Note that in the following discussions of the Freundlich and Temkin isotherms, a similar procedure can be applied by using the activity of practical solutions.

The coefficient b_α of (5.112) and (5.115) represents the equilibrium constant of adsorption/desorption as shown by (5.108). Then, by applying (E.141), we have the following interpretation:

$$b_\alpha = B_\alpha\, e^{-(\Delta H^\ominus)_\alpha/RT}, \qquad B_\alpha = e^{(\Delta S^e)^\ominus_\alpha/R}. \tag{5.117}$$

We can outline the procedure used to solve the diffusion problem under the Langmuir equilibrium adsorption isotherm. A diffusion field in the solution (5.29) together with Fick's law gives

$$n\rho\frac{dc_\alpha}{dt} + \nabla \cdot \boldsymbol{q}_\alpha = n\gamma_\alpha, \tag{5.118}$$

$$\boldsymbol{q}_\alpha = -k_\alpha\, \nabla\, c_\alpha \qquad (!\alpha) \tag{5.119}$$

where the adsorption term ζ^*_α is ignored since it is evaluated in the adsorption isotherm, $\boldsymbol{q}_\alpha$ is the mass flux of the species α in the solution, and k_α is a diffusion coefficient of species α in the solution. On the other hand, the diffusion field in the solid is given by (5.42) together with Fick's law:

$$(1-n)\rho^*\frac{d_* c^*_\alpha}{dt} + \nabla \cdot \boldsymbol{q}^*_\alpha = (1-n)\gamma^*_\alpha, \tag{5.120}$$

$$\boldsymbol{q}^*_\alpha = -k^*_\alpha\, \nabla\, c^*_\alpha \qquad (!\alpha). \tag{5.121}$$

Note that even if the adsorption reaction is in equilibrium, the diffusion field is time-dependent. Here $\boldsymbol{q}^*_\alpha$ is the mass flux of the species α in the solid, and k^*_α is a diffusion coefficient of a species α in the solid. Differentiating (5.112) of the Langmuir equilibrium adsorption isotherm with respect to time gives

$$\frac{\partial c^*_\alpha}{\partial t} = K^d_\alpha \frac{\partial c_\alpha}{\partial t}, \qquad \nabla\, c^*_\alpha = K^d_\alpha\, \nabla\, c_\alpha, \qquad K^d_\alpha = \frac{b_\alpha}{(1 + b_\alpha\, c_\alpha)^2} \qquad (!\alpha). \tag{5.122}$$

Substituting these into (5.120) and (5.121) and summing with (5.118) yields the following diffusion equation in terms of the concentration c_α of the solution:

$$n\rho R_\alpha^d \frac{\partial c_\alpha}{\partial t} + \left[n\rho \boldsymbol{v}_\alpha + (1-n)\rho^* K_\alpha^d \boldsymbol{v}_\alpha^* \right] \cdot \nabla c_\alpha - \nabla \cdot \left(\widetilde{k}_\alpha \nabla c_\alpha \right) = \widetilde{\gamma}_\alpha \quad (!\alpha), \tag{5.123}$$

$$R_\alpha^d = 1 + \frac{(1-n)\rho^*}{n\rho} K_\alpha^d, \quad \widetilde{k}_\alpha = k_\alpha + k_\alpha^* K_\alpha^d, \quad \widetilde{\gamma}_\alpha = n\gamma_\alpha + (1-n)\gamma_\alpha^* \quad (!\alpha). \tag{5.124}$$

Since the diffusion velocity in the solid is usually slow, we have $\widetilde{k}_\alpha \simeq k_\alpha$. The particle velocity $\boldsymbol{v}_\alpha^*$ is also slow. Therefore (5.123) can be approximated by the result

$$n\rho \left[R_\alpha^d \frac{\partial c_\alpha}{\partial t} + \boldsymbol{v}_\alpha \cdot \nabla c_\alpha \right] - \nabla \cdot \left(k_\alpha \nabla c_\alpha \right) = \widetilde{\gamma}_\alpha \quad (!\alpha). \tag{5.125}$$

5.5.2 Freundlich's Equilibrium Adsorption Isotherm

On the surface of a crystal, defects related to dislocation of the crystal structure are commonly observed. On these nonuniform surfaces a formulation adopted by Freundlich (1926) fits with experimental data: i.e.,

$$\theta = k(P_\alpha)^{1/m} \tag{5.126}$$

which gives a nonlinear equilibrium adsorption isotherm. Here k is constant and m is also a constant greater than unity.

Equation 5.126 can be rewritten in terms of the concentrations c_α^* and c_α as

$$c_\alpha^* = k_\alpha (c_\alpha)^{1/m_\alpha} \quad (!\alpha). \tag{5.127}$$

5.5.3 Temkin's Equilibrium Adsorption Isotherm

Temkin's equilibrium isotherm corresponds to a logarithmic form of a nonlinear adsorption isotherm given by

$$\theta = c_1 \ln(c_2 P_\alpha) \tag{5.128}$$

where c_1 and c_2 are constants. In this case, the enthalpy of adsorption is known to be linear with respect to the partial pressure P_α.

Equation 5.128 can be rewritten in terms of the concentrations c_α^* and c_α as follows:

$$c_\alpha^* = c_{1\alpha} \ln(c_{2\alpha} c_\alpha) \qquad (!\alpha). \tag{5.129}$$

5.5.4 *Langmuir's Nonequilibrium Adsorption Isotherm*

For the case of nonequilibrium adsorption where the rate of reaction cannot be ignored, Langmuir's nonequilibrium adsorption isotherm corresponding to (5.112) is given by

$$\frac{\partial c_\alpha^*}{\partial t} = k_\alpha \left(\frac{b_\alpha\, c_\alpha}{1 + b_\alpha\, c_\alpha} - c_\alpha^* \right) \qquad (!\alpha) \tag{5.130}$$

where k_α and b_α are constants (Hendricks 1972). If $b_\alpha\, c_\alpha \gg$, $(!\alpha)$, it reduces to a linear form of the nonequilibrium adsorption isotherm

$$\frac{\partial c_\alpha^*}{\partial t} + k_\alpha\, c_\alpha^* = 0 \qquad (!\alpha) \tag{5.131}$$

where k_α is a constant. Thus, (5.114) can be rewritten as

$$\frac{\partial c_\alpha^*}{\partial t} = k_\alpha\, (k_{1\alpha}\, c_\alpha + k_{2\alpha} - c_\alpha^*) \tag{5.132}$$

where $k, k_\alpha, k_{1\alpha}, k_{2\alpha}$ are constants (Lapidus and Amundson 1952).

5.5.5 *Freundlich's Nonequilibrium Adsorption Isotherm*

A nonequilibrium adsorption isotherm corresponding to (5.127) is given by

$$\frac{\partial c_\alpha^*}{\partial t} = k_\alpha \left(k_{1\alpha}\, (c_\alpha)^{1/m_\alpha} - c_\alpha^* \right) \tag{5.133}$$

where $k_\alpha, k_{1\alpha}, m_\alpha$ are constants.

5.6 Transport Equations and Similitude Laws

The Navier-Stokes equation (2.238) for an incompressible Newtonian viscous fluid can be written as

$$\frac{\partial \boldsymbol{v}}{\partial t} + \nabla \boldsymbol{v} \cdot \boldsymbol{v} = -\frac{1}{\rho} \nabla p + \mu^{\ddagger} \Delta \boldsymbol{v} + \boldsymbol{b} \tag{5.134}$$

$$\nabla \cdot \boldsymbol{v} = 0 \tag{5.135}$$

where $\Delta = \nabla \cdot \nabla$ is the Laplacian. The diffusion equation (5.125) together with Fick's law is given by

$$R_d \frac{\partial c}{\partial t} + \nabla c \cdot \boldsymbol{v} = D \Delta c, \qquad D = \frac{k_D}{n\rho} \tag{5.136}$$

where k_D is the diffusion coefficient. The thermal conduction equation together with Fourier's law is given by

$$\frac{\partial T}{\partial t} + \nabla T \cdot \boldsymbol{v} = \alpha \Delta T, \qquad \alpha = \frac{k_T}{C_v} \tag{5.137}$$

where k_T is the thermal conductivity. We will discuss similitude laws for these transport equations.

Let τ, L and V be the standard time, length and velocity, respectively, then the normalized time, length and velocity are given by

$$t^* = \frac{t}{\tau}, \qquad \boldsymbol{x}^* = \frac{\boldsymbol{x}}{L}, \qquad \boldsymbol{v}^* = \frac{\boldsymbol{v}}{V} \tag{5.138}$$

Note that in most cases we can set $\tau = L/V$. Using the above definitions of time-space normalization we can introduce the following normalized pressure p^*, concentration c^* and temperature T^*:

$$p - p_0 = \rho V^2 p^*, \quad c - c_0 = (c_0 - \bar{c})\, c^*, \quad T - T_0 = (T_0 - \overline{T})\, T^*. \tag{5.139}$$

where p_0, c_0, T_0 are the initial values for the pressure, concentration and temperature, respectively, and $\bar{c}$, $\overline{T}$ are boundary values of concentration and temperature.

Substituting (5.138) and (5.139) into (5.134)–(5.137), we obtain the normalized transport equations as[3]

$$\mathrm{St} \frac{\partial \boldsymbol{v}^*}{\partial t^*} + \nabla^* \boldsymbol{v}^* \cdot \boldsymbol{v}^* = -\nabla^* p^* + \frac{1}{\mathrm{Re}} \Delta^* \boldsymbol{v}^* + \frac{1}{\mathrm{Fr}} \frac{\boldsymbol{b}}{\rho g}, \tag{5.140}$$

$$\nabla^* \cdot \boldsymbol{v}^* = 0, \tag{5.141}$$

$$\mathrm{St}_D \frac{\partial c^*}{\partial t^*} + \nabla^* c^* \cdot \boldsymbol{v}^* = \frac{1}{\mathrm{Pe}_D} \Delta^* c^*, \tag{5.142}$$

[3] Note that under the base vectors $\{\boldsymbol{e}_i^*\}$ differential operators in the normalized space are defined by

$$\mathrm{grad}^* = \nabla^* = \boldsymbol{e}_i^* \frac{\partial}{\partial x_i^*}, \qquad \mathrm{div}^* = \nabla^* \cdot, \qquad \Delta^* = \nabla^* \cdot \nabla^*.$$

$$\mathrm{St}\frac{\partial T^*}{\partial t^*} + \nabla^* T^* \cdot v^* = \frac{1}{\mathrm{Pe}_T}\Delta^* T^* \tag{5.143}$$

where g is the gravitational acceleration, St is the Strouhal number, Re is the Reynolds number, Fr is the Froude number, St_D is the Strouhal number for diffusion, Pe_D is the Péclet number for diffusion and Pe_T is the Péclet number for temperature. Note that we have the following physical interpretations:

$$\mathrm{St} = \frac{L/V}{\tau} = \frac{\text{time scale for flow}}{\text{time scale for unsteady flow}}, \tag{5.144}$$

$$\mathrm{Re} = \frac{V^2/L}{\mu^\ddagger V/L^2} = \frac{VL}{\mu^\ddagger} = \frac{\text{inertia force}}{\text{viscous force}}, \tag{5.145}$$

$$\mathrm{Fr} = \frac{\rho V^2/L}{\rho g} = \frac{\text{inertia force}}{\text{gravitational force}}, \tag{5.146}$$

$$\mathrm{St}_D = \frac{R_d L/V}{\tau} = \frac{\text{flow time scale including absorption}}{\text{unsteady time scale}}, \tag{5.147}$$

$$\mathrm{Pe}_D = \frac{VL}{D} = \frac{\text{mass transport by convection}}{\text{mass transport by diffusion}}, \tag{5.148}$$

$$\mathrm{Pe}_T = \frac{VL}{\alpha} = \frac{\text{heat transport by convection}}{\text{heat transport by diffusion}}. \tag{5.149}$$

Chapter 6
Classical Theory of Consolidation for Saturated Porous Media

Soil is a mixture of a solid phase, a liquid phase (i.e., water) and a gaseous phase. Saturated soil is a two-phase material which consists of a solid phase and a liquid phase. The formulation of a theory for a saturated soil commenced with the concept of the effective stress of Terzaghi at the beginning of the twentieth century, and Terzaghi's concept was extended by Biot to a three dimensional consolidation theory.

In this Chapter, we review the classical theory of consolidation for saturated soils under the assumption that both the intrinsic part of the solid phase and the fluid are incompressible, and that the deformation of the soil results from a nonlinear deformation characteristic of the skeleton and the seepage of fluid.

6.1 Mass Conservation Law and Seepage Equation

As described in Sect. 5.2, the fluid flow problem in a porous medium is given by a seepage equation, which we again show here together with Darcy's law, assuming that the intrinsic solid particles and fluid are incompressible. Note that we will not consider diffusive processes.

A mass conservation law for water together with the incompressibility condition results in the following seepage equation:

$$\operatorname{tr} \boldsymbol{D}^* + \operatorname{div} \tilde{\boldsymbol{v}} = 0 \tag{6.1}$$

where $\tilde{\boldsymbol{v}}$ is the seepage velocity which is defined by (see (5.55))

$$\tilde{\boldsymbol{v}} = n\,(\boldsymbol{v} - \boldsymbol{v}^*). \tag{6.2}$$

The driving force associated with the seepage velocity $\tilde{\boldsymbol{v}}$ is the reduced Bernoulli potential, which consists of the pressure head and the datum head ζ; thus Darcy's law takes the form

Y. Ichikawa and A.P.S. Selvadurai, *Transport Phenomena in Porous Media*,
DOI 10.1007/978-3-642-25333-1_6, © Springer-Verlag Berlin Heidelberg 2012

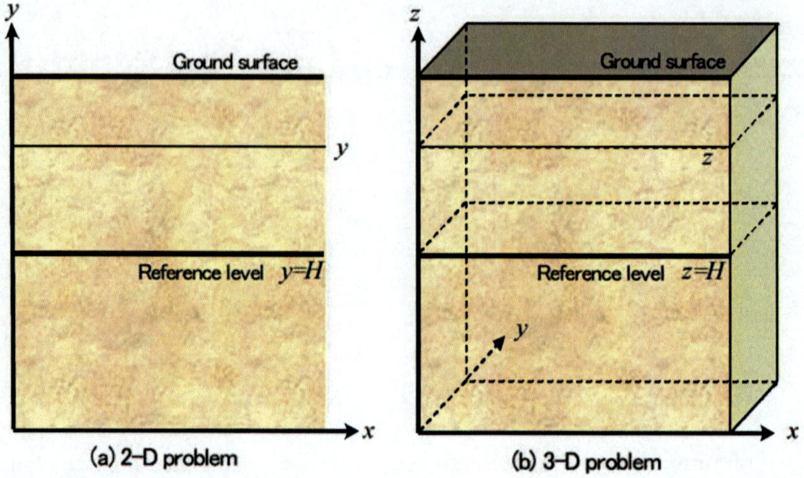

Fig. 6.1 Reference levels

$$\tilde{v} = -\boldsymbol{k}\,\mathrm{grad}\,\phi, \qquad \phi = \frac{p}{\rho g} + \zeta \tag{6.3}$$

where ρ is the mass density of water, $\boldsymbol{k}$ is the hydraulic conductivity tensor, g is the gravity constant, ϕ is the total potential, and p is the pore pressure. The datum potential is measured from a reference level H; for a two-dimensional problem as shown in Fig. 6.1a and for a three-dimensional problem as shown in Fig. 6.1b, we have, respectively

$$\zeta = y - H, \qquad \zeta = z - H. \tag{6.4}$$

Note that in conventional geotechnical engineering problems, the pore fluid is water and the influence of temperature is neglected. If the soil is considered to be isotropic, the second order tensor $\boldsymbol{k}$ reduces to $\boldsymbol{k} = k\,\boldsymbol{i}$ where $\boldsymbol{i}$ is the unit tensor of the coordinate system of the current configuration, and k is referred to as the hydraulic conductivity.

By substituting (6.2) and (6.3) into the mass conservation equation (6.1), and considering the incompressibility of water (ρ = constant), we have the following seepage equation, which includes the effect of the volume change of the skeleton $d\varepsilon_v/dt = \mathrm{tr}\,\boldsymbol{D}^*$:

$$\frac{d\varepsilon_v}{dt} + \nabla \cdot (\boldsymbol{k}\,\nabla\phi) = 0. \tag{6.5}$$

6.2 Conservation of Linear Momentum, Effective Stress and Biot's Consolidation Theory

The conservation law of linear momentum for a saturated porous medium can be written as

$$\frac{d}{dt}\int_{\Omega} n\rho v\, dv + \frac{d}{dt}\int_{\Omega}(1-n)\rho^{*}v^{*}\, dv = \int_{\Omega}(\nabla\cdot\sigma + \bar{\rho}b)dv \tag{6.6}$$

where ρ^* is the mass density of the solid phase, v^* is the particle velocity of the solid phase, σ is the total stress, and b is the body force vector, which is given as $b = (0, 0, -g)$ if the z-axis is directed away from the ground surface (g is the gravitational constant). Note that the mean mass density $\bar{\rho}$ is calculated as

$$\bar{\rho} = n\,\rho + (1-n)\,\rho^{*}. \tag{6.7}$$

We introduce Terzaghi's effective stress by

$$\sigma = \sigma' - p i\,. \tag{6.8}$$

Then, (6.6) is written as

$$\rho\frac{d(nv)}{dt} + \rho^{*}\frac{d(1-n)v^{*}}{dt} = \nabla\cdot\sigma' - \nabla p + \bar{\rho}b \tag{6.9}$$

where σ' is the effective stress, which acts in the solid skeleton, and p is the pore fluid pressure. We introduce the lower-case p in order to distinguish it from the pore fluid pressure for the current configuration in a finite strain theory as shown in Sect. 6.3. It should be noted that σ and σ' are positive for tension, and p is positive for compression.

For static problems (i.e., in the absence of inertia terms), (6.9) is written as

$$\nabla\cdot\sigma' - \nabla p + \bar{\rho}b = 0. \tag{6.10}$$

This is combined with the seepage equation (6.5), giving Biot's consolidation equation.

The pore pressure p is usually set as the hydrostatic pressure p_0 plus the excess pore pressure dp. Then, the effective stress is written in an incremental form as

$$\sigma + d\sigma = \sigma' + d\sigma' - (p_0 + dp)i \quad \Rightarrow \quad \dot{\sigma} = \dot{\sigma}' - \dot{p}i\,. \tag{6.11}$$

The equation of equilibrium (6.10) is then given as

$$\nabla\cdot\dot{\sigma}' - \nabla\dot{p} = 0. \tag{6.12}$$

6.3 Finite Strain Theory of Consolidation ♣

In this section we reconstruct the theory of consolidation by introducing the concept of a finite strain and a nominal stress rate, which are given in Chap. 2. Note that in Chap. 5 a mixture theory was developed for a porous medium with multiple chemical species, while in this chapter we will only discuss a complete fluid phase and solid phase in which the behavior of each species is not considered. We can therefore introduce the seepage equations as shown in Sects. 5.2.3 and 5.2.4. That is, in the mixture theory for a porous medium we introduce, e.g., for a fluid phase, a particle velocity $\boldsymbol{v}_\alpha$ and a deformation gradient $\boldsymbol{F}_\alpha$ for each species, and define the mass-averaged velocity $\boldsymbol{v} = \sum_\alpha \rho_\alpha \boldsymbol{v}_\alpha / \rho$, which gives the velocity gradient $\boldsymbol{L}$, the stretching $\boldsymbol{D}$ and the spin tensor $\boldsymbol{W}$, therefore we cannot introduce a reference configuration and a deformation gradient for the whole fluid phase. In this section we examine the entire fluid phase and solid phase without discussing the behavior of individual chemical species. In this case the reference configuration is given for the solid phase, since it is possible to measure the deformation of the solid phase. Then we can introduce a finite strain theory of consolidation for a porous medium.

6.3.1 *Seepage Equation of Consolidation in a Lagrangian Form*

By introducing the reference configuration for the entire solid phase, we define the basis $\boldsymbol{E}_K$ for the reference configuration, and the reference point is given by $\boldsymbol{X}^* = X_K^* \boldsymbol{E}_K$. For the current configuration the basis is $\boldsymbol{e}_i$, and the current point of the solid phase is given by $\boldsymbol{x} = x_i\, \boldsymbol{e}_i$ (which is the same as the current point of the fluid phase). A deformation gradient of the solid phase is $\boldsymbol{F}^*$. Then we have

$$\boldsymbol{F}^* = \text{Grad}^* \boldsymbol{x} = \frac{\partial x_i}{\partial X_K^*} \boldsymbol{e}_i \otimes \boldsymbol{E}_K, \quad \text{Grad}^* = \boldsymbol{E}_K \frac{\partial}{\partial X_K^*}, \quad J^* = \det \boldsymbol{F}^*. \tag{6.13}$$

As shown by (6.3), Darcy's law for the current configuration is given for the seepage velocity $\tilde{\boldsymbol{v}}$ as

$$\tilde{\boldsymbol{v}} = -\boldsymbol{k} \left(\text{grad} p - \rho \boldsymbol{b} \right), \qquad \text{grad} = \boldsymbol{e}_i \frac{\partial}{\partial x_i} \tag{6.14}$$

where $\rho = \sum_\alpha \rho_\alpha$ is the total mass density of the fluid phase defined by $(5.20)_2$, $\boldsymbol{k}$ is the hydraulic conductivity tensor for the current configuration, $\boldsymbol{b}$ is the body force vector, which is $\boldsymbol{b} = (0,\, 0,\, -g)$ if gravity is working in $-x_3$-direction.

A surface element ds for the current configuration is related to the surface element dS for the reference configuration of the solid phase by Nanson's formula (2.86) as

$$\boldsymbol{n}\, ds = J^* (\boldsymbol{F}^*)^{-T} \boldsymbol{N}\, dS. \tag{6.15}$$

Note that the mass density of the fluid phase is ρ for the current configuration, and we introduce the mass density of the fluid phase for the reference configuration as

$$\rho_0^{\dagger} = \rho J^*. \tag{6.16}$$

The seepage velocity $\tilde{v}$ for the current configuration is related to the seepage velocity $\tilde{V}$ for the reference configuration as

$$\tilde{\boldsymbol{v}} \cdot \boldsymbol{n}\, ds = \tilde{\boldsymbol{V}} \cdot \boldsymbol{N}\, dS. \tag{6.17}$$

We substitute (6.17) into (6.15), and obtain

$$\tilde{\boldsymbol{V}} = J^*(\boldsymbol{F}^*)^{-1}\tilde{\boldsymbol{v}} = -\boldsymbol{K}\left(\text{Grad}^* P - \rho_0^{\dagger}\boldsymbol{b}^0\right), \tag{6.18}$$

$$\boldsymbol{K} = J^*(\boldsymbol{F}^*)^{-1}\boldsymbol{k}\,(\boldsymbol{F}^*)^{-T}, \quad \text{Grad}^* P = (\boldsymbol{F}^*)^T \text{grad} p, \quad \boldsymbol{b}^0 = (J^*)^{-1}(\boldsymbol{F}^*)^T\boldsymbol{b} \tag{6.19}$$

where $\boldsymbol{K}$ is the hydraulic conductivity tensor for the reference configuration, P is the nominal pore pressure, which is written as

$$P = J^*\frac{\partial X_K^*}{\partial x_i}\delta_{iK}\, p = J^* \text{tr}\,(\boldsymbol{F}^*\mathbb{I})^{-1} p, \qquad \mathbb{I} = \delta_{Ki}\, \boldsymbol{E}_K \otimes \boldsymbol{e}_i. \tag{6.20}$$

If $\boldsymbol{i} = \delta_{ij}\, \boldsymbol{e}_i \otimes \boldsymbol{e}_j$ is the identity tensor for the current configuration, the pore pressure tensor $\boldsymbol{p}$ and the nominal pore pressure tensor $\boldsymbol{P}$ are introduced as

$$\boldsymbol{p} = p\,\boldsymbol{i} = p\,\delta_{ij}\, \boldsymbol{e}_i \otimes \boldsymbol{e}_j, \qquad \boldsymbol{P} = P\,\mathbb{I} = P\,\delta_{Ki}\, \boldsymbol{E}_K \otimes \boldsymbol{e}_i = J^*(\boldsymbol{F}^*)^{-1}\boldsymbol{p} \tag{6.21}$$

where $\mathbb{I}$ is the shifter, which transfers the basis $\boldsymbol{e}_i$ for the current configuration to the basis $\boldsymbol{E}_K$ for the reference configuration.

We now introduce incompressibility conditions for the fluid phase and the solid phase (ρ = constant, ρ^* = constant). Under these conditions we have (5.55) and (5.56). Since $\text{tr}\,\boldsymbol{D}^* = \text{div}\,\boldsymbol{v}^*$, we can apply Nanson's formula to (5.55), and obtain the following seepage equation for consolidation in the reference configuration:

$$\text{tr}\,\tilde{\boldsymbol{D}}^* + \text{Div}^*\,\tilde{\boldsymbol{V}} = 0 \tag{6.22}$$

$$\tilde{\boldsymbol{D}}^* = \frac{1}{2}\Big[\text{Grad}^*\,\boldsymbol{V}^* + (\text{Grad}^*\,\boldsymbol{V}^*)^T\Big], \quad \boldsymbol{V}^* = J^*(\boldsymbol{F}^*)^{-1}\boldsymbol{v}^*. \tag{6.23}$$

It may be noted that Darcy's law (6.18) is applied to the term $\tilde{\boldsymbol{V}}$ of (6.22).

6.3.2 Lagrangian Equation of Equilibrium

As shown by (2.116), the equation of equilibrium in terms of the total stress for the reference configuration can be written as

$$\mathrm{Div}^* \boldsymbol{\Pi}^T(\boldsymbol{X}^*) + \bar{\rho}_0 \boldsymbol{b}(\boldsymbol{X}^*) = \boldsymbol{0} \tag{6.24}$$

where

$$\boldsymbol{\Pi} = J^*(\boldsymbol{F}^*)^{-1}\boldsymbol{\sigma} \tag{6.25}$$

is the first Piola-Kirchhoff stress corresponding to the total stress $\boldsymbol{\sigma}$, and we have

$$\bar{\rho}_0 = \bar{\rho} J^*, \qquad \bar{\rho} = \rho n + \rho^*(1-n). \tag{6.26}$$

Let us decompose $\boldsymbol{\Pi}$ into an effective stress $\boldsymbol{\Pi}'$ and a pore pressure $\boldsymbol{P} = P\,\mathbb{I}$ such that

$$\boldsymbol{\Pi} = \boldsymbol{\Pi}' - \boldsymbol{P}, \qquad \boldsymbol{\Pi}' = J^*(\boldsymbol{F}^*)^{-1}\boldsymbol{\sigma}', \quad \boldsymbol{P} = J^*(\boldsymbol{F}^*)^{-1}\boldsymbol{p}. \tag{6.27}$$

Note that $\boldsymbol{P}$ is already given in (6.21). By substituting (6.27) into (6.24), we obtain the following Lagrangian equation of equilibrium under a finite strain field:

$$\mathrm{Div}^* \boldsymbol{\Pi}'^T - \mathbb{I}^T \mathrm{Grad}^* P + \bar{\rho}_0 \boldsymbol{b} = \boldsymbol{0}. \tag{6.28}$$

In indicial form (6.28) can be written as

$$\frac{\partial \Pi'_{Ki}}{\partial X^*_K} - \frac{\partial P}{\partial X^*_K}\delta_{Ki} + \bar{\rho}_0 b_i = 0. \tag{6.29}$$

6.3.3 *Incremental Form of the Equation of Equilibrium*

By taking a time-differentiation of (6.28), we have

$$\frac{\partial \dot{\Pi}'_{Ki}}{\partial X^*_K} - \frac{\partial \dot{P}}{\partial X^*_K}\delta_{Ki} + \dot{\bar{\rho}}_0 b_i = 0 \tag{6.30}$$

where $\boldsymbol{b}$ is a constant. This is referred to as the *total Lagrangian incremental equilibrium equation*. The vector form of (6.30) is given as

$$\mathrm{Div}^* \dot{\boldsymbol{\Pi}}'^T - \mathbb{I}^T \mathrm{Grad}^* \dot{P} + \dot{\bar{\rho}}_0 \boldsymbol{b} = \boldsymbol{0}. \tag{6.31}$$

As shown by (2.128), we can introduce a rate of the effective nominal stress $\overset{\circ}{\boldsymbol{\Pi}}{}'$ and a rate of the nominal pore pressure $\overset{\circ}{\boldsymbol{P}}$ as

$$\overset{\circ}{\boldsymbol{\Pi}}{}' = (J^*)^{-1}\boldsymbol{F}^*\dot{\boldsymbol{\Pi}}' = \dot{\boldsymbol{\sigma}}' - \boldsymbol{L}^*\boldsymbol{\sigma}' + \boldsymbol{\sigma}' \,\mathrm{tr}\, \boldsymbol{D}^*, \tag{6.32}$$

$$\overset{\circ}{\boldsymbol{P}} = (J^*)^{-1}\boldsymbol{F}^*\dot{\boldsymbol{p}} = \dot{\boldsymbol{p}} - \boldsymbol{L}^*\boldsymbol{p} + \boldsymbol{p}\,\mathrm{tr}\,\boldsymbol{D}^*. \tag{6.33}$$

Then, we have

$$\frac{\partial \overset{\circ}{\Pi}{}'_{ji}}{\partial x_j} = (J^*)^{-1}\frac{\partial \dot{\Pi}'_{Ki}}{\partial X^*_K}, \qquad \frac{\partial \overset{\circ}{P}}{\partial x_j}\delta_{ji} = (J^*)^{-1}\frac{\partial \dot{P}}{\partial X^*_K}\delta_{Ki}.$$

Thus, (6.30) can be rewritten as

$$\frac{\partial \overset{\circ}{\Pi}{}'_{ji}}{\partial x_j} - \frac{\partial \overset{\circ}{P}}{\partial x_i} + \dot{\bar{\rho}} b_i = 0 \tag{6.34}$$

where $\bar{\rho} = (J^*)^{-1}\bar{\rho}_0$. This is referred to as the *updated Lagrangian incremental equilibrium equation*. In a vector form we have

$$\mathrm{div}\left(\overset{\circ}{\boldsymbol{\Pi}}{}'\right)^T - \mathrm{grad}\,\overset{\circ}{P} + \dot{\bar{\rho}}\,\boldsymbol{b} = \boldsymbol{0}. \tag{6.35}$$

6.4 A Weak Form of Biot's Consolidation Equations and Finite Element Analysis

The analytical solution of these three-dimensional coupled equations can be accomplished for only very specialized states of deformation (Selvadurai 2007). For this reason, computational approaches have been developed for the solution of poroelasticity problems; a finite element scheme is used here to solve this problem.

6.4.1 Strong Form

The strong form of Biot's consolidation equations is given as follows:

Governing equations

$$\frac{\partial \Delta\sigma'_{ij}}{\partial x_j} - \frac{\partial \Delta p}{\partial x_i} = 0 \qquad \text{in } \Omega \tag{6.36}$$

$$\frac{\partial \Delta\varepsilon_v}{\partial t} + \frac{\partial}{\partial x_i}\left(\frac{k_{ij}}{\gamma_w}\frac{\partial \Delta p}{\partial x_j}\right) = 0 \quad \text{in } \Omega \tag{6.37}$$

Strain–displacement relation

$$\Delta\varepsilon_{ij} = \frac{1}{2}\left(\frac{\partial \Delta u_i}{\partial x_j} + \frac{\partial \Delta u_j}{\partial x_i}\right) \tag{6.38}$$

Boundary conditions (BC)[1]

$$\Delta u_i = \Delta\bar{u}_i \quad \text{on} \quad \partial\Omega_u \tag{6.39}$$

$$\left(\Delta\sigma'_{ij} - \Delta p\,\delta_{ij}\right)n_j = \Delta\bar{t}_i \quad \text{on} \quad \partial\Omega_t \tag{6.40}$$

$$\Delta p = \Delta\bar{p} \quad \text{on} \quad \partial\Omega_p \tag{6.41}$$

$$-\frac{k_{ij}}{g}\frac{\partial \Delta p}{\partial x_j}n_i = \Delta\bar{q}^w \quad \text{on} \quad \partial\Omega_q \tag{6.42}$$

Initial conditions (IC)

$$\Delta u_i(\boldsymbol{x}, t=0) = \Delta u_0(\boldsymbol{x}), \quad \Delta p(\boldsymbol{x}, t=0) = \Delta p_0(\boldsymbol{x}) \quad \text{in } \Omega \tag{6.43}$$

Here σ'_{ij} is the effective stress, which acts on the solid skeleton, p is the pore pressure, ε_v is the volumetric strain, k_{ij} is the hydraulic conductivity tensor, and $\gamma_w = \rho g$ is the bulk weight of the liquid phase. It should be noted that the mass flux is described as $\boldsymbol{q}^w = -\boldsymbol{K}\nabla p/g$, and the boundary condition is given by (6.42). We use a convention where σ'_{ij} is positive for tension, and p is positive for compression.

We assume that in Biot's consolidation equations the porous medium is fully saturated, and the fluid phase is incompressible. The excess pore pressure Δp is an increase (or a decrease) of pore pressure from the hydrostatic pressure; therefore the equation of equilibrium (6.36) is given in an incremental form.

6.4.2 Weak Form

The weak form of Biot's consolidation equations is obtained as

$$\int_{\partial\Omega_t} \Delta\bar{t}_i\, v_i\, ds - \int_{\Omega}\left(\Delta\sigma'_{ij} - \Delta p\delta_{ij}\right)\frac{\partial v_i}{\partial x_j}\, dv = 0, \tag{6.44}$$

---

[1]The load boundary condition (6.40) is given for a total stress. If it is given for an effective stress, it is modified as

$$\Delta\sigma'_{ij}n_j = \Delta\bar{t}'_i.$$

The weak form is modified because of this condition.

$$\int_{\partial\Omega_q} \frac{\Delta\bar{q}^w}{\rho}\eta\, ds + \int_{\Omega}\left(\frac{\partial\Delta\varepsilon_v}{\partial t}\eta - \frac{k_{ij}}{\gamma_w}\frac{\partial\Delta p}{\partial x_j}\frac{\partial\eta}{\partial x_i}\right) dv = 0. \tag{6.45}$$

6.4.3 Finite Element Analysis

For purposes of illustration we restrict our attention to the two-dimensional problem, and introduce a Galerkin finite element approximation to the weak form (6.44) and (6.45). The unknown functions are the increment of displacement $\Delta\boldsymbol{u}$ and the incremental excess pore pressure Δp. Introducing the global shape functions $N_\alpha(x, y)$ $(\alpha = 1, 2, \cdots, N)$ and $\bar{N}_\alpha(x, y)$ $(\alpha = 1, 2, \cdots, \bar{N})$ corresponding to each variable, we have

$$\Delta\boldsymbol{u} \simeq \Delta\boldsymbol{u}_h = \boldsymbol{N}(x, y)\Delta\boldsymbol{U}(t), \tag{6.46}$$

$$\Delta p \simeq \Delta p_h = \bar{\boldsymbol{N}}(x, y)\Delta\boldsymbol{P}(t), \tag{6.47}$$

where

$$\boldsymbol{N}(x, y) = \begin{bmatrix} N_1 & 0 & N_2 & 0 & \cdots & N_\alpha & 0 & \cdots & N_N & 0 \\ 0 & N_1 & 0 & N_2 & \cdots & 0 & N_\alpha & \cdots & 0 & N_N \end{bmatrix} \tag{6.48}$$

$$\Delta\boldsymbol{U}(t) = \left[\Delta U_x^1\ \Delta U_y^1\ \Delta U_x^2\ \Delta U_y^2 \cdots \Delta U_x^\alpha\ \Delta U_y^\alpha \cdots \Delta U_x^N\ \Delta U_y^N\right]^T \tag{6.49}$$

$$\bar{\boldsymbol{N}}(x, y) = \left[\bar{N}_1\ \bar{N}_2 \cdots \bar{N}_\alpha \cdots \bar{N}_{\bar{N}}\right] \tag{6.50}$$

$$\Delta\boldsymbol{P}(t) = \left[\Delta P_1\ \Delta P_2 \cdots \Delta P_\alpha \cdots \Delta P_{\bar{N}}\right]^T \tag{6.51}$$

where N and $\bar{N}$ are the numbers of nodes of the displacement and pore pressure, respectively. Note that in a finite element analysis of the consolidation problem we commonly use a shape function for the pore pressure that is one order less than the shape function for the displacement due to the stability of a numerical scheme. Details can be found in Zienkiewicz et al. (1999) and Ichikawa (1990).

Under the interpolation of an incremental displacement (6.46) the strain increment (6.38) is represented as

$$\Delta\boldsymbol{\varepsilon} \simeq \Delta\boldsymbol{\varepsilon}_h = \boldsymbol{B}\Delta\boldsymbol{U} \tag{6.52}$$

where

$$\boldsymbol{B} = \left[\boldsymbol{B}_1\ \boldsymbol{B}_2 \cdots \boldsymbol{B}_\alpha \cdots \boldsymbol{B}_N\right] \tag{6.53}$$

$$\boldsymbol{B}_\alpha = \begin{bmatrix} \dfrac{\partial N_\alpha}{\partial x} & 0 \\ 0 & \dfrac{\partial N_\alpha}{\partial y} \\ \dfrac{\partial N_\alpha}{\partial y} & \dfrac{\partial N_\alpha}{\partial x} \end{bmatrix}$$

The increment of volumetric strain $\Delta\varepsilon_v$ for a two-dimensional plane strain problem is

$$\Delta\varepsilon_v = \Delta\varepsilon_{xx} + \Delta\varepsilon_{yy} = \frac{\partial \Delta u_x}{\partial x} + \frac{\partial \Delta u_y}{\partial y}, \tag{6.54}$$

and by substituting the representation (6.46) we have

$$\Delta\varepsilon_v \simeq \Delta\varepsilon_{vh} = \boldsymbol{B}_v \Delta \boldsymbol{U} \tag{6.55}$$

where

$$\boldsymbol{B}_v = \left[\frac{\partial N_1}{\partial x} \ \frac{\partial N_1}{\partial y} \ \frac{\partial N_2}{\partial x} \ \frac{\partial N_2}{\partial y} \cdots \frac{\partial N_\alpha}{\partial x} \ \frac{\partial N_\alpha}{\partial y} \cdots \frac{\partial N_N}{\partial x} \ \frac{\partial N_N}{\partial y} \right] \tag{6.56}$$

By using the interpolation function (6.47) of the increment of pore pressure, its gradient $\nabla(\Delta p)$ is represented as

$$\nabla(\Delta p) \simeq \nabla(\Delta p)_h = \boldsymbol{G} \Delta \boldsymbol{P} \tag{6.57}$$

where

$$\boldsymbol{G} = \begin{bmatrix} \dfrac{\partial N_1}{\partial x} & \dfrac{\partial N_2}{\partial x} & \cdots & \dfrac{\partial N_\alpha}{\partial x} & \cdots & \dfrac{\partial N_{\bar{N}}}{\partial x} \\ \dfrac{\partial N_1}{\partial y} & \dfrac{\partial N_2}{\partial y} & \cdots & \dfrac{\partial N_\alpha}{\partial y} & \cdots & \dfrac{\partial N_{\bar{N}}}{\partial y} \end{bmatrix}. \tag{6.58}$$

We use the same interpolation functions for arbitrary functions $\boldsymbol{v}$ and η as $\Delta\boldsymbol{u}$ and Δp, which is known as Galerkin's method:

$$\boldsymbol{v} \simeq \boldsymbol{v}_h = \boldsymbol{N}(x, y)\boldsymbol{V}, \tag{6.59}$$

$$\eta \simeq \eta_h = \bar{\boldsymbol{N}}(x, y)\boldsymbol{H}. \tag{6.60}$$

We apply the above interpolations into the weak forms (6.44) and (6.45). Since vectors $\boldsymbol{V}$ and $\boldsymbol{H}$ are arbitrary, we have the following vector form of the coupled ordinary differential equations:

$$\begin{bmatrix} \boldsymbol{K} & -\boldsymbol{C} \\ -\boldsymbol{C}^T & \boldsymbol{0} \end{bmatrix} \begin{bmatrix} \Delta\dot{\boldsymbol{U}} \\ \Delta\dot{\boldsymbol{P}} \end{bmatrix} + \begin{bmatrix} \boldsymbol{0} & \boldsymbol{0} \\ \boldsymbol{0} & -\bar{\boldsymbol{K}} \end{bmatrix} \begin{bmatrix} \Delta\boldsymbol{U} \\ \Delta\boldsymbol{P} \end{bmatrix} = \begin{bmatrix} \Delta\boldsymbol{F} \\ \Delta\boldsymbol{Q} \end{bmatrix} \tag{6.61}$$

where

$$\boldsymbol{K} = \int_\Omega \boldsymbol{B}^T \boldsymbol{D}\boldsymbol{B}\, dv, \qquad \boldsymbol{C} = \int_\Omega \boldsymbol{B}_v^T \bar{\boldsymbol{N}}\, dv, \qquad \Delta\boldsymbol{F} = \int_{\partial\Omega_t} \boldsymbol{N}^T \Delta\bar{\boldsymbol{t}}\, ds, \tag{6.62}$$

$$\bar{\boldsymbol{K}} = \int_\Omega \frac{k}{\gamma_w} \boldsymbol{G}^T \boldsymbol{G}\, dv, \qquad \Delta\boldsymbol{Q} = \int_{\partial\Omega_q} \bar{\boldsymbol{N}}^T \frac{\Delta\bar{q}^w}{\rho}\, dv, \tag{6.63}$$

and k is assumed to be isotropic (i.e. $k_{ij} = k\,\delta_{ij}$).

Let us write (6.61) as

$$\boldsymbol{A}\dot{\boldsymbol{X}} + \boldsymbol{B}\boldsymbol{X} = \boldsymbol{Y} \tag{6.64}$$

$$\boldsymbol{A} = \begin{bmatrix} \boldsymbol{K} & \boldsymbol{C} \\ \boldsymbol{C}^T & \boldsymbol{0} \end{bmatrix}, \quad \boldsymbol{B} = \begin{bmatrix} \boldsymbol{0} & \boldsymbol{0} \\ \boldsymbol{0} & \bar{\boldsymbol{K}} \end{bmatrix}, \quad \boldsymbol{X} = \begin{bmatrix} \Delta\boldsymbol{U} \\ \Delta\boldsymbol{P} \end{bmatrix}, \quad \boldsymbol{Y} = \begin{bmatrix} \Delta\boldsymbol{F} \\ \Delta\boldsymbol{Q} \end{bmatrix}.$$

Since we introduced the Galerkin finite element approximations in space, the result (6.64) is an ordinary differential equation in the time-domain.

We solve the ordinary differential equation (6.64) using a finite difference scheme, referred to as the θ-method. Let the value $\boldsymbol{X}^{n-1}$ at a discretized time-step $n-1$ be known, then we can write the unknown value $\boldsymbol{X}^n$ at the time-step n as follows:

$$\boldsymbol{X}^n = \boldsymbol{X}^{n-1} + \Delta t \left[(1-\theta)\dot{\boldsymbol{X}}^{n-1} + \theta\dot{\boldsymbol{X}}^n \right] \tag{6.65}$$

where Δt is a time difference between the step $n-1$ and the step n, while the parameter θ is given as

$$0 \le \theta \le 1.$$

The cases $\theta = 0$, $\theta = 1$ and $\theta = 0.5$ correspond to the explicit, implicit and Crank-Nicolson finite difference schemes, respectively. Since (6.64) is satisfied both for the $(n-1)$th and nth steps, we have

$$\boldsymbol{A}\dot{\boldsymbol{X}}^{n-1} + \boldsymbol{B}\boldsymbol{X}^{n-1} = \boldsymbol{Y}^{n-1}, \qquad \boldsymbol{A}\dot{\boldsymbol{X}}^n + \boldsymbol{B}\boldsymbol{X}^n = \boldsymbol{Y}^n. \tag{6.66}$$

We multiply $(6.66)_1$ by $(1-\theta)$ and $(6.66)_2$ by θ and add both equations to obtain

$$\boldsymbol{A}\left[(1-\theta)\dot{\boldsymbol{X}}^{n-1} + \theta\dot{\boldsymbol{X}}^n \right] + (1-\theta)\boldsymbol{B}\boldsymbol{X}^{n-1} + \theta\boldsymbol{B}\boldsymbol{X}^n = (1-\theta)\boldsymbol{Y}^{n-1} + \theta\boldsymbol{Y}^n. \tag{6.67}$$

By substituting (6.65) into (6.67), the successive form of the θ-method is given as

$$\left[\frac{1}{\Delta t}\boldsymbol{A} + \theta\boldsymbol{B} \right] \boldsymbol{X}^n = (1-\theta)\boldsymbol{Y}^{n-1} + \theta\boldsymbol{Y}^n + \left[\frac{1}{\Delta t}\boldsymbol{A} - (1-\theta)\boldsymbol{B} \right] \boldsymbol{X}^{n-1} \tag{6.68}$$

We then obtain $\Delta\boldsymbol{U}^n$ and $\Delta\boldsymbol{P}^n$ at each time-step by solving (6.68).

6.5 The Cam Clay Model

Constitutive developments in the description of geomaterials have been the subject of extensive research over the past six decades. The scope of these developments is given by Desai and Siriwardane (1984), Darve (1990), Davis and Selvadurai (2002), and Pietruszczak (2010), etc. The *Cam clay model* (also known as the *Cambridge model*) is one that has been used widely to describe geomaterial behavior (Schofield and Wroth 1968; Parry 1971; Houlsby and Schofield 1993; Oka 1995).

In this Section we present an outline of the Cam clay model. Note that we assume a conventional triaxial stress state ($\sigma_2 = \sigma_3$), and the following variables for stress and strain are introduced:

$\sigma_1, \sigma_2, \sigma_3$	: principal values of the total stress $\boldsymbol{\sigma}$ (tension positive)
$\sigma_1', \sigma_2', \sigma_3'$	: principal values of the effective stress $\boldsymbol{\sigma}'$ (tension positive)
$\varepsilon_1, \varepsilon_2, \varepsilon_3$	: principal values of the strain $\boldsymbol{\varepsilon}$ (tension positive)
$p' = -(\sigma_1' + 2\sigma_3')/3$	: effective pressure (i.e., isotropic stress) under biaxial state (compression positive)
$q = -(\sigma_1 - \sigma_3)$	: shear stress
$\varepsilon_v = -(\varepsilon_1 + 2\varepsilon_3)$	: volumetric strain under a biaxial state (compression positive)
$\varepsilon_s = -2(\varepsilon_1 - \varepsilon_3)/3$	: shear strain

6.5.1 *Normally Consolidated Clay*

If we perform undrained triaxial tests (namely, void ratio e=constant) on isotropically and normally consolidated clay at several confining pressures (p_0'), the stress-strain behavior is schematically shown in Fig. 6.2. It is observed that at the final stage of loading (namely, the failure state) the stress ratio $\eta = q/p'$ becomes constant ($\eta_f = (q/p')_f = M$), which is referred to as the *critical state*, which means that the deformation is developed under a constant volumetric plastic strain and a constant shear stress at the critical state $q = Mp'$.

This result is similar to drained triaxial tests (namely, σ_3' = constant, $dp' = -(1/3)d\sigma_1'$, $dq = -d\sigma_1'$, $dq/dp' = 3$). As shown in Fig. 6.2d, if an undrained stress path B-B′ under a consolidation pressure p_{0B}' intersects with a drained stress path A-A″ under a consolidation pressure p_{0A}' at a point C, the void ratio obtained from the undrained test is the same as the void ratio obtained from the drained test. We can draw these states in a space (p', q, e) as illustrated in Fig. 6.2f, which shows that the critical state is reached after travelling the surface referred to as the *state boundary surface* or *Roscoe surface* in both the undrained and drained tests. The line of failure is known as the *critical state line* (CSL). It is noted that the projection of CLS in the space (p', q, e) onto the space (p', q) is given as $q = Mp'$, and the surface formed by CLS and its projection onto (q, e) is referred to as *Hvoslev surface*, which is a failure surface found by Hvoslev in 1937 through a series of direct shear tests conducted on Vienna clay.

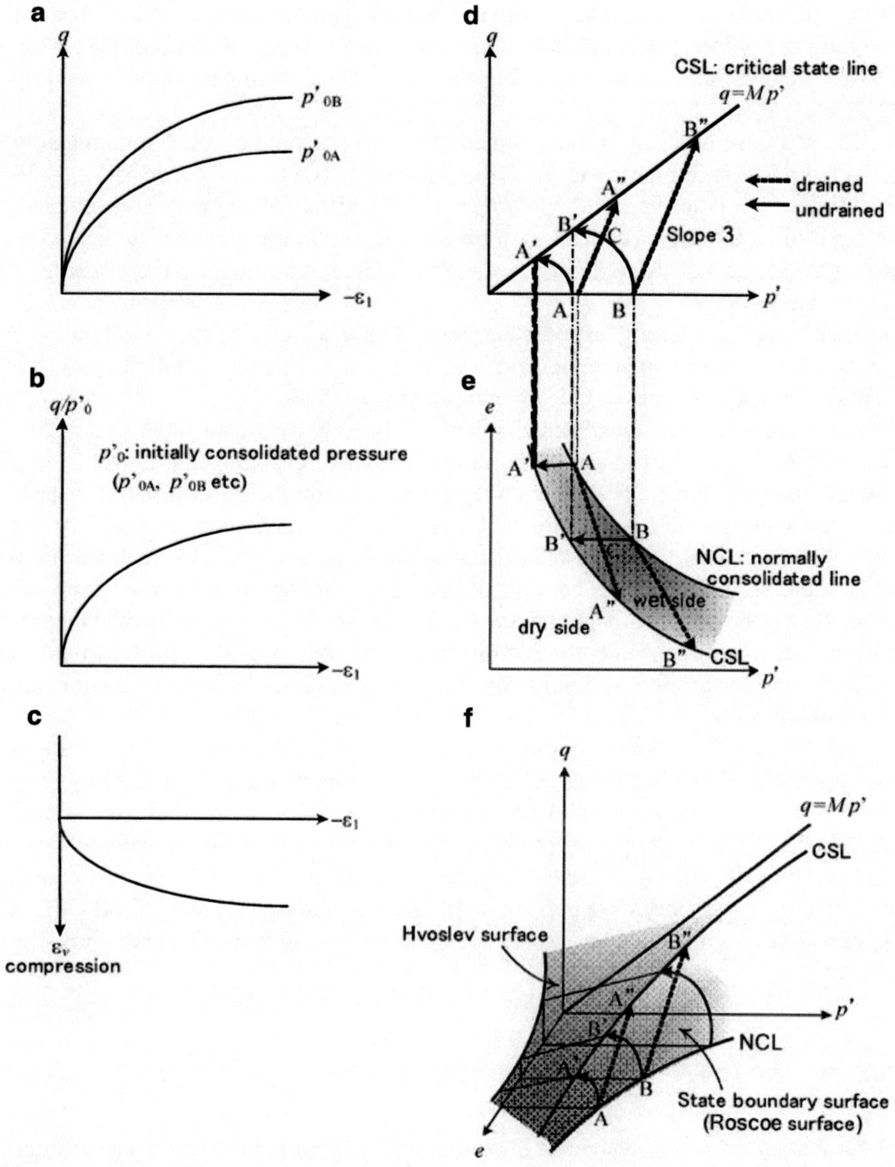

Fig. 6.2 Response of normally consolidated clay

6.5.2 Over-consolidated Clay

If the current stress state of volumetric effective stress is lower than the stress state experienced in the past, the clay is said to be *over-consolidated*, and we define the over-consolidated ratio as OCR= p_0'/p_i' where p_0' is the preload pressure and p_i'

is the current effective pressure swelled from p'_0. Note that for a natural clay the swelling curve does not clearly turn at the preloading curve, and therefore the over-consolidated ratio is defined by using the value of the turning point p'_e, namely OCR$= p'_e / p'_i$.

The response of an over-consolidated clay depends on the initial state, namely from the position in relation to the critical state line (CSL). As shown in Fig. 6.3d the clay, which is located at the left side of the CSL is heavily over-consolidated, referred to as the *dry side* or *drier than critical*, while the clay that is located at the right side of the CSL is lightly over-consolidated, referred to as the *wet side* or *wetter than critical*.

We observe a softening response, as shown Fig. 6.3a, b for a drained test, and the stress reaches a residual state described as R in Fig. 6.3a after a peak point Q; the volumetric strain changes at Q from compression to dilation.

Let us consider the shearing behavior of a heavily over-consolidated clay. We start to make it swell from a point A on the normally consolidated line (NCL) to a point B shown in Fig. 6.3d, e. When we perform an undrained triaxial test, the shear stress q attains the yield point P on the Hvoslev surface (namely the maximum point for the stress ratio $\eta = q/p'$), and moves on the Hvoslev surface to reach the point C on the CSL. On the other hand, if a drained triaxial test is performed from the point B, the shear stress q attains the peak strength Q, then the stress reaches a residual state R through a softening process. Note that in practical experiments a shearing slip surface is commonly observed, and the whole specimen cannot reach the residual state.

The shearing behavior of a lightly over-consolidated clay is as follows: initially it swells from a point A′ on the NCL to a point D, as shown in Fig. 6.3d, e. Then, after performing an undrained triaxial test, the shear stress q directly attains the point C on the CSL. Thus, the behavior of a lightly over-consolidated clay is different from that of a heavily over-consolidated clay, and is similar to the normally consolidated clay. On the other hand, if we perform a drained test, the stress q reaches the CSL at a point F, after which the clay experiences plastic flow under a constant volumetric plastic strain.

6.5.3 The Original Cam Clay Model

A hardening/softening behavior is shown in Figs. 6.2e or 6.3d. It has been rewritten in the space $(e, \ln p')$, to give Fig. 6.4 where $v = 1 + e$ is the volume ratio. Figure 6.4a gives the normally consolidated state, and Fig. 6.4b gives a general stress state where the shear stress is generated by moving along the state boundary surface shown in Fig. 6.2.

Let the compressive strain be positive, then the increment of the volumetric strain is given as

$$d\varepsilon_v = -\frac{dv}{v} = -\frac{de}{1+e}. \tag{6.69}$$

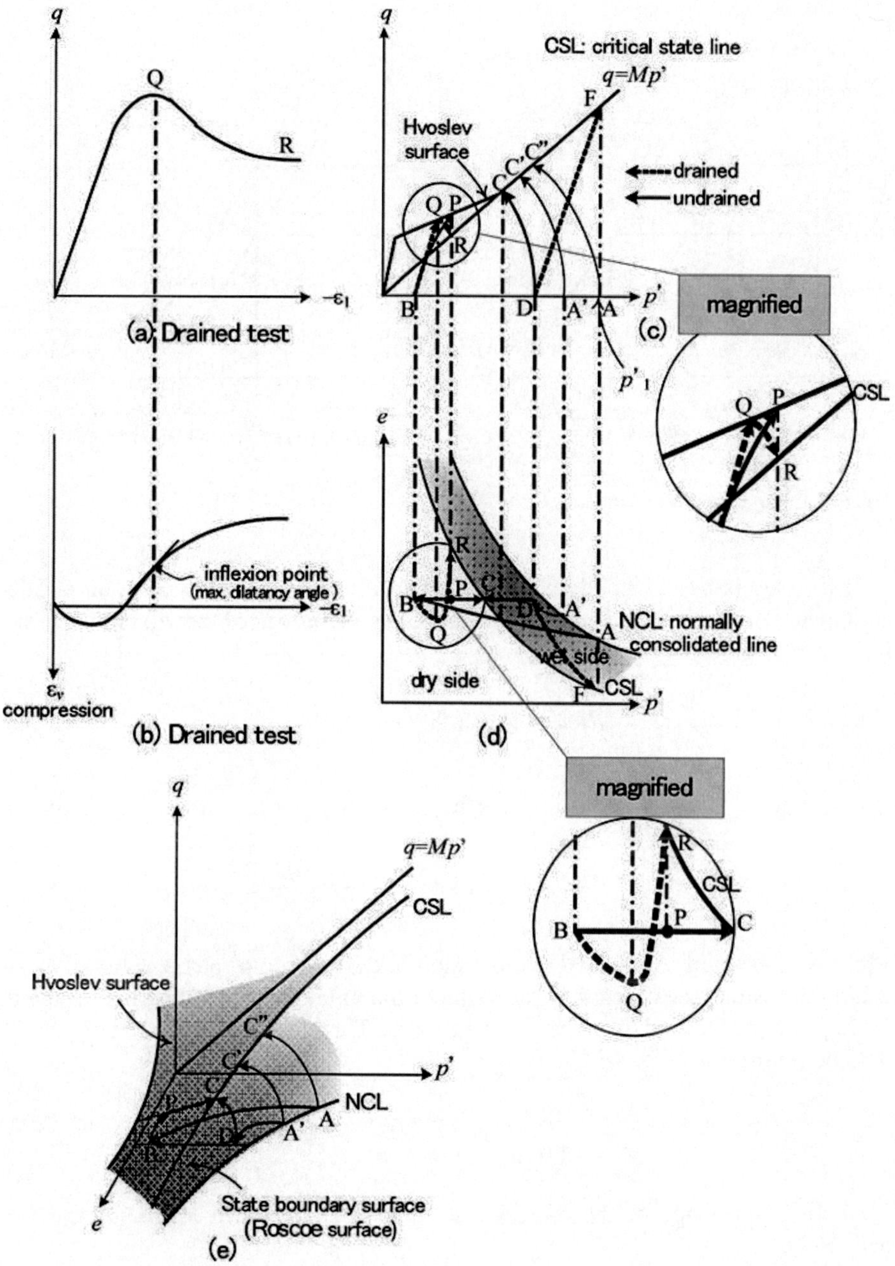

Fig. 6.3 Response of over-consolidated clay

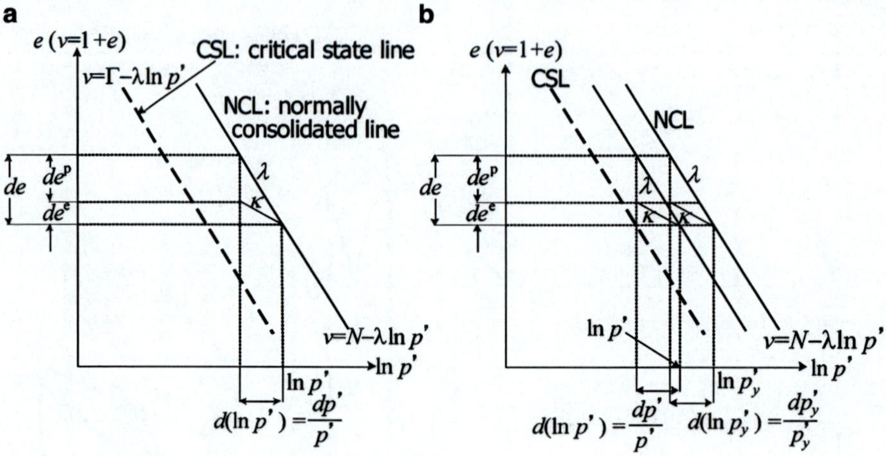

Fig. 6.4 Compression characteristics

If the clay is in a normally consolidated state, referring to Fig. 6.4a, an elastic component de^e and plastic component de^p of an increment of the void ratio $de = de^e + de^p$ is given by

$$de = -\lambda\, d(\ln p') = -\lambda \frac{dp'}{p'} \tag{6.70}$$

$$de^e = -\kappa\, d(\ln p') = -\kappa \frac{dp'}{p'} \tag{6.71}$$

$$de^p = de - de^e = -(\lambda - \kappa) \frac{dp'}{p'} \tag{6.72}$$

where λ is the slope of normal compression in the e vs. $\ln p'$ plane, referred to as the compression index, and κ is the slope of the unloading-reloading line in the e vs. $\ln p'$ plane, referred to as the swelling index. Therefore the increment of plastic volumetric strain $d\varepsilon_v^p$ is written as

$$d\varepsilon_v^p = -\frac{de^p}{1+e} = \frac{\lambda - \kappa}{1+e} \frac{dp'}{p'} = MD \frac{dp'}{p'} \tag{6.73}$$

Here M is the slope of the critical state line CSL shown in Fig. 6.2d, and we define D as

$$D = \frac{\lambda - \kappa}{M(1+e)}. \tag{6.74}$$

If we have a general stress state as shown in Fig. 6.4b, the results (6.69)–(6.74) are satisfied. In addition we can introduce the stress state on the normally consolidated line $(p' = p'_y, q = 0)$ under the same condition of void ratio as observed in Fig. 6.4b, and we have

$$d\varepsilon_v^p = -\frac{de^p}{1+e} = \frac{\lambda - \kappa}{1+e}\frac{dp'_y}{p'_y} = MD\frac{dp'_y}{p'_y}. \tag{6.75}$$

A normality rule is assumed on the yield locus $f(p', q) = 0$ (namely, the associated flow rule) as shown in Fig. 6.5, so that we have

$$\frac{d\varepsilon_v^p}{d\varepsilon_s^p} = -\frac{dq}{dp'} \tag{6.76}$$

where $d\varepsilon_s^p$ is an increment of the plastic shear strain.

We introduce the stress–dilatancy relation presented by Taylor (1948), which is schematically shown in Fig. 6.6. Taylor visualized direct shearing along jagged surfaces, and obtained an increment of the work done due to a normal force P and shear force Q on the respective displacements, as follows:

$$dW = P\,du_y + Q\,du_x \tag{6.77}$$

where du_x and du_y are the increments of normal and shearing displacements. Let us assume that all external work is consumed by friction (namely, no elastic energy), and the Coulomb friction is generated by the action of the normal force and a friction coefficient μ. Then we have

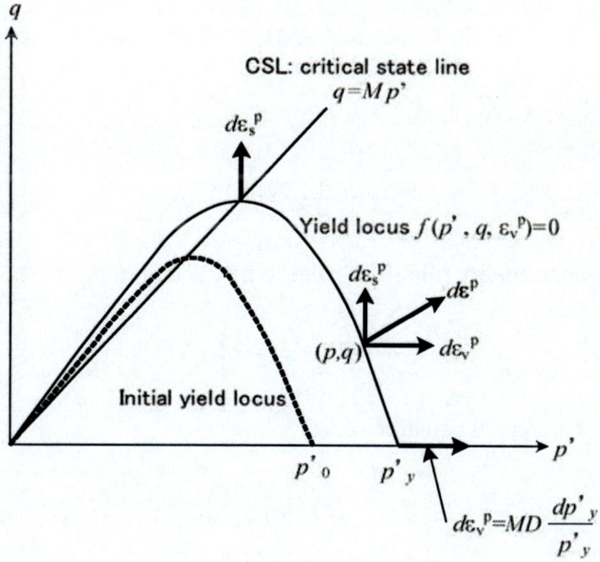

Fig. 6.5 Cam clay model

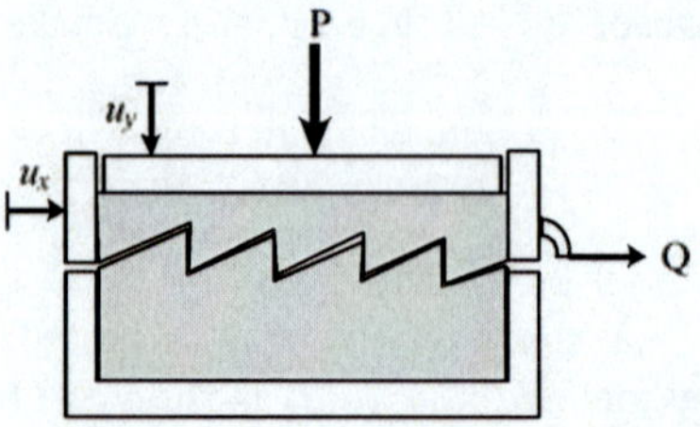

Fig. 6.6 Taylor's stress–dilatancy model

$$dW = dW^p = \mu P\, du_x. \tag{6.78}$$

In this case we have the following *stress-dilatancy relation* of Taylor:

$$P\, du_y + Q\, du_x = \mu P\, du_x. \tag{6.79}$$

We rewrite (6.79) in terms of stress and strain as follows:

$$dW^p = p'\, d\varepsilon_v^p + q\, d\varepsilon_s^p = Mp'\, d\varepsilon_s^p. \tag{6.80}$$

Note that at the critical state we have

$$\frac{d\varepsilon_v^p}{d\varepsilon_s^p} = 0$$

and (6.80) gives the following relationship:

$$\frac{q}{p'} = M.$$

By modifying (6.80) we have

$$\frac{q}{p'} + \frac{d\varepsilon_v^p}{d\varepsilon_s^p} = M. \tag{6.81}$$

We substitute the normality rule (6.76) into (6.81), and obtain

$$\frac{dq}{dp'} - \frac{q}{p'} = -M. \tag{6.82}$$

By solving the differential equation (6.82) we have

$$q = -Mp' \ln\left(\frac{p'}{p'_0}\right).$$

This corresponds to a stress part $f_1(\boldsymbol{\sigma})$ of the yield function (2.297):

$$f_1(\boldsymbol{\sigma}) = \frac{q}{Mp'} + \ln\left(\frac{p'}{p'_0}\right) = 0 \qquad (\text{for } \kappa = 0) \tag{6.83}$$

where p'_0 is the value of p'_y on the initial yielding surface (Fig. 6.5).

For simplicity, we introduce the following isotropic hardening rule with the strain hardening parameter:

$$f(\boldsymbol{\sigma},\, \kappa(\varepsilon^p_v)) = f_1(\boldsymbol{\sigma}) - K(\kappa) = 0, \tag{6.84}$$

$$d\kappa = |d\varepsilon^p_v| = \left(d\varepsilon^p_v d\varepsilon^p_v\right)^{1/2}. \tag{6.85}$$

We must satisfy the following condition, which corresponds to the state on the normally consolidated line ($p' = p'_y, q = 0$):

$$f(p'{=}p'_y, q{=}0, \kappa) = \ln\left(\frac{p'_y}{p'_0}\right) - K(\kappa) = 0.$$

Then the consistency condition gives

$$df = d(\ln p'_y) - \frac{dK}{d\kappa} d\kappa = 0. \tag{6.86}$$

We substitute (6.75) into (6.86), and using (6.85) we obtain

$$\left(\frac{d\varepsilon^p_v}{MD}\right)^2 = \left(\frac{dK}{d\kappa}\right)^2 \left(d\varepsilon^p_v\right)^2. \tag{6.87}$$

Under the conditions $K > 0,\ \kappa > 0$ we solve (6.87):

$$\frac{dK}{d\kappa} = \frac{1}{MD} \quad \Rightarrow \quad K = \frac{1}{MD}\int d\kappa.$$

Finally we obtain the following yield function, which is referred to as the original *Cam clay model*:

$$f = MD \ln \frac{p'}{p'_0} + D\eta - \int d\kappa = 0, \tag{6.88}$$

$$\eta = \frac{q}{p'} : \text{ stress ratio.} \tag{6.89}$$

When deriving (6.89), we have assumed that MD is a constant. However, referring to (6.74), we understand $MD = (\lambda - \kappa)/(1 + e)$, which shows that MD is a function of the void ratio e. In this sense, the model (6.89) is relevant to a small strain theory.

6.5.4 Modified Cam Clay Model

Recall that Taylor's stress–dilatancy model is introduced under the hypothesis that only the friction term of shearing displacement due to the normal force contributes to the plastic work done, as given in (6.78). Here this is modified by considering the component of volumetric displacement as

$$dW^p = p'\,d\varepsilon_v^p + q\,d\varepsilon_s^p = p'\Big[(d\varepsilon_v^p)^2 + (M\,d\varepsilon_s^p)^2\Big]^{1/2}. \tag{6.90}$$

This is rewritten as

$$\frac{d\varepsilon_v^p}{d\varepsilon_s^p} = \frac{(p')^2M^2 - q^2}{2p'q}. \tag{6.91}$$

We substitute the normality rule (6.76) into (6.91), and obtain

$$\frac{dq}{dp'} + \frac{M^2-\eta^2}{2\eta} = 0 \tag{6.92}$$

where $\eta = q/p'$. We solve (6.92) and obtain

$$f_1(\sigma) = \frac{p'}{p_0'} - \frac{M^2}{M^2+\eta^2} = 0 \quad (\text{for } \kappa = 0). \tag{6.93}$$

Note that as shown in Fig. 6.7, (6.93) gives an ellipse in the space (p', q).

The hardening rule is introduced under an isotropic hardening model with a strain hardening parameter, which is the same as the original Cam clay model:

$$f(\sigma, \kappa(\varepsilon_v^p)) = f_1(\sigma) - K(\kappa) = 0, \qquad d\kappa = |d\varepsilon_v^p| = \left(d\varepsilon_v^p d\varepsilon_v^p\right)^{1/2} \tag{6.94}$$

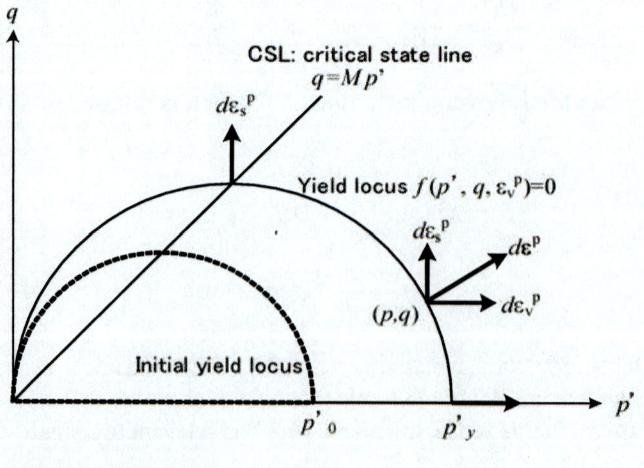

Fig. 6.7 Modified Cam clay model

We use the condition

$$f(p'=p'_y, q=0, \kappa) = \frac{p'_y}{p'_0} - 1 - K = 0 \quad \Rightarrow \quad p'_y = (1+K)p'_0 \tag{6.95}$$

and apply the consistency condition

$$df = \frac{dp'_y}{p'_0} - \frac{dK}{d\kappa}d\kappa = 0 \quad \Rightarrow \quad \frac{dp'_y}{p'_y} = \frac{p'_0}{p'_y}\frac{dK}{d\kappa}d\kappa. \tag{6.96}$$

We then substitute (6.72) into (6.96), and by using (6.85) and (6.95) we have

$$\left(\frac{d\varepsilon_v^p}{MD}\right)^2 = \left(\frac{1}{1+K}\right)^2\left(\frac{dK}{d\kappa}\right)^2(d\kappa)^2$$

Since $K > 0$, $\kappa > 0$, we obtain

$$\int \frac{dK}{1+K} = \ln(1+K) = \frac{1}{MD}\int d\kappa,$$

which results in

$$K = \exp\left(\frac{1}{MD}\int d\kappa\right) - 1. \tag{6.97}$$

The yield function is finally given as

$$f = \frac{p'}{p'_0} - \frac{M^2}{M^2+\eta^2} + 1 - \exp\left(\frac{1}{MD}\int d\kappa\right) = 0. \tag{6.98}$$

6.5.5 *Elasto-plastic Constitutive Law*

By applying the procedure described in Sect. 2.10.3, the elasto-plastic constitutive law is obtained, which relates the increment of effective stress $d\boldsymbol{\sigma}'$ to the strain increment $d\boldsymbol{\varepsilon}$ as

$$d\sigma'_{ij} = D^{ep}_{ijkl}d\varepsilon_{kl}, \tag{6.99}$$

$$D^{ep}_{ijkl} = D^e_{ijkl} - \frac{1}{H}\left(D^e_{ijmn}\frac{\partial g}{\partial \sigma'_{mn}}\right)\left(D^e_{klst}\frac{\partial f}{\partial \sigma'_{st}}\right), \quad H = h + \frac{\partial f}{\partial \sigma'_{ij}}\left(D^e_{ijkl}\frac{\partial g}{\partial \sigma'_{kl}}\right) \tag{6.100}$$

Since both the original and modified Cam clay models employ an associated flow rule, the plastic potential function g is the same as the yield function f. We use (6.88) and (6.98), and obtain the hardening coefficient h as follows:

$$\text{Cam clay model:} \qquad h = \frac{\partial K}{\partial \kappa}\left|\frac{\partial f}{\partial \boldsymbol{\sigma}'}\right| = \frac{1}{MD}\left|\frac{\partial f}{\partial \boldsymbol{\sigma}'}\right|, \tag{6.101}$$

$$\text{Modified Cam clay model:} \qquad h = \frac{\partial K}{\partial \kappa}\left|\frac{\partial f}{\partial \boldsymbol{\sigma}'}\right| = \frac{1}{MD}\exp\left(\frac{1}{MD}\int d\kappa\right)\left|\frac{\partial f}{\partial \boldsymbol{\sigma}'}\right|. \tag{6.102}$$

Chapter 7
Introduction to Homogenization Analysis

We outline the essential features of a multiscale homogenization analysis. A problem of a one-dimensional elastic bar is given as an example.

The topic of homogenization has been extensively discussed in the literature in the context of multiphase components and heterogeneous media that are either microscopically periodic or non-periodic (Sanchez-Palencia 1980; Bakhvalov & Panasenko 1984; Ene and Polisevski 1987; Hornung 1997; Milton 2002; Torquato 2002; Dormieux et al. 2006).

7.1 One-dimensional Problem of an Elastic Bar

We consider the problem of an elastic bar where the elastic modulus is inhomogeneous. The bar is subjected to a state of uniaxial force F at the end point. The equation of equilibrium for the bar is given by

$$\frac{d}{dx}\left(E\frac{du^{\varepsilon}(x)}{dx}\right) = f \tag{7.1}$$

where $u^{\varepsilon}(x)$ is the displacement (φ^{ε} shows that the function φ rapidly changes in the local scale), E is the elastic modulus, and f is the body force acting on a unit volume.

7.2 Micro/Macro Coordinates

As shown in Fig. 7.1, we consider a one-dimensional bar with a periodic structure at the microscale level.

Y. Ichikawa and A.P.S. Selvadurai, *Transport Phenomena in Porous Media*,
DOI 10.1007/978-3-642-25333-1_7, © Springer-Verlag Berlin Heidelberg 2012

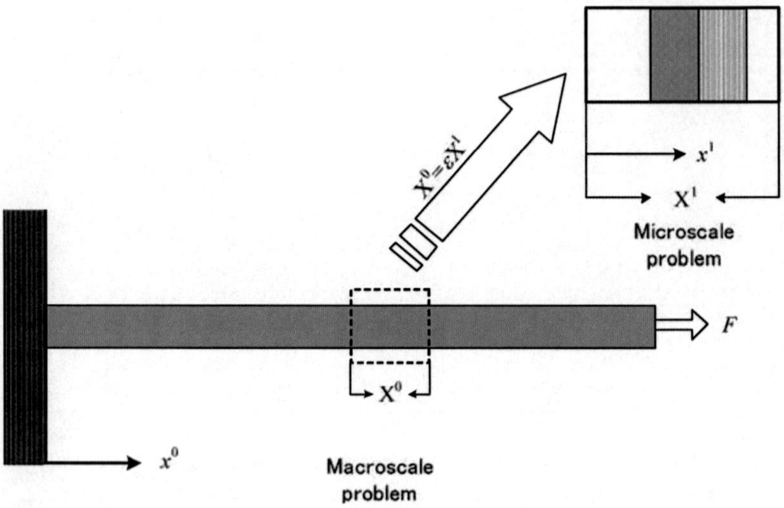

Fig. 7.1 The problem of a one-dimensional elastic bar with periodic microstructure

We introduce a macroscale coordinate system x^0 and a microscale one x^1 for the general system x as

$$x = x(x^0, x^1). \tag{7.2}$$

We assume the systems x^0 and x^1 are related, with

$$x^1 = \frac{x^0}{\varepsilon} \tag{7.3}$$

where ε is a (small) parameter. If we consider the limit condition $\varepsilon \to 0$, the differential operation in (7.1) should be changed to

$$\frac{d}{dx} = \frac{\partial}{\partial x^0} + \frac{\partial}{\partial x^1}\frac{\partial x^1}{\partial x^0} = \frac{\partial}{\partial x^0} + \frac{1}{\varepsilon}\frac{\partial}{\partial x^1}. \tag{7.4}$$

The periodic microstructure is referred to as a *unit cell*. In homogenization analysis we assume that the size of the unit cell is sufficiently small, and by taking the limit $\varepsilon \to 0$ we derive a system of differential equations that relates the microscale behavior to the microscale characteristics.

7.3 Microscale and Macroscale Problems

We assume that the displacement $u^\varepsilon(x)$ can be represented as a power series in the small parameter ε in the form

$$u^\varepsilon(x) = u_0(x^0, x^1) + \varepsilon u_1(x^0, x^1) + \varepsilon^2 u_2(x^0, x^1) + \cdots \tag{7.5}$$

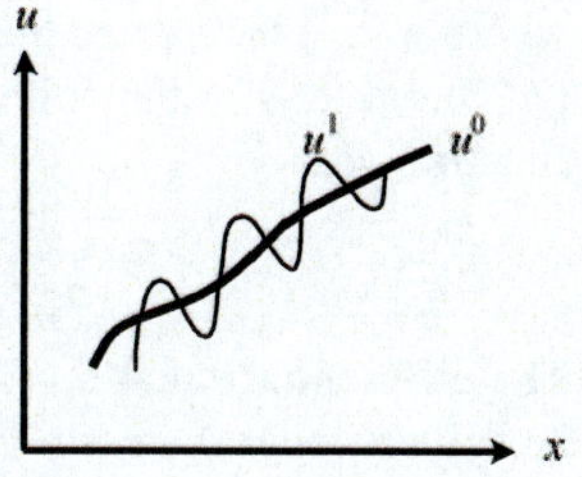

Fig. 7.2 A microscale periodic function

where $u_0(x^0, x^1)$, $u_1(x^0, x^1)$, $u_2(x^0, x^1)$, $\cdots$ are periodic functions in terms of the size of the unit cell X^1 of the microscale system x^1:

$$u_i(x^0, x^1) = u_i(x^0, x^1 + X^1), \quad i = 0, 1, 2, \cdots \tag{7.6}$$

$u_i(x^0, x^1)$ are referred to as X^1-periodic functions. The expansion (7.5) implies that the displacement u^ε consists of a globally changing part, which is related to x^0 and a locally changing one that is related to x^1, see Fig. 7.2.

By substituting (7.4) and (7.5) into the governing equation (7.1), we have

$$\frac{1}{\varepsilon^2}\frac{\partial}{\partial x^1}\left(E\frac{\partial u_0}{\partial x^1}\right) + \frac{1}{\varepsilon}\left[\frac{\partial}{\partial x^1}\left(E\frac{\partial u_0}{\partial x^0}\right) + \frac{\partial}{\partial x^1}\left(E\frac{\partial u_1}{\partial x^1}\right)\right]$$
$$+\left[\frac{\partial}{\partial x^0}\left(E\frac{\partial u_0}{\partial x^0}\right) + \frac{\partial}{\partial x^1}\left(E\frac{\partial u_1}{\partial x^0}\right) + \frac{\partial}{\partial x^0}\left(E\frac{\partial u_1}{\partial x^1}\right) + \frac{\partial}{\partial x^1}\left(E\frac{\partial u_2}{\partial x^1}\right) - f(x^0)\right]$$
$$+ \cdots = 0. \tag{7.7}$$

We require that as $\varepsilon \to 0$, all terms with respect to ε must be null, and we then obtain the following equations corresponding to each ε-term:

$$\varepsilon^{-2}\text{-term}: \quad \frac{\partial}{\partial x^1}\left(E\frac{\partial u_0}{\partial x^1}\right) = 0. \tag{7.8}$$

This equation suggests that u_0 is a function of only x^0:

$$u_0 = u_0(x^0). \tag{7.9}$$

$$\varepsilon^{-1}\text{-term}: \quad \frac{\partial}{\partial x^1}\left(E\frac{\partial u_1}{\partial x^1}\right) = -\frac{\partial}{\partial x^1}\left(E\frac{\partial u_0}{\partial x^0}\right). \tag{7.10}$$

This gives a differential equation of $u_1(x^0, x^1)$ with respect to the coordinate system x^1 if the term $u_0(x^0)$ is known. Then, we introduce a separation of variables as

$$u_1(x^0, x^1) = N(x^1)\frac{\partial u_0(x^0)}{\partial x^0} \tag{7.11}$$

where $N(x^1)$ is referred to as the *characteristic function*. By substituting (7.11) into (7.10), we obtain the following differential equation to identify the characteristic function $N(x^1)$:

$$\frac{d}{dx^1}\left(E(x^1)\frac{d}{dx^1}(N(x^1)+x^1)\right)=0. \tag{7.12}$$

The differential equation (7.12) is referred to as the *microscale equation*, which can be solved under the periodic boundary condition (7.6).

$$\varepsilon^0\text{-term}: \quad \frac{\partial}{\partial x^1}\left(E\frac{\partial u_2}{\partial x^1}\right)=f(x^0)-\frac{\partial}{\partial x^0}\left(E\frac{\partial u_0}{\partial x^0}\right)-\frac{\partial}{\partial x^1}\left(E\frac{\partial u_1}{\partial x^0}\right)-\frac{\partial}{\partial x^0}\left(E\frac{\partial u_1}{\partial x^1}\right) \tag{7.13}$$

We introduce an averaging operation for a function $\phi(x^1)$ for the unit cell as

$$\langle\phi\rangle=\frac{1}{|X^1|}\int_0^{X^1}\phi\,dx^1$$

where $|X^1|$ is the length of the unit cell. If ϕ is an X^1-periodic function, we have

$$\left\langle\frac{d\phi}{dx^1}\right\rangle=\frac{1}{|X^1|}\left(\phi(X^1)-\phi(0)\right)=0.$$

Thus the average of (7.13) is written as

$$\frac{d}{\partial x^0}\left(E^*\frac{du_0}{dx^0}\right)=f(x^0) \tag{7.14}$$

where

$$E^*=\left\langle E(x^1)+E(x^1)\frac{dN}{dx^1}\right\rangle \tag{7.15}$$

is the averaged or homogenized elastic modulus. Equation 7.14 is referred to as the *macroscale equation.*

In conclusion, the homogenization analysis procedure can be stated as follows: (1) The microscale equation (7.12) is first solved under the periodic boundary condition, which gives the characteristic function $N(x^1)$. (2) Using the characteristic function $N(x^1)$ we then calculate the averaged elastic modulus E^*. (3) The macroscale equation (7.14) can then be solved, giving the first perturbed term $u_0(x^0)$. Since $u_1(x^0,x^1)$ is calculated by (7.11), the first order approximation of $u^\varepsilon(x)$ can be represented as

$$u^\varepsilon(x)\simeq u_0(x^0)+\varepsilon\,u_1(x^0,x^1). \tag{7.16}$$

The strain, that includes a microscale effect, is then calculated as

$$\varepsilon_x = \frac{du^{\varepsilon}(x)}{dx} \simeq \frac{\partial u_0}{\partial x^0} + \varepsilon \frac{\partial u_1}{\partial x^0} + \frac{\partial u_1}{\partial x^1}. \tag{7.17}$$

The stress is given by

$$\sigma = E\varepsilon_x. \tag{7.18}$$

Chapter 8
Homogenization Analysis and Permeability of Porous Media

The Navier-Stokes' (NS) equations can be used to describe problems of fluid flow. Since these equations are scale-independent, flow in the microscale structure of a porous medium can also be described by a NS field. If the velocity on a solid surface is assumed to be null, the velocity field of a porous medium problem with a small pore size rapidly decreases (see Sect. 5.3.2). We describe this flow field by omitting the convective term $\boldsymbol{v} \cdot \nabla \boldsymbol{v}$, which gives rise to the classical Stokes' equation[1]. We recall that Darcy's theory is usually applied to describe seepage in a porous medium, where the scale of the solid skeleton does not enter the formulation as an explicit parameter. The scale effect of a solid phase is implicitly included in the permeability coefficient, which is specified through experiments. It should be noted that Kozeny-Carman's formula (5.88) involves a parameter of the solid particle; however, it is not applicable to a geometrical structure at the local pore scale.

If a homogenization analysis (HA) is applied to porous media flow, which is described by the Stokes' equation, we can immediately obtain Darcy's formula and the seepage equation in a macroscale field while in the microscale field the distributions of velocity and pressure are specified (Sanchez-Palencia 1980). We can also apply HA for a problem with a locally varying viscosity.

In this Section we first show that a local variation of viscosity in the pore water of a saturated smectitic clay such as montmorillonite or beidellite, which is a platelet crystal of about one nanometer ($=10^{-9}$ m) thickness, can be calculated by a molecular dynamic (MD) simulation. Then, by applying the HA with the locally distributed viscosity, we can calculate the seepage field of the smectitic clay, which consists of stacks of clay minerals. Consequently, we apply a three-scale analysis of homogenization for a bentonite clay with quartz grains of about 10 μm (1 μm $= 10^{-6}$ m).

[1]Note that even if a multiscale procedure is started with a NS equation with the convective term, this nonlinear term is dropped as a higher order term.

Y. Ichikawa and A.P.S. Selvadurai, *Transport Phenomena in Porous Media*,
DOI 10.1007/978-3-642-25333-1_8, © Springer-Verlag Berlin Heidelberg 2012

8.1 Micro-inhomogeneous Porous Media and Stokes' Equation

Stokes' equation for an incompressible fluid is given as follows:

$$-\frac{\partial p^\varepsilon}{\partial x_i} + \frac{\partial}{\partial x_j}\left(\eta \frac{\partial v_i^\varepsilon}{\partial x_j}\right) + f_i = 0 \qquad \text{in } \Omega_f, \tag{8.1}$$

$$\frac{\partial v_i^\varepsilon}{\partial x_i} = 0 \qquad \text{in } \Omega_f \tag{8.2}$$

where v_i^ε is the velocity, p^ε is the pressure, f_i is the body force, η is the viscosity, and Ω_f is the fluid domain. The superscript ε indicates that the corresponding function rapidly varies in the local field because of the micro-inhomogeneous property. We recall that (8.1) is based on the conservation law of linear momentum, and (8.2) is the mass conservation law for an incompressible viscous fluid. The convective term $v_j\, \partial v_i / \partial x_j$ is disregarded during the procedure of homogenization as a second order of small value. The velocity at a solid-fluid interface is assumed to be null:

$$v_i^\varepsilon = 0 \qquad \text{on } \Gamma. \tag{8.3}$$

8.2 Seepage Theory for Two-scale Porous Media

We consider the micro-/macro-problem for a saturated two-scale porous medium as shown in Fig. 8.1 with the macro-domain Ω_0 and the micro-domain Ω_1, and Ω_{1f} is the fluid phase in the micro-domain, Ω_{1s} is the solid phase in the micro-domain, Γ is the fluid/solid interface, and $\partial\Omega_{1f}$ is the periodic boundary of the fluid phase in the micro-domain.

Let us introduce a macroscale coordinate system $\boldsymbol{x}^0$ and a microscale coordinate system $\boldsymbol{x}^1$ as

$$\boldsymbol{x} = \boldsymbol{x}(\boldsymbol{x}^0, \boldsymbol{x}^1). \tag{8.4}$$

Both systems are assumed to be related by

$$\boldsymbol{x}^1 = \frac{\boldsymbol{x}^0}{\varepsilon} \tag{8.5}$$

where ε is a scaling parameter. As shown in Fig. 8.1, if the size of a unit cell in the macro-domain is given by $(X_1^0 = \varepsilon X_1^1, X_2^0 = \varepsilon X_2^1)$, the size in the micro-domain is (X_1^1, X_2^1). In taking the limit $\varepsilon \to 0$, the differential operation can be changed to

$$\frac{\partial}{\partial x_i} = \frac{\partial}{\partial x_i^0} + \frac{1}{\varepsilon}\frac{\partial}{\partial x_i^1}. \tag{8.6}$$

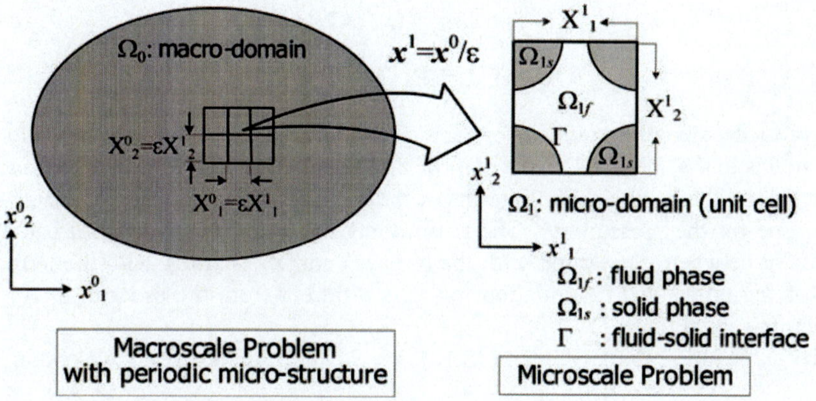

Fig. 8.1 A micro-/macro-problem for two-scale porous media

Note 8.1 (On the size of a unit cell). We can determine the size of the micro-domain based on (8.5). If the dimensions of the unit cell are given by X_1^1, X_2^1, X_3^1 in each direction as shown Fig. 8.1, we have the following relations in the macro-domain:

$$X_1^0 = \varepsilon\, X_1^1, \quad X_2^0 = \varepsilon\, X_2^1, \quad X_3^0 = \varepsilon\, X_3^1. \tag{8.7}$$

Thus for a non-cubic unit cell ($X_1^1 \neq X_2^1 \neq X_3^1$) we have

$$\varepsilon = L = X_1^0 \ll 1; \qquad X_1^1 = 1, \quad X_2^1 = \frac{X_2^0}{L}, \quad X_3^1 = \frac{X_3^0}{L} \tag{8.8}$$

where $L = X_1^0$. Under this condition the geometry of the unit cell is maintained as $\varepsilon \to 0$. ■

8.2.1 Homogenization Analysis and Seepage Problem of Porous Media

Let us introduce a perturbation in terms of both the $\boldsymbol{x}^0$ and $\boldsymbol{x}^1$ coordinate systems for the velocity v_i^ε and the pressure p^ε:

$$v_i^\varepsilon(\boldsymbol{x}) = \varepsilon^2 v_i^0(\boldsymbol{x}^0, \boldsymbol{x}^1) + \varepsilon^3 v_i^1(\boldsymbol{x}^0, \boldsymbol{x}^1) + \cdots, \tag{8.9}$$

$$p^\varepsilon(\boldsymbol{x}) = p^0(\boldsymbol{x}^0, \boldsymbol{x}^1) + \varepsilon p^1(\boldsymbol{x}^0, \boldsymbol{x}^1) + \cdots, \tag{8.10}$$

where $v_i^\alpha(\boldsymbol{x}^0, \boldsymbol{x}^1)$ and $p^\alpha(\boldsymbol{x}^0, \boldsymbol{x}^1)$ ($\alpha = 0, 1, 2, \cdots$) are periodic functions, which satisfy the following periodic boundary conditions in the micro-domain:

$$v_i^{\alpha}(\mathbf{x}^0, \mathbf{x}^1) = v_i^{\alpha}(\mathbf{x}^0, \mathbf{x}^1 + \mathbf{X}^1), \tag{8.11}$$

$$p^{\alpha}(\mathbf{x}^0, \mathbf{x}^1) = p^{\alpha}(\mathbf{x}^0, \mathbf{x}^1 + \mathbf{X}^1). \tag{8.12}$$

$\mathbf{X}^1$ gives the size of a unit cell for each direction. The perturbation of the velocity v_i^{ε} implies that v_i^{ε} consists of the sum of the global change in terms of the macro-domain and the local change in terms of the micro-domain, and the conditions are the same for the pressure p^{ε}. The reason why the initial order of perturbation is different between v_i^{ε} (starting with the term ε^2) and p^{ε} (starting with the term ε^0) is that the order of differentiation for v_i^{ε} is different from the order of p^{ε} in the governing equation (8.1).

By substituting (8.6), (8.9) and (8.10) into Stokes' equation (8.1), we obtain

$$\begin{aligned}
&\left\{-\left(\frac{\partial p^0}{\partial x_i^0} + \frac{1}{\varepsilon}\frac{\partial p^0}{\partial x_i^1}\right) - \varepsilon\left(\frac{\partial p^1}{\partial x_i^0} + \frac{1}{\varepsilon}\frac{\partial p^1}{\partial x_i^1}\right) - \cdots\right\}\\
&+\varepsilon^2\left\{\frac{\partial}{\partial x_j^0}\left(\eta\frac{\partial v_i^0}{\partial x_j^0}\right) + \frac{2}{\varepsilon}\frac{\partial}{\partial x_j^0}\left(\eta\frac{\partial v_i^0}{\partial x_j^1}\right) + \frac{1}{\varepsilon^2}\frac{\partial}{\partial x_j^1}\left(\eta\frac{\partial v_i^0}{\partial x_j^1}\right)\right\}\\
&+\varepsilon^3\left\{\frac{\partial}{\partial x_j^0}\left(\eta\frac{\partial v_i^1}{\partial x_j^0}\right) + \frac{2}{\varepsilon}\frac{\partial}{\partial x_j^0}\left(\eta\frac{\partial v_i^1}{\partial x_j^1}\right) + \frac{1}{\varepsilon^2}\frac{\partial}{\partial x_j^1}\left(\eta\frac{\partial v_i^1}{\partial x_j^1}\right)\right\} + \cdots + f_i\\
&= -\frac{1}{\varepsilon}\frac{\partial p^0}{\partial x_i^1} + \varepsilon^0\left\{-\frac{\partial p^0}{\partial x_i^0} - \frac{\partial p^1}{\partial x_i^1} + \frac{\partial}{\partial x_j^1}\left(\eta\frac{\partial v_i^0}{\partial x_j^1}\right) + f_i\right\}\\
&+\varepsilon^1\left\{-\frac{\partial p^1}{\partial x_i^0} - \frac{\partial p^2}{\partial x_i^1} + 2\frac{\partial}{\partial x_j^0}\left(\eta\frac{\partial v_i^0}{\partial x_j^1}\right) + \frac{\partial}{\partial x_j^1}\left(\eta\frac{\partial v_i^1}{\partial x_j^1}\right)\right\}\\
&+\varepsilon^2\left\{-\frac{\partial p^2}{\partial x_i^0} - \frac{\partial p^3}{\partial x_i^1} + \frac{\partial}{\partial x_j^0}\left(\eta\frac{\partial v_i^0}{\partial x_j^0}\right) + 2\frac{\partial}{\partial x_j^0}\left(\eta\frac{\partial v_i^1}{\partial x_j^1}\right) + \frac{\partial}{\partial x_j^1}\left(\eta\frac{\partial v_i^2}{\partial x_j^1}\right)\right\} + \cdots\\
&= 0.
\end{aligned}$$

As $\varepsilon \to 0$, we have the following perturbed equations:

$$\varepsilon^{-1}\text{-term:} \quad \frac{\partial p^0}{\partial x_i^1} = 0 \tag{8.13}$$

$$\varepsilon^{0}\text{-term:} \quad -\frac{\partial p^1}{\partial x_i^1} + \frac{\partial}{\partial x_j^1}\left(\eta\frac{\partial v_i^0}{\partial x_j^1}\right) = \frac{\partial p^0}{\partial x_i^0} - f_i \tag{8.14}$$

We estimate the term ε^{-1} as

$$\frac{\partial p^0}{\partial x_i^1} = 0 \quad \Rightarrow \quad p^0(\mathbf{x}^0, \mathbf{x}^1) = p^0(\mathbf{x}^0). \tag{8.15}$$

This implies that p^0 is a function of only $\mathbf{x}^0$.

By substituting (8.6) and (8.9) into the mass conservation equation (8.2), we have

$$\varepsilon^2\left(\frac{\partial v_i^0}{\partial x_i^0}+\frac{1}{\varepsilon}\frac{\partial v_i^0}{\partial x_i^1}\right)+\varepsilon^3\left(\frac{\partial v_i^1}{\partial x_i^0}+\frac{1}{\varepsilon}\frac{\partial v_i^1}{\partial x_i^1}\right)+\cdots$$
$$=\varepsilon^1\frac{\partial v_i^0}{\partial x_i^1}+\varepsilon^2\left(\frac{\partial v_i^0}{\partial x_i^0}+\frac{\partial v_i^1}{\partial x_i^1}\right)+\cdots$$
$$=0.$$

Then for the terms ε^1 and ε^2, we obtain

$$\varepsilon^1\text{-term:}\quad \frac{\partial v_i^0}{\partial x_i^1}=0 \tag{8.16}$$

$$\varepsilon^2\text{-term:}\quad \frac{\partial v_i^0}{\partial x_i^0}+\frac{\partial v_i^1}{\partial x_i^1}=0. \tag{8.17}$$

The r.h.s. terms of (8.14) are functions of only $\boldsymbol{x}^0$, and we introduce a separation of variables for the r.h.s. terms $p^1(\boldsymbol{x}^0,\boldsymbol{x}^1)$ and $v_i^0(\boldsymbol{x}^0,\boldsymbol{x}^1)$ as

$$v_i^0(\boldsymbol{x}^0,\boldsymbol{x}^1)=-\left(\frac{\partial p^0(\boldsymbol{x}^0)}{\partial x_j^0}-f_j(\boldsymbol{x}^0)\right)v_i^j(\boldsymbol{x}^1) \tag{8.18}$$

$$p^1(\boldsymbol{x}^0,\boldsymbol{x}^1)=-\left(\frac{\partial p^0(\boldsymbol{x}^0)}{\partial x_j^0}-f_j(\boldsymbol{x}^0)\right)p^j(\boldsymbol{x}^1). \tag{8.19}$$

By subsituting (8.18) and (8.19) into (8.14) and (8.16), we obtain the following incompressible flow equations in the micro-domain, referred to as the *microscale equations for Stokes' flow*:

$$-\frac{\partial p^k}{\partial x_i^1}+\frac{\partial}{\partial x_j^1}\left(\eta\frac{\partial v_i^k}{\partial x_j^1}\right)+\delta_{ik}=0 \quad \text{in } \Omega_{1f} \tag{8.20}$$

$$\frac{\partial v_i^k}{\partial x_i^1}=0 \quad \text{in } \Omega_{1f} \tag{8.21}$$

where $v_i^k(\boldsymbol{x}^1)$ and $p^k(\boldsymbol{x}^1)$ are *characteristic functions for velocity and pressure*, respectively, and δ_{ik} is Kronecker's delta. If we solve (8.20) and (8.21) under the microscale periodic boundary conditions, we obtain the characteristic functions $v_i^k(\boldsymbol{x}^1)$ and $p^k(\boldsymbol{x}^1)$, which involve the microscale inhomogeneity in geometry and material properties.

We recall that Darcy's law gives a relationship between the seepage velocity and the gradient of the total head. Based on the homogenization theory, we obtain the following homogenized form of Darcy's law by averaging (8.18) in the micro-domain:

$$\tilde{v}_i^0(\boldsymbol{x}^0) = \langle v_i^0(\boldsymbol{x}^0, \boldsymbol{x}^1)\rangle = -K_{ji}\left(\frac{\partial p^0(\boldsymbol{x}^0)}{\partial x_j^0} - f_j(\boldsymbol{x}^0)\right), \tag{8.22}$$

$$K_{ij} \equiv \langle v_j^i(\boldsymbol{x}^1)\rangle = \frac{1}{|\Omega_1|}\int_{\Omega_{1f}} v_j^i(\boldsymbol{x}^1)\, d\boldsymbol{x}^1 \tag{8.23}$$

where $|\Omega_1|$ is the volume of the unit cell, and $\langle\cdot\rangle$ gives an averaging operation.

Let us introduce a weak form for the microscale equations for Stokes' flow (8.20):

$$\left\langle -\eta\frac{\partial v_i^k}{\partial x_j^1} + p^k\delta_{ij},\ \frac{\partial w_i}{\partial x_j^1}\right\rangle + \left\langle \delta_{ki},\ w_i\right\rangle = 0, \qquad \forall \boldsymbol{w} \in V_{X^1} \tag{8.24}$$

where V_{X^1} is defined by

$$V_{X^1} = \left\{\boldsymbol{u} \in \left(H^1(\Omega_f)\right)^3,\ \partial u_i/\partial x_i^1 = 0,\ X^1\text{-periodic}\right\}. \tag{8.25}$$

The space of functions $(H^1(\Omega_f))^3$ is a subspace of the Hilbert space $(L_2(\Omega_f))^3$, and we introduce an L_2-inner product $\langle\cdot,\cdot\rangle$[2] as

$$\langle f, g\rangle = \int_{\Omega_1} f(\boldsymbol{x}^1)\, g(\boldsymbol{x}^1)\, d\boldsymbol{x}^1.$$

Let us assume that $w_i = v_i^k$ in (8.24) under the microscale incompressibility condition (8.21), and we obtain

$$\left\langle -\eta\frac{\partial v_i^k}{\partial x_j^1},\ \frac{\partial v_i^k}{\partial x_j^1}\right\rangle + \left\langle \delta_{ki},\ v_i^k\right\rangle = \int_{\Omega_1}\left(-\eta\frac{\partial v_i^k}{\partial x_j^1}\frac{\partial v_i^k}{\partial x_j^1} + \delta_{ki}v_i^k\right)d\boldsymbol{x}^1 = 0. \tag{8.26}$$

Equation 8.26 shows that v_i^j is symmetric, which results in the symmetry and positive definiteness of K_{ji}:

$$v_i^j = v_j^i, \qquad K_{ji} = K_{ij}, \qquad K_{ij}\,\xi_i\xi_j \geq \xi_i\xi_i \quad \forall\boldsymbol{\xi}. \tag{8.27}$$

By averaging (8.17) in the micro-domain, the second term of the l.h.s. vanishes because of the periodic boundary condition in the micro-domain; substituting

[2]The space of functions $(H^1(\Omega_f))^3$ implies that a function $f \in (H^1(\Omega_f))^3$ and its first order differential are bounded:

$$\langle f, f\rangle + \langle\boldsymbol{\nabla} f, \boldsymbol{\nabla} f\rangle < +\infty.$$

Details are given in, e.g., Sanchez-Palencia (1980).

Darcy's law (8.22) into the averaged (8.17), we obtain the following *homogenized seepage equation*, which gives the macroscale incompressibility condition:

$$\frac{\partial \tilde{v}_i^0}{\partial x_i^0} = 0 \quad \Rightarrow \quad -\frac{\partial}{\partial x_i^0}\left[-K_{ji}\left\{\frac{\partial p^0(\boldsymbol{x}^0)}{\partial x_j^0} - f_j(\boldsymbol{x}^0)\right\}\right] = 0. \tag{8.28}$$

The true velocity and pressure are given in their first order approximation as

$$v_i^\varepsilon(\boldsymbol{x}^0, \boldsymbol{x}^1) \simeq \varepsilon^2 v_i^0(\boldsymbol{x}^0, \boldsymbol{x}^1), \qquad p^\varepsilon(\boldsymbol{x}^0) \simeq p^0(\boldsymbol{x}^0). \tag{8.29}$$

Thus by applying the HA, we can obtain the true distributions of velocity and pressure.

The conventional theory of soil mechanics states that Darcy's law is introduced as

$$\tilde{v}_i^* = -K_{ij}^* \frac{\partial \phi}{\partial x_j^0}, \qquad \phi = \frac{p}{\rho g} + \zeta \tag{8.30}$$

where ϕ is the total head, $p/\rho g$ is the pressure head, ζ is the elevation head, ρ is the mass density of the fluid, and g is the gravitational constant. K_{ij}^* of (8.30) is here referred to as the C-permeability, and if we recall the relationship

$$\tilde{v}_i^* = \tilde{v}_i^\varepsilon \simeq \varepsilon^2 \tilde{v}_i^0,$$

we have the following relationship between the HA-permeability and the C-permeability:

$$K_{ij}^* = \varepsilon^2 \rho g K_{ij}. \tag{8.31}$$

8.2.2 *Analytical Solution for a Microscale Poiseulle Flow*

A pure smectitic clay consists of stacks of clay minerals such as montmorillonite or beidellite. One mineral is a platelet of about 100 × 100 × 1 nm, and several crystals stack parallelly as shown in Fig. 1.7. Keeping this fact in mind, we consider a microscale structure, i.e., a unit cell, with flow between two parallel platelets as shown in Fig. 8.2c. If the viscosity η of the fluid is constant, the solutions of microscale equations (8.20) and (8.21) are given by

$$v_1^1 = \frac{1}{8\eta}\left(d^2 - 4(x_2^1)^2\right), \qquad p^1 = C_1, \qquad p^2 = x_2^1 + C_2$$

where C_1 and C_2 are constants, and other characteristic functions are null. Thus the HA-permeability can be calculated as

$$K_{11} = \frac{d^3}{12\eta X_2^1}. \tag{8.32}$$

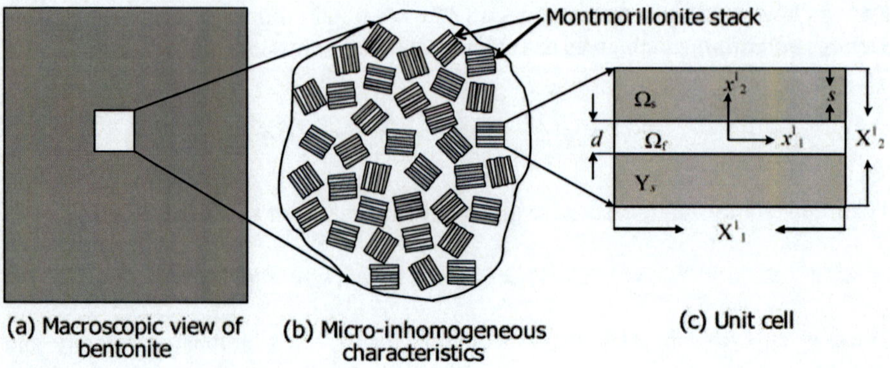

Fig. 8.2 Multiscale structures of pure smectitic clay

If we use the values of the water viscosity $\eta = 0.8 \times 10^{-3}$ Pa·s at temperature 300 K, and the geometry of a unit cell $X_1^1 = 100$ nm, $X_2^1 = 0.02$ nm, $d = 1$ nm, $\varepsilon = 10^{-5}$, we obtain the HA-permeability as $K_{11} = 5.208 \times 10^{-4}$ s cm³/g. Since the density of water is $\rho = 0.99651$ Mg/m³, the C-permeability (8.23) is $K_{11}^* = 5.086 \times 10^{-11}$ cm/s.

As shown in Fig. 8.2b, the stacks of clay minerals are randomly distributed in most cases, and then the permeability is isotropic. From (8.31) and (8.32) we can assume that the permeability is proportional to $\varepsilon^2 d^2$, since d is the same dimension as X_2^1. The permeability of clay is extremely small, since the value $\varepsilon^2 d^2$ is very small. Note that this is the case if the viscosity η is constant. However, by MD simulations we know that the viscosity η varies widely in the neighborhood of the smectitic clay surface.

8.2.3 Finite Element Analysis for the Mass Conservation Equation in the Micro-domain

The major mineral of bentonite is a smectitic clay (see Fig. 1.7). Since the surface of smectitic clay is negatively charged, the surface molecules of interlayer water will exhibit different characteristics in diffusion and viscosity from that of normal water. The swelling property of beidellite, which is calculated by a Molecular Dynamic (MD) simulation is given in Figs. 8.3 and 8.4 (see Kawamura and Ichikawa 2001). Distributions of self-diffusivity and viscosity of water in the neighborhood of the clay surface are shown in Fig. 8.5; the viscosity of water becomes extremely high, which, together with the size effect described in Sect. 8.2.2, leads to the low permeability of bentonite.

When the viscosity η is distributed inhomogeneously, an analytical solution of the microscale equations (8.20) and (8.21) is not possible; we therefore perform a finite element calculation. We first give a penalized weak form of (8.20) with the

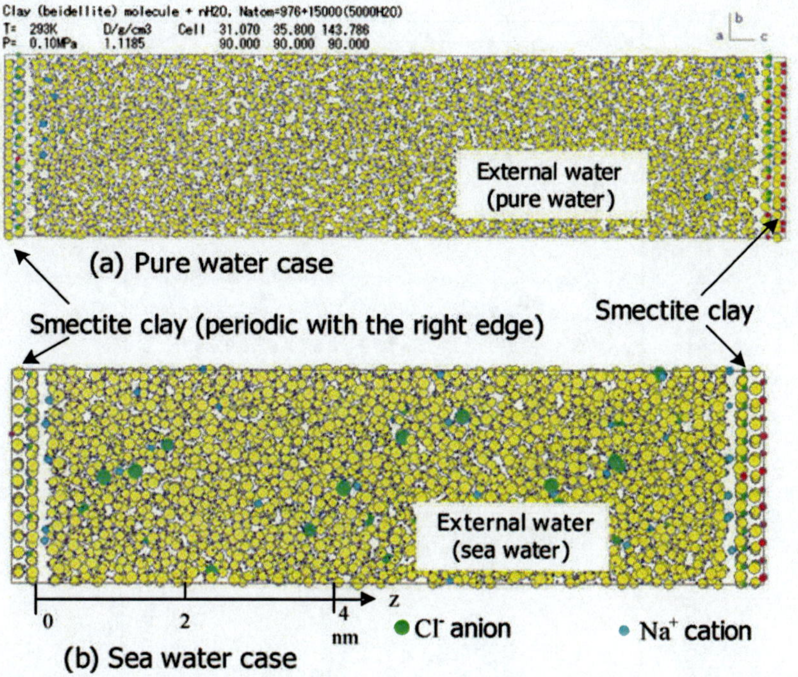

Fig. 8.3 Snapshots of hydrated beidellite calculated using an MD simulation (Courtesy of Prof. K. Kawamura)

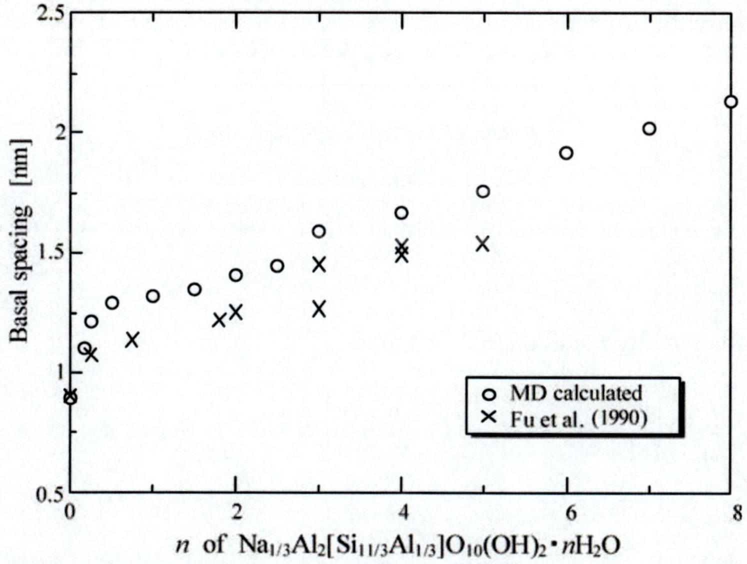

Fig. 8.4 Swelling property of bentonite: ○ calculated using an MD simulation for beidellite, × experiments for Wyoming clay (Courtesy of Prof. K. Kawamura)

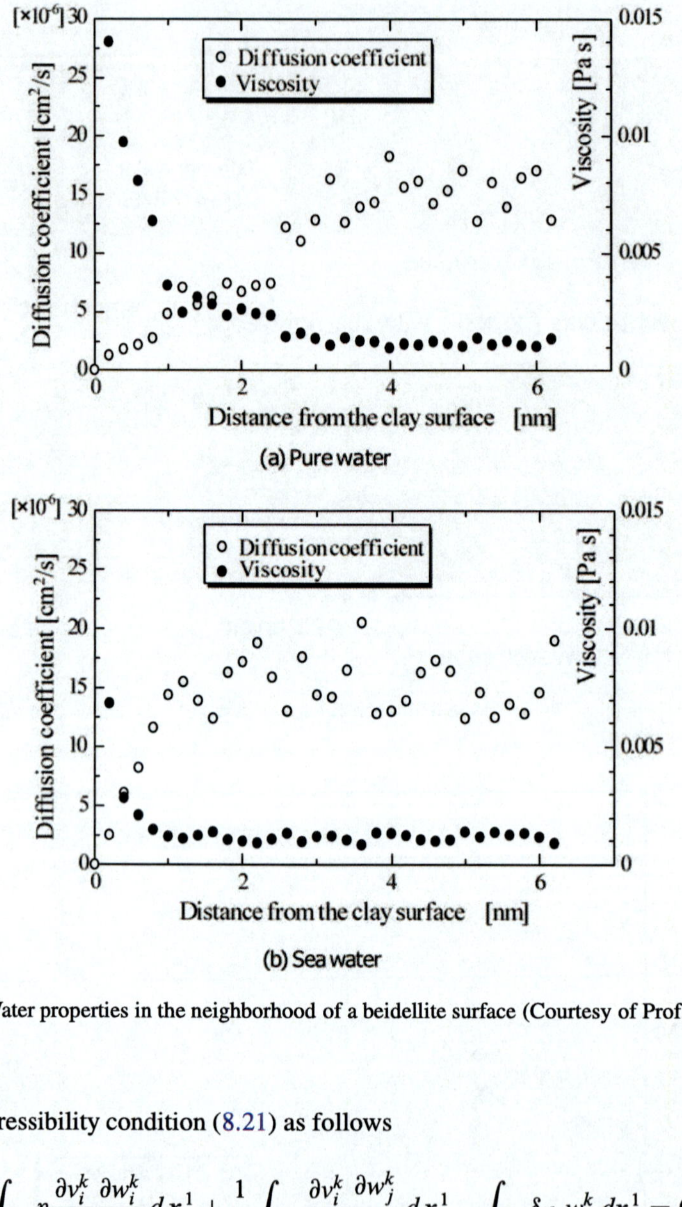

Fig. 8.5 Water properties in the neighborhood of a beidellite surface (Courtesy of Prof. K. Kawamura)

incommpressibility condition (8.21) as follows

$$\int_{\Omega_f} \eta \frac{\partial v_i^k}{\partial x_j^1} \frac{\partial w_i^k}{\partial x_j^1} \, d\boldsymbol{x}^1 + \frac{1}{\lambda} \int_{\Omega_f} \frac{\partial v_i^k}{\partial x_i^1} \frac{\partial w_j^k}{\partial x_j^1} \, d\boldsymbol{x}^1 - \int_{\Omega_f} \delta_{ik} \, w_i^k \, d\boldsymbol{x}^1 = 0$$

$$\forall w_i^k \;\; (w_i^k = 0 \;\; \text{on} \;\; \Gamma) \tag{8.33}$$

where λ $(0 < \lambda \ll 1)$ is a penalty constant, and Γ is a solid/fluid interface in the micro-domain.

We next introduce finite element interpolations for v_i^k and w_i^k using the shape functions Φ_r:

$$v_i^k \cong \sum_{r=1}^{m} \Phi_r V_{ir}^k, \qquad w_i^k \cong \sum_{s=1}^{m} \Phi_s \delta V_{is}^k \tag{8.34}$$

where V_{ir}^k and δV_{is}^k are nodal vectors corresponding to v_i^k and w_i^k. By substituting (8.34) into (8.33), we have

$$\sum_{r=1}^{m}\sum_{s=1}^{m} \delta V_{is}^k (K_{rs}^V V_{ir}^k + K_{rsji}^{VP} V_{jr}^k) = \sum_{r=1}^{m} \delta V_{is}^k F_{is}^k, \tag{8.35}$$

$$K_{rs}^V = \int_{\Omega_f} \eta \frac{\partial \Phi_r}{\partial x_j^1} \frac{\partial \Phi_s}{\partial x_j^1} d\boldsymbol{x}^1, \quad K_{rsji}^{VP} = \frac{1}{\lambda} \int_{\Omega_f} \frac{\partial \Phi_r}{\partial x_j^1} \frac{\partial \Phi_s}{\partial x_i^1} d\boldsymbol{x}^1, \quad F_{is}^k = \int_{\Omega_f} \Phi_s \delta_{ik}\, d\boldsymbol{x}^1. \tag{8.36}$$

Since δV_{is}^k are arbitrary in Ω_f, they can be omitted to obtain the following simultaneous equation:

$$\left[\boldsymbol{K}^v + \boldsymbol{K}^{VP}\right]\left\{\boldsymbol{V}^k\right\} = \left\{\boldsymbol{F}^k\right\} \qquad (k = 1, 2, 3). \tag{8.37}$$

We solve this to obtain $\{V\}^k$. In our analysis we set the penalty constant as $\lambda = 10^{-1}$, and use a reduced integration technique of the Gaussian integral in order to prevent over-constraining (see Hughes 1987).

8.2.4 *Numerical Results of Seepage Analysis for Pure Smectitic Clay*

8.2.4.1 Flow Between Two Platelets Under Distributed Water Viscosity for Pure Water

The water in the neighborhood of the smectitic clay surface is structured due to the hydrogen bond, and the viscosity varies inversely with distance from the surface. In this case we can apply the finite element method to solve the microscale equations (8.20) and (8.21), as described previously.

Figure 8.6a shows the finite element mesh of the micro-domain. We use the distribution of viscosity given by Fig. 8.5a, which is calculated by MD, and the result for the distribution of the characteristic function v_1^1 is shown in Fig. 8.6b. The analytical solution under a constant viscosity is represented in Fig. 8.6c, which is very different from Fig. 8.6b. This shows the strong influence of the surface on the water viscosity.

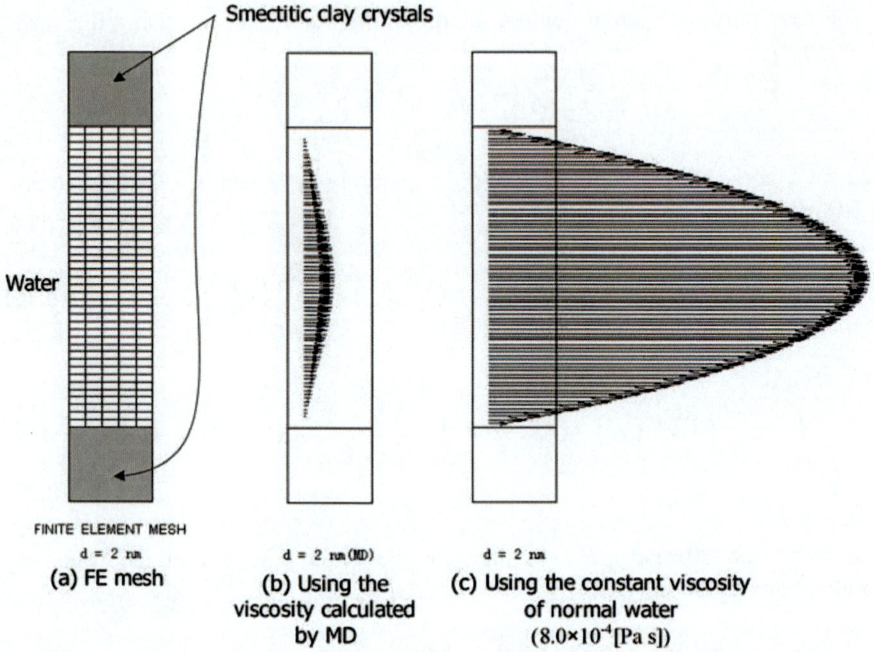

Fig. 8.6 Distribution of the characteristic velocity function v_1^1 for pure water

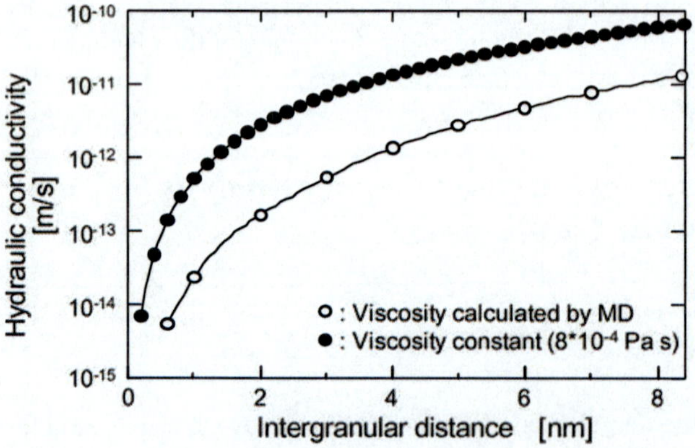

Fig. 8.7 C-permeability for pure water as a function of distance between platelets

The C-permeability can then be calculated as a function of the distance between two platelets (Fig. 8.7); it can clearly been seen that the permeability obtained from the MD/HA procedure is considerably smaller than when using constant viscosity.

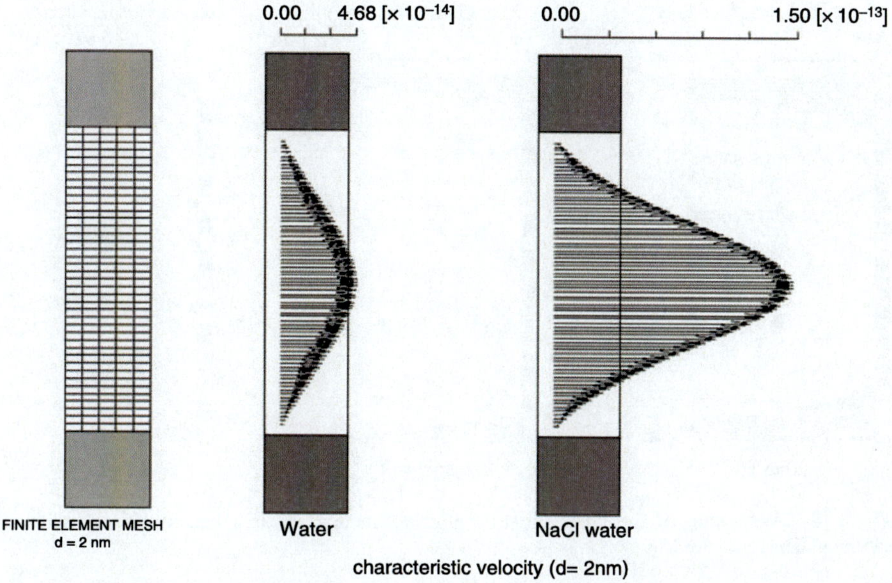

Fig. 8.8 Distribution of the characteristic velocity function v_1^1 for pure water and salt water

Fig. 8.9 C-permeability for salt water and pure water as a function of distance between platelets

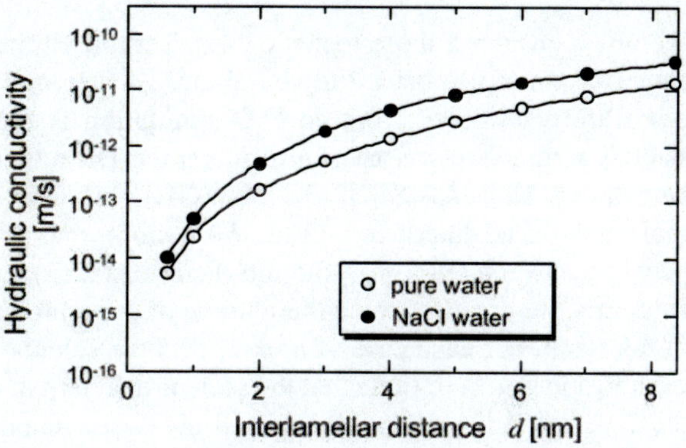

8.2.4.2 Flow Between Two Platelets Under Distributed Water Viscosity for Salt Water

By using the MD result for salt water (where the salt content is similar to that of the sea water) shown in Fig. 8.5b, we can calculate the distribution of the characteristic function v_1^1 (Fig. 8.8) and the C-permeability (Fig. 8.9), which are compared with those of pure water.

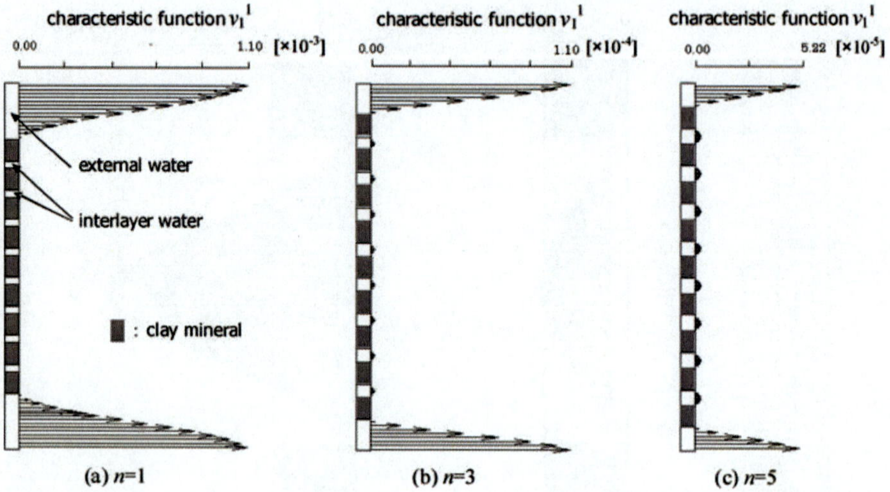

Fig. 8.10 Distribution of the characteristic velocity function v_1^1 of a beidellite stack with a saturated density of 1.8 Mg/m^3

8.2.4.3 Flow in a Stack of Clay Minerals with Pure Water

We now consider a more realistic case. Let the saturated density of pure smectitic clay (for example, beidellite) be about 1.8 [Mg/m^3]. The crystal density of the beidellite determined from an MD simulation is found to be 2.901 [Mg/m^3]. A stack is assumed to consist of nine minerals. The molecular formula of the hydrated beidellite is $Na_{1/3}Al_2[Si_{11/3}Al_{1/3}]O_{10}(OH)_2\ nH_2O$ where n is the number of water molecules in an interlayer space. We assume that $n = 1, 3, 5$, and the distance between two minerals (i.e., the interlayer distance) can be obtained from Fig. 8.4; from this, we can determine the volume of external water that exists on the outside of the stack. For each case of $n = 1, 3, 5$ we calculate the characteristic functions as shown in Fig. 8.10 (note that the scale is different in each case). Then we compute the C-permeability as shown in Fig. 8.11. Based on numerous experimental results, Pusch (1994) obtained the permeability characteristics of clays as a function of density as shown in Fig. 8.12. We recall that the permeability of the saturated smectitic clay is not only a function of the density but also of the ratio of interlayer water to the external water, which indicates that there exists a distribution of permeability for the same density. The range of permeability given in Fig. 8.12 with a saturated density of 1.8 Mg/m^3 corresponds well to our calculated results, which were obtained using the MD/HA procedure.

8.2.4.4 Flow in a Stack of Clay Minerals with Salt Water

The property of viscosity distribution for salt water, which was obtained using MD (Fig. 8.5b) is used to calculate the characteristic function of HA. The characteristic

Fig. 8.11 Hydraulic conductivity of a beidellite stack with a saturated density of 1.8 Mg/m^3

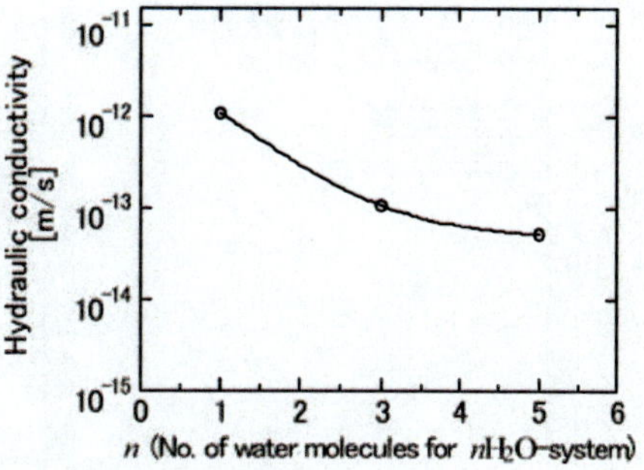

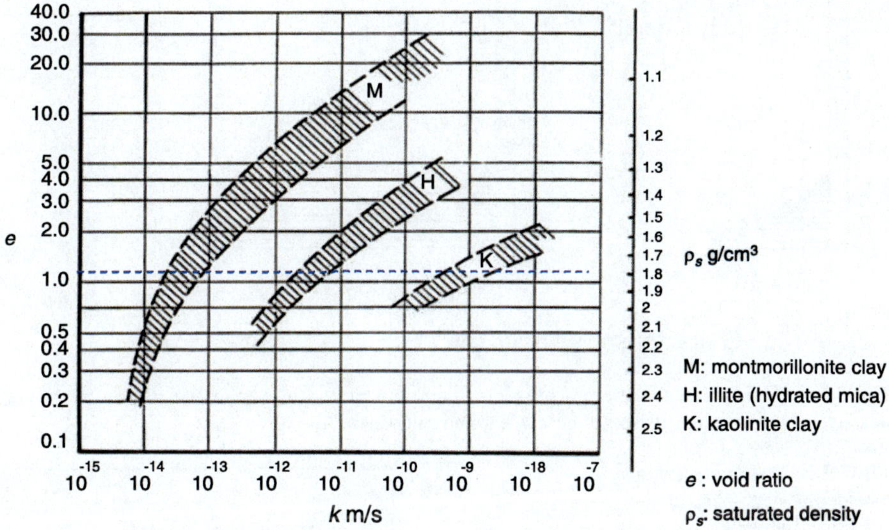

Fig. 8.12 Hydraulic conductivity for several pure clays (Pusch 1994)

function v_1^1 for the beidellite stack with a saturated density of 1.8 Mg/m^3 is shown in Fig. 8.13 for cases of both pure water and salt water. The C-permeability of a beidellite stack for salt water is shown in Fig. 8.14, and is dependent on the molecular number of the interlayer water.

8.2.5 Three-dimensional Seepage Analysis of Sand

8.2.5.1 Permeability by Homogenization Analysis

We consider a simple geometrical model of sand in the micro-domain as shown in Fig. 8.15, and apply the HA.

As described previously, the permeability calculated by the HA is written as

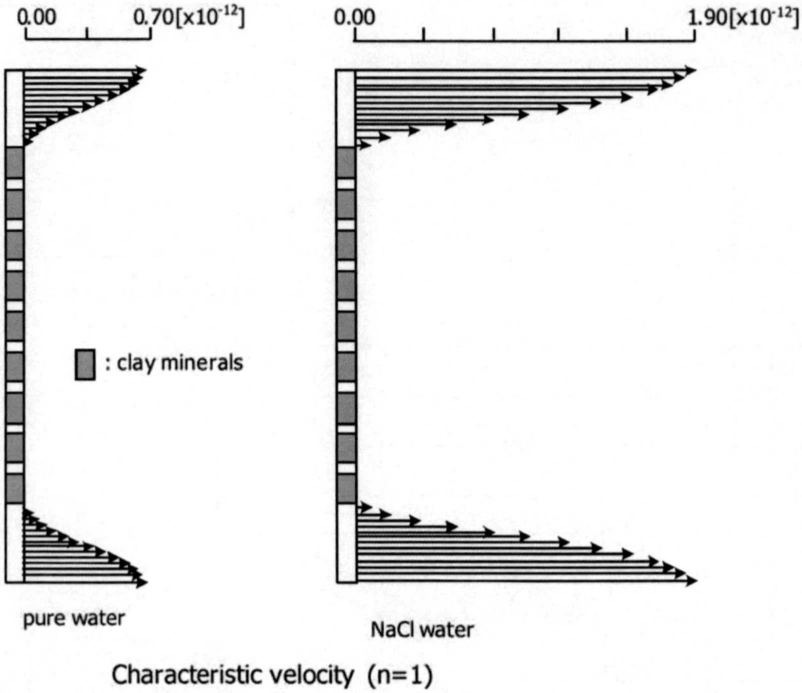

Fig. 8.13 Distribution of the characteristic velocity function v_1^1 of a beidellite stack of saturated density 1.8 Mg/m^3: comparison of pure water and salt water

Fig. 8.14 C-permeability of a beidellite stack of saturated density 1.8 Mg/m^3: comparison between results for pure water and for salt water

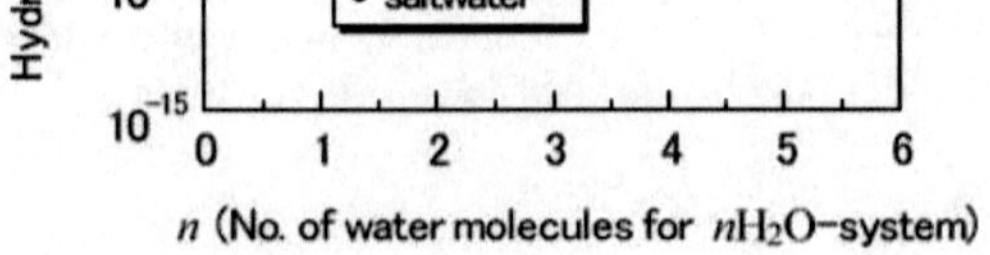

$$K_{ij}^* = \varepsilon^2 \rho g K_{ij}, \qquad K_{ij} = \frac{1}{|\Omega_1|} \int_{\Omega_1} v_j^i \, d\boldsymbol{x}^1$$

where K_{ij}^* is the C-permeability, and K_{ij} is the H-permeability. Since it is difficult to obtain an analytical solution for this problem, we apply a finite element approximation.

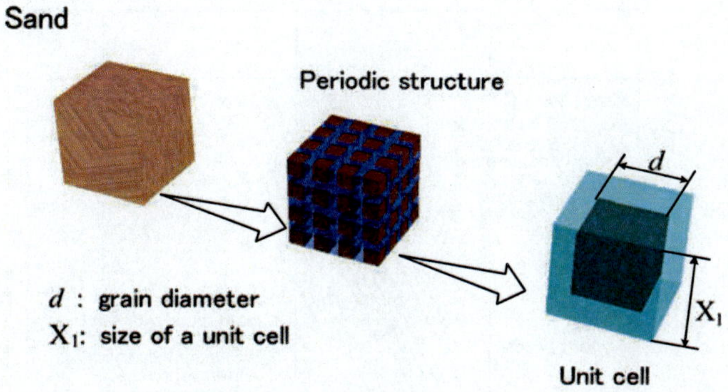

Fig. 8.15 A micro-domain model of sand and a unit cell

8.2.5.2 Emperical Formulae and HA-Calculated Values

Several empirical formulae were proposed for sand, which are given as follows:

$$\text{Hazen (Hazen 1892):} \quad k = C_K(0.7 + 0.03T)D_w^2,$$

$$\text{Terzaghi (Terzaghi and Peck 1964):} \quad k = \frac{C_t}{\eta}\left(\frac{n - 0.13}{\sqrt[3]{1-n}}\right)^2 D_{10}^2,$$

$$\text{Zunker (Zunker 1930):} \quad k = \frac{C_z}{\eta}\left(\frac{n}{1-n}\right)^2 D_w^2,$$

$$\text{Kozeny-Donat (Kozeny 1927; Donat 1929):} \quad k = \frac{C_k}{\eta}\left(\frac{n^3}{(1-n)^2}\right)^2 D_w^2$$

where C_K, C_t, C_z, C_k are empirical constants, T is the temperature, D_w is the average grain size, D_{10} is the effective grain size, n is the porosity, and η is the dynamic viscosity of water. It should be noted that the permeability k increases with porosity n, it increases with the square of the grain size D_{10} or D_w (which relates to the pore size), and it increases inversely with the fluid viscosity η. These experimental observations are schematically shown in Fig. 8.16.

In Fig. 8.17 we show the HA-calculated results (•) using the regression formula with a viscosity $\eta = 1.14 \times 10^{-3}$ Pa·s (water at $T = 15°$C). A comparison with Fig. 8.16 indicates that both are reconcilable.

In the empirical formulae, the permeability is proportional to the porosity and inversely proportional to the viscosity. We validate this through the HA calculation by changing the porosity n and the viscosity η. Figures 8.18–8.20 illustrate the results for viscosity changes at $T = 10$, 20 and 30°C respectively, and show good agreement between the HA results and the empirical relationships.

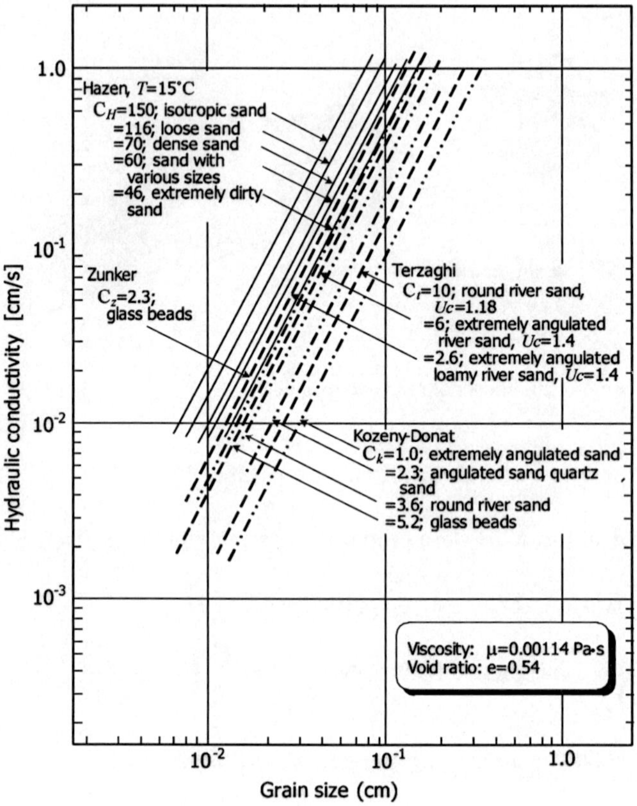

Fig. 8.16 Hydraulic conductivity of sand with various grain sizes (Kubota et al. 1976)

8.3 A Permeability Theory for Multiscale Porous Media

We can extend the two-scale HA theory to a multiscale porous medium, which is schematically shown in Fig. 8.21 where Ω_f is the fluid phase, Ω_s is the solid phase, Ω_{sf} is the mixture phase of fluid and solid (i.e., a mixture of stacks of beidellite and fluid) in the meso-domain, Γ_{sf} is the solid/fluid interface, and Γ_f is the periodic boundary for the fluid phase.

The three scales are as follows: the clay mineral is the microscale medium, which has a thickness of 10^{-9} m and a width of about 10^{-7} m; the quartz grain, which constitutes the mesoscale, has dimensions of 10^{-4}–10^{-5} m; the triaxial specimen used in the experiments has dimensions 10^{-1} m. Considering these scales, the parameter ε for a smectitic bentonite is about 10^{-4}. Then we can introduce the following relations between the coordinate systems:

$$\boldsymbol{x}^1 = \frac{\boldsymbol{x}^0}{\varepsilon}, \qquad \boldsymbol{x}^2 = \frac{\boldsymbol{x}^1}{\varepsilon}. \tag{8.38}$$

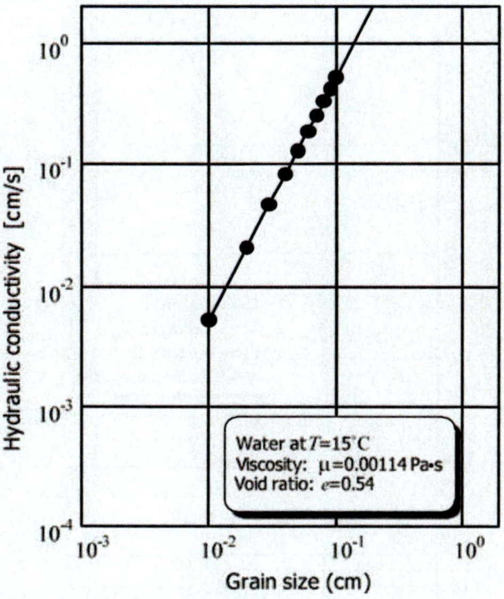

Fig. 8.17 HA-calculated hydraulic conductivity (•) and the regression line ($e = 0.54,\ T = 15°\mathrm{C}$)

These relations are approximate; however, if the size of each domain is widely different, we need to introduce another form. The size of a unit cell of the meso-domain is $|\boldsymbol{X}^1| = \varepsilon^{-1}|\boldsymbol{X}^2|$, and the size of a unit cell of the micro-domain is $|\boldsymbol{X}^2|$.

Since we introduce the multiscale systems $\boldsymbol{x}^0$, $\boldsymbol{x}^1$ and $\boldsymbol{x}^2$, the original coordinates should be changed to

$$\boldsymbol{x} = \boldsymbol{x}(\boldsymbol{x}^0, \boldsymbol{x}^1, \boldsymbol{x}^2), \tag{8.39}$$

and the differentiation is transformed to

$$\frac{\partial}{\partial x_i} = \frac{\partial}{\partial x_i^0} + \frac{1}{\varepsilon}\frac{\partial}{\partial x_i^1} + \frac{1}{\varepsilon^2}\frac{\partial}{\partial x_i^2}. \tag{8.40}$$

8.3.1 Multiscale Permeability Theory

As mentioned previously, the flow field of water in a porous medium can be represented by Stokes' equation (8.3).

We introduce perturbations of the velocity v_i^ε and the pressure p^ε such that (8.9) and (8.10) now become

$$\boldsymbol{v}^\varepsilon(\boldsymbol{x}) = \varepsilon^4\boldsymbol{v}^0(\boldsymbol{x}^0, \boldsymbol{x}^1, \boldsymbol{x}^2) + \varepsilon^5\boldsymbol{v}^1(\boldsymbol{x}^0, \boldsymbol{x}^1, \boldsymbol{x}^2) + \varepsilon^6\boldsymbol{v}^2(\boldsymbol{x}^0, \boldsymbol{x}^1, \boldsymbol{x}^2) + \cdots, \tag{8.41}$$

$$p^\varepsilon(\boldsymbol{x}) = p^0(\boldsymbol{x}^0) + \varepsilon p^1(\boldsymbol{x}^0, \boldsymbol{x}^1, \boldsymbol{x}^2) + \varepsilon^2 p^2(\boldsymbol{x}^0, \boldsymbol{x}^1, \boldsymbol{x}^2) + \cdots \tag{8.42}$$

where $\boldsymbol{v}^\alpha(\boldsymbol{x}^0, \boldsymbol{x}^1, \boldsymbol{x}^2)$, $p^\alpha(\boldsymbol{x}^0, \boldsymbol{x}^1, \boldsymbol{x}^2)$ $(\alpha = 0, 1, \dots)$ are X^1-/X^2-periodic functions:

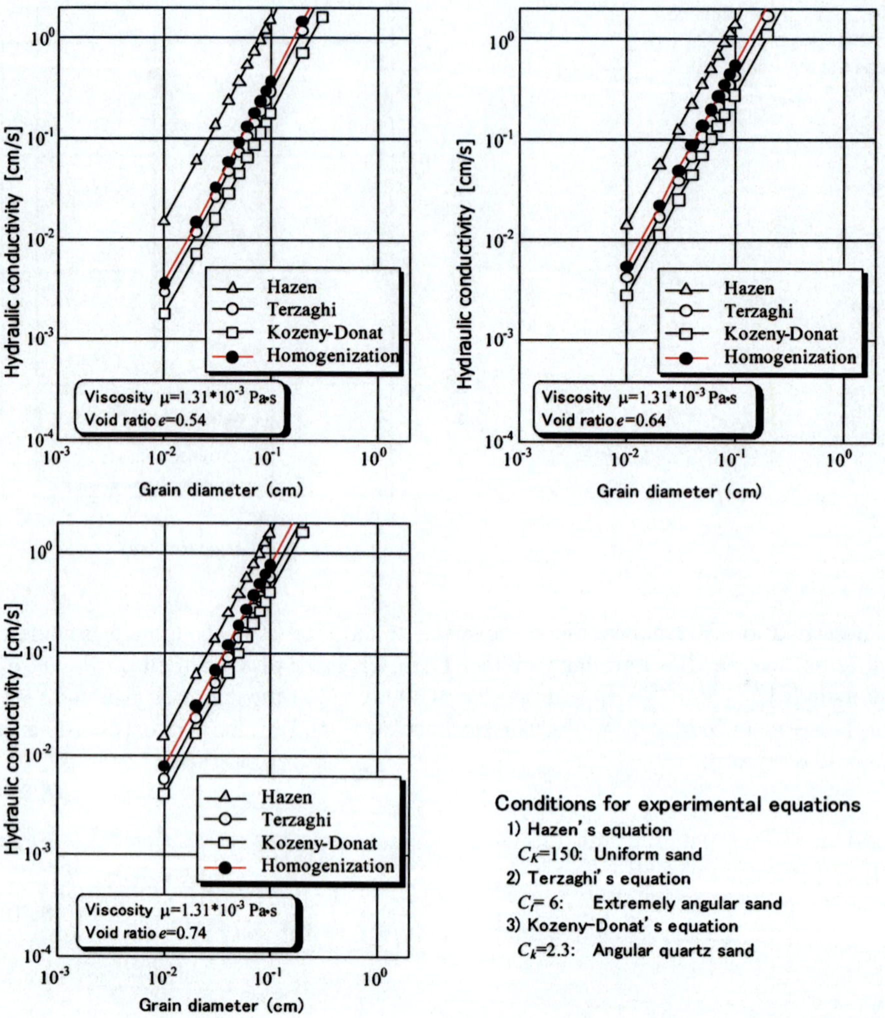

Fig. 8.18 Hydraulic conductivity by HA and empirical formulae: $T = 10°\text{C}$

$$v^{\alpha}(\boldsymbol{x}^0, \boldsymbol{x}^1, \boldsymbol{x}^2) = v^{\alpha}(\boldsymbol{x}^0, \boldsymbol{x}^1 + \boldsymbol{X}^1, \boldsymbol{x}^2), \quad v^{\alpha}(\boldsymbol{x}^0, \boldsymbol{x}^1, \boldsymbol{x}^2) = v^{\alpha}(\boldsymbol{x}^0, \boldsymbol{x}^1, \boldsymbol{x}^2 + \boldsymbol{X}^2)$$
$$p^{\alpha}(\boldsymbol{x}^0, \boldsymbol{x}^1, \boldsymbol{x}^2) = p^{\alpha}(\boldsymbol{x}^0, \boldsymbol{x}^1 + \boldsymbol{X}^1, \boldsymbol{x}^2), \quad p^{\alpha}(\boldsymbol{x}^0, \boldsymbol{x}^1, \boldsymbol{x}^2) = p^{\alpha}(\boldsymbol{x}^0, \boldsymbol{x}^1, \boldsymbol{x}^2 + \boldsymbol{X}^2).$$

The reason why the perturbation of $\boldsymbol{v}^{\varepsilon}(\boldsymbol{x})$ starts with a ε^4-term is to ensure reduction to the corresponding Stokes' equation in the micro-domain as a microscale equation (to be discussed later). We assume that the first-order term of pressure p^0 is a function of only the macroscale coordinate system $\boldsymbol{x}^0$.

By substituting (8.40), (8.41) and (8.42) into (8.1), we have

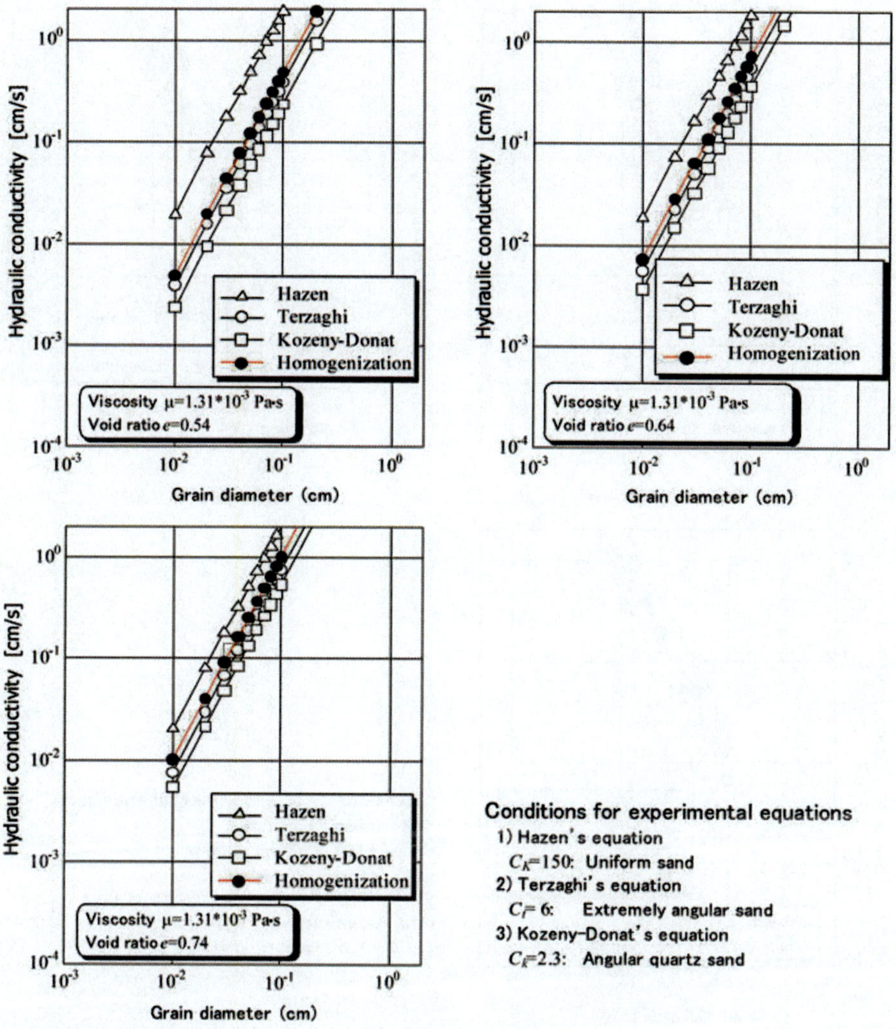

Fig. 8.19 Hydraulic conductivity by HA and empirical formulae: $T = 20°\mathrm{C}$

$$-\varepsilon^{-2}\frac{\partial p^0}{\partial x_i^2} - \varepsilon^{-1}\left(\frac{\partial p^0}{\partial x_i^1} + \frac{\partial p^1}{\partial x_i^2}\right) - \left(\frac{\partial p^0}{\partial x_i^0} + \frac{\partial p^1}{\partial x_i^1} + \frac{\partial p^2}{\partial x_i^2}\right) + \cdots$$

$$+\frac{\partial}{\partial x_j^2}\left(\eta\frac{\partial v_i^0}{\partial x_j^2}\right) + \varepsilon\left[\frac{\partial}{\partial x_j^1}\left(\eta\frac{\partial v_i^0}{\partial x_i^2}\right) + \frac{\partial}{\partial x_j^2}\left\{\eta\left(\frac{\partial v_i^0}{\partial x_j^1} + \frac{\partial v_i^1}{\partial x_j^2}\right)\right\}\right] + \cdots + f_i = 0,$$

and as $\varepsilon \to 0$, we obtain the following:

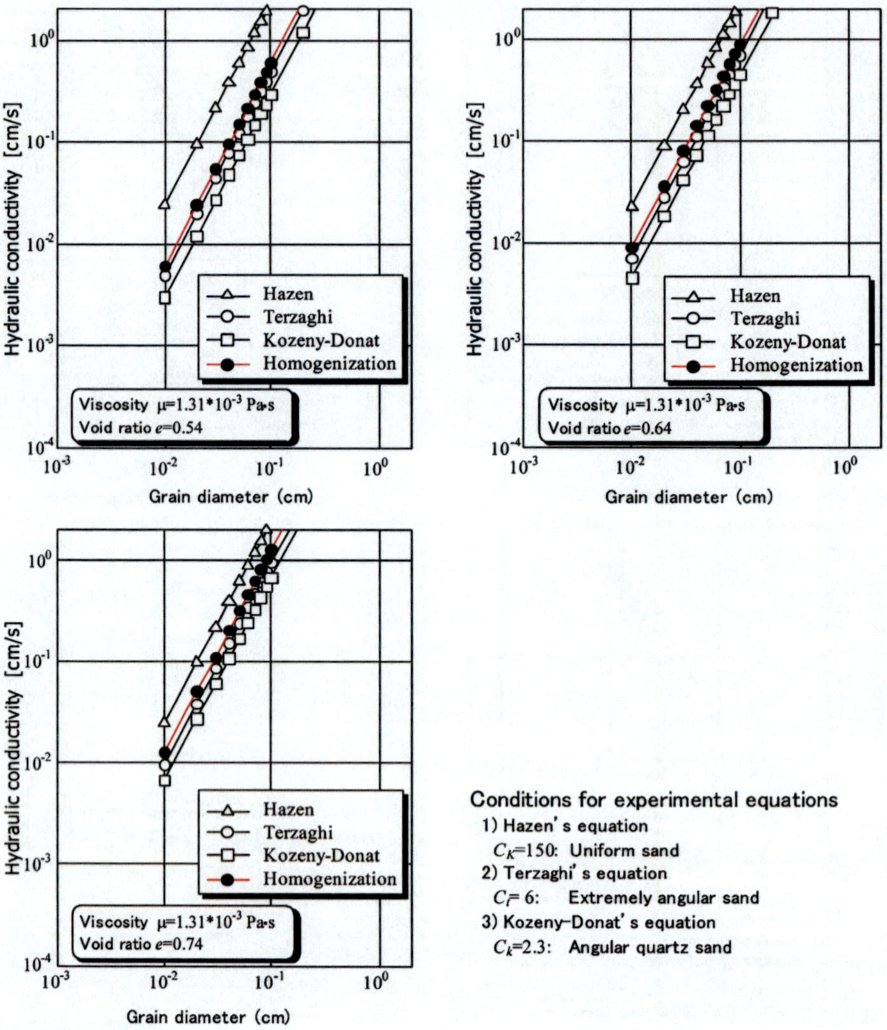

Fig. 8.20 Hydraulic conductivity by HA and empirical formulae: $T = 30°\mathrm{C}$

$$\varepsilon^{-2}\text{-term:} \quad \frac{\partial p^0}{\partial x_i^2} = 0 \tag{8.43}$$

$$\varepsilon^{-1}\text{-term:} \quad \frac{\partial p^0}{\partial x_i^1} + \frac{\partial p^1}{\partial x_i^2} = 0 \tag{8.44}$$

$$\varepsilon^{0}\text{-term:} \quad \frac{\partial}{\partial x_j^2}\left(\eta \frac{\partial v_i^0}{\partial x_j^2}\right) - \frac{\partial p^2}{\partial x_i^2} = \frac{\partial p^0}{\partial x_i^0} + \frac{\partial p^1}{\partial x_i^1} - f_i \tag{8.45}$$

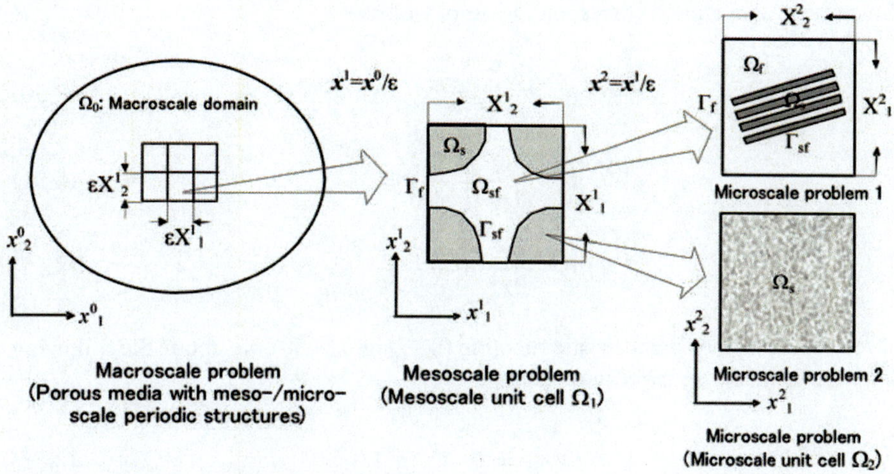

Fig. 8.21 Multiscale problem for porous media

Since p^0 is a function of only $\boldsymbol{x}^0$, (8.43) is automatically satisfied. By substituting (8.43) into (8.44), we have

$$p^1 = p^1(\boldsymbol{x}^0, \boldsymbol{x}^1). \tag{8.46}$$

The r.h.s. terms of (8.45) are functions of $\boldsymbol{x}^0$ and $\boldsymbol{x}^1$, since f_i is a function of only $\boldsymbol{x}^0$. Thus we can introduce normalized characteristic functions $v_i^k(\boldsymbol{x}^2)$ and $p^k(\boldsymbol{x}^2)$ such that

$$v_i^0 = -\left(\frac{\partial p^0}{\partial x_k^0} + \frac{\partial p^1}{\partial x_k^1} - f_k\right) v_i^k(\boldsymbol{x}^2) \tag{8.47}$$

$$p^2 = -\left(\frac{\partial p^0}{\partial x_k^0} + \frac{\partial p^1}{\partial x_k^1} - f_k\right) p_2^k(\boldsymbol{x}^2) \tag{8.48}$$

where $v_i^k(\boldsymbol{x}^2)$ and $p_2^k(\boldsymbol{x}^2)$ are referred to as the *velocity characteristic function* and the *pressure characteristic function*. By substituting (8.47) and (8.48) into (8.45), we obtain the following *microscale Stokes' equation*:

$$\frac{\partial}{\partial x_j^2}\left(\eta \frac{\partial v_i^k}{\partial x_j^2}\right) - \frac{\partial p_2^k}{\partial x_i^2} + \delta_{ki} = 0 \qquad \text{in } \Omega_2. \tag{8.49}$$

Under the transformation (8.40) we substitute the perturbation (8.41) into the incompressibility condition (8.2), and obtain

$$\frac{\partial v_i^\varepsilon}{\partial x_i} = \varepsilon^2 \frac{\partial v_i^0}{\partial x_i^2} + \varepsilon^3\left(\frac{\partial v_i^0}{\partial x_i^1} + \frac{\partial v_i^1}{\partial x_i^2}\right) + \varepsilon^4\left(\frac{\partial v_i^0}{\partial x_i^0} + \frac{\partial v_i^1}{\partial x_i^1} + \frac{\partial v_i^2}{\partial x_i^2}\right) + \cdots . = 0$$

Since each term should correspond to zero, we have

$$\varepsilon^2\text{-term:}\quad \frac{\partial v_i^0}{\partial x_i^2} = 0 \tag{8.50}$$

$$\varepsilon^3\text{-term:}\quad \frac{\partial v_i^0}{\partial x_i^1} + \frac{\partial v_i^1}{\partial x_i^2} = 0 \tag{8.51}$$

$$\varepsilon^4\text{-term:}\quad \frac{\partial v_i^0}{\partial x_i^0} + \frac{\partial v_i^1}{\partial x_i^1} + \frac{\partial v_i^2}{\partial x_i^2} = 0. \tag{8.52}$$

By substituting the characteristic function (8.47) into (8.50), we obtain the following *microscale incompressibility condition*:

$$\frac{\partial v_i^k}{\partial x_i^2} = 0 \qquad \text{in } \Omega_2. \tag{8.53}$$

Under the X^2-periodicity condition we can solve the microscale equations (8.49) and (8.53), and obtain the characteristic functions $v_i^k(\boldsymbol{x}^2)$, $p_2^k(\boldsymbol{x}^2)$.

If we introduce an integration average $\langle\cdot\rangle_2$ in the macro-domain Ω_2 for the characteristic function v_i^j, we have the *HA-meso-permeability* K_{ji}^2 as

$$K_{ji}^2 = \left\langle v_i^j \right\rangle_2 = \frac{1}{|\Omega_2|}\int_{\Omega_2} v_i^j \, d\boldsymbol{x}^2. \tag{8.54}$$

Let us introduce a weak form of the microscale Stokes' equation (8.49) as

$$\left\langle -\eta\frac{\partial v_i^k}{\partial x_j^2} + p_2^k\delta_{ij},\ \frac{\partial w_i}{\partial x_j^2}\right\rangle + \left\langle \delta_{ki},\ w_i\right\rangle = 0, \qquad \forall \boldsymbol{w} \in V_{X^2} \tag{8.55}$$

where, similar to (8.25), V_{X^2} is defined as

$$V_{X^2} = \left\{\boldsymbol{u} \in \left(H^1(\Omega_{2f})\right)^3,\ \partial u_i/\partial x_i^2 = 0,\ X^2\text{-periodic}\right\}.$$

If we set $w_i = v_i^k$ in (8.55) and take into account the microscale incompressibility condition (8.53), we obtain

$$\left\langle -\eta\frac{\partial v_i^k}{\partial x_j^2},\ \frac{\partial v_i^k}{\partial x_j^2}\right\rangle + \left\langle \delta_{ki},\ v_i^k\right\rangle = 0. \tag{8.56}$$

The result (8.56) implies that v_i^j is symmetric, and so is K_{ji}^2, which is positive definite in the sense that

$$v_i^j = v_j^i, \qquad K_{ji}^2 = K_{ij}^2, \qquad K_{ij}^2 \xi_i \xi_j \geq \xi_i \xi_i \quad \forall \boldsymbol{\xi}. \tag{8.57}$$

By introducing integral averaging $\langle \cdot \rangle_2$ in the micro-domain for (8.47), we have

$$\begin{aligned}
\left\langle v_i^0 \right\rangle_2 &= -K_{ji}^2 \left(\frac{\partial p^0}{\partial x_j^0} - f_j \right) - \frac{1}{|\Omega_2|} \int_{\Omega_2} \frac{\partial p^1}{\partial x_j^1} v_i^j \, d\boldsymbol{x}^2 \\
&= -K_{ji}^2 \left(\frac{\partial p^0}{\partial x_j^0} - f_j \right) - \frac{1}{|\Omega_2|} \int_{\Gamma_2} p^1 v_i^j n_j \, ds^2 + \frac{1}{|\Omega_2|} \int_{\Omega_2} p^1 \frac{\partial v_i^j}{\partial x_j^1} \, d\boldsymbol{x}^2
\end{aligned}$$

where the second term of the r.h.s. disappears because of the periodicity condition, and the third term of the r.h.s. also disappears because of the symmetry of v_i^j and the incompressibility condition. Thus we obtain the following *mesoscale Darcy's law*:

$$\left\langle v_i^0 \right\rangle_2 = -K_{ji}^2 \left(\frac{\partial p^0}{\partial x_j^0} - f_j \right). \tag{8.58}$$

By introducing integral averaging $\langle \cdot \rangle_2$ in the micro-domain for (8.52) and setting the contribution from the periodic terms to zero, we have

$$-\frac{\partial}{\partial x_i^0} \left\{ K_{ji}^2 \left(\frac{\partial p^0}{\partial x_j^0} - f_j \right) \right\} + \frac{\partial}{\partial x_i^1} \left\langle v_i^1 \right\rangle_2 = 0.$$

Furthermore, if we introduce an integral averaging $\langle \cdot \rangle_1$ in the micro-domain Ω_1, we obtain the following *macroscale seepage equation*:

$$-\frac{\partial}{\partial x_i^0} \left\{ K_{ji} \left(\frac{\partial p^0}{\partial x_j^0} - f_j \right) \right\} = 0 \qquad \text{in } \Omega_0 \tag{8.59}$$

where the *HA-macro-permeability* K_{ji} is defined as

$$K_{ji} = \left\langle K_{ji}^2 \right\rangle_1 = \frac{1}{|\Omega_1|} \int_{\Omega_1} K_{ji}^2 \, d\boldsymbol{x}^1, \tag{8.60}$$

and the *macroscale Darcy's law* is given as

$$\tilde{v}_i^0 = \left\langle \left\langle v_i^0 \right\rangle_2 \right\rangle_1 = -K_{ji} \left(\frac{\partial p^0}{\partial x_j^0} - f_j \right) \tag{8.61}$$

where $\tilde{v}_i^0$ is the macro-averaging value of v_i^0.

The pressure p^ε and velocity v_i^ε are approximated as

$$v_i^\varepsilon(\boldsymbol{x}) \simeq \varepsilon^4 v_i^0(\boldsymbol{x}^0, \boldsymbol{x}^1, \boldsymbol{x}^2), \qquad p^\varepsilon(\boldsymbol{x}) \simeq p^0(\boldsymbol{x}^0). \tag{8.62}$$

In order to solve the multiscale HA seepage problem, we proceed as follows: First we solve the microscale equations (8.49) and (8.53), and obtain the characteristic functions v_i^k and p^k. We next determine the permeabilities K_{ij}^2 and K_{ij} in the meso-domain and macro-domain using (8.54) and (8.60). Then we can solve the macroscale equation (8.59), and obtain the pressure p^0. By substituting these into (8.62), we can determine the velocity and pressure fields.

In geotechnical engineering, we commonly use the empirical form of Darcy's law:

$$\tilde{v}_i^* = -K_{ij}^* \frac{\partial \phi}{\partial x_j}, \qquad \phi = \frac{p}{\rho g} + \zeta \tag{8.63}$$

where $\tilde{v}_i^*$ is the mean velocity, ϕ is the total head, ζ is the elevation head, ρ is the mass density of the fluid, which is assumed to be constant because of the incompressibility of the fluid, and g is the gravitational constant. By comparing (8.63) with (8.61) and (8.62), we obtain the following relationship:

$$\tilde{v}_i^* = \left\langle \left\langle v_i^\varepsilon \right\rangle_2 \right\rangle_1 \simeq \varepsilon^4 \tilde{v}_i^0. \tag{8.64}$$

We can therefore see that the HA-permeability K_{ij} is related to the C-permeability K_{ij}^* as

$$K_{ij}^* = \varepsilon^4 \rho g K_{ij}. \tag{8.65}$$

Once again it should be noted that the velocity v_i^ε is sensitive to the sizes of the mesoscale and microscale grains, whereas p^ε is relatively insensitive to these scales.

8.3.2 *Seepage Analysis of Bentonite*

As described previously, the viscosity of water in the interlayer space of smectite clay stacks is strictly influenced by the processes at the clay surface. If the viscosity is location-dependent, the seepage problem is usually not solvable through an analytical technique. In this case, we solve the microscale equations (8.49) and (8.53) using a finite element method, and determine the HA-permeability K_{ij} and C-permeability K_{ij}^*.

The bentonite analysed here is Kunigel V1®, which is a candidate buffer material for the Japanese high-level nuclear waste management program. This bentonite is made up of 50% smectite minerals by weight, while the remainder is mainly grains of quartz. The intrinsic mass density of a clay mineral is almost the same as that for quartz, which is about 2.7 Mg/m^3, and the porosity of Kunigel V1® at a dry density of 1.8 Mg/m^3 is approximately one-third (i.e., the pore ratio $e = 0.5$).

From the measured data of the quartz grains, we can see that the diameter varies from about 5–100 μm. Based on the peak value of the data and the fact that the permeability characteristics are mainly controlled by smaller grains, we assume

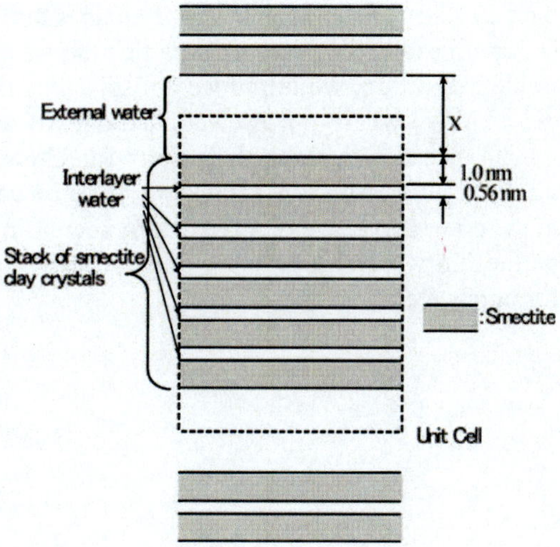

Fig. 8.22 Unit cell of a clay stack in the micro-domain for Kunigel V1®

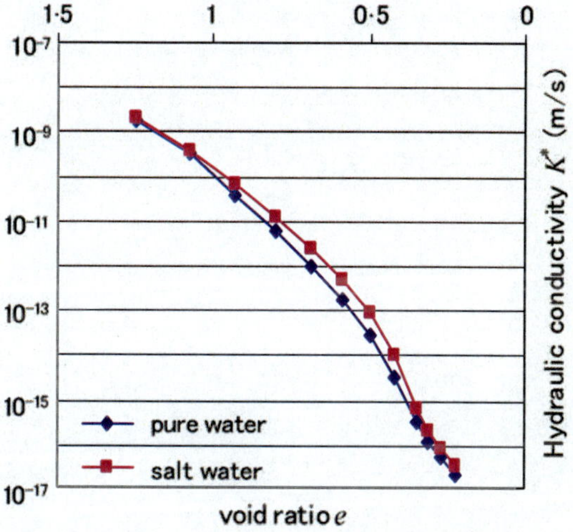

Fig. 8.23 C-permeability (i.e., hydraulic conductivity) of Kunigel V1® calculated by multiscale HA for pure water and salt water

that the quartz grains are approximately 15 μm, and the size of the meso-domain is 45 μm.

One crystal of the smectite clay is a platelet of about $100 \times 100 \times 1$ nm, as mentioned previously, and several crystals form a stack. Here we assume that six crystals form a stack. The interlayer distance is 0.56 nm, which corresponds to the state of hydration by two water layers. The distance between stacks, shown as X in Fig. 8.22, is determined by the given saturated density of the bentonite, and if the distance X becomes smaller than 0.56 nm, we assume that both X and the interlayer distance of the stack are equally reduced. Since the smectite mineral is very thin, one of the micro-domains is pseudo-one-dimensional, which represents the clay stack, and another micro-domain is fully occupied by a quartz crystal, which is impermeable. Since we assume that the stacks of clays are randomly distributed in their direction, we introduce an isotropic HA-meso-permeability K^2 by setting $K^2 = K_{11}^2/3$ where K_{11}^2 is calculated directly using (8.54).

In Fig. 8.23 we show the results for C-permeability using the distribution of viscosity obtained by MD (Fig. 8.6), for pure water and for salt water, as a function of the pore ratio e. We observe that by increasing the density (i.e., decreasing e), the permeability decreases rapidly, and the permeability for salt water is larger than that for pure water.

Chapter 9
Homogenization Analysis of Diffusion in Porous Media

We examine the problem of diffusion in a porous medium using a homogenization analysis (HA). Diffusion problems have important applications in environmental geosciences. We clarify the mechanism of diffusion, convective transport and adsorption in porous media at both the microscale and macroscale levels. Attention is particularly focused on diffusion processes in bentonite, which is an engineered geological barrier to be used to buffer the transport of radionuclides from deep geologic repositories.

9.1 Micro-inhomogeneous Porous Media and Diffusion Problems

As mentioned in Sect. 5.5, in the classical diffusion theory for a porous medium, adsorption is described by a distribution coefficient K_d resulting from the transfer of the species from the fluid phase to the solid phase through the linearized equation of equilibrium adsorption isotherm (5.113).

We conjecture that the actual adsorption mechanism results in a 'source' term rather than a 'storage' that provides a coefficient to the term $\partial c/\partial t$. This perspective, however, is relevant only if the theory is developed through a formulation that couples the microscale phenomena with the macroscale behavior. This is because in the classical theory, it is simple to evaluate an experimental result macroscopically due to adsorption as K_d. Here we will develop an alternative adsorption/desorption/diffusion theory, which is based on MD (molecular dynamics) and HA (homogenization analysis).

Let us consider the problem of the diffusion of a multi-component fluid (see e.g., Drew and Passman 1998) in a non-deformable porous medium which is saturated with the solution. Note that in ensuring adsorption, desorption is also taken into account unless otherwise specified. Referring to (3.246), the mass conservation equation for the multi-component fluid is

Y. Ichikawa and A.P.S. Selvadurai, *Transport Phenomena in Porous Media*,
DOI 10.1007/978-3-642-25333-1_9, © Springer-Verlag Berlin Heidelberg 2012

$$\int_{\Omega_f}\left[\frac{\partial(\rho c_\alpha)}{\partial t}+\frac{\partial}{\partial x_i}(\rho c_\alpha v_i^\alpha)-\gamma_\alpha^*\right]dv-\int_{\Omega_{ad}}\gamma_\alpha^{ad}\,dv=0 \tag{9.1}$$

where c_α is the mass concentration of the α-th component, ρ is the mean mass density defined by (3.242), $\boldsymbol{v}^\alpha$ is the particle velocity of the α-th component, γ_α^* is the source term due to, e.g., chemical reaction, Ω_{ad} is the microscale domain where some chemical species are adsorbed or desorbed, γ_α^{ad} is the source term due to the adsorption in the domain Ω_{ad} (positive in adsorption and negative in desorption).

The mean velocity $\boldsymbol{v}$ of the fluid mixture is defined by (3.251), and the diffusing mass fluxes $\boldsymbol{j}_\alpha$ are introduced by (3.256) as follows:

$$\boldsymbol{v}=\frac{1}{\rho}\sum_{\alpha=1}^{n^*}\rho_\alpha\boldsymbol{v}_\alpha=\sum_{\alpha=1}^{n^*}c_\alpha\boldsymbol{v}_\alpha, \tag{9.2}$$

$$\boldsymbol{j}_\alpha=\rho_\alpha\bar{\boldsymbol{v}}_\alpha=\rho_\alpha(\boldsymbol{v}_\alpha-\boldsymbol{v}). \tag{9.3}$$

By applying Fick's law to the second term on the l.h.s. of (9.1), we have

$$\int_{\Omega_f}\left[\frac{\partial(\rho c_\alpha)}{\partial t}+v_i\frac{\partial(\rho c_\alpha)}{\partial x_i}-\frac{\partial}{\partial x_i}\left(\sum_{\beta=1}^{n^*}\rho D_{ij}^{\alpha\beta}\frac{\partial c_\beta}{\partial x_j}\right)-\gamma_\alpha^*\right]dv-\int_{\Omega_{ad}}\gamma_\alpha^{ad}\,dv=0. \tag{9.4}$$

Thus we obtain the following advective transport equation in the domain Ω_f:

$$\frac{\partial(\rho c_\alpha)}{\partial t}+v_i\frac{\partial(\rho c_\alpha)}{\partial x_i}-\frac{\partial}{\partial x_i}\left(\sum_{\beta=1}^{n^*}\rho D_{ij}^{\alpha\beta}\frac{\partial c_\beta}{\partial x_j}\right)-\gamma_\alpha=0 \quad \text{in } \Omega_f. \tag{9.5}$$

where

$$\gamma_\alpha=\gamma_\alpha^*+\gamma_\alpha^{ad}\chi_{ad}, \tag{9.6}$$

$$\chi_{ad}=\begin{cases}1 & \text{if } \boldsymbol{x}\in\Omega_{ad}\\ 0 & \text{if } \boldsymbol{x}\notin\Omega_{ad}\end{cases}. \tag{9.7}$$

We consider the problem of edge adsorption, which can occur at the edges of clay minerals and is schematically shown in Fig. 9.1. Considering molecular based simulations, we can establish that for smectitic clay minerals, cations are adsorbed at the edges, and they diffuse into the interlayer space forming a hydrated structure, which can be different for each cation species.

At the edge region of the clay Ω_{ad} we can consider a condition for adsorption of the species α in a multi-component fluid with n^*-species in the following form:

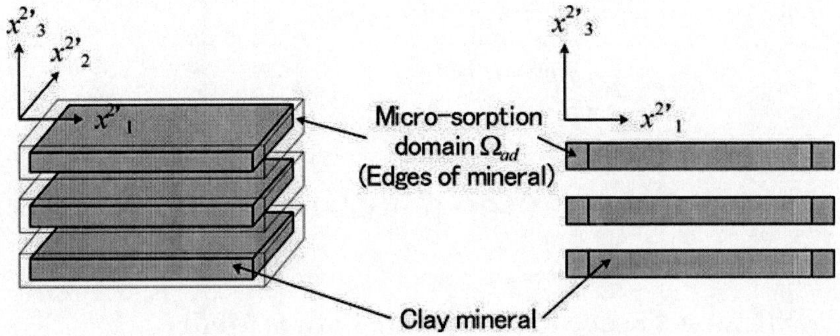

Fig. 9.1 Schematic micro-structure of adsorption/desorption of clay edges

$$\gamma_\alpha^{ad} = \begin{cases} \sum_{\beta=1}^{n^*} s_{\alpha\beta}(c_\beta^l - c_\beta), & \text{if } c_\beta < c_\beta^l \\ 0, & \text{if } c_\beta \geq c_\beta^l \end{cases} \quad \text{in } \Omega_{ad} \tag{9.8}$$

where $s_{\alpha\beta}$ is an adsorption coefficient of species β related to species α, which can be a function of the hydrogen-ion concentration pH, etc., and c_β^l denotes the limit value of adsorption, which is also a function of pH, etc.

On the interlayer surface Γ_i we can observe both adsorption (mainly physisorption) and retardation of diffusion, which is not accompanied by adsorption. If there is adsorption, we can introduce the same condition as imposed on the edge domain Ω_{ad}. If not, we have $\gamma_\alpha^{ad} = 0$.

In the ensuring we assume that the interaction between species α and β is reciprocal, i.e. $D_{ij}^{\alpha\beta} = D_{ij}\delta_{\alpha\beta}$ where $\delta_{\alpha\beta}$ is Kronecker's delta. For simplicity we treat only one species, and set the concentration as $c^\varepsilon = c_\alpha$, the source as $\rho f^\varepsilon = \gamma_\alpha$ where the superscript $^\varepsilon$ implies that this value is rapidly changed in a scale of the micro-domain as mentioned previously. Then, if we assume incompressibility of fluid (ρ = constant, $\partial v_i^\varepsilon/\partial x_i = 0$), we have the following initial-boundary value problem governing the advective-diffusive movement of the species:

Governing equation

$$\frac{\partial c^\varepsilon}{\partial t} + v_j^\varepsilon \frac{\partial c^\varepsilon}{\partial x_j} - \frac{\partial}{\partial x_i}\left(D_{ij}^\varepsilon \frac{\partial c^\varepsilon}{\partial x_j}\right) - f^\varepsilon = 0 \quad \text{in } \Omega \tag{9.9}$$

Boundary conditions

$$\text{Dirichlet boundary:} \quad c^\varepsilon(\boldsymbol{x}, t) = \bar{c}^\varepsilon(t) \quad \text{on } \partial\Omega_c \tag{9.10}$$

$$\text{Neumann boundary:} \quad -D_{ij}^\varepsilon \frac{\partial c^\varepsilon}{\partial x_j} n_i = \bar{q}(t) \quad \text{on } \partial\Omega_q \tag{9.11}$$

Adsorption/desorption condition in the micro-domain

$$\gamma_\alpha^{ad} = \begin{cases} s(c^l - c^\varepsilon), & \text{if } c^\varepsilon < c^l \\ 0, & \text{if } c^\varepsilon \geq c^l \end{cases} \quad \text{in } \Omega_{ad} \tag{9.12}$$

Initial condition

$$c^\varepsilon(\boldsymbol{x}, t{=}t_0) = c_0^\varepsilon(\boldsymbol{x}) \tag{9.13}$$

9.2 Diffusion Theory for Two-scale Porous Media

Referring to Fig. 8.1, we consider a diffusion problem for a two-scale porous medium in which Ω_f is the fluid-phase domain, Ω_s is the solid-phase domain, Γ is the solid/fluid interface, and Γ_f is the periodic boundary of the fluid-phase in the micro-domain.

Following the HA theory we introduce the coordinate systems $\boldsymbol{x}^0$ and $\boldsymbol{x}^1$ for the macro-domain and the micro-domain, respectively, and we have

$$\boldsymbol{x} = \boldsymbol{x}(\boldsymbol{x}^0, \boldsymbol{x}^1), \qquad \boldsymbol{x}^1 = \frac{\boldsymbol{x}^0}{\varepsilon} \tag{9.14}$$

$$\frac{\partial}{\partial x_i} = \frac{\partial}{\partial x_i^0} + \frac{1}{\varepsilon}\frac{\partial}{\partial x_i^1} \tag{9.15}$$

where $\varepsilon(\ll 1)$ is a scaling parameter.

9.2.1 *HA for Diffusion Problems in Porous Media*

We introduce a perturbation scheme for the concentration $c^\varepsilon(\boldsymbol{x}, t)$ as

$$c^\varepsilon(\boldsymbol{x}, t) = c^0(\boldsymbol{x}^0, \boldsymbol{x}^1, t) + \varepsilon\, c^1(\boldsymbol{x}^0, \boldsymbol{x}^1, t) + \varepsilon^2 c^2(\boldsymbol{x}^0, \boldsymbol{x}^1, t) + \cdots \tag{9.16}$$

where c^i is an X^1-periodic function in the sense that

$$c^i(\boldsymbol{x}^0, \boldsymbol{x}^1 + \boldsymbol{X}^1, t) = c^i(\boldsymbol{x}^0, \boldsymbol{x}^1, t) \tag{9.17}$$

where X_i^1 is the size of a unit cell in the x_i^1-direction.

In addition to the conventional source term $f^0(\boldsymbol{x}^0)$, we introduce the following *micro-sorption term* $\Gamma_i(\boldsymbol{x}^1)$ in order to represent microscale adsorption/desorption phenomena in the micro-domain Ω_{ad}:

$$f^\varepsilon = f^0(\boldsymbol{x}^0) + \varepsilon^{-1}\gamma^{ad}(\boldsymbol{x}^1, t), \qquad \gamma^{ad}(\boldsymbol{x}^1, t) = \Gamma_i(\boldsymbol{x}^1)\frac{\partial c^0}{\partial x_i^0}. \tag{9.18}$$

From (9.15) and (9.16) we have the spatial derivative of the concentration c^ε as

$$\begin{aligned}\frac{\partial c^\varepsilon}{\partial x_j^0} &= \left(\frac{\partial}{\partial x_j^0} + \frac{1}{\varepsilon}\frac{\partial}{\partial x_j^1}\right)\left(c^0 + \varepsilon c^1 + \varepsilon^2 c^2 + \cdots\right) \\ &= \frac{1}{\varepsilon}\frac{\partial c^0}{\partial x_j^1} + \left(\frac{\partial c^0}{\partial x_j^0} + \frac{\partial c^1}{\partial x_j^1}\right) + \varepsilon\left(\frac{\partial c^1}{\partial x_j^0} + \frac{\partial c^2}{\partial x_j^1}\right) + \cdots\end{aligned}$$

The governing equation (9.9) can be represented in the expanded form

$$\begin{aligned}&\left(\frac{\partial c^0}{\partial t} + \varepsilon\frac{\partial c^1}{\partial t} + \varepsilon^2\frac{\partial c^2}{\partial t} + \cdots\right) \\ &\quad + \left[\frac{1}{\varepsilon}v_j^\varepsilon\frac{\partial c^0}{\partial x_j^1} + v_j^\varepsilon\left(\frac{\partial c^0}{\partial x_j^0} + \frac{\partial c^1}{\partial x_j^1}\right) + \varepsilon v_j^\varepsilon\left(\frac{\partial c^1}{\partial x_j^0} + \frac{\partial c^2}{\partial x_j^1}\right) + \cdots\right] \\ &\quad - \left[\frac{1}{\varepsilon^2}\left\{\frac{\partial}{\partial x_i^1}\left(D_{ij}^\varepsilon\frac{\partial c^0}{\partial x_j^1}\right)\right\} + \frac{1}{\varepsilon}\left\{\frac{\partial}{\partial x_i^0}\left(D_{ij}^\varepsilon\frac{\partial c^0}{\partial x_j^1}\right) + \frac{\partial}{\partial x_i^1}\left(D_{ij}^\varepsilon\left(\frac{\partial c^0}{\partial x_j^0} + \frac{\partial c^1}{\partial x_j^1}\right)\right)\right\}\right. \\ &\quad \left. + \left\{\frac{\partial}{\partial x_i^0}\left(D_{ij}^\varepsilon\left(\frac{\partial c^0}{\partial x_j^0} + \frac{\partial c^1}{\partial x_j^1}\right)\right) + \frac{\partial}{\partial x_i^1}\left(D_{ij}^\varepsilon\left(\frac{\partial c^1}{\partial x_j^0} + \frac{\partial c^2}{\partial x_j^1}\right)\right)\right\} + \varepsilon\{\cdots\} + \cdots\right] \\ &\quad - f^0(\boldsymbol{x}^0) - \frac{1}{\varepsilon}\Gamma_i(\boldsymbol{x}^1)\frac{\partial c^0}{\partial x_i^0} = 0.\end{aligned} \tag{9.19}$$

Considering terms in the series, we have the following relations:

For terms $O(\varepsilon^{-2})$:

$$\frac{\partial}{\partial x_i^1}\left(D_{ij}^\varepsilon\frac{\partial c^0}{\partial x_j^1}\right) = 0, \tag{9.20}$$

which implies that

$$c^0(\boldsymbol{x}^0, \boldsymbol{x}^1, t) = c^0(\boldsymbol{x}^0, t). \tag{9.21}$$

That is, the first term of the perturbation c^0 is not a function of the micro-domain coordinates $\boldsymbol{x}^1$, but the macro-domain coordinates $\boldsymbol{x}^0$.

For terms $O(\varepsilon^{-1})$: Microscale equation

$$v_j^\varepsilon\frac{\partial c^0}{\partial x_j^1} - \frac{\partial}{\partial x_i^0}\left(D_{ij}^\varepsilon\frac{\partial c^0}{\partial x_j^1}\right) - \frac{\partial}{\partial x_i^1}\left\{D_{ij}^\varepsilon\left(\frac{\partial c^0}{\partial x_j^0} + \frac{\partial c^1}{\partial x_j^1}\right)\right\} - \Gamma_i(\boldsymbol{x}^1)\frac{\partial c^0}{\partial x_i^0} = 0. \tag{9.22}$$

By substituting (9.21) into (9.22), we have

$$\frac{\partial}{\partial x_i^1}\left\{D_{ij}^\varepsilon\left(\frac{\partial c^0}{\partial x_j^0}+\frac{\partial c^1}{\partial x_j^1}\right)\right\}+\Gamma_i(\boldsymbol{x}^1)\frac{\partial c^0}{\partial x_i^0}=0. \tag{9.23}$$

Since $c^0(\boldsymbol{x}^0,t)$ is constant in the micro-domain, we can normalize (9.23) by $\partial c^0/\partial x_i^0$ to obtain

$$c^1(\boldsymbol{x}^0,\boldsymbol{x}^1,t)=-N_1^k(\boldsymbol{x}^1)\frac{\partial c^0(\boldsymbol{x}^0,t)}{\partial x_k^0}+C_1(\boldsymbol{x}^0,t) \tag{9.24}$$

where N_1^k is the *characteristic function*. By substituting (9.24) into (9.23), we have

$$\left[\frac{\partial}{\partial x_i^1}\left\{D_{ij}^\varepsilon\left(\delta_{kj}-\frac{\partial N_1^k}{\partial x_j^1}\right)\right\}+\Gamma_k\right]\frac{\partial c^0}{\partial x_k^0}=0$$

$$\Rightarrow \quad \frac{\partial}{\partial x_i^1}\left[D_{ij}^\varepsilon\left(\delta_{kj}-\frac{\partial N_1^k}{\partial x_j^1}\right)\right]+\Gamma_k=0. \tag{9.25}$$

The result (9.25) is referred to as the *microscale equation* for the two-scale diffusion problem in porous media. The boundary condition (9.25) is the X^1-periodic condition (9.17).

By taking into account the condition (9.17), we introduce a weak form of (9.25) as follows:

$$\int_{\Omega_1} D_{ij}^\varepsilon \frac{\partial N_1^k}{\partial x_j^1}\frac{\partial V_k}{\partial x_i^1}d\boldsymbol{x}^1-\int_{\Omega_1}\left(D_{ik}^\varepsilon\frac{\partial V_k}{\partial x_i^1}+\Gamma_k V_k\right)d\boldsymbol{x}^1=0 \tag{9.26}$$

where $V_k(\boldsymbol{x}^1)$ are arbitrary functions with the X^1-periodic conditions. Thus the microscale problem is reduced to the weak form problem (9.26), and we can obtain the solution N_1^k under the periodic condition (9.17). Note that if we consider the adsorption/desorption condition (9.12) in the micro-domain, the microscale equation (9.25) is not completely independent from the macroscale problem, therefore we need to introduce an iteration procedure, which is treated later.

For terms $O(\varepsilon^0)$: Macroscale equation

$$\begin{aligned}\frac{\partial c^0}{\partial t}&+v_j^\varepsilon\left(\frac{\partial c^0}{\partial x_j^0}+\frac{\partial c^1}{\partial x_j^1}\right)-\frac{\partial}{\partial x_i^0}\left\{D_{ij}^\varepsilon\left(\frac{\partial c^0}{\partial x_j^0}+\frac{\partial c^1}{\partial x_j^1}\right)\right\}\\&-\frac{\partial}{\partial x_i^1}\left\{D_{ij}^\varepsilon\left(\frac{\partial c^1}{\partial x_j^0}+\frac{\partial c^2}{\partial x_j^1}\right)\right\}-f^0=0.\end{aligned} \tag{9.27}$$

When introducing a volume integral average in (9.27), the fourth term of the l.h.s. can be reduced to a surface integral: i.e.,

$$\frac{1}{|\Omega_1|}\int_{\Omega_1}\frac{\partial}{\partial x_i^1}\left\{D_{ij}^{\varepsilon}\left(\frac{\partial c^1}{\partial x_j^0}+\frac{\partial c^2}{\partial x_j^1}\right)\right\}d\boldsymbol{x}^1$$
$$=\frac{1}{|\Omega_1|}\int_{\partial\Omega_1}\left\{D_{ij}^{\varepsilon}\left(\frac{\partial c^1}{\partial x_j^0}+\frac{\partial c^2}{\partial x_j^1}\right)\right\}\,n_i\,ds^1=0 \tag{9.28}$$

where $|\Omega_1|$ is the volume of the unit cell; we used the conditions that c^1 and c^2 are X^1-periodic functions and that n_i gives inverse values at both sides of a unit cell. Thus from the volume integral average of (9.27), we have

$$\frac{1}{|\Omega_1|}\int_{\Omega_1}\frac{\partial c^0}{\partial t}d\boldsymbol{x}^1+\frac{1}{|\Omega_1|}\int_{\Omega_1}v_j^{\varepsilon}\left(\frac{\partial c^0}{\partial x_j^0}+\frac{\partial c^1}{\partial x_j^1}\right)d\boldsymbol{x}^1$$
$$-\frac{1}{|\Omega_1|}\int_{\Omega_1}\frac{\partial}{\partial x_i^0}\left\{D_{ij}^{\varepsilon}\left(\frac{\partial c^0}{\partial x_j^0}+\frac{\partial c^1}{\partial x_j^1}\right)\right\}d\boldsymbol{x}^1-\frac{1}{|\Omega_1|}\int_{\Omega_1}f^0\,d\boldsymbol{x}^1=0. \tag{9.29}$$

By substituting the characteristic function given by (9.24) into (9.29), we obtain

$$\frac{\partial c^0}{\partial t}+v_j^H\frac{\partial c^0}{\partial x_j^0}-\frac{\partial}{\partial x_i^0}\left(D_{ij}^H\frac{\partial c^0}{\partial x_j^0}\right)-f^H=0 \tag{9.30}$$

$$D_{ij}^H\equiv\frac{1}{|\Omega_1|}\int_{\Omega_1}D_{ik}^{\varepsilon}\left(\delta_{kj}-\frac{\partial N_1^j}{\partial x_k^1}\right)d\boldsymbol{x}^1, \tag{9.31}$$

$$v_j^H\equiv\frac{1}{|\Omega_1|}\int_{\Omega_1}v_k^{\varepsilon}\,d\boldsymbol{x}^1,\qquad f^H\equiv\frac{1}{|\Omega_1|}\int_{\Omega_1}f^0 d\boldsymbol{x}^1. \tag{9.32}$$

The result (9.30) is the *macroscale transport equation* of the diffusion problem in porous media, and D_{ij}^H is the homogenized diffusion coefficient.

Under the boundary conditions (9.10) and (9.11) we can develop a weak form of (9.30) as

$$\int_{\Omega_0}\frac{\partial c^0}{\partial t}w\,d\boldsymbol{x}^0+\int_{\Omega_0}v_j^H\frac{\partial c^0}{\partial x_j^0}w\,d\boldsymbol{x}^0+\int_{\Omega_0}D_{ij}^H\frac{\partial c^0}{\partial x_j^0}\frac{\partial w}{\partial x_i^0}d\boldsymbol{x}^0-\int_{\Omega_0}f^H w\,d\boldsymbol{x}^0$$
$$=\int_{\partial\Omega_q}\bar{q}\,w\,ds^0\qquad\forall w(\boldsymbol{x}^0,t)\quad(w=0\ \text{ on }\partial\Omega_c). \tag{9.33}$$

Thus the macroscale equation in the macro-domain is reduced to the problem of obtaining c^0 by solving (9.33) under the homogenized D_{ij}^H, v_j^H and f^H, which are given by (9.31) and (9.32). The local distribution of concentration is calculated as

$$c^{\varepsilon}(\boldsymbol{x},t)\simeq c^0(\boldsymbol{x}^0,t)+\varepsilon c^1(\boldsymbol{x}^0,\boldsymbol{x}^1,t)=c^0(\boldsymbol{x}^0,t)-\varepsilon N_1^k(\boldsymbol{x}^1)\frac{\partial c^0(\boldsymbol{x}^0,t)}{\partial x_k^0}. \tag{9.34}$$

9.2.2 Simulation of a Through-diffusion Test

In order to verify the validity of the HA theory, we compare the numerical work with the result of a through-diffusion test for a pure smectitic clay. The HA macro-model and micro-model, i.e., the finite elements in the macro-domain and in the micro-domain, are shown in Fig. 9.2.

We consider a diffusion/adsorption test of cesium (Cs), and the data provided are as follows: The size of a clay platelet is $100 \times 100 \times 1$. Based on experimental data of Baeyens and Bradbury (1997) the cross-section area of clay edges, which are adsorption sites, are assumed to be 35 m^2/g. Since the atomic radius of cesium is 3.34 Å, and because of the monolayer adsorption, we set the layer thickness of the edge domain where cesium ions are adsorbed as 0.67 nm. The maximum amount of cesium adsorbed in this domain is 6.51×10^{-5} mol/g. The molecular number of interlayer water is given as $n{=}2.5$. The diffusion coefficient of cesium ions in the bulk water is $2{\times}10^{-5}$ cm^2/s, and in the interlayer water is $2.62{\times}10^{-6}$ cm^2/sec, which is obtained from the MD simulation. The concentration of cesium at the upstream boundary, i.e., the l.h.s. surface of Fig. 9.2 (Dirichlet boundary condition) is given as: Case (1) 10^{-2} mol/l, Case (2) 10^{-3} mol/l and Case (3) 10^{-4} mol/l.

The calculated break-through curves for Case (1) at each point of the macro-domain are shown in Fig. 9.3, where the curves corresponding to a situation without adsorption are also presented. Similarly, the curves for Cases (2) and (3) are shown

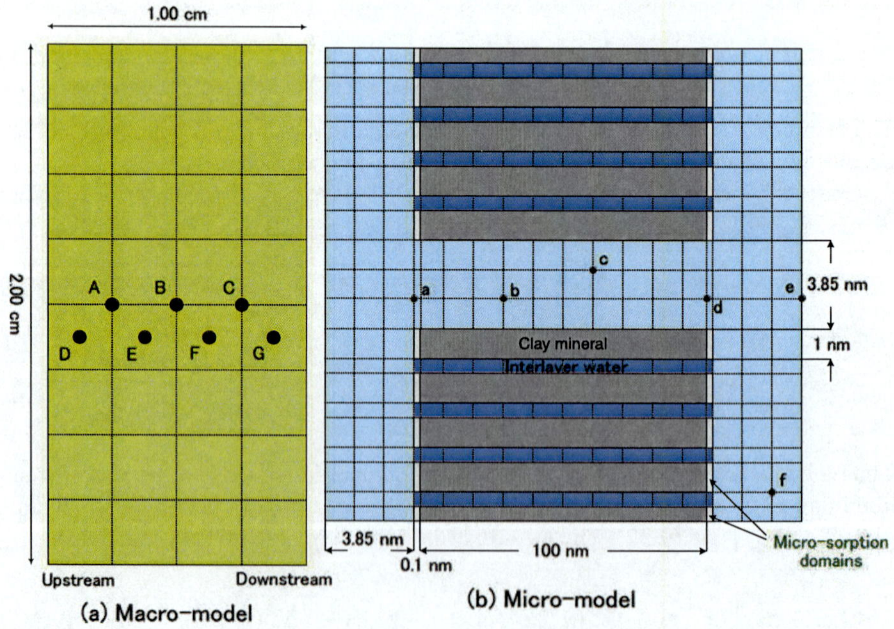

Fig. 9.2 Macro-model (**a**) and micro-model (**b**) of a through-diffusion test for a pure smectitic clay

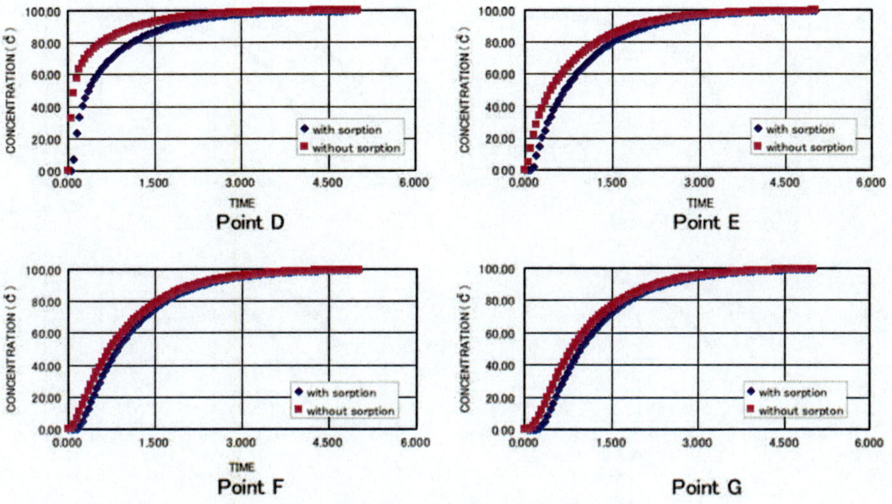

Fig. 9.3 Break-through curves at each macro-domain point: Case (1)

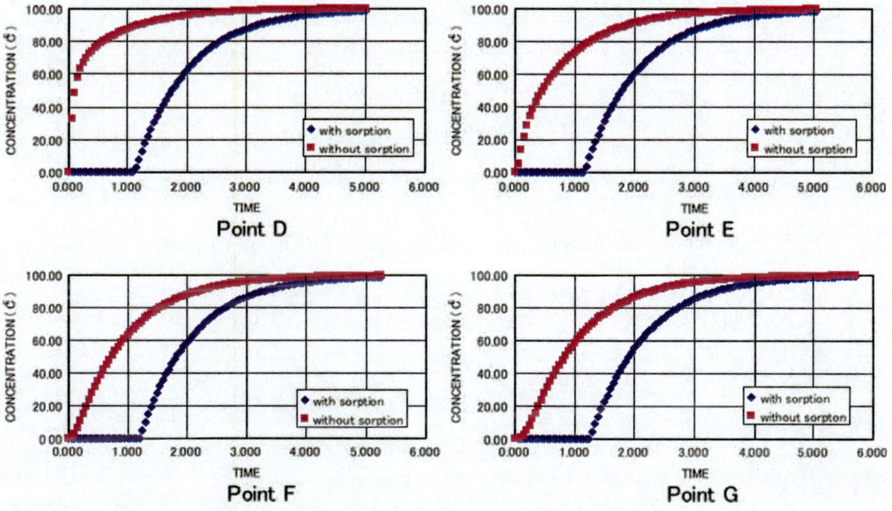

Fig. 9.4 Break-through curves at each macro-domain point: Case (2)

in Figs. 9.4 and 9.5, respectively. We observe that in Case (1), where a higher concentration is prescribed at the upstream side, the break-through curve rises at the early stage, since the adsorbed edge domains are saturated relatively quickly, while in Case (3), which has a lower concentration at the boundary, we observe a delayed break-through curve. These are expected results.

The distribution of concentration in the micro-domain for Case (1) after a sufficiently long elapsed time, is shown in Fig. 9.6. Similarly, the results for Cases (2) and (3) are given in Figs. 9.7 and 9.8, respectively.

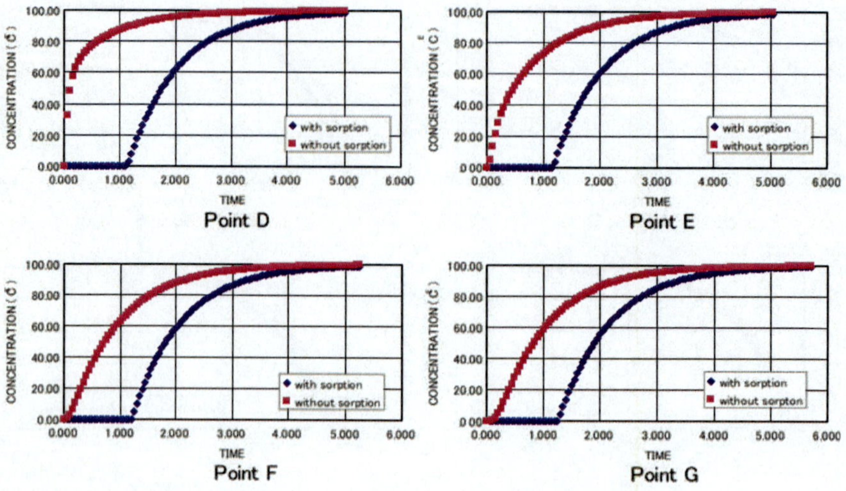

Fig. 9.5 Break-through curves at each macro-domain point: Case (3)

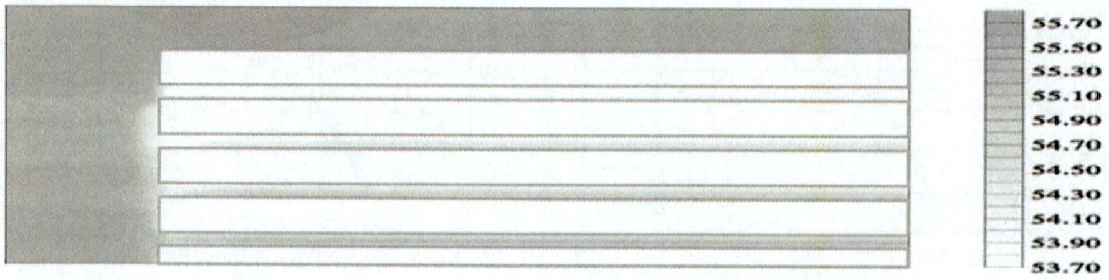

Fig. 9.6 Distribution of concentration in the micro-model (1/4 domain): Case (1)

Fig. 9.7 Distribution of concentration in the micro-model (1/4 domain): Case (2)

9.2.3 HA Diffusion Equation with Higher Order Derivatives

In certain situations it is useful to have an approximation to the problem with convection and adsorption in the micro-domain by including a term c^2. If we take the unknown c^2 of (9.27), and substitute the characteristic function of (9.24), we obtain the following higher (i.e., second) order microscale equation:

Fig. 9.8 Distribution of concentration in the micro-model (1/4 domain): Case (3)

$$\frac{\partial c^0}{\partial t} + v_k^\varepsilon\left(\delta_{jk} - \frac{\partial N_1^j}{\partial x_k^1}\right)\frac{\partial c^0}{\partial x_j^0} - \frac{\partial}{\partial x_i^0}\left\{D_{ij}^\varepsilon\left(\delta_{jk} - \frac{\partial N_1^k}{\partial x_j^1}\right)\frac{\partial c^0}{\partial x_k^0}\right\}$$

$$+\frac{\partial}{\partial x_i^1}\left\{D_{ij}^\varepsilon\frac{\partial}{\partial x_j^0}\left(N_1^k\frac{\partial c^0}{\partial x_k^0}\right)\right\} - \frac{\partial}{\partial x_i^1}\left(D_{ij}^\varepsilon\frac{\partial c^2}{\partial x_j^1}\right) - f^\varepsilon = 0. \tag{9.35}$$

We then set c^2 as

$$c^2(\boldsymbol{x}^0, \boldsymbol{x}^1, t) = N_{21}^{kl}(\boldsymbol{x}^1)\frac{\partial^2 c^0(\boldsymbol{x}^0, t)}{\partial x_k^0 \partial x_l^0} + N_{22}^k(\boldsymbol{x}^1)\frac{\partial c^0(\boldsymbol{x}^0, t)}{\partial x_k^0}$$

$$+N_{23}(\boldsymbol{x}^1)\frac{\partial c^0(\boldsymbol{x}^0, t)}{\partial t} + N_{24}(\boldsymbol{x}^1) + \bar{c}^2(\boldsymbol{x}^0, t) \tag{9.36}$$

where $N_{21}^{kl}(\boldsymbol{x}^1)$, $N_{22}^k(\boldsymbol{x}^1)$, $N_{23}(\boldsymbol{x}^1)$ and $N_{24}(\boldsymbol{x}^1)$ are the characteristic functions with respect to a small disturbance of the position in the micro-domain, with respect to convection, time-dependence and adsorption, respectively, and $\bar{c}^2(\boldsymbol{x}^0, t)$ is an integration constant.

Substituting (9.36) into (9.35), we derive a weak form as

$$\frac{\partial^2 c^0}{\partial x_k^0 \partial x_l^0}\left\{\int_{\Omega_1} D_{ij}^\varepsilon \frac{\partial N_{21}^{kl}}{\partial x_j^1}\frac{\partial V}{\partial x_i^1} d\boldsymbol{x}^1 - \int_{\Omega_1} D_{il}^\varepsilon N_1^k \frac{\partial V}{\partial x_i^1} d\boldsymbol{x}^1 - \int_{\Omega_1} D_{lj}^\varepsilon\left(\delta_{jk} - \frac{\partial N_1^k}{\partial x_j^1}\right) V\, d\boldsymbol{x}^1\right\}$$

$$+\frac{\partial c^0}{\partial x_k^0}\left\{\int_{\Omega_1} D_{ij}^\varepsilon \frac{\partial N_{22}^k}{\partial x_j^1}\frac{\partial V}{\partial x_i^1} d\boldsymbol{x}^1 + \int_{\Omega_1} v_j^\varepsilon\left(\delta_{jk} - \frac{\partial N_1^k}{\partial x_j^1}\right) V\, d\boldsymbol{x}^1\right\}$$

$$+\frac{\partial c^0}{\partial t}\left\{\int_{\Omega_1} D_{ij}^\varepsilon \frac{\partial N_{23}}{\partial x_j^1}\frac{\partial V}{\partial x_i^1} d\boldsymbol{x}^1 + \int_{\Omega_1} V\, d\boldsymbol{x}^1\right\}$$

$$+\left\{\int_{\Omega_1} D_{ij}^\varepsilon \frac{\partial N_{24}}{x_j^1}\frac{\partial V}{\partial x_i^1} d\boldsymbol{x}^1 - \int_{\Omega_1} f^\varepsilon V\, d\boldsymbol{x}^1\right\} = 0. \tag{9.37}$$

Since each term of (9.37) must be null because of the arbitrariness of c^0, we obtain the following equations for determining $N_{21}^{kl}(\boldsymbol{x}^1)$, $N_{22}^k(\boldsymbol{x}^1)$, $N_{23}(\boldsymbol{x}^1)$ and $N_{24}(\boldsymbol{x}^1)$:

$$\int_{\Omega_1} D_{ij}^{\varepsilon} \frac{\partial N_{21}^{kl}}{\partial x_j^1} \frac{\partial V}{\partial x_i^1} d\boldsymbol{x}^1 = \int_{\Omega_1} D_{lj}^{\varepsilon} \left(\delta_{jk} - \frac{\partial N_1^k}{\partial x_j^1} \right) V \, d\boldsymbol{x}^1 + \int_{\Omega_1} D_{il}^{\varepsilon} N_1^k \frac{\partial V}{\partial x_i^1} d\boldsymbol{x}^1 \tag{9.38}$$

$$\int_{\Omega_1} D_{ij}^{\varepsilon} \frac{\partial N_{22}^{k}}{\partial x_j^1} \frac{\partial V}{\partial x_i^1} d\boldsymbol{x}^1 = -\int_{\Omega_1} v_j^{\varepsilon} \left(\delta_{jk} - \frac{\partial N_1^k}{\partial x_j^1} \right) V \, d\boldsymbol{x}^1 \tag{9.39}$$

$$\int_{\Omega_1} D_{ij}^{\varepsilon} \frac{\partial N_{23}}{\partial x_j^1} \frac{\partial V}{\partial x_i^1} d\boldsymbol{x}^1 = -\int_{\Omega_1} V \, d\boldsymbol{x}^1 \tag{9.40}$$

$$\int_{\Omega_1} D_{ij}^{\varepsilon} \frac{\partial N_{24}}{\partial x_j^1} \frac{\partial V}{\partial x_i^1} d\boldsymbol{x}^1 = \int_{\Omega_1} f^{\varepsilon} V \, d\boldsymbol{x}^1 \tag{9.41}$$

where $V(\boldsymbol{x}^1)$ is an arbitrary X^1-periodic function. From Onsager's relationships (Kondepudi and Prigogine 1998) it can be shown that the diffusion matrix is symmetric, and, in addition, the assumption of isotropy gives $D_{ij}^{\varepsilon} = D\delta_{ij}$. It should be noted that if we substitute N_1^k into (9.38), which is determined by (9.26), we obtain N_{21}^{kl}. Similarly, we can obtain N_{22}^{kl} using (9.39). We also understand that (9.41) involves the local adsorption term f^{ε}. Thus we can see that for the second order microscale problem, (9.38)–(9.41) can be solved to obtain N_{21}^{kl}, N_{22}^{k}, N_{23} and N_{24}.

The diffusion problem with two scale domains including second order differentials can be solved as follows:

1. Obtain the characteristic function N_l^k by solving (9.26).
2. Specify the homogenized diffusion coefficient D_{ij}^H by substituting N_l^k into (9.31).
3. Obtain c^0 by solving the macroscale equation (9.33).
4. Obtain N_{21}^{kl}, N_{22}^{k}, N_{23} and N_{24} by solving the second order microscale equations (9.38)–(9.41).
5. The local distribution of concentration can be specified as follows:

$$\begin{aligned} c^{\varepsilon}(\boldsymbol{x}^0, t) &\simeq c^0(\boldsymbol{x}^0, \boldsymbol{x}^1, t) + \varepsilon c^1(\boldsymbol{x}^0, \boldsymbol{x}^1, t) + \varepsilon^2 c^2(\boldsymbol{x}^0, \boldsymbol{x}^1, t) \\ &= c^0(\boldsymbol{x}^0, t) - \varepsilon N_1^k(\boldsymbol{x}^1) \frac{\partial c^0(\boldsymbol{x}^0, t)}{\partial x_k^0} \\ &+ \varepsilon^2 \left\{ N_{21}^{kl}(\boldsymbol{x}^1) \frac{\partial^2 c^0(\boldsymbol{x}^0, t)}{\partial x_k^0 \partial x_l^0} + N_{22}^{k}(\boldsymbol{x}^1) \frac{\partial c^0(\boldsymbol{x}^0, t)}{\partial x_k^0} + N_{23}(\boldsymbol{x}^1) \frac{\partial c^0(\boldsymbol{x}^0, t)}{\partial t} \right. \\ &\left. + N_{24}(\boldsymbol{x}^1, t) \right\} \end{aligned} \tag{9.42}$$

9.3 Diffusion Problem for Multiscale Porous Media

Bentonite is a composite material with micron sized grains, composed mainly of quartz, and nanometer sized clay minerals as described previously. We present here a multiscale HA theory for diffusion in such a composite porous medium.

We consider a material that has macro-/meso-/micro-structures as shown in Fig. 8.21, where Ω_f and Ω_s are the domains of the fluid phase and solid phase, respectively, Γ_{sf} is the solid/fluid interface, Γ_f is the periodic boundary of the fluid phase, and Ω_{sf} is the mixture domain of smectite stacks and fluid in the meso-domain,

Let $\boldsymbol{x}^0$, $\boldsymbol{x}^1$ and $\boldsymbol{x}^2$ be coordinate systems introduced in the macro-/meso-/micro-domains, respectively. We consider a triaxial test on the typical bentonite of Japan, Kunigel V1®, using a specimen of about 10^{-1} m in diameter, which is the macro-domain, where the size of the quartz grains are about $\varepsilon = 10^{-4}$, which is the meso-domain, and the size of clay stacks is about 10^{-8} m. Then we can set $\varepsilon=10^{-4}$, and introduce the following relations:

$$\boldsymbol{x} = \boldsymbol{x}(\boldsymbol{x}^0, \boldsymbol{x}^1, \boldsymbol{x}^2); \qquad \boldsymbol{x}^1 = \frac{\boldsymbol{x}^0}{\varepsilon}, \qquad \boldsymbol{x}^2 = \frac{\boldsymbol{x}^1}{\varepsilon}, \tag{9.43}$$

$$\frac{\partial}{\partial x_i} = \frac{\partial}{\partial x_i^0} + \frac{1}{\varepsilon}\frac{\partial}{\partial x_i^1} + \frac{1}{\varepsilon^2}\frac{\partial}{\partial x_i^2} \tag{9.44}$$

9.3.1 *Multiscale HA for Diffusion Problems in Porous Media*

Let us introduce a perturbation of $c^\varepsilon(\boldsymbol{x}, t)$ for the macro-/meso/micro-coordinate systems $\boldsymbol{x}^0$, $\boldsymbol{x}^1$, $\boldsymbol{x}^2$ as

$$c^\varepsilon(\boldsymbol{x}, t) = c^0(\boldsymbol{x}^0, t) + \varepsilon c^1(\boldsymbol{x}^0, \boldsymbol{x}^1, \boldsymbol{x}^2, t) + \varepsilon^2 c^2(\boldsymbol{x}^0, \boldsymbol{x}^1, \boldsymbol{x}^2, t) + \cdots \tag{9.45}$$

where c^α s ($\alpha = 0, 1, 2, \ldots$) are X^1-/X^2-periodic functions:

$$c^\alpha(\boldsymbol{x}^0, \boldsymbol{x}^1, \boldsymbol{x}^2, t) = c^\alpha(\boldsymbol{x}^0, \boldsymbol{x}^1 + \boldsymbol{X}^1, \boldsymbol{x}^2, t),$$
$$c^\alpha(\boldsymbol{x}^0, \boldsymbol{x}^1, \boldsymbol{x}^2, t) = c^\alpha(\boldsymbol{x}^0, \boldsymbol{x}^1, \boldsymbol{x}^2 + \boldsymbol{X}^2, t).$$

From (9.44) the differential of the concentration c^ε is given by

$$\frac{\partial c^\varepsilon}{\partial x_j} = \varepsilon^{-2}\frac{\partial c^0}{\partial x_j^2} + \varepsilon^{-1}\left(\frac{\partial c^0}{\partial x_j^1} + \frac{\partial c^1}{\partial x_j^2}\right) + \left(\frac{\partial c^0}{\partial x_j^0} + \frac{\partial c^1}{\partial x_j^1} + \frac{\partial c^2}{\partial x_j^2}\right) + \varepsilon\left(\cdots\right) + \cdots. \tag{9.46}$$

Thus the governing equation (9.9) is transformed into

$$
\begin{aligned}
&\left(\frac{\partial c^0}{\partial t}+\varepsilon\frac{\partial c^1}{\partial t}+\varepsilon^2\frac{\partial c^2}{\partial t}+\cdots\right)\\
&+\left[\varepsilon^{-2}v_j^\varepsilon\frac{\partial c^0}{\partial x_j^2}+\varepsilon^{-1}v_j^\varepsilon\left(\frac{\partial c^0}{\partial x_j^1}+\frac{\partial c^1}{\partial x_j^2}\right)+v_j^\varepsilon\left(\frac{\partial c^0}{\partial x_j^0}+\frac{\partial c^1}{\partial x_j^1}+\frac{\partial c^2}{\partial x_j^2}\right)+\cdots\right]\\
&-\left[\varepsilon^{-4}\left\{\frac{\partial}{\partial x_i^2}\left(D_{ij}^\varepsilon\frac{\partial c^0}{\partial x_j^2}\right)\right\}+\varepsilon^{-3}\left\{\frac{\partial}{\partial x_i^1}\left(D_{ij}^\varepsilon\frac{\partial c^0}{\partial x_j^2}\right)+\frac{\partial}{\partial x_i^2}\left(D_{ij}^\varepsilon\left(\frac{\partial c^0}{\partial x_j^1}+\frac{\partial c^1}{\partial x_j^2}\right)\right)\right\}\right.\\
&+\varepsilon^{-2}\left\{\frac{\partial}{\partial x_i^0}\left(D_{ij}^\varepsilon\frac{\partial c^0}{\partial x_j^2}\right)+\frac{\partial}{\partial x_i^1}\left(D_{ij}^\varepsilon\left(\frac{\partial c^0}{\partial x_j^1}+\frac{\partial c^1}{\partial x_j^2}\right)\right)\right.\\
&\left.+\frac{\partial}{\partial x_i^2}\left(D_{ij}^\varepsilon\left(\frac{\partial c^0}{\partial x_j^0}+\frac{\partial c^1}{\partial x_j^1}+\frac{\partial c^2}{\partial x_j^2}\right)\right)\right\}\\
&+\varepsilon^{-1}\left\{\frac{\partial}{\partial x_i^0}\left(D_{ij}^\varepsilon\left(\frac{\partial c^0}{\partial x_j^1}+\frac{\partial c^1}{\partial x_j^2}\right)\right)+\frac{\partial}{\partial x_i^1}\left(D_{ij}^\varepsilon\left(\frac{\partial c^0}{\partial x_j^0}+\frac{\partial c^1}{\partial x_j^1}+\frac{\partial c^2}{\partial x_j^2}\right)\right)\right.\\
&\left.+\frac{\partial}{\partial x_i^2}\left(D_{ij}^\varepsilon\left(\frac{\partial c^1}{\partial x_j^0}+\frac{\partial c^2}{\partial x_j^1}+\frac{\partial c^3}{\partial x_j^2}\right)\right)\right\}\\
&+\left\{\frac{\partial}{\partial x_i^0}\left(D_{ij}^\varepsilon\left(\frac{\partial c^0}{\partial x_j^0}+\frac{\partial c^1}{\partial x_j^1}+\frac{\partial c^2}{\partial x_j^2}\right)\right)+\frac{\partial}{\partial x_i^1}\left(D_{ij}^\varepsilon\left(\frac{\partial c^1}{\partial x_j^0}+\frac{\partial c^2}{\partial x_j^1}+\frac{\partial c^3}{\partial x_j^2}\right)\right)\right.\\
&\left.+\frac{\partial}{\partial x_i^2}\left(D_{ij}^\varepsilon\left(\frac{\partial c^2}{\partial x_j^0}+\frac{\partial c^3}{\partial x_j^1}+\frac{\partial c^4}{\partial x_j^2}\right)\right)\right\}\\
&\left.+\varepsilon\left\{\cdots\right\}+\cdots\right]-f^\varepsilon=0
\end{aligned}
$$

Therefore, for each power series of ε we obtain the following equations:

For terms $O(\varepsilon^{-4})$:

$$
\frac{\partial}{\partial x_i^2}\left(D_{ij}^\varepsilon\frac{\partial c^0}{\partial x_j^2}\right)=0. \tag{9.47}
$$

c^0 is a function of only $\boldsymbol{x}^0$, thus (9.47) is automatically satisfied.

For terms $O(\varepsilon^{-3})$: First order characteristic function

$$
\frac{\partial}{\partial x_i^1}\left(D_{ij}^\varepsilon\frac{\partial c^0}{\partial x_j^2}\right)+\frac{\partial}{\partial x_i^2}\left\{D_{ij}^\varepsilon\left(\frac{\partial c^0}{\partial x_j^1}+\frac{\partial c^1}{\partial x_j^2}\right)\right\}=0.
$$

The terms $\partial c^0/\partial x_j^2$ and $\partial c^0/\partial x_j^1$ are null, and c^1 is a function of only $\boldsymbol{x}^0$ and $\boldsymbol{x}^1$. This suggests that for c^1 we can introduce the following *first order characteristic function* $N_1^k(\boldsymbol{x}^1)$:

$$
c^1=c^1(\boldsymbol{x}^0,\boldsymbol{x}^1,t)=-N_1^k(\boldsymbol{x}^1)\frac{\partial c^0(\boldsymbol{x}^0,t)}{\partial x_k^0}+\bar{c}^1(\boldsymbol{x}^0,t) \tag{9.48}
$$

where $\bar{c}^1(\boldsymbol{x}^0,t)$ is the integral constant.

For terms $O(\varepsilon^{-2})$: Second order characteristic function and microscale equation

Since c^0 is a function of only $\boldsymbol{x}^0$ and c^1 is a function of only $\boldsymbol{x}^0$ and $\boldsymbol{x}^1$, and we have the following equation in the micro-domain:

$$\frac{\partial}{\partial x_i^2}\left\{D_{ij}^{\varepsilon}\left(\frac{\partial c^0}{\partial x_j^0}+\frac{\partial c^1}{\partial x_j^1}+\frac{\partial c^2}{\partial x_j^2}\right)\right\}=0. \tag{9.49}$$

The term $\partial c^0/\partial x_j^0 + \partial c^1/\partial x_j^1$ is a function of only $\boldsymbol{x}^0$ and $\boldsymbol{x}^1$, therefore we can introduce the *second order characteristic function* $N_2^k(\boldsymbol{x}^2)$ with the intergral constant $\bar{c}^2(\boldsymbol{x}^0,\boldsymbol{x}^1,t)$ as

$$c^2(\boldsymbol{x}^0,\boldsymbol{x}^1,\boldsymbol{x}^2,t)=-N_2^k(\boldsymbol{x}^2)\left(\delta_{kl}-\frac{\partial N_1^l}{\partial x_k^1}\right)\frac{\partial c^0}{\partial x_l^0}+\bar{c}^2(\boldsymbol{x}^0,\boldsymbol{x}^1,t) \tag{9.50}$$

where we have used (9.48). Then we have

$$D_{ij}^{\varepsilon}\left(\frac{\partial c^0}{\partial x_j^0}+\frac{\partial c^1}{\partial x_j^1}+\frac{\partial c^2}{\partial x_j^2}\right)=D_{ij}^{\varepsilon}\left(\delta_{jk}-\frac{\partial N_2^k}{\partial x_j^2}\right)\left(\delta_{kl}-\frac{\partial N_1^l}{\partial x_k^1}\right)\frac{\partial c^0}{\partial x_k^0}. \tag{9.51}$$

Substituting (9.51) into (9.49), we obtain the following *microscale equation* in the micro-domain:

$$\frac{\partial}{\partial x_i^2}\left\{D_{ij}^{\varepsilon}\left(\delta_{jk}-\frac{\partial N_2^k}{\partial x_j^2}\right)\right\}=0 \quad \text{in } \Omega_2. \tag{9.52}$$

It is clear that (9.52) can be solved under the X^2-periodic boundary condition.

For terms $O(\varepsilon^{-1})$: Mesoscale equation

As mentioned previously, c^0 is a function of only $\boldsymbol{x}^0$ and c^1 is a function of only $\boldsymbol{x}^0$ and $\boldsymbol{x}^1$; thus we have

$$\frac{\partial}{\partial x_i^1}\left\{D_{ij}^{\varepsilon}\left(\frac{\partial c^0}{\partial x_j^0}+\frac{\partial c^1}{\partial x_j^1}+\frac{\partial c^2}{\partial x_j^2}\right)\right\}+\frac{\partial}{\partial x_i^2}\left\{D_{ij}^{\varepsilon}\left(\frac{\partial c^1}{\partial x_j^0}+\frac{\partial c^2}{\partial x_j^1}+\frac{\partial c^3}{\partial x_j^2}\right)\right\}=0. \tag{9.53}$$

If we introduce an integral average $\langle\cdot\rangle_2$ in the micro-domain to (9.53), then the last term is reduced. By substituting (9.51) into this, we have the following *mesoscale equation* under the X^1-periodic condition:

$$\frac{\partial}{\partial x_i^1}\left\{D_{ik}^{H_2}\left(\delta_{kl}-\frac{\partial N_1^l}{\partial x_k^1}\right)\right\}=0 \quad \text{in } \Omega_1 \tag{9.54}$$

$$D_{ik}^{H_2}=\frac{1}{|\Omega_2|}\int_{\Omega_2}D_{ij}^{\varepsilon}\left(\delta_{jk}-\frac{\partial N_2^k}{\partial x_j^2}\right)d\boldsymbol{x}^2. \tag{9.55}$$

where $D_{ik}^{H_2}$ is the meso-diffusivity.

For terms $O(\varepsilon^0)$: Macroscale equation
Since c^0 is a function of only $\boldsymbol{x}^0$ and c^1 is a function of only $\boldsymbol{x}^0$ and $\boldsymbol{x}^1$, for the terms ε^0 we introduce an integration average $\langle\cdot\rangle_2$ in the micro-domain Ω_2 under the relation (9.51), and obtain

$$\frac{\partial c^0}{\partial t} + v_k^{H_2}\left(\delta_{kl} - \frac{\partial N_1^l}{\partial x_k^1}\right)\frac{\partial c^0}{\partial x_l^0} - \frac{\partial}{\partial x_i^0}\left\{D_{ik}^{H_2}\left(\delta_{kl} - \frac{\partial N_1^l}{\partial x_k^1}\right)\frac{\partial c^0}{\partial x_l^0}\right\}$$
$$-\frac{\partial}{\partial x_i^1}\left\{D_{ij}^{\varepsilon}\left(\frac{\partial c^1}{\partial x_j^0} + \frac{\partial c^2}{\partial x_j^1} + \frac{\partial c^3}{\partial x_j^2}\right)\right\} - f^{H_2} = 0, \tag{9.56}$$

$$v_k^{H_2} = \frac{1}{|\Omega_2|}\int_{\Omega_2} v_j^{\varepsilon}\left(\delta_{jk} - \frac{\partial N_2^k}{\partial x_j^2}\right)d\boldsymbol{x}^2, \quad f^{H_2} = \frac{1}{|\Omega_2|}\int_{\Omega_2} f^{\varepsilon}\, d\boldsymbol{x}^2. \tag{9.57}$$

Furthermore, we introduce an integration average $\langle\cdot\rangle_1$ in the meso-domain Ω_1, and obtain the following *macroscale equation*:

$$\frac{\partial c^0}{\partial t} + v_l^H\frac{\partial c^0}{\partial x_l^0} - \frac{\partial}{\partial x_i^0}\left(D_{il}^H\frac{\partial c^0}{\partial x_l^0}\right) - f^H = 0 \quad \text{in } \Omega_0 \tag{9.58}$$

$$D_{il}^H = \frac{1}{|\Omega_1|}\int_{\Omega_1} D_{ik}^{H_2}\left(\delta_{kl} - \frac{\partial N_1^l}{\partial x_k^1}\right)d\boldsymbol{x}^1, \tag{9.59}$$

$$v_l^H = \frac{1}{|\Omega_1|}\int_{\Omega_1} v_k^{H_2}\left(\delta_{kl} - \frac{\partial N_1^l}{\partial x_k^1}\right)d\boldsymbol{x}^1, \quad f^H = \frac{1}{|\Omega_1|}\int_{\Omega_2} f^{H_2}\, d\boldsymbol{x}^1 \tag{9.60}$$

where D_{il}^H is the macro-diffusivity. We can solve the macroscale equation (9.58) using the boundary conditions (9.10) and (9.11) and the initial condition (9.13).

We summarize the procedure to solve the diffusion problem of a multiscale porous media using the HA method as follows:

1. Obtain the second order characteristic function N_2^k by solving the microscale equation (9.52) under the X^2-periodic condition.
2. Determine the meso-diffusivity $D_{ik}^{H_2}$ using (9.55), and obtain the first order characteristic function N_1^k by solving the mesoscale equation (9.54) under the X^1-periodic condition.
3. Determine the mesoscale velocity $v_k^{H_2}$ and the mesoscale source f^{H_2} from (9.56).
4. Determine the macro-diffusivity D_{il}^H, macro-velocity v_l^H, macro-source f^H by (9.59), and obtain c^0 by solving the macroscale equation (9.58) under the boundary conditions (9.10) and (9.11) and under the initial condition (9.13).
5. The first order approximation of the true concentration c^ε is obtained as follows:

$$c^{\varepsilon}(\boldsymbol{x},t) \simeq c^0(\boldsymbol{x}^0,t) + \varepsilon c^1(\boldsymbol{x}^0,\boldsymbol{x}^1,t) + \varepsilon^2 c^2(\boldsymbol{x}^0,\boldsymbol{x}^1,\boldsymbol{x}^2,t)$$
$$= c^0 - \varepsilon\left\{N_1^l + \varepsilon N_2^k\left(\delta_{kl} - \frac{\partial N_1^l}{\partial x_k^1}\right)\right\}\frac{\partial c^0}{\partial x_l^0}. \tag{9.61}$$

9.4 Diffusivity of Compacted Bentonite Considering Micro-structure

We can find many reports of diffusion experiments for bentonite (Gillham et al. 1983; Miyahara et al. 1991; Sato et al. 1992; Oscarson and Hume 1994; Sato and Shibutani 1994; Kozaki et al. 1999). These types of experimental data of diffusion and distribution coefficients will be used for the analysis of the long-term safety of nuclear waste disposal facilities that use bentonite clay barriers. The data, however, have not yet been estimated systematically in the sense of a theory of physical chemistry; therefore it is necessary to re-assess these data not only by experiments but also from the perspective of a molecular- and multiscale-based theory.

In this section we present the results of the MD/HA method for bentonite diffusivity under the assumption of the periodic geometry of bentonite in the micro-domain. That is, we provide the data of diffusivity of a micro-domain by MD, then introduce a multiscale finite element method. Note that detailed results for this section are given in Fujii (2002).

9.4.1 Experimental Data of Bentonite Diffusivity

Diffusion experiments of bentonite are classified into transient methods and steady-state methods (Sato and Shibutani 1994; Kozaki et al. 1999).

A typical transient method is the in-diffusion method (i.e., concentration profile method) shown in Fig. 9.9: After the species is allowed to diffuse into the one-dimensional specimen, it is sliced into segments of approximate thickness of $0.5 \sim 1$ mm, and the amount of the species in each slice is measured. The diffusivity is then determined using this concentration profile. The diffusivity determined by this method is referred to as the *apparent diffusion coefficient*, which is denoted by D_a.

A typical steady-state method is the through-diffusion method shown in Fig. 9.10, where the concentration of species from the sampling cell is measured at each time interval. At its steady-state the concentration is linearly increased, and the time-rate corresponds to the mass flux, thus we obtain the diffusivity from the slope of the mass flux. This diffusivity is referred to as the *effective diffusion coefficient*, which is denoted by D_e.

The apparent values of the diffusion coefficient obtained by the in-diffusion method are shown in Tables 9.1 and 9.2, and the effective diffusion coefficients obtained by the through-diffusion method are shown in Table 9.3. The effective

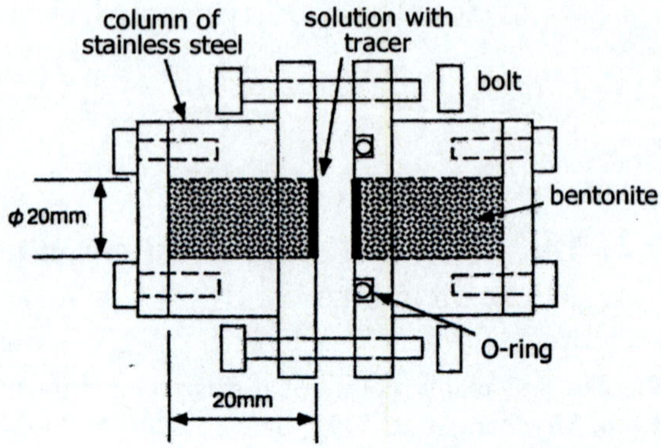

Fig. 9.9 Apparatus for the in-diffusion test (Sato and Shibutani 1994)

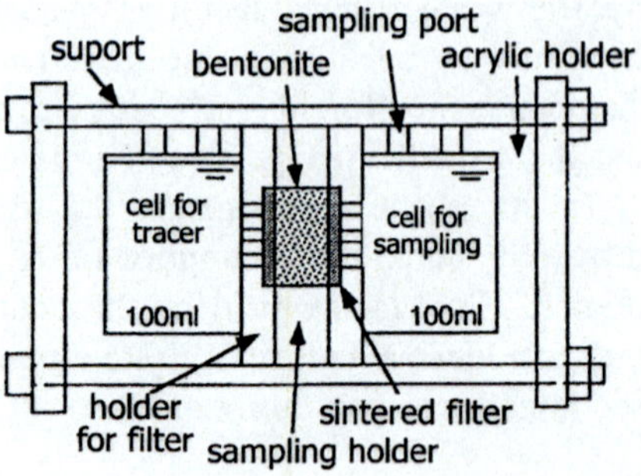

Fig. 9.10 Apparatus for the through-diffusion test (Sato and Shibutani 1994)

diffusion coefficients are in the range applicable to cations, neutral species and anions. The reason may be that in the neighborhood of the clay surface the cations are diffusing faster because of their surface diffusion characteristics, while the anions diffuse more slowly because of their ion exclusion characteristics.

Table 9.1 shows the results for a refined bentonite (i.e., Kunipia F®), and Tables 9.2 and 9.3 the results for an unrefined bentonite (i.e., Kunigel V1®). The chemical composition of these compounds was given in Table 1.1. By comparing Tables 9.1 and 9.2 we can see that Kunipia F®, which consists mainly of smectitic clays, gives a lower value for the apparent diffusion coefficient than the Kunigel V1®.

From Tables 9.2 and 9.3 we recognize a difference in the apparent diffusion coefficient D_a and the effective one D_e. It is known that both are represented in terms of the diffusion coefficient in bulk water D_0 as

$$D_a = \frac{n}{n + (1-n)\rho K_d} \frac{\delta}{\tau} D_0, \tag{9.62}$$

$$D_e = n \frac{\delta}{\tau} D_0 \tag{9.63}$$

Table 9.1 Apparent diffusion coefficient in a refined bentonite (Kunipia F®; Sato 1998)

Species	Dry density [Mg/m³]							
	0.8	1.0	1.2	1.4	1.5	1.6	1.8	2.0
HTO	4.8E−10		1.8E−10			9.5E−11		
^{90}Sr		7.6E−12	7.6E−12		5.4E−12			4.3E−12
^{99}Tc		1.3E−10		3.2E−11			1.2E−11	1.0E−11
^{129}I	2.6E−10			8.5E−11				2.4E−11
^{137}Cs	6.6E−12	4.8E−12	3.2E−12	2.2E−12		1.3E−12	7.9E−13	4.0E-13
^{237}Np	5.0E−12	2.0E−12	1.0E−12	3.0E−13		2.0E−13	1.0E−13	3.0E−14

Unit: [m²/s]

Table 9.2 Apparent diffusion coefficient in an unrefined bentonite (Kunigel V1®; Sato 1998)

Species	Dry density [Mg/m³]					
	0.8	1.0	1.4	1.6	1.8	2.0
HTO	9.9E−10		5.8E−10	4.2E−10		2.3E−10
^{99}Tc	7.2E−10	4.7E−10		1.8E−10	1.2E−10	4.4E−11
^{137}Cs	1.4E−11		7.8E−12			5.2E−12
^{237}Np	7.0E−12		1.5E−12			2.0E−13

Unit: [m²/s]

where n is the porosity, ρ is the intrinsic mass density of the smectitic clay (2.7 Mg/m³), δ is the constrictivity of pores, and τ is the tortuosity. From (9.62) and (9.63), we note that the apparent diffusion coefficient D_a is related to the effective one D_e as

$$D_a = \frac{1}{n + (1-n)\rho K_d} D_e. \tag{9.64}$$

The apparent diffusion coefficient includes the term for the distribution coefficient K_d; that is, the apparent diffusion coefficient, which includes the effect of adsorption, represents the diffusivity in a simplified manner, whereas the effective diffusion coefficient is not influenced by the adsorption.

Since tritium water HTO is not adsorptive (see later), we can set K_d=0. Then (9.64) is given as

$$D_a = \frac{1}{n} D_e. \tag{9.65}$$

9.4.2 *Microscale Problem of HA for Bentonite*

We present here the HA results for diffusion in bentonite and compare these with the experimental results. For a pure smectitic bentonite, Kunipai F®, we apply the two-scale HA: By solving the microscale equation (9.26) using a finite element method

Table 9.3 Effective diffusion coefficient in an unrefined bentonite (Kunigel V1®; Sato 1998)

Species	Dry density [Mg/m^3]						
	0.8	1.0	1.3	1.4	1.5	1.8	2.0
HTO		3.6E−10			1.4E−10		4.2E−11
		5.6E−10			1.9E−10		5.1E−11
^{99}Tc			4.4E−11			9.3E−12	1.4E−12
			4.9E−11				1.9E−12
^{137}Cs	8.6E−10			3.9E−10		2.5E−10	3.2E−11
					1.6E−10	4.3E−11	
^{237}Np	1.2E−10			2.5E−11		2.5E−12	

Unit: [m^2/s]

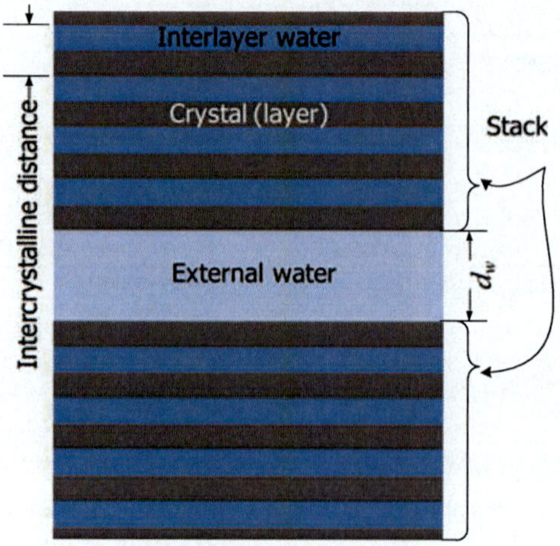

Fig. 9.11 Unit cell for a pure smectitic clay with eight layers of clay minerals (Fujii 2002)

we obtain the homogenized diffusivity coefficient given by (9.31). For an unrefined bentonite, Kunigel V1®, we apply the three-scale HA because it incorporates quartz grains of about 10 μm: By solving the microscale equation (9.52) and the mesoscale equation (9.54) we obtain the homogenized diffusivity coefficient given by (9.59). When solving the microscale equations in both cases, we use the distributed viscosity characteristics, which are obtained from MD calculations.

9.4.2.1 Diffusion of Tritium in Pure Smectitic Bentonite

First we consider a pure smectitic bentonite, Kunipia F®, which consists of stacks of clay minerals including interlayer water, and external water. We assume a simple stack with eight clay minerals in the micro-domain (Fig. 9.11). From MD results, we know the diffusivity and viscosity as a function of the distance from the surface of a clay mineral.

Tritium water HTO (^3H_2O) consists of a tritium atom, which is an isotope of hydrogen, an oxygen atom and a normal hydrogen atom. The chemical property is almost same as the normal water H_2O, and in a bentonite saturated with normal water, the tritium behaves as a non-adsorbed species. Tritium is a radioactive isotope with a half-life of 12.33 years, and it can be used as a tracer for identifying the diffusivity in bentonite because it is non-adsorptive. We can directly compare the experimental results with the HA calculated ones. The properties used in the HA calculation for a pure smectitic bentonite are given in Table 9.4.

The water in the pure smectitic bentonite is classified as interlayer water, which exists in the interlayer space between the clay stacks, and external water, which exists externally to the clay stacks. Both the physical and chemical properties of each type of water are strongly affected by the charged state of the mineral surface, and the interlayer water is also affected by the interlayer cations. In Fig. 9.12 we show the distribution of the self-diffusion coefficient of water as a function of the distance from the clay surface.

We assume that the profile of diffusivity of tritium water HTO is the same as the normal water shown in Fig. 9.12; however, its diffusivity in the bulk water is specified by experiments as $2.44 \times 10^{-5}\,cm^2/s$ at 25°C (Klitzsche et al. 1976), therefore the maximum value in Fig. 9.12 is replaced by this experimental value. The diffusivity in the clay mineral is, of course, null.

In the unit cell model of HA, the distances between each clay mineral (i.e., the interlayer distance) and between the two stacks (i.e., the thickness of external water) strongly affects the calculated results. Note that the thickness of a clay mineral is

Table 9.4 Properties for HA calculation for pure smectitic bentonite (Fujii 2002)

Bentonite	Kunipia F®
Dry density	1.6, 2.0 (Mg/m^3)
Porosity (dry density)	0.41(1.6), 0.26(2.0)
Saturation	100(%)
Temperature	25°C
Species	HTO (tritium water)
Interlayer water	Normal water

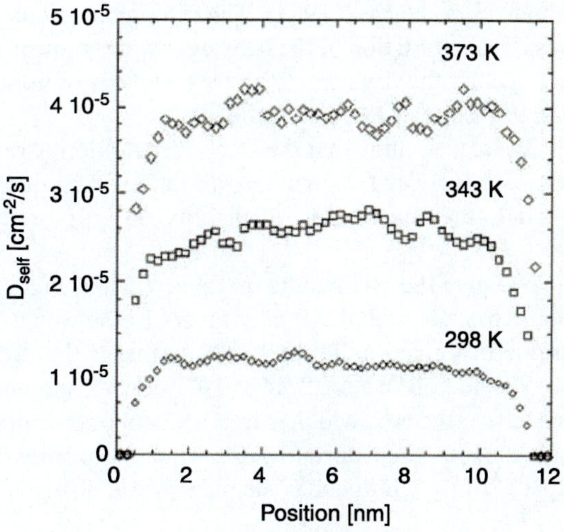

Fig. 9.12 Diffusivity of interlayer water obtained by MD (courtesy of Prof. K. Kawamura)

Table 9.5 HA calculated diffusivity of HTO in pure smectitic bentonite (Fujii 2002)

No. of layers	Dry density [Mg/m^3]					
	1.6			2.0		
	Thick. ext. water	Diffusivity D^H [m^2/s]		Thick. ext. water	Diffusivity D^H [m^2/s]	
	d_w[nm]	Without interlayer diffusion	With interlayer diffusion	d_w[nm]	Without interlayer diffusion	With interlayer diffusion
4	2.48	1.10×10^{-10}	1.11×10^{-10}	4.87	3.57×10^{-11}	4.82×10^{-11}
6	3.54	1.42×10^{-10}	1.52×10^{-10}	6.96	4.94×10^{-11}	6.25×10^{-11}
8	4.61	1.79×10^{-10}	1.90×10^{-10}	9.05	6.02×10^{-11}	7.37×10^{-11}
10	5.67	2.07×10^{-10}	2.18×10^{-10}	11.14	7.13×10^{-11}	8.49×10^{-11}
For bulk water diffusivity		3.31×10^{-10}			2.11×10^{-10}	

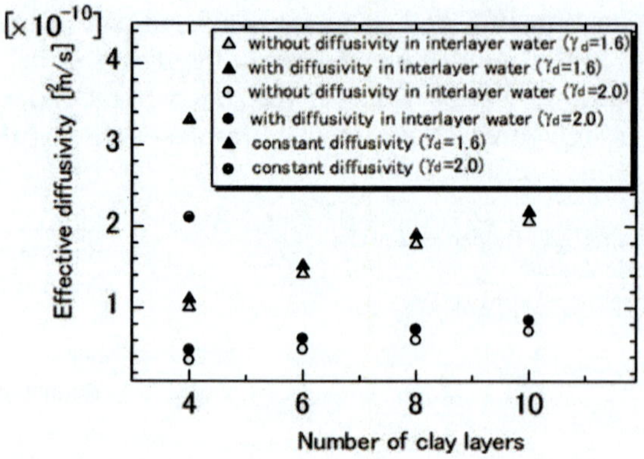

Fig. 9.13 Diffusivity in pure smectitic bentonite (Fujii 2002)

about 1 nm (Nakano 1991). The interlayer distance is determined by using the MD results as a function of the interlayer water molecules n, and here we assume $n=2$. If we specify the dry density and the number of minerals in a stack, we can determine the thickness of external water d_w.

We also assume that the stacks are randomly oriented; therefore for the homogenized diffusivity D_{11}^H in the x_1-direction of the quasi-one dimensional micro-domain model, the macroscale diffusivity D^H is assumed to be isotropic, given by $D^H = D_{11}^H/3$.

We give the HA results in Table 9.5 and Fig. 9.13, which were calculated using stack models with 4–10 clay layers in the micro-domain model Fig. 9.11, and the properties given in Table 9.4. Note that if the diffusivity is constant and the same as for the bulk water (2.44×10^{-5} cm^2/s), the homogenized diffusivity D^H is the same for all cases, which is independent of the number of minerals in a stack and the thickness of the external water. Thus the profile of the diffusivity in the interlayer space strongly influences the macroscale diffusivity.

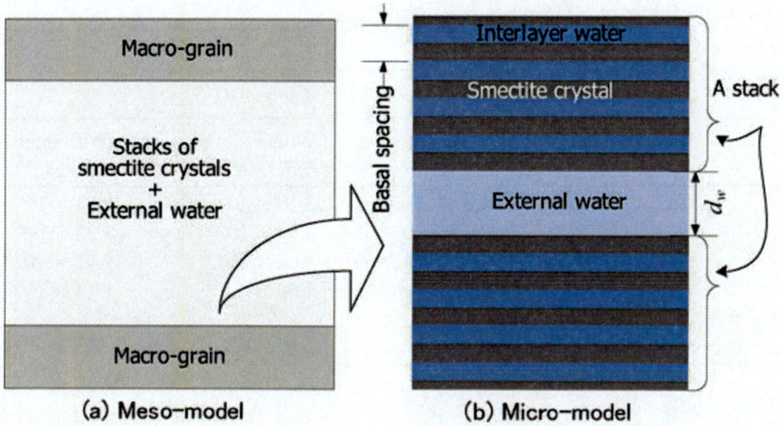

Fig. 9.14 Meso-/micro-models of the multiscale HA (Fujii 2002)

Since the interlayer distance is small, the water in that space is strongly affected by the surface of the clay minerals. Thus, the diffusivity in that space becomes very small, which is the same mechanism shown for permeability (Chap. 8). In order to verify this fact we calculated the results for both cases (i.e., with/without diffusion in the interlayer space) and these are shown in Table 9.5. The difference between both cases is significant, therefore the major transport path is in the external water.

9.4.2.2 Multiscale HA for Bentonite

We use a three-scale HA for Kunigel V1® with the same properties as indicated in Table 9.4. Since Kunigel V1® includes the same percent for the quartz grains and the smectitic clays by weight, this can be simplified to give 2/3 by volume as stacks and 1/3 by volume of quartz grains. The schematic diagram shown in Fig. 9.14a is the unit cell in the meso-domain, which consists of a domain of quartz and a mixture of stacks and external water. The diffusivity in the quartz is null. The unit cell in the micro-domain shown in Fig. 9.14b is the same as that for the pure smectitic clay. We assume that the pore structure of Kunigel V1® is intensively represented in the micro-domain, and there exist only quartz grains and stacks in the meso-domain.

We calculate the model of Kunigel V1® described above. The number of layers of clay minerals in a one stack is given as 4–10, the same as for the case of the pure smectite, Kunipia F®. The results are shown in Table 9.6 and Fig. 9.15 together with the results for Kunipia F®. We also show the effective diffusion coefficient obtained by the through-diffusion method, which is same as the homogenized diffusivity D^H in this HTO self-diffusion problem.

We observe that the HA-calculated results are close to the experimentally derived results for Kunigel V1®. The diffusivity of Kunipia F® is less than for Kunigel V1® for the same dry density of 2.0 Mg/m^3. The experimental data of Sato (1998) show that the diffusivity of Kunigel V1® is several fold that of Kunigel V1® for a

Table 9.6 Diffusivities of HTO in bentonite (Fujii 2002)

No. of clay minerals	Diffusivity D^H [m²/s]			
	Kunipia F®		Kunigel V1®	
	Without interlayer diffusion	With interlayer diffusion	Without interlayer diffusion	With interlayer diffusion
4	3.57×10^{-11}	4.82×10^{-11}	5.10×10^{-11}	5.81×10^{-11}
6	4.94×10^{-11}	6.25×10^{-11}	6.62×10^{-11}	7.38×10^{-11}
8	6.02×10^{-11}	7.37×10^{-11}	9.00×10^{-11}	9.77×10^{-11}
10	7.13×10^{-11}	8.49×10^{-11}	1.06×10^{-10}	1.14×10^{-10}
Experiment (Sato and Shibutani 1994)			4.2×10^{-11} 5.1×10^{-11}	

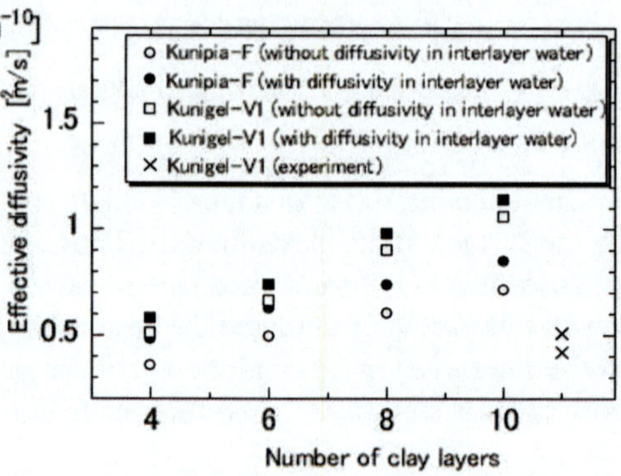

Fig. 9.15 Effective diffusivity of bentonite with a dry density of 2.0 Mg/m³ and the number of clay minerals in a stack (Fujii 2002)

dry density of 1.6 Mg/m³, while the HA results show ratios of 1.2:1.3 times. This may be due to the high tortuosity of Kunipia F®, and the possible consequences of a smaller interlayer space.

It is reported that the number of layers in a stack in smectitic clays is on average eight (Fujii and Nakano 1984). If we observe the numerical results and consider the effects of tortuosity, we can conclude that the number of layers in a stack can be around eight.

9.5 HA and Similitude for the Seepage/Diffusion Problem

By using the micro-/macro-model shown in Fig. 9.16 and Table 9.7, we can calculate the flow field in pure smectitic bentonite. We assume that nine layers of clay minerals are stacked in this micro-model, and the stacks are distributed in a random orientation. The height of a macro-model is set as $l=1$ m, which gives the parameter $\varepsilon=10^{-7}$. In the microscale model of Fig. 9.16, d_{ext} is the height of external water, and d_{int} is the height of interlayer water, which is determined depending on the dry density ρ_d as shown in Table 9.7. The height of the micro-model is

$$X_2^1 = 9 + 8d_{int} + d_{ext} \text{ [nm]}.$$

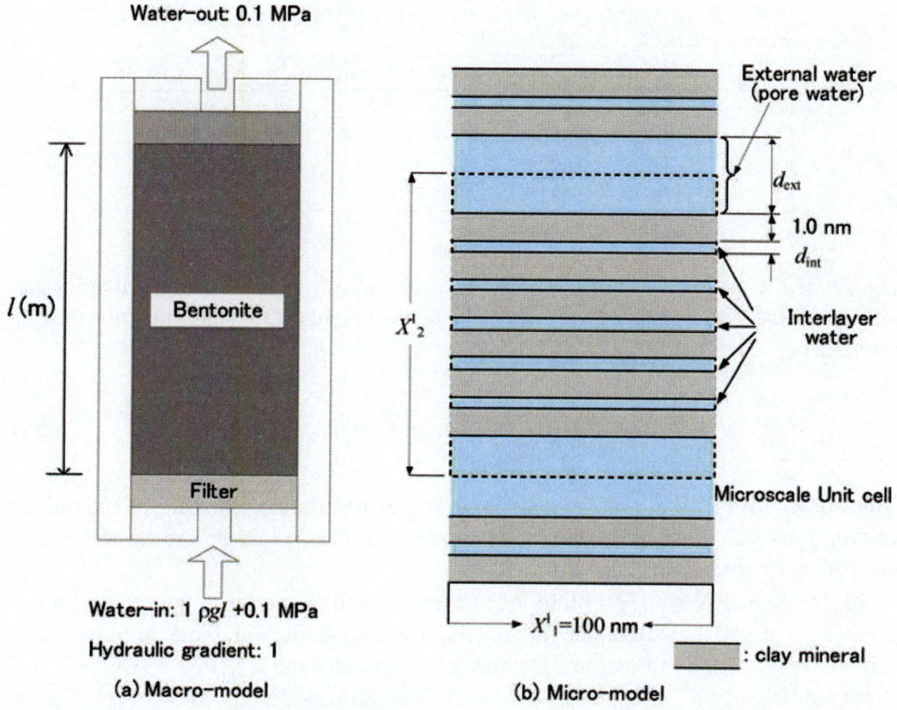

Fig. 9.16 Macro-model and micro-model for a hydrated pure smectite

Table 9.7 Calculated properties and permeability for a hydrated pure smectite

ρ_d Mg/m^3	Void ratio e	d_{int} nm	d_{ext} nm	d_{int}/d_{ext}	K_{11} $\times 10^{-15}$m/s
1.2	1.24	1.12	2.20	0.80	23.70
1.4	0.93	0.91	1.10	0.87	7.82
1.6	0.59	0.59	–	1.00	2.28
1.8	0.50	0.50	–	1.00	1.42

Note that the external space disappears for the case $\rho_d > 1.6\,\text{g/cm}^3$. The viscosity η is given as a function of distance from the surface of the clay mineral. The boundary condition in the macro-model is set as the hydraulic gradient between the upper and lower surface, which is 1.0. The stacks are randomly oriented, therefore the permeability can be regarded as isotropic; $K = K_{11}^*/3$ where K_{11}^* is the C-permeability calculated by (8.31) and K_{11} is the H-permeability in x_1^1-direction. The homogenized velocity is given by $v_i^H = \langle v_i \rangle$. Since the stack orientation is random, the averaged velocity is $v^H = v_i^H/3$, and the calculated velocities are given for a hydrated pure smectite in Table 9.8.

As discussed in Sect. 5.6, the Peclét number in the diffusion Pe_D denotes the ratio of the mass transported by convection relative to the mass transported by diffusion. In the HA for the pure smectitic bentonite, the normalized length L of (5.142) is

Table 9.8 Averaged velocity, diffusivity and Peclét number for a hydrated pure smectite

ρ_d Mg/m^3	L nm	v^H ×10^{-17}m/s	D^H ×10^{-10}m^2/s	Pe$_D$ ×10^{-14}
1.2	4.49	7.260	1.089	2.993
1.4	4.17	2.390	0.767	1.299
1.6	3.78	0.698	0.508	0.520
1.8	3.67	0.433	0.412	0.386

chosen as the size of the unit cell, and the normalized velocity and diffusivity are chosen as the homogenized velocity v^H and diffusivity D^H; therefore the Peclét number Pe_D is calculated as

$$\mathrm{Pe}_D = \frac{v^H/L}{D^H/L^2}, \qquad L = \varepsilon(X_1^1\, X_2^1\, X_3^1)^{1/3}. \tag{9.66}$$

Thus, using the HA method, we can obtain the Peclét number correctly. Note that in conventional calculations the normalizing length L can be chosen arbitrarily, which can influence the calculation.

In the self-diffusion problem of water molecules under a given hydraulic gradient, we can calculate the local velocity distribution and the homogenized diffusivity of water molecules for the micro-/macro-model (Fig. 9.16), and the averaged velocity v^H is given in Table 9.8. The homogenized diffusive coefficient D^H is also given in Table 9.8. The Peclét number Pe_D can also be calculated correctly, as in Table 9.8, and, since the Peclét number for this problem is extremely small ($\mathrm{Pe}_D \ll 1$), the problem is diffusion-dominant.

Chapter 10
Long-Term Consolidation of Bentonite and a Homogenization Analysis of the Flow Field

Proposals for the geological disposal of heat emitting high-level radioactive wastes (HLW) have been put forward by many countries including Japan, Canada, Sweden, Switzerland, USA, Spain and others. The disposal concepts invariably involve underground multi-barrier schemes where bentonite clay is chosen for a number of desirable attributes including its swelling potential and ability to trap the majority of released radionuclides (JNC 1999; Chapman and McCombie 2003).

The Japanese proposals for the disposal of HLW also advocate a multi-barrier system with an engineered barrier system (EBS) and a natural geological barrier, i.e., the surrounding rock mass. The EBS consists of the vitrified waste, a metal overpack and compacted bentonite. As a physical and chemical buffer, the bentonite plays an important role in preventing the long-term transport of radionuclides from a repository. Therefore, it is necessary to understand correctly the performance of bentonite in its role as an engineered geological barrier. However, in laboratory consolidation experiments on bentonite specimens over a long-time period, we frequently observe a secondary consolidation, and the true mechanism of this secondary consolidation is not yet clearly understood; researchers have applied viscoelastic and elasto-visco-plastic models (Gibson and Lo 1961; Schiffman et al. 1964; Singh and Mitchell 1968), while others have proposed a conceptual model of macro-pores and micro-pores (Kamon et al. 1985), but these cannot be easily applied to practical engineering problems, particularly those involving geological disposal activities.

The experimental results of long-term consolidation tests on samples of unrefined bentonite, Kunigel V1®, show that the secondary consolidation is caused by the micro-inhomogeneous behavior of a seepage process in the bentonite that occurs due to the compression of the external pores and interlayer spaces. In this analysis the change in permeability is provided by the results of the MD/HA method, which is described in Chap. 8. The deformation behavior is analyzed by using the standard Cam clay model (Schofield and Wroth 1968; Atkinson and Bransby 1978; Atkinson 1993; Davis and Selvadurai 2002; Pietruszczak 2010).

Y. Ichikawa and A.P.S. Selvadurai, *Transport Phenomena in Porous Media*, DOI 10.1007/978-3-642-25333-1_10, © Springer-Verlag Berlin Heidelberg 2012

10.1 Long-term Consolidation Test on Bentonite

Many of the nuclear waste disposal concepts advocate the use of bentonite as a buffer between the host rock and the metal overpack, which contains HLW, i.e., the spent fuel or the vitrified waste, but little is known about the long-term behavior of bentonite.

A test result on Kunigel V1®, which was performed over a period of about 3 years will be described. The Cam clay data for the compression index λ and a swelling index κ defined by (6.70) and (6.71), respectively, is utilized in short-duration tests involving consolidation and unloading before long-term consolidation tests are performed. The swelling due to suction of water in the interlayer space of smectitic clay minerals has to be distinguished from the swelling due to elasto-plastic characteristics; the former is called 'suction swelling' and the latter is 'elasto-plastic swelling'.

In this test, water was first imbibed by a specimen of Kunigel V1® that was constrained from movement at both ends, allowing the developed swelling pressure to reach a constant value. Next, a short-term test sequence of consolidation and unloading was performed to obtain the compression index λ and the swelling index.κ. After this procedure, a constant stress was applied and the displacement was measured. The procedure employed in the experiments is outlined in Table 10.1 and a schematic diagram of the apparatus is shown in Fig. 10.1.

Results for the long-term consolidation are shown in Fig. 10.2, which indicates that a secondary consolidation process takes place and then, after about 100 days, a tertiary consolidation.

Table 10.1 Outline of the long-term consolidation of Kinigel V1®

Item	Content
Type of experiment	1D consolidation
Apparatus	Standard consolidation apparatus with functions measuring the swelling pressure and the amount of imbibed water
Specimens	Unrefined bentonite (Kinigel V1®)
Dimension	Diameter 60 mm, height 10 mm
Initial dry density ρ_d	1.37 and 1.60 Mg/m^3
Procedure	(1) Compacting the powder bentonite in a cell.
	(2) Absorbing water under constant volume (measuring the suction pressure).
	(3) Performing the short-time consolidating and unloading to obtain λ and κ.
	(4) Continuing the long-term consolidation under a given stress.
Applied stress	Max 3.25 MPa
No. of tests	Two tests under five different conditions (total ten tests)
Temperature	Constant (21°C)
Pore water	Pure water and NaCl water

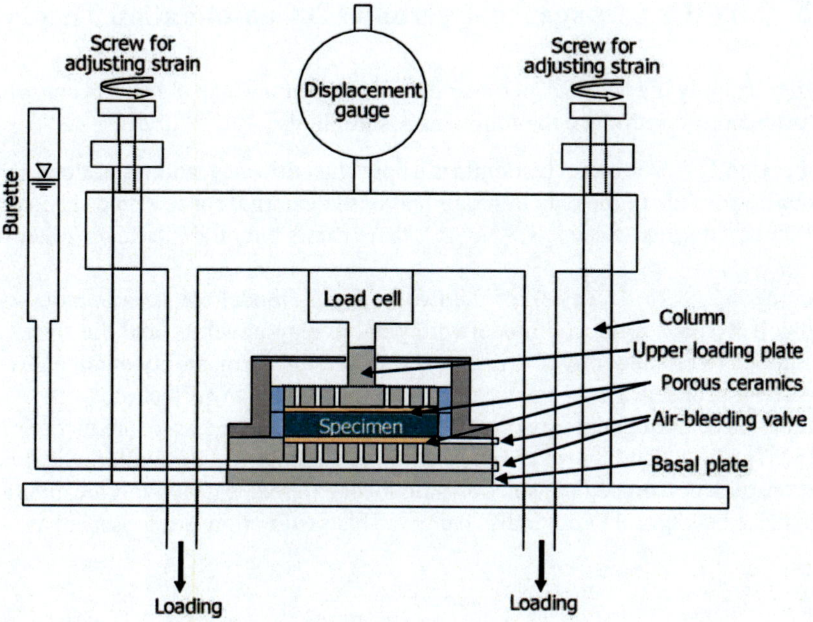

Fig. 10.1 Schematic diagram of the 1D consolidation apparatus used for long-term consolidation tests

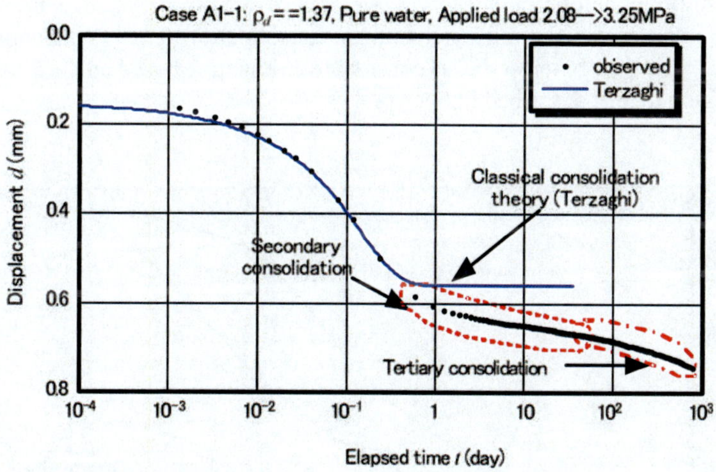

Fig. 10.2 Secondary/tertiary consolidation for compacted bentonite

10.2 MD/HA Seepage Analysis and 1D Consolidation Theory

In order to apply the MD/HA permeability developed in Chap. 8 to a 1D consolidation problem, we introduce the following assumptions:

Assumption 1: The target bentonite is a pure smectitic clay, and water flows in the interlayer space of the clay minerals and/or the external space of the clay stacks. If the dry density exceeds 1.8 Mg/m^3, there exists only the interlayer water (see Fig. 10.3).

Assumption 2: In a unit cell of the two-scale HA model we assume a quasi-1D flow if we have a stacked model with parallel clay minerals, and the stacks are assumed to be randomly distributed. Therefore the permeability obtained by the MD/HA procedure is isotropic, and water can flow in any direction.

Assumption 3: Both the clay crystal (i.e., solid part) and the water are incompressible. The permeability can change due to the volumetric deformation $\Delta\varepsilon_v$ during a consolidation procedure, since an amount of external water and/or the interlayer distance between clay minerals changes. This assumption is represented as

$$\Delta\varepsilon_v = -\frac{\Delta e}{1+e}$$

where we use the convention that the compression is negative.

Gibson et al. (1967) proposed an alternative 1D consolidation theory under a finite strain condition; however, if we ignore a volume change due to the deformation process of the solid skeleton, the consolidation theory is not completely correct. We present a finite strain 1D consolidation equation based on the finite strain consolidation theory described in Sect. 6.3.

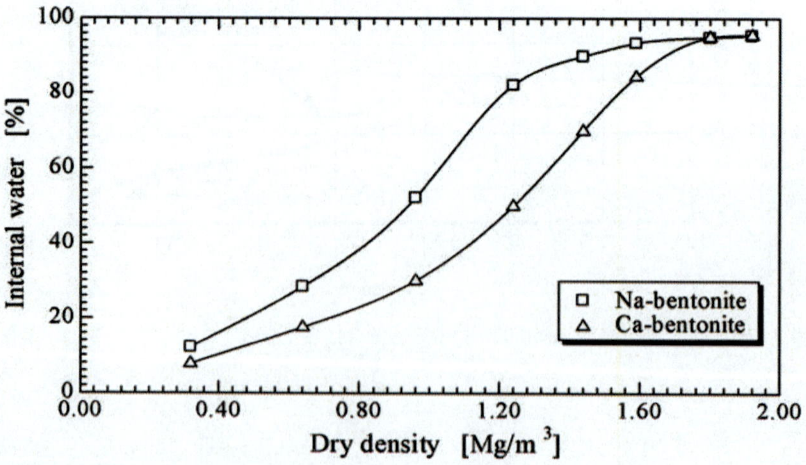

Fig. 10.3 Dry density and interlayer water for bentonite (Pusch 1994)

10.2.1 1D Finite Strain Consolidation Theory

Let $\boldsymbol{X}^* = X_K^* \boldsymbol{E}_K$ and $\boldsymbol{x} = x_i \boldsymbol{e}_i$ be position vectors in the reference and current configurations (Fig. 10.4), respectively. We assume that in our 1D problem the deformation exists only in the x_3-direction, i.e.,

$$x_1 = X_1^*, \qquad x_2 = X_2^*, \qquad x_3 = x_3(X_3^*, t). \tag{10.1}$$

Thus the deformation gradient and its inverse are obtained as

$$\left(F_{iK}^*\right) = \begin{pmatrix} 1\,0\,0 \\ 0\,1\,0 \\ 0\,0\,\delta \end{pmatrix}, \quad \left(F_{iK}^*\right)^{-1} = \begin{pmatrix} 1\;0\;0 \\ 0\;1\;0 \\ 0\;0\;\delta^{-1} \end{pmatrix}, \quad \delta = \frac{\partial x_3}{\partial X_3^*}, \quad \boldsymbol{F}^* = (\boldsymbol{F}^*)^T. \tag{10.2}$$

The Jacobian is given as

$$J^* = \det \boldsymbol{F}^* = \delta. \tag{10.3}$$

Note that gravity acts in the negative x_3 direction, and we have $\boldsymbol{b} = (0,\, 0,\, -g)$.

The differential terms with respect to X_1^* and X_2^* are all null. The components with respect to X_3^* of the nominal effective Cauchy stress $\boldsymbol{\Pi}'$ and the nominal pore pressure P given by (6.27) are obtained as

$$\Pi_{33}' = \left(J^*\,(\boldsymbol{F}^*)^{-1}\boldsymbol{\sigma}'\right)_{33} = J^*\,(F^*)_{3i}^{-1}\sigma_{i3}' = \sigma_{33}', \quad \left(P\boldsymbol{I}\right)_{33} = \left(J^*(\boldsymbol{F}^*)^{-1}p\right)_{33} = p \tag{10.4}$$

where $\boldsymbol{\sigma}'$ and p are the effective stress and the pore pressure in the current configuration, respectively.

Following the sign convention used in soil mechanics, we can write the mean effective stress as $p' = -\sigma_{33}'$. Then the equilibrium equation (6.29) is given by

$$\frac{\partial p}{\partial X_3^*} = \frac{\partial \sigma_{33}'}{\partial X_3^*} - mg = -\frac{\partial p'}{\partial X_3^*} - mg. \tag{10.5}$$

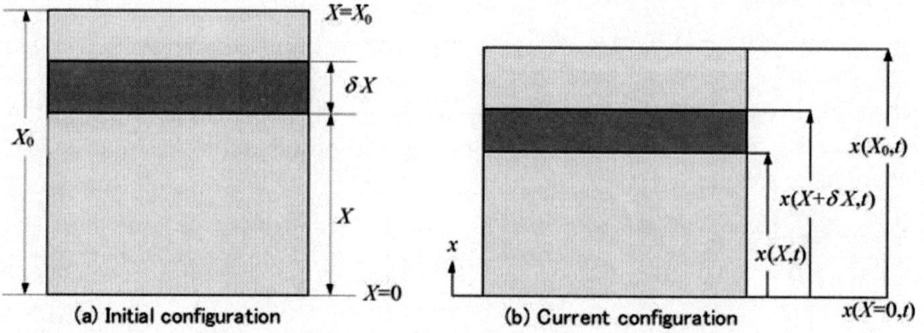

Fig. 10.4 1D consolidation model under finite strain

Also, if we assume that the permeability is isotropic, i.e., $\boldsymbol{k} = (k\,\delta_{ij})$, we obtain the permeability and mass flux in the reference state as

$$\boldsymbol{K} = J^*(\boldsymbol{F}^*)^{-1}\boldsymbol{k}\,(\boldsymbol{F}^*)^{-T} = \begin{pmatrix} k\delta & 0 & 0 \\ 0 & k\delta & 0 \\ 0 & 0 & k\delta^{-1} \end{pmatrix},$$
$$\tilde{V} = -\boldsymbol{K}\left(\text{Grad}^* p - \rho \boldsymbol{b}^0\right) = -k \begin{pmatrix} 0 \\ 0 \\ \delta^{-1}\dfrac{\partial p}{\partial X_3^*} - \rho g \end{pmatrix}. \tag{10.6}$$

The intrinsic solid part (i.e., clay crystals) and water are assumed to be incompressible, i.e., $\dot{\rho}^* = 0$, $\dot{\rho} = 0$. Then the mass of the solid part is given as $m^* = \rho^*(1-n)J^*$, and the mass conservation law implies $\dot{m}^* = 0$, which suggests that

$$\dot{n}\,\delta = (1-n)\,\dot{\delta} \tag{10.7}$$

where n is the porosity. By substituting (10.6) and (10.7) into (6.22), the seepage equation under finite strain is obtained as

$$\dot{\delta} - \frac{\partial}{\partial X_3^*}\left[k\left(\delta^{-1}\frac{\partial p}{\partial X_3^*} - \rho g\right)\right] = 0. \tag{10.8}$$

Let e be the void ratio, and e_0 be its initial value. Then we have

$$J^* = \frac{1+e}{1+e_0} = \delta. \tag{10.9}$$

The classical theory of soil mechanics states that under normal consolidation conditions the relationship between e and $\ln\, p'$ can be represented in the form

$$e = -\lambda\left(\ln\, p' - \ln\, p_0'\right) \tag{10.10}$$

where p_0' is the initial mean effective stress. Then we have

$$de = -\lambda\frac{dp'}{p'} = -(1+e)\,m_v\,dp', \qquad m_v = \frac{de}{(1+e)\,dp'} = \frac{\lambda}{(1+e)\,p'} \tag{10.11}$$

where m_v is the volume compressibility coefficient. By substituting (10.5), (10.9) and (10.11) into (10.8), we obtain the following 1D consolidation equation applicable for finite strains:

$$\frac{1}{1+e_0}\frac{de}{dt} - \frac{\partial}{\partial X_3^*}\left[\frac{k}{m_v}\frac{1+e_0}{(1+e)^2}\frac{\partial e}{\partial X_3^*} + m'g\right] = 0, \qquad m' = \frac{kg}{1+e}(\rho^* - \rho) \tag{10.12}$$

where the term $m'g$ represents the influence of buoyancy.

For simplicity we set $X_3^* = X$, and assume that the initial void ratio is uniformly distributed. Thus we have the following initial boundary value problem (IBVP):

Governing equation

$$\frac{\partial e}{\partial t} - \frac{\partial}{\partial X}\left(C_F \frac{\partial e}{\partial X} + \zeta\right) = \frac{de}{dt} - \frac{\partial}{\partial X}\left(C_F \frac{\partial e}{\partial X}\right) - C_K \frac{\partial e}{\partial X} = 0, \tag{10.13}$$

$$C_F = \frac{k}{m_v}\left(\frac{1+e}{1+e_0}\right)^{-2}, \quad \zeta = (1+e_0)m'g = (\rho^* - \rho)kg\left(\frac{1+e}{1+e_0}\right)^{-1} \tag{10.14}$$

$$C_K = \frac{\partial \zeta}{\partial e} = (\rho^* - \rho)g\left(\frac{1+e}{1+e_0}\right)^{-1}\left(\frac{\partial k}{\partial e} - \frac{k}{1+e}\right) \tag{10.15}$$

Boundary conditions

Dirichlet condition: $e(X,t) = \hat{e}(t) \quad \text{at } X = a$ (10.16)

Neumann condition: $-\rho\left(C_F \frac{\partial e}{\partial X} + \zeta\right)n = \hat{Q}_w(t) \quad \text{at } X = b$ (10.17)

Initial condition

$$e(X, t = t_0) = e_0(X) \tag{10.18}$$

where n is the unit outward normal at $X = b$, and $\hat{Q}_w$ is the measured value of mass flux.

The third term of the l.h.s. of (10.13) is caused by buoyancy, as mentioned previously; if the permeability is small and if the nonlinear term that depends on the void ratio e can be ignored, the IBVP system is given as follows:

Governing equation: $C_K = 0$

$$\frac{\partial e}{\partial t} - \frac{\partial}{\partial X}\left(C_F \frac{\partial e}{\partial X}\right) = 0 \tag{10.19}$$

Boundary conditions: $C_K = 0$

Dirichlet condition: $e(X,t) = \hat{e}(t) \quad \text{at } X = a$ (10.20)

Neumann condition: $-\rho C_F \frac{\partial e}{\partial X} n = \hat{Q}_w(t) \quad \text{at } X = b$ (10.21)

Initial condition: $C_K = 0$

$$e(X, t = t_0) = e_0(X) \tag{10.22}$$

10.2.2 Weak Form of the 1D Finite Strain Consolidation Equation

To develop a finite element scheme we introduce a weak form of the strong form system (10.13)–(10.18). We integrate (10.13) after multiplying an arbitrary function $w(X)$, which is null at $X = a$, and obtain the following *weak form of the 1D finite strain consolidation equation*:

$$\left\langle \frac{\partial e}{\partial t}, w \right\rangle + \left\langle C_F \frac{\partial e}{\partial X}, \frac{\partial w}{\partial X} \right\rangle + \left\langle C_K \frac{\partial e}{\partial X}, w \right\rangle - \hat{Q}_w \, w(b,t) = 0,$$

$$^\forall w (w = 0 \text{ at } X = a) \qquad (10.23)$$

where we integrated the second term by parts, and introduced an inner product as

$$\left\langle f, g \right\rangle = \int_a^b f(X)\, g(X)\, dX.$$

10.2.3 FEM for 1D Finite Strain Consolidation Equation

We discretize the weak form (10.23) by FEM, and introduce a θ-method for the time discretization.

10.2.3.1 Finite Element Galerkin Method

We apply the finite element Galerkin method to (10.23). That is, we introduce FE interpolation functions N_α $(\alpha = 1, 2, \cdots, N)$ for the void ratio e and N_β $(\beta = 1, 2, \cdots, N)$ for the arbitrary function w:

$$e(X,t) = N_\alpha(X)\, e_\alpha(t), \qquad w(X,t) = N_\beta(X)\, w_\beta(t). \qquad (10.24)$$

where $e_\alpha(t)$ $(\alpha = 1, 2, \cdots, N)$ and w_β $(\beta = 1, 2, \cdots, N)$ are node vectors of e and w, respectively. We substitute the approximations (10.24) into the weak form (10.23), and eliminate the arbitrary vector $\{w_\beta\}$ to obtain

$$M_{\alpha\beta} \frac{\partial e_\beta}{\partial t} + K_{\alpha\beta}\, e_\beta + R_{\alpha\beta}\, e_\beta = Q_\alpha \qquad (10.25)$$

where we have

$$M_{\alpha\beta} = \left\langle N_\alpha, N_\beta \right\rangle, \qquad (10.26)$$

$$K_{\alpha\beta} = \left\langle \frac{dN_\alpha}{dX},\ C_F(e)\frac{dN_\beta}{dX} \right\rangle, \tag{10.27}$$

$$R_{\alpha\beta} = \left\langle N_\alpha,\ C_K(e)\frac{dN_\beta}{dX} \right\rangle, \tag{10.28}$$

$$Q_\alpha = \left\langle \hat{Q}_w,\ N_\alpha(X{=}b) \right\rangle. \tag{10.29}$$

The result (10.25) is a nonlinear equation. We can introduce a modified Newton-Raphson method (Owen and Hinton 1980) for solving (10.25). Then we rewrite (10.25), and (10.26)–(10.29) by introducing variables with superscripts k and $k-1$, which implies the values of the variables at each iteration step, and we have

$$M_{\alpha\beta}\frac{\partial e^k_\beta}{\partial t} + K^{k-1}_{\alpha\beta} e^k_\beta + R^{k-1}_{\alpha\beta} e^k_\beta = Q^k_\alpha, \tag{10.30}$$

$$M_{\alpha\beta} = \left\langle N_\alpha,\ N_\beta \right\rangle, \tag{10.31}$$

$$K^{k-1}_{\alpha\beta} = \left\langle \frac{dN_\alpha}{dX},\ C_F(e^{k-1})\frac{dN_\beta}{dX} \right\rangle, \qquad e^{k-1} = N_\gamma(X) e^{k-1}_\gamma \tag{10.32}$$

$$R^{k-1}_{\alpha\beta} = \left\langle N_\alpha,\ C_K(e^{k-1})\frac{dN_\beta}{dX} \right\rangle, \tag{10.33}$$

$$Q^k_\alpha = \left\langle \hat{Q}^k_w,\ N_\alpha(X{=}b) \right\rangle, \tag{10.34}$$

10.2.3.2 Time Discretization: θ-method

The node vector e_β ($\beta = 1, 2, \cdots, N$) is a function of time t, since we introduced a discretization in space using the FEM. We use a θ-method for time discretization.

Let e^{N-1}_β, which is a node vector at the time step $N-1$ be known. Then the unknown vector $e^{k,N}_\beta$, which is given at the time step N and at the nonlinear iteration step k is assumed to be

$$e^{k,n}_\beta = e^{k,N-1}_\beta + \Delta t\left[(1-\theta)\dot{e}^{k,N-1}_\beta + \theta\dot{e}^{k,N}_\beta\right] \tag{10.35}$$

where Δt is the time increment, and θ is a constant given by

$$0 \le \theta \le 1. \tag{10.36}$$

The procedure with $\theta = 0$ corresponds to an explicit difference scheme; $\theta = 1$ is an implicit difference scheme; and $\theta = 0.5$ is the Crank-Nicolson scheme.

Since (10.30) is satisfied both for the steps $N-1$ and N, we have

$$M_{\alpha\beta}\dot{e}^{k,N-1}_\beta + K^{k-1}_{\alpha\beta} e^{k,N-1}_\beta + R^{k-1}_{\alpha\beta} e^{k,N-1}_\beta = Q^{k,N-1}_\alpha, \tag{10.37}$$

$$M_{\alpha\beta}\dot{e}_{\beta}^{k,N} + K_{\alpha\beta}^{k-1} e_{\beta}^{k,N} + R_{\alpha\beta}^{k-1} e_{\beta}^{k,N} = Q_{\alpha}^{k,N}. \tag{10.38}$$

Then after multiplying (10.37) by $1-\theta$ and (10.38) by θ, and adding the resulting equations we obtain

$$\breve{K}_{\alpha\beta}\, e_{\beta}^{k,N} = \breve{Q}_{\alpha}^{k,N} \tag{10.39}$$

$$\breve{K}_{\alpha\beta} = \frac{M_{\alpha\beta}}{\Delta t} + \theta\big(K_{\alpha\beta}^{k-1} + R_{\alpha\beta}^{k-1}\big),$$

$$\breve{Q}_{\alpha}^{k,N} = (1-\theta)Q_{\alpha}^{k,N-1} + \theta\, Q_{\alpha}^{k,N} + \left[\frac{M_{\alpha\beta}}{\Delta t} - (1-\theta)\big(K_{\alpha\beta}^{k-1} + R_{\alpha\beta}^{k-1}\big)\right] e_{\beta}^{k,N-1}$$

where we used (10.35). If we solve (10.39) by an iteration procedure, we can determine the node vector $e_{\beta}^{k,N}$ at each time step N.

10.2.4 Relation Between Permeability and Void Ratio for Compacted Bentonite

We use a model for a pure smectitic bentonite, Kunipia-F®, which is compacted initially to a dry density of 1.4 Mg/m^3. The material properties of this bentonite are shown in Table 10.2.

We assume that there is no external water when the dry density exceeds 1.6 Mg/m^3. We consider a unit cell in the two scale HA, which involves a stack of clay minerals as shown in Fig. 10.5. Note that the following analysis is due to Theramast (2003).

Assuming that the number of clay minerals involved in a stack is 10, the interlayer distance is arranged between $d = 0.2 \sim 0.85$ corresponding to the void ratio $e = 0.2 \sim 1.2$. We use the relationship between the interlayer distance and the permeability, which is obtained by the MD/HA calculation as shown in Fig. 10.6, which is given as

$$e = \alpha_H + \beta_H \ln k \tag{10.40}$$

where we determine the coefficients by a least squares approximation as follows:

Table 10.2 Material properties of a compacted bentonite, Kunipia-F®

Elastic coeff. [MPa]	Poisson's ratio	Wet density [Mg/m^3]	Dry density [Mg/m^3]	Int. frict. angle [°]
1,100	–	2.0	1.4	30
Vol. comp. coeff. m_v [MPa]$^{-1}$	λ	κ	M	p_0' [MPa]
4.10×10^{-4}	0.1403	–	–	0.1

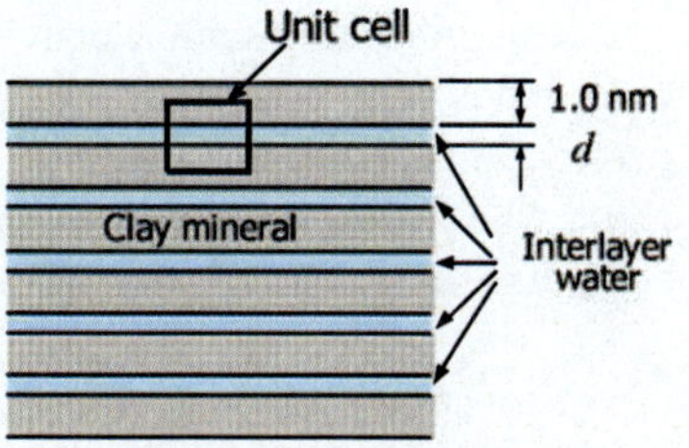

Fig. 10.5 A unit cell model of a stack of clay minerals

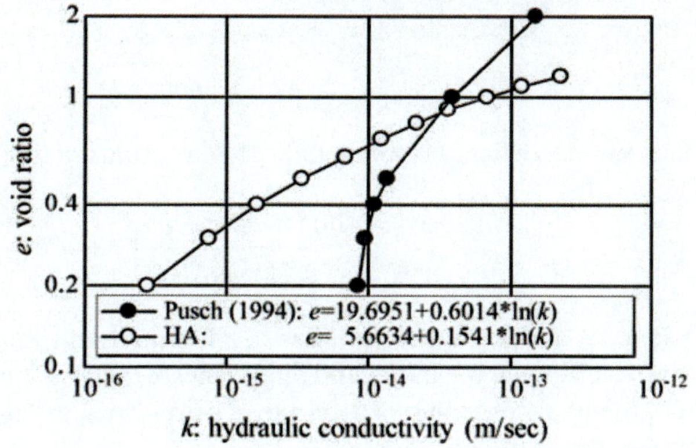

Fig. 10.6 Relationship between void ratio and permeability

$$\alpha_H = 5.663, \qquad \beta_H = 0.1541.$$

Using previous experimental results, Pusch (1994) schematically showed the relationships between the void ratio and permeability for three kinds of clays (Fig. 8.12). From these results we obtain a relationship between the void ratio and permeability for montmorillonite clay as

$$e = \alpha_P + \beta_P \ln(k) \tag{10.41}$$

$$\alpha_P = 19.695, \qquad \beta_P = 0.6013.$$

This result is shown in Fig. 10.6. We understand that the permeability obtained by the MD/HD procedure is different from the experimental data at these values of low void ratio. However, as we will discuss later, if we use both sets of data for the 1D consolidation analysis, the results in both cases are similar, and we cannot reproduce the secondary consolidation response. Conducting permeability experiments on these low permeable materials is extremely difficult. Thus we cannot conclude that the MD/HD results are incompatible with the experimental results.

In soil mechanics it is common practice to adopt a viscous model to describe the process of secondary consolidation (see Mitchell 1993); however, by using an MD simulation it can be shown that a smectite clay stack does not possess a high-enough viscosity to account for the secondary consolidation process (Ichikawa et al. 2004). Therefore, it is necessary to consider an alternative mechanism to represent the secondary consolidation.

Let us introduce such a concept where changing the state of consolidation also contributes to a change in the permeability. That is, at the first stage $t = 0$ the permeability is homogeneous everywhere in a specimen with a value k_0. At stage $t = t$, when consolidation is in progress, the permeability varies at each point. Then at the final stage $t = t_f$, the permeability again becomes homogeneous with a value of k_f.

As consolidation progresses, the permeability changes, therefore if we introduce the state of consolidation at $t = t$ in terms of the void changes as

$$u = \frac{e_0 - e}{e_0 - e_f}, \tag{10.42}$$

then we can define the permeability at any arbitrary time

$$\left(\frac{\log k - \log k_f}{\log k_0 - \log k_f}\right)^{\alpha} + u^{\alpha} = 1 \tag{10.43}$$

where e_0 and e_f are void ratios at the initial and final stages, respectively. The permeabilities at the initial and final stages k_0 and k_f are determined from Fig. 10.6. The material constant α is determined by referring to Fig. 10.7. Note that u is known as the degree of consolidation.

Thus, we have an alternative form of (10.43):

$$\log\left(\frac{k}{k_f}\right) = \log\left(\frac{k_0}{k_f}\right)(1 - u^{\alpha})^{\frac{1}{\alpha}}.$$

If we set

$$\chi = (1 - u^{\alpha})^{\frac{1}{\alpha}},$$

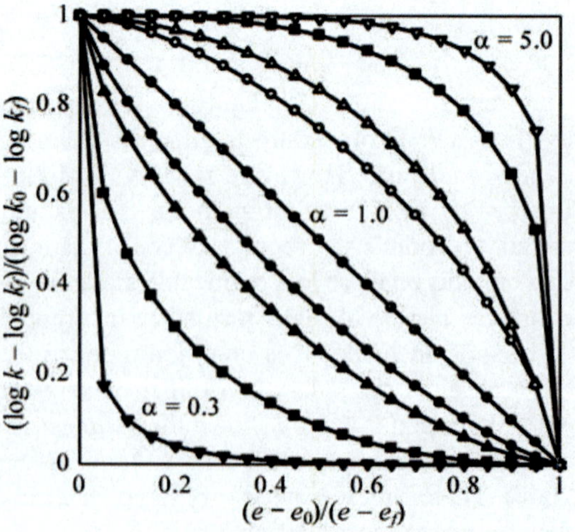

Fig. 10.7 A function which represents the inhomogeneous consolidation

then the permeability k is obtained as

$$k = k_0 \left(\frac{k_f}{k_0} \right)^{1-\chi} . \tag{10.44}$$

10.2.5 Consolidation Experiment and Inhomogeneous 1D Analysis

We use a model of double-sided (both top and bottom) drainage consolidation for our 1D analysis as shown in Fig. 10.8.

Using the material parameters given in Table 10.2, and applying loads $F^{ext} = 0 \sim 2.05\,\text{MPa}$ at the top surface, we obtain the results shown in Fig. 10.9.

The curves shown in Fig. 10.9 indicate the following:

Solid line: Data fitted curve for a bentonite consolidation experiment.

Dashed line: Consolidation curve under a constant permeability $k = k_0$ for all time-steps and all parts of the specimen.

Line with ∘: Consolidation curve using the void ratio-permeability relationship (10.40), which is obtained by the MD/HA simulation. The permeability is constant in the specimen at each time-step (i.e., homogeneous permeability).

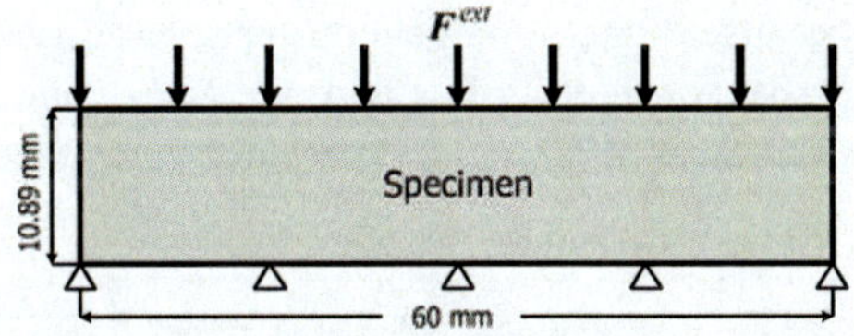

Fig. 10.8 Model for inhomogeneous 1D consolidation analysis

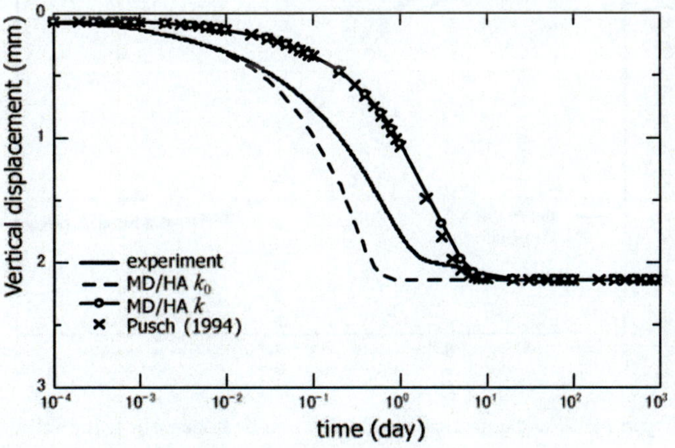

Fig. 10.9 Experimental and numerical results: For homogeneous permeability cases

Line with ×: Consolidation curve using the void ratio-permeability relationship (10.41), which is based on the work of Pusch (1994). The permeability is constant in the specimen at each time-step (i.e., homogeneous permeability).

Note that in our analysis the initial void ratio is given as $e_0 = 1.2$, and therefore the initial permeability from Fig. 10.6 is $k_0 = 3.0 \times 10^{-13}$ m/s.

From the results shown in Fig. 10.9, it can be seen that for the case of a constant permeability $k = k_0$ for all time-steps and in all regions of the specimen (dashed line) the consolidation is terminated at an early stage. If the permeabilities given by the MD/HA simulation (line with ∘) and the results of Pusch 1994 (line with ×) are used, the consolidation terminates later. However, none of these cases correctly represents the secondary consolidation.

By comparing the permeabilities given by the MD/HA simulation (line with ∘) and the results given by Pusch 1994 (line with ×) we can see that both cases give similar results; this suggests that for very low permeability, the relative difference in permeability is not particularly important.

We assume that at the early stage of consolidation, water is drained from the space external to the stacks in the specimen of bentonite; at the intermediate stage, the water is drained from both the external space and the interlayer space, and at the final stage, the water is drained from the interlayer space. Bentonite has a double space structure and shows secondary consolidation behavior because of this structure. From this point of view it is clear that even for permeability changes over time, the secondary consolidation behavior is not represented in these analyses because of the homogeneous permeability in the specimen.

By considering this fact, we can derive a model given by (10.42), in which the permeability varies in a specimen since the void ratio is nonhomogeneous.

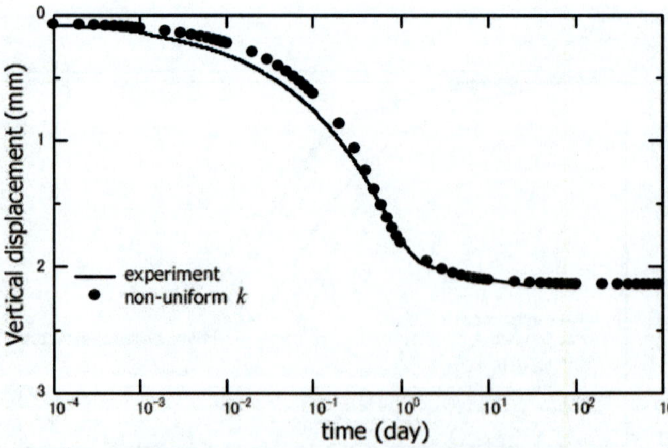

Fig. 10.10 Experimental and numerical results: For non-homogeneous permeability case

The result obtained using this model with non-homogeneous permeability in the specimen is shown in Fig. 10.10, which represents the secondary consolidation of bentonite more accurately.

The most important observation derived from this analysis is that the secondary consolidation is not caused by creep of the solid skeleton. This was confirmed using X-ray diffraction: Ichikawa et al. (2004) have used an X-ray diffraction apparatus to measure the crystalline structure at each point of the specimen during consolidation. From this X-ray diffraction measurement, the above conclusion was verified. It is important to carefully plan experiments when using materials with extremely low permeability, since a non-homogeneous state is easily achieved; in the experiments cited, the specimen was only about 10 mm in height.

Note that we use (10.42) for deciding on the local permeability, where we introduce a parameter α to represent the internal pore structure schematically. Further work is needed to establish the correct permeability model for bentonite with a multiscale pore structure.

Appendix A
Introduction to Vectors

We show some details of the algebra required to study the continuum mechanics of porous media. Further details can be found in the introductory chapters of Malvern (1969), Fung (1969), Mase (1970), Chadwick (1976), Selvadurai (2000a), and Spencer (2004).

A.1 Vectors in $\mathbb{R}^3$

Let us introduce an orthogonal coordinate system (i.e., Euclidean coordinates) x_1, x_2, x_3 (instead of x, y, z) in the three-dimensional real number space $\mathbb{R}^3$. The Euclidean basis is $\{\boldsymbol{e}_1, \boldsymbol{e}_2, \boldsymbol{e}_3\}$. Then, a vector $\boldsymbol{v}$ in $\mathbb{R}^3$ can be given as

$$\boldsymbol{v} = v_1 \boldsymbol{e}_1 + v_2 \boldsymbol{e}_2 + v_3 \boldsymbol{e}_3$$

(Fig. A.1). By using Einstein's summation convention, we write this as

$$\boldsymbol{v} = v_i \, \boldsymbol{e}_i \, . \tag{A.1}$$

That is, the index i in the r.h.s. term of (A.1) appears twice, which is referred to as the *dummy index* and implies that the index is added from 1 to 3 in $\mathbb{R}^3$. On the other hand, we have a *free index* that appears only once in a term. For example, in the following equation:

$$v'_i = Q_{ij} \, v_j \, , \tag{A.2}$$

the index i appears once in each term, and it implies $i = 1$ or 2 or 3. Note that in (A.2) the repeated index j is dummy, which must be summed up from 1 to 3.

Y. Ichikawa and A.P.S. Selvadurai, *Transport Phenomena in Porous Media*,
DOI 10.1007/978-3-642-25333-1, © Springer-Verlag Berlin Heidelberg 2012

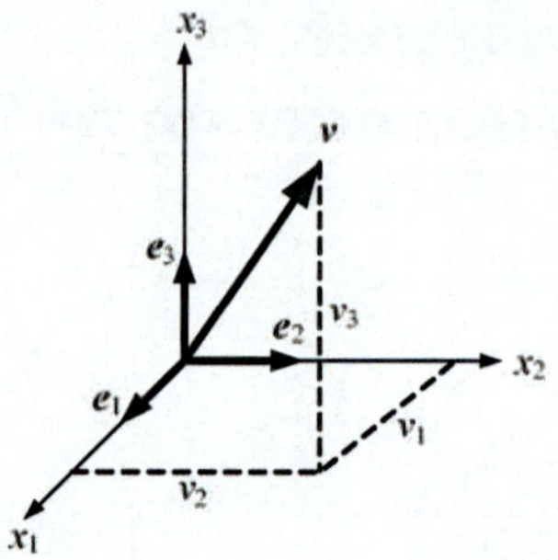

Fig. A.1 A vector in the three-dimensional real number space $\mathbb{R}^3$

A.2 Inner Product and the Length of Vectors

An inner product of the Euclidean bases $\boldsymbol{e}_i$ and $\boldsymbol{e}_j$ gives the following ortho-normal condition:

$$\boldsymbol{e}_i \cdot \boldsymbol{e}_j = \delta_{ij} \tag{A.3}$$

$$\delta_{ij} = \begin{cases} 1 & \text{if} \quad i = j \\ 0 & \text{if} \quad i \neq j \end{cases} \tag{A.4}$$

where δ_{ij} is Kronecker's delta, which is equivalent to the unit tensor $\boldsymbol{I}$:

$$\left((\boldsymbol{I})_{ij}\right) = \left(\delta_{ij}\right) = \begin{pmatrix} 1 & 0 & 0 \\ 0 & 1 & 0 \\ 0 & 0 & 1 \end{pmatrix}. \tag{A.5}$$

By using (A.3), an inner product of vectors $\boldsymbol{u}$ and $\boldsymbol{v}$ is written as

$$\boldsymbol{u} \cdot \boldsymbol{v} = u_i\, v_j\, \boldsymbol{e}_i \cdot \boldsymbol{e}_j = u_i\, v_j\, \delta_{ij} = u_i\, v_i\,. \tag{A.6}$$

Note that the coefficient v_i of a vector $\boldsymbol{v} = v_i\, \boldsymbol{e}_i$ is obtained by an inner product between $\boldsymbol{v}$ and $\boldsymbol{e}_i$:

$$v_i = \boldsymbol{v} \cdot \boldsymbol{e}_i\,. \tag{A.7}$$

Assume a linear transformation $\boldsymbol{A}$ that maps a vector $\boldsymbol{u}$ into another vector $\boldsymbol{v}$ (i.e., $\boldsymbol{v} = \boldsymbol{A}\, \boldsymbol{u}$) where the coefficients are represented by a matrix $(\, A_{ij}\,)$.[1] Then, an inner product between this $\boldsymbol{v}$ and a vector $\boldsymbol{w}$ is given as

[1] Let $\mathcal{U}$ and $\mathcal{V}$ be linear vector spaces (see Appendix C for the linear vector space). A transformation $\boldsymbol{v} = \boldsymbol{A}\boldsymbol{u}$ ($\boldsymbol{u} \in \mathcal{U}$, $\boldsymbol{v} \in \mathcal{V}$) is linear if

$$\boldsymbol{A}(\boldsymbol{u}_1 + \boldsymbol{u}_2) = \boldsymbol{A}\boldsymbol{u}_1 + \boldsymbol{A}\boldsymbol{u}_2, \qquad \forall \boldsymbol{u}_1, \boldsymbol{u}_2,$$
$$\boldsymbol{A}(k\boldsymbol{u}) = k\, \boldsymbol{A}\boldsymbol{u} \qquad \forall k, \boldsymbol{u}$$

where k is a scalar number.

$$\boldsymbol{w} \cdot \boldsymbol{v} = \boldsymbol{w} \cdot \boldsymbol{A}\,\boldsymbol{u} = w_i\, A_{ij} u_j = A_{ji}\, w_j\, u_i = \boldsymbol{A}^T \boldsymbol{w} \cdot \boldsymbol{u} \tag{A.8}$$

where $\boldsymbol{A}^T$ represents the transpose of $\boldsymbol{A}$, and the coefficient matrix of $\boldsymbol{A}^T$ is

$$\left(A_{ij}^T\right) = \begin{pmatrix} A_{11} & A_{21} & A_{31} \\ A_{12} & A_{22} & A_{32} \\ A_{13} & A_{23} & A_{33} \end{pmatrix}.$$

The length of a vector $\boldsymbol{v}$ is written as $|\,\boldsymbol{v}\,|$, which is obtained by using an inner product as

$$|\,\boldsymbol{v}\,| = \sqrt{\boldsymbol{v} \cdot \boldsymbol{v}}. \tag{A.9}$$

The inner product of two vectors $\boldsymbol{u}$ and $\boldsymbol{v}$ is then rewritten as

$$\boldsymbol{u} \cdot \boldsymbol{v} = |\,\boldsymbol{u}\,|\,|\,\boldsymbol{v}\,| \cos\theta \tag{A.10}$$

where θ is the angle between $\boldsymbol{u}$ and $\boldsymbol{v}$.

A.3 Coordinate Transformation

Consider the coordinate transformation from a system $\{x_i\}$ into another system $\{x_i'\}$, which is shown in Fig. A.2. The base vectors are $\{\boldsymbol{e}_i\}$ and $\{\boldsymbol{e}_i'\}$ corresponding to each coordinate system. A transformation can be written as

$$\boldsymbol{e}_i' = Q_{ij}\, \boldsymbol{e}_j \tag{A.11}$$

where the coefficient matrix of $\boldsymbol{Q}$ is

$$(Q_{ij}) = \begin{pmatrix} Q_{11} & Q_{12} & Q_{13} \\ Q_{21} & Q_{22} & Q_{23} \\ Q_{31} & Q_{32} & Q_{33} \end{pmatrix}, \tag{A.12}$$

which is a coordinate transformation matrix with the following properties:

$$\boldsymbol{Q}\,\boldsymbol{Q}^T = \boldsymbol{Q}^T\,\boldsymbol{Q} = \boldsymbol{I}, \qquad \boldsymbol{Q}^{-1} = \boldsymbol{Q}^T \tag{A.13}$$

($\boldsymbol{Q}^{-1}$ is the inverse of $\boldsymbol{Q}$, and $\boldsymbol{I}$ is the unit tensor). An index notation of (A.13) is represented as

$$Q_{ik}\, Q_{jk} = Q_{ki}\, Q_{kj} = \delta_{ij}. \tag{A.14}$$

Let an angle between $\boldsymbol{e}_i'$ and $\boldsymbol{e}_j$ be θ_{ij} as shown in Fig. A.2a for the three-dimensional case, and we have

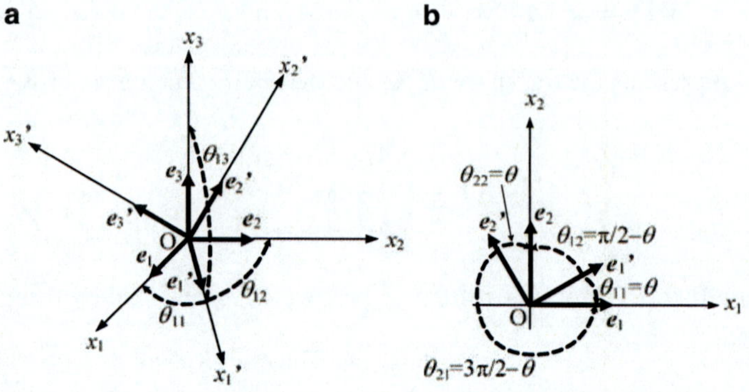

Fig. A.2 Coordinate transformation; (**a**) three-dimensional case, (**b**) two-dimensional case

$$\boldsymbol{e}'_i \cdot \boldsymbol{e}_j = |\,\boldsymbol{e}'_i\,|\,|\,\boldsymbol{e}_j\,| \cos\theta_{ij} = \cos\theta_{ij}.$$

This results in

$$Q_{ij} = \cos\theta_{ij}. \tag{A.15}$$

(Q_{ij}) is then referred to as the matrix of direction cosines. We have the following identity:

$$\boldsymbol{e}'_i \cdot \boldsymbol{e}'_j = \delta_{ij} = Q_{ik}\,\boldsymbol{e}_k \cdot Q_{jl}\,\boldsymbol{e}_l = Q_{ik}\,Q_{jl}\,\delta_{kl} = Q_{ik}\,Q_{jk},$$

which proves the result (A.13) or (A.14).

For the two-dimensional case shown in Fig. A.2b, we have $\cos(\pi/2-\theta) = \sin\theta$, $\cos(3\pi/2–\theta) = -\sin\theta$, so that we have the following simplified relationship:

$$(Q_{ij}) = \begin{pmatrix} Q_{11} & Q_{12} \\ Q_{21} & Q_{22} \end{pmatrix} = \begin{pmatrix} \cos\theta & \sin\theta \\ -\sin\theta & \cos\theta \end{pmatrix}. \tag{A.16}$$

If a vector $\boldsymbol{v}$ is transformed into $\boldsymbol{v}'$ by $\boldsymbol{Q}$, we have a similar rule to (A.11):

$$\boldsymbol{v}' = \boldsymbol{Q}\,\boldsymbol{v}. \tag{A.17}$$

The relationship of a linear transformation $\boldsymbol{v} = \boldsymbol{A}\,\boldsymbol{u}$ can be similarly represented in a new coordinate system as

$$\boldsymbol{v}' = \boldsymbol{A}'\,\boldsymbol{u}'; \qquad \boldsymbol{v}' = \boldsymbol{Q}\,\boldsymbol{v}, \quad \boldsymbol{u}' = \boldsymbol{Q}\,\boldsymbol{u},$$

which gives the following transformation rule for the second order tensor $\boldsymbol{A}$:

$$\begin{aligned} v = A\,u = Q^T\,v' &= A\,Q^T\,u' \\ \Rightarrow \quad A' &= Q\,A\,Q^T. \end{aligned} \tag{A.18}$$

A.4 Outer Product

An *outer product* or *vector product* of two base vectors $\boldsymbol{e}_i$ and $\boldsymbol{e}_j$ is given as

$$\boldsymbol{e}_i \wedge \boldsymbol{e}_j = e_{ijk}\,\boldsymbol{e}_k \tag{A.19}$$

where

$$e_{ijk} = \begin{cases} 1, & \text{if} \quad (i,j,k) = \{(1,2,3),(2,3,1),(3,1,2)\} \\ -1, & \text{if} \quad (i,j,k) = \{(1,3,2),(3,2,1),(2,1,3)\} \\ 0, & \text{otherwise} \end{cases} \tag{A.20}$$

is the permutation symbol. An outer product of two vectors $\boldsymbol{a}$ and $\boldsymbol{b}$ is then given as

$$\boldsymbol{a} \wedge \boldsymbol{b} = a_i\,\boldsymbol{e}_i \wedge b_j\,\boldsymbol{e}_j = e_{ijk}\,a_i\,b_j\,\boldsymbol{e}_k. \tag{A.21}$$

Note that by using the permutation symbol e_{ijk}, the determinant of a (3×3)-matrix (A_{ij}) can be written as

$$\det A = \begin{vmatrix} A_{11} & A_{12} & A_{13} \\ A_{21} & A_{22} & A_{23} \\ A_{31} & A_{32} & A_{33} \end{vmatrix} = e_{ijk}\,A_{i1}\,A_{j2}\,A_{k3}. \tag{A.22}$$

Appendix B
Partial Differentiation and Integral Theorems

Several fundamental results concerning partial differentials and integral theorems are summarized. Further expositions are given by Spiegel (1971), Greenberg (1998), Riley et al. (1998), and Selvadurai (2000a).

B.1 Calculus of Partial Differentials

If a function f is defined in a one-dimensional domain x, its differential is represented as an ordinary differential:

$$\frac{df}{dx} = \lim_{\Delta x \to 0} \frac{f(x + \Delta x) - f(x)}{\Delta x}.$$

In a two-dimensional domain (x, y) the partial differential of a function $g(x, y)$ in each direction is given as

$$\frac{\partial g}{\partial x} = \lim_{\Delta x \to 0} \frac{g(x + \Delta x, y) - g(x, y)}{\Delta x}, \qquad \frac{\partial g}{\partial y} = \lim_{\Delta y \to 0} \frac{g(x, y + \Delta y) - g(x, y)}{\Delta y}.$$

A geometrical interpretation of the above is shown in Fig. B.1. For a function $h(x, y, z)$ in a three-dimensional domain (x, y, z) we can have similar partial differentials.

Let $\{\boldsymbol{e}_i\}_{i=1}^3$ be an orthogonal basis in $\mathbb{R}^3$. The gradient of a scalar-valued function $h(\boldsymbol{x})$ is defined by

$$\nabla h = \frac{\partial h}{\partial x_i} \boldsymbol{e}_i$$

where $x_1 = x$, $x_2 = y$, $x_3 = z$ and Einstein's summation convention is used. The symbol ∇ is 'nabla', which is represented as

$$\nabla = \boldsymbol{e}_i \frac{\partial}{\partial x_i}.$$

Y. Ichikawa and A.P.S. Selvadurai, *Transport Phenomena in Porous Media*, DOI 10.1007/978-3-642-25333-1, © Springer-Verlag Berlin Heidelberg 2012

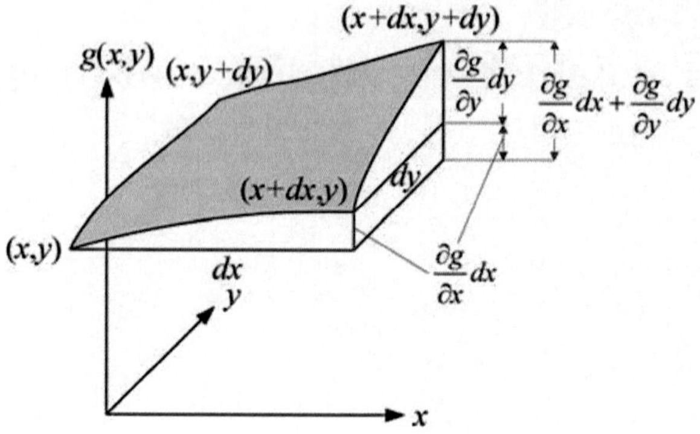

Fig. B.1 Geometrical interpretation of partial differentials

Let $\boldsymbol{h}(\boldsymbol{x}) = h_i \boldsymbol{e}_i$ be a vector-valued function, e.g., a displacement $\boldsymbol{u}(\boldsymbol{x})$. The gradient of $\boldsymbol{h}(\boldsymbol{x})$ is given as a second order tensor, and the components are written in a matrix form as

$$((\nabla \boldsymbol{h})_{ij}) = \begin{pmatrix} \partial h_x/\partial x & \partial h_x/\partial y & \partial h_x/\partial z \\ \partial h_y/\partial x & \partial h_y/\partial y & \partial h_y/\partial z \\ \partial h_z/\partial x & \partial h_z/\partial y & \partial h_z/\partial z \end{pmatrix}. \tag{B.1}$$

In (B.1) we used the right form, which is given in Note 2.2, p. 14.

The gradient $\nabla \boldsymbol{h}$ is uniquely decomposed into a symmetric part $(\nabla \boldsymbol{h})_S$ and an anti-symmetric part $(\nabla \boldsymbol{h})_A$ as follows:

$$\nabla \boldsymbol{h} = (\nabla \boldsymbol{h})_S + (\nabla \boldsymbol{h})_A, \tag{B.2}$$

$$(\nabla \boldsymbol{h})_S = \frac{1}{2}\Big[(\nabla \boldsymbol{h}) + (\nabla \boldsymbol{h})^T\Big], \qquad (\nabla \boldsymbol{h})_A = \frac{1}{2}\Big[(\nabla \boldsymbol{h}) - (\nabla \boldsymbol{h})^T\Big]. \tag{B.3}$$

The divergence of $\boldsymbol{h}(\boldsymbol{x})$ is introduced as a scalar function by

$$\nabla \cdot \boldsymbol{h} = \frac{\partial h_i}{\partial x_i}. \tag{B.4}$$

Note that the divergence decreases the order of the tensor, while the gradient increases its order.

The rotation of a vector-valued function $\boldsymbol{h}(x, y, z)$ is given by

$$\nabla \wedge \boldsymbol{h} = \left(\frac{\partial h_z}{\partial y} - \frac{\partial h_y}{\partial z}\right)\boldsymbol{e}_1 + \left(\frac{\partial h_x}{\partial z} - \frac{\partial h_z}{\partial y}\right)\boldsymbol{e}_2 + \left(\frac{\partial h_y}{\partial x} - \frac{\partial h_x}{\partial z}\right)\boldsymbol{e}_3. \tag{B.5}$$

B.2 Gauss-Green Theorem

Theorem B.1 (Gauss' Theorem). *Let S be the surface of $G \subset \mathbb{R}^3$, then we have*

$$\int_G \frac{\partial f}{\partial x_i} dv = \int_S f\, n_i\, ds \tag{B.6}$$

Proof. A domain G with a surface S_0, which is parallel to the z-axis is shown in Fig. B.2. The upper surface of G is S_1, and the lower one is S_2. The surfaces S_1 and S_2 are represented by the following equations:

$$z = \varphi_1(x, y), \quad \text{on } S_1; \qquad z = \varphi_2(x, y), \quad \text{on } S_2.$$

Let a projection of G onto the (x, y)-surface be A. We have a unit outward normal $\boldsymbol{n} = (n_1, n_2, n_3)$ at a point P on S_2, and an angle between $\boldsymbol{n}$ and the z-axis is assumed to be γ. Similarly, an angle γ' can be introduced on S_1. Then we have

$$dA = dx\, dy = \begin{cases} \cos\gamma\, ds, & \text{on } S_2 \\ \cos\gamma'\, ds, & \text{on } S_1 \end{cases}$$

and

$$n_3 = \begin{cases} \cos\gamma, & \text{on } S_2 \\ -\cos\gamma', & \text{on } S_1 \\ 0, & \text{on } S_0 \end{cases}$$

Since the total surface of G is designated as $S = S_0 + S_1 + S_2$, we have

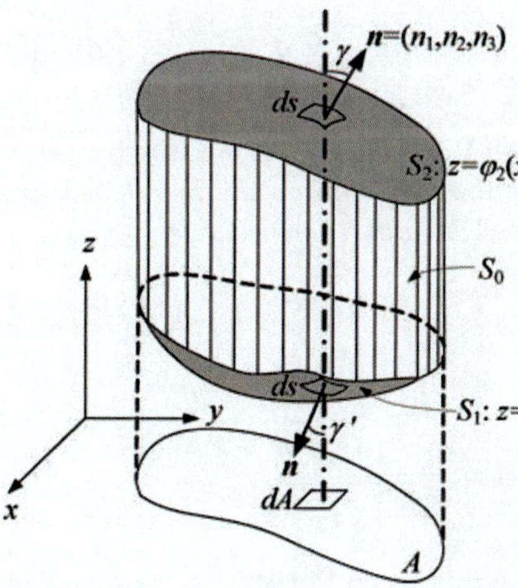

Fig. B.2 Gauss theorem

$$\begin{aligned}\int_G \frac{\partial f}{\partial z} dv &= \iint_A \left[\int_{\varphi_1(x,y)}^{\varphi_2(x,y)} \frac{\partial f}{\partial z} dz\right] dx\, dy \\ &= \iint_A \left[f(x,y,\varphi_2(x,y)) - f(x,y,\varphi_1(x,y))\right] dx\, dy \\ &= \int_{S_2} f \cos\gamma\, ds - \int_{S_1} f \cos\gamma'\, ds = \int_S f\, n_3\, ds.\end{aligned}$$

Here we have assumed that the function f is sufficiently smooth. We can obtain the same results for the x and y directions, so that we have Gauss' Theorem (B.6). ■

Theorem B.2 (Divergence Theorem) .

$$\int_G \nabla \cdot \boldsymbol{\psi}\, dv = \int_{\partial G} \boldsymbol{\psi} \cdot \boldsymbol{n}\, ds \tag{B.7}$$

Proof. In Gauss' Theorem (B.6), we set $f = \psi_i$, and prove (B.7). Note that $\partial\psi_i/\partial x_i = \nabla \cdot \boldsymbol{\psi}$. ■

Theorem B.3 (Green's Theorem) .

$$\int_G (\Delta u)\, v\, dv = \int_S v \frac{\partial u}{\partial n} ds - \int_G \nabla u \cdot \nabla v\, dv \tag{B.8}$$

where Δu is the Laplacian of the function u, and we have $\sum_i (\partial u/\partial x_i)\, n_i = \partial u/\partial n$.

Proof. We set $\psi_i = v\, \partial u/\partial x_i$, and by using $\nabla \cdot (v\, \nabla u) = (\Delta u)\, v + \nabla u \cdot \nabla v$, we can obtain (B.8). ■

We note that in the one-dimensional case, (B.8) is equivalent to the following integration-by-parts:

$$\int_{x_1}^{x_2} \frac{d^2 u}{dx^2} v\, dx = \left[\frac{du}{dx} v\right]_{x_1}^{x_2} - \int_{x_1}^{x_2} \frac{du}{dx}\frac{dv}{dx} dx.$$

Theorem B.4 (Gauss' Double Integral Theorem) . *Let $\boldsymbol{F} = F_1 \boldsymbol{i} + F_2 \boldsymbol{j}$ be a vector-valued function in the two-dimensional space, and $R \subset \mathbb{R}^2$ be a simply-connected domain. Then we have*

$$\int_R (\nabla \wedge \boldsymbol{F}) \cdot \boldsymbol{k}\, dx\, dy = \oint_C \boldsymbol{F} \cdot d\boldsymbol{r} \quad \text{or} \tag{B.9}$$

$$\int_R \left(\frac{\partial F_2}{\partial x} - \frac{\partial F_1}{\partial y}\right) dx\, dy = \oint_C (F_1\, dx + F_2\, dy) \tag{B.10}$$

where we set $\boldsymbol{r} = x\, \boldsymbol{i} + y\, \boldsymbol{j} + z\, \boldsymbol{k}$.

Proof. As shown in Fig. B.3, we assume that the domain R is given in each x and y direction as

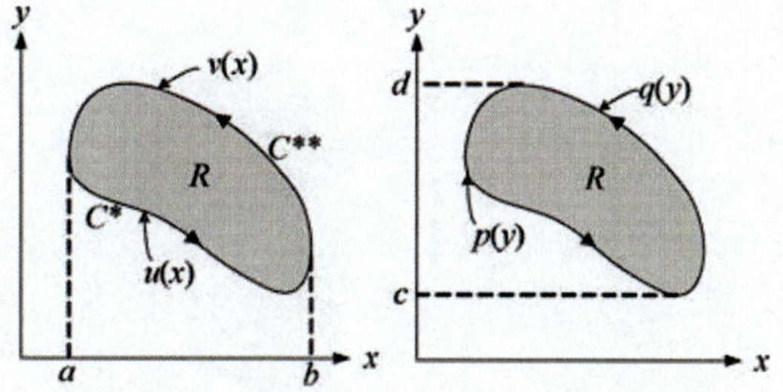

Fig. B.3 Integral path

$$a \le x \le b, \quad u(x) \le y \le v(x),$$
$$c \le y \le d, \quad p(y) \le x \le q(y).$$

Then we have

$$\int_R \frac{\partial F_1}{\partial y} dx\, dy = \int_a^b \left[\int_{u(x)}^{v(x)} \frac{\partial F_1}{\partial y} dy \right] dx = -\int_a^b F_1\,(x, u(x))\; dx - \int_b^a F_1\,(x, v(x))\; dx$$
$$= -\int_{C^*} F_1\,(x, u(x))\; dx - \int_{C^{**}} F_1\,(x, v(x))\; dx = -\oint F_1\, dx. \qquad \text{(B.11)}$$

where C^* and C^{**} are paths through the upper and lower parts of S, respectively. Similarly, we obtain

$$\int_R \frac{\partial F_2}{\partial x} dx\, dy = \oint F_2\, dy. \qquad \text{(B.12)}$$

By combining (B.11) and (B.12), we obtain (B.10). ■

B.3 Stokes' Theorem and Exact Differentiability

Let S be a curved surface in $\mathbb{R}^3$, and $\boldsymbol{r} = x\,\boldsymbol{i} + y\,\boldsymbol{j} + z\,\boldsymbol{k}$ be a position vector on the surface S. Since the surface S is two-dimensional, we can write the position vector $\boldsymbol{r}$ in terms of two parameters $(u,\, v)$, so that

$$\boldsymbol{r}(u,\, v) = x(u,\, v)\,\boldsymbol{i} + y(u,\, v)\,\boldsymbol{j} + z(u,\, v)\,\boldsymbol{k}$$

where $(u,\, v)$ is a coordinate system defined in a two-dimensional domain R, which is a projection of S onto a flat plane (see Fig. B.4a). A trajectory C on S is represented as $\boldsymbol{r}\,(u(t),\, v(t))$ where t is an arbitrary parameter.

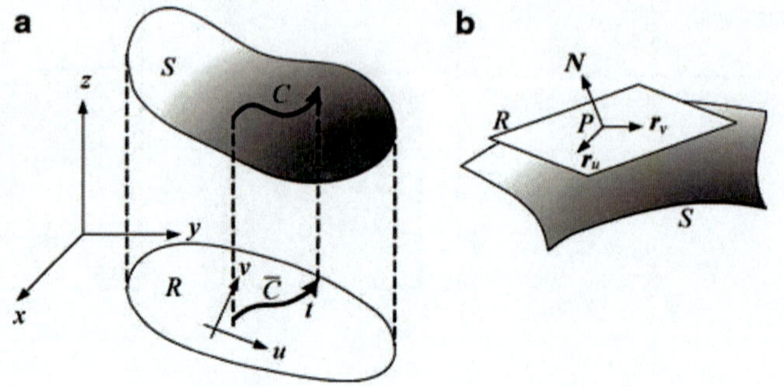

Fig. B.4 Curved surface and curved line in $\mathbb{R}^3$

A tangent vector of the trajectory C is obtained as

$$\frac{d\boldsymbol{r}}{dt} = \boldsymbol{r}_u \frac{du}{dt} + \boldsymbol{r}_v \frac{dv}{dt}, \qquad \boldsymbol{r}_u = \frac{\partial \boldsymbol{r}}{\partial u}, \quad \boldsymbol{r}_v = \frac{\partial \boldsymbol{r}}{\partial v}.$$

The vectors $\boldsymbol{r}_u$ and $\boldsymbol{r}_v$ span a tangential plane at a point P on the surface S (cf. Fig. B.4b). Therefore a vertical vector $\boldsymbol{N}$ at the point P and its normalized unit vector $\boldsymbol{n}$ are given as

$$\boldsymbol{N} = \boldsymbol{r}_u \wedge \boldsymbol{r}_v, \qquad \boldsymbol{n} = \frac{\boldsymbol{N}}{|\boldsymbol{N}|}.$$

Thus the integral of a vector-valued function $\boldsymbol{F}(x,y,z) = F_1(x,y,z)\,\boldsymbol{i} + F_2(x,y,z)\,\boldsymbol{j} + F_3(x,y,z)\,\boldsymbol{k}$ on S can be written as an integral on R:

$$\int_S \boldsymbol{F} \cdot d\boldsymbol{s} = \int_R \boldsymbol{F}(\boldsymbol{r}(u,v)) \cdot \boldsymbol{N}\, du\, dv \qquad (d\boldsymbol{s} = \boldsymbol{n}\, ds = \boldsymbol{N}\, du\, dv). \tag{B.13}$$

From the above results, we can prove the following Stokes' theorem.

Theorem B.5 (Stokes' Theorem) . *Let S be a sufficiently smooth surface in $\mathbb{R}^3$, and C a piece-wise smooth curve defined on S (cf. Fig. B.5). Then we have*

$$\int_S (\nabla \wedge \boldsymbol{F}) \cdot d\boldsymbol{s} = \oint_C \boldsymbol{F} \cdot d\boldsymbol{r}. \tag{B.14}$$

Proof. From (B.13) the l.h.s. term of (B.14) is written as

$$\int_S (\nabla \wedge \boldsymbol{F}) \cdot d\boldsymbol{s} = \int_R \left[\left(\frac{\partial F_3}{\partial y} - \frac{\partial F_2}{\partial z} \right) N_1 + \left(\frac{\partial F_1}{\partial z} - \frac{\partial F_3}{\partial x} \right) N_2 + \left(\frac{\partial F_2}{\partial x} - \frac{\partial F_1}{\partial y} \right) N_3 \right] du\, dv.$$

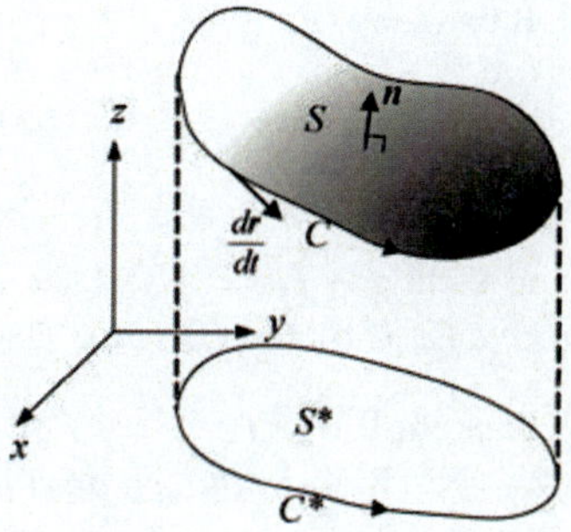

Fig. B.5 Integrating domain for Stokes' theorem

Therefore, if we can prove the following:

$$\int_R \left(\frac{\partial F_1}{\partial z} N_2 - \frac{\partial F_1}{\partial y} N_3 \right) du\, dv = \oint_{\bar{C}} F_1\, dx \tag{B.15a}$$

$$\int_R \left(\frac{\partial F_2}{\partial y} N_3 - \frac{\partial F_2}{\partial z} N_1 \right) du\, dv = \oint_{\bar{C}} F_2\, dy \tag{B.15b}$$

$$\int_R \left(\frac{\partial F_2}{\partial z} N_1 - \frac{\partial F_1}{\partial x} N_2 \right) du\, dv = \oint_{\bar{C}} F_3\, dz, \tag{B.15c}$$

it proves Stokes' theorem.

First, we will prove (B.15a). Let S be denoted as

$$z = f(x, y).$$

Assume that S^* is the projection of S onto the (x, y)-plane. In this case, we have $u = x$, $v = y$, so that $R = S^*$, $\bar{C} = C^*$, and from the position vector $\boldsymbol{r}$, we can decide the tangential vectors $\boldsymbol{r}_u$, $\boldsymbol{r}_v$ and the normal vector $\boldsymbol{N}$ as follows:

$$\boldsymbol{r}(u, v) = x\,\boldsymbol{i} + y\,\boldsymbol{j} + f(x, y)\,\boldsymbol{k},$$

$$\boldsymbol{r}_u = \boldsymbol{r}_x = \boldsymbol{i} + f_x\,\boldsymbol{k}, \quad \boldsymbol{r}_v = \boldsymbol{r}_y = \boldsymbol{j} + f_y\,\boldsymbol{k},$$

$$\boldsymbol{N} = \boldsymbol{r}_u \wedge \boldsymbol{r}_v = -f_x\,\boldsymbol{i} - f_y\,\boldsymbol{j} + \boldsymbol{k}.$$

Therefore the l.h.s. term of (B.15a) is written as

$$\int_{S^*} \left(\frac{\partial F_1}{\partial z}(-f_y) - \frac{\partial F_1}{\partial y} \right) dx\, dy.$$

On the other hand, if we set $F_2 = 0$ for (B.10), we have

$$\oint_{\bar{C}} F_1\, dx = \int_{S^*} -\frac{\partial F_1}{\partial y}\, dx\, dy.$$

Let us recall that

$$-\frac{\partial F_1(x,y,f(x,y))}{\partial y} = -\frac{\partial F_1}{\partial y} - \frac{\partial F_1}{\partial z} f_y,$$

and setting $-f_y = N_2$, $1 = N_3$, we obtain (B.15a). In a similar manner we can prove (B.15b) and (B.15c). Therefore, Stokes' theorem (B.14) is proved. ■

Theorem B.6 *Let $\boldsymbol{F} = F_1\,\boldsymbol{i} + F_2\,\boldsymbol{j} + F_3\,\boldsymbol{k}$ be a vector-valued function defined on $D \subset \mathbb{R}^3$. We introduce a path integral by*

$$\int_C \boldsymbol{F} \cdot d\boldsymbol{r} = \int_C (F_1\,dx + F_2\,dy + F_3\,dz)\,. \tag{B.16}$$

Then, the integral (B.16) has a path-independent value if and only if there exists a scalar function f such that

$$\boldsymbol{F} = \nabla f. \tag{B.17}$$

Proof. ($\Leftarrow$) Let us assume that there is a function f defined on D, which satisfies (B.17). In the domain D we introduce a path C with its parameter t, which moves from a point A to a point B. If the path C is represented as

$$\boldsymbol{r}(t) = x(t)\,\boldsymbol{i} + y(t)\,\boldsymbol{j} + z(t)\,\boldsymbol{k}, \qquad a \le t \le b,$$

the integral of (B.16) can be calculated as

$$\int_A^B \boldsymbol{F} \cdot d\boldsymbol{r} = \int_A^B \nabla f \cdot d\boldsymbol{r} = \int_a^b \frac{df}{dt} dt = f(B) - f(A),$$

which shows that the integral is path-independent.

($\Rightarrow$) Assume that the integral (B.16) is path-independent. For a given point $A(x_0, y_0, z_0) \in D$ we choose a point $B(x, y, z) \in D$, and we define f as

$$f(x,y,z) = f_0 + \int_A^B \boldsymbol{F} \cdot d\boldsymbol{r}^*, \qquad d\boldsymbol{r}^* = dx^*\boldsymbol{i} + dy^*\boldsymbol{j} + dz^*\boldsymbol{k} \tag{B.18}$$

where f_0 is an arbitrary constant. Since the integral (B.18) is path-independent from the above assumption, we change the integration path into $A(x_0, y_0, z_0) \to B_1(x_1, y, z) \to B(x, y, z)$ such that along the first path $A(x_0, y_0, z_0) \to B_1(x_1, y, z)$ we vary x from x_0 to x_1 under the constant y and z, and along the second path $B_1(x_1, y, z) \to B(x, y, z)$ we vary y and z under the constant $x = x_1$. Since y and z are constant along the first path, we have $F_2\,dy = F_3\,dz = 0$. Similarly, since x is constant along the second path, we have $F_1\,dx = 0$. Therefore, we obtain

$$f(x, y, z) = f_0 + \int_A^{B_1} F_1 \, dx^* + \int_{B_1}^{B} (F_2 \, dy^* + F_3 \, dz^*). \tag{B.19}$$

By operating a differentiation $\partial / \partial x$ to (B.19), the result is

$$\frac{\partial f}{\partial x} = F_1.$$

Similar results can be obtained for the y and z directions. ■

Theorem B.7 (Integrability Condition) . *The integral (B.16) is path-independent if and only if we have*

$$\nabla \wedge \boldsymbol{F} = \mathbf{0}. \tag{B.20}$$

Proof. (⇒) Assume that the integral (B.16) is path-independent. Then there exists a scalar function f due to Theorem B.6. Therefore we have

$$\nabla \wedge \boldsymbol{F} = \nabla \wedge (\nabla f) \equiv \mathbf{0}.$$

(⇐) Assume that $\nabla \wedge \boldsymbol{F} = \mathbf{0}$. If the domain D is simply connected and C is an arbitrary path in D, then due to Stokes' theorem (B.14), we have

$$\oint_C \boldsymbol{F} \cdot d\boldsymbol{r} = \int_S (\nabla \wedge \boldsymbol{F}) \cdot d\boldsymbol{s} = 0.$$

Thus, the integral (B.16) is path-independent. ■

Appendix C
A Summary of Linear Vector Spaces

Here we present a general concept of the vector space in which the n-dimensional real number space $\mathbb{R}^n$ is an example. The central idea of the vector space is to introduce a 'vector addition' and a 'scalar multiplication'. Further expositions are given by Protter and Morrey (1977), Friedman (1982) and Keener (1988).

C.1 Algebraic System

Let $\mathcal{S} = \{ G, * \}$ be an algebraic system where G is a set and $*$ is an operation. We can then introduce the following hierarchy for $\mathcal{S}$:

1. $\mathcal{S} = \{ G, * \}$: Algebraic system (G: a set, $*$: an operation)
   - Groupoid: the operation $*$ is closed; i.e., $a * b \in G, \quad \forall a, b \in G$
   - Semi-group: (the above +) $*$ is associative; i.e., $a * (b * c) = (a * b) * c, \quad \forall a, b, c \in G$
   - Monoid: (the above +) $\exists$ an identity element e s.t. $a * e = e * a = a, \quad \forall a \in G$
   - Group: (the above +) $\forall\, a \in G\ \exists$ an inverse element b s.t. $a * b = b * a = e$
   - Abelian group: (the above +) $*$ is commutative; i.e., $a * b = b * a, \quad \forall a, b \in G$
2. $\mathcal{S} = \{ G, +, * \}$: Algebraic system ($G$: a set, $+$ & $*$: operations)

(2.a) Ring

(1) $\{ G, + \}$ is an Abelian group,
(2) $\{ G, * \}$ is a semi-group,
(3) $*$ is distributive w.r.t. $+$;

Y. Ichikawa and A.P.S. Selvadurai, *Transport Phenomena in Porous Media*,
DOI 10.1007/978-3-642-25333-1, © Springer-Verlag Berlin Heidelberg 2012

$$a * (b + c) = a * b + a * c, \quad \forall a,\, b,\, c \in G$$
$$(a + b) * c = a * c + b * c, \quad \forall a,\, b,\, c \in G$$

Example 1. Set of integers $\{\mathbb{Z},\ +,\ \times\}$
Example 2. Set of all n-th order polynomials $P^n[f]$ where the coefficients are in a field $\mathcal{K}$.

$$P^n[f] = \left\{ f : \ f(x) = a_0 + a_1 \times x + \cdots + a_n \times x^n;\ a_0, a_1, \cdots, a_n \in \mathcal{K} \right\}$$

(2.b) Field

(1) $\{G,\ +\}$ is an Abelian group,
(2) $\{G,\ *\}$ is an Abelian group,
(3) $*$ is distributive w.r.t. $+$;

$$a * (b + c) = a * b + a * c, \quad \forall a,\, b,\, c \in G$$
$$(a + b) * c = a * c + b * c, \quad \forall a,\, b,\, c \in G$$

Example 3. Set of rational numbers $\mathbb{Q}$, set of real numbers $\mathbb{R}$, set of complex numbers $\mathbb{C}$. The addition $+$ and multiplication $\times$ follow the known operation rule.

C.2 A Linear Vector Space $\mathcal{V} = \{X, +, *\}$

A linear vector space (LVS) is a mathematical system where the following operations of a vector addition and a scalar multiplication are defined with a scalar field $\mathcal{K}$:

1. Vector addition $+$:

(1-1) Commutative; $\boldsymbol{u} + \boldsymbol{v} = \boldsymbol{v} + \boldsymbol{u}, \quad \forall \boldsymbol{u}, \boldsymbol{v} \in X$
(1-2) Associative; $\boldsymbol{u} + (\boldsymbol{v} + \boldsymbol{w}) = (\boldsymbol{u} + \boldsymbol{v}) + \boldsymbol{w}, \quad \forall \boldsymbol{u}, \boldsymbol{v}, \boldsymbol{w} \in X$
(1-3) Zero element; $\exists !\, \boldsymbol{0} \in X$ s.t. $\boldsymbol{u} + \boldsymbol{0} = \boldsymbol{u}, \quad \forall \boldsymbol{u} \in X$
(1-4) Inverse element; $\exists -\boldsymbol{u} \in X$ s.t. $\boldsymbol{u} + (-\boldsymbol{u}) = \boldsymbol{0}, \quad \forall \boldsymbol{u} \in X$

2. Scalar multiplication $*$:

(2-1) For $1 \in \mathcal{K}$, $1 * \boldsymbol{u} = \boldsymbol{u}, \quad \forall \boldsymbol{u} \in X$
(2-2) $(l \times m) * \boldsymbol{u} = l * (m * \boldsymbol{u}), \quad \forall \boldsymbol{u} \in X, \quad \forall l, m \in \mathcal{K}$
(2-3) $(l + m) * \boldsymbol{u} = l * \boldsymbol{u} + m * \boldsymbol{u}, \quad \forall \boldsymbol{u} \in X, \quad \forall l, m \in \mathcal{K}$
(2-4) $l * (\boldsymbol{u} + \boldsymbol{v}) = l * \boldsymbol{u} + l * \boldsymbol{v}, \quad \forall \boldsymbol{u}, \boldsymbol{u} \in X, \quad \forall l \in \mathcal{K}$

Example 1. Set of all n-th order polynomials:

$$P^n[f] = \left\{ f : \ f(x) = a_0 + a_1 \times x + \cdots + a_n \times x^n;\ a_0, a_1, \cdots, a_n \in \mathcal{K} \right\}$$

Example 2. Set of all $(m \times n)$-matrices:

$$M^{(m,n)} = \left\{ A : A = \left(a_{ij}\right),\ i = 1, \ldots, m;\ j = 1, \ldots, n \right\}$$

Example 3. Set of all linear transformations from a LVS $\mathcal{U} = \{X, +, *\}$ to a LVS $\mathcal{V} = \{Y, +, *\}$:

$$\mathcal{L}(\mathcal{U}, \mathcal{V}) = \Big\{ T : v = T(u)\ \ \forall u \in X,\ \forall v \in Y \text{ s.t.}$$
$$T(u_1 + u_2) = T(u_1) + T(u_2),\ T(ku) = kT(u),\ \ u, u_1, u_2 \in X,\ k \in \mathcal{K} \Big\}$$

Example 4. Set of all continuous functions $C(0,\ 1)$ defined on $(0, 1)$:
The vector addition is defined by $f + g \ \Leftarrow\ (f + g)(x) = f(x) + g(x)$
The scalar multiplication is defined by $k * f \ \Leftarrow\ (k * f)(x) = k(f(x))$

C.3 Basis and Dimension

A set $\{ \boldsymbol{e}_i \}_{i=1}^{n}$ is a *(Hamel) basis* of $\mathcal{V}$ if

1. $\{ \boldsymbol{e}_i \}$ are linearly independent,[1] i.e., $a_1 \boldsymbol{e}_1 + a_2 \boldsymbol{e}_2 + \cdots + a_n \boldsymbol{e}_n = \boldsymbol{0}\ \ \forall a_1, a_2, \cdots,$ $a_n \in \mathcal{K}$ iff $a_1 = a_2 = \cdots = a_n = 0$, and
2. $\{ \boldsymbol{e}_i \}$ spans $\mathcal{V}$, i.e., all vectors $\boldsymbol{v} \in \mathcal{V}$ can be represented by a linear combination of $\{ \boldsymbol{e}_i \}$ $(\boldsymbol{v} = v_1 \boldsymbol{e}_1 + v_2 \boldsymbol{e}_2 + \cdots + v_n \boldsymbol{e}_n)$.[2]

A basis $\{ \boldsymbol{e}_i \}$ is *orthonormal* if

$$\left\langle \boldsymbol{e}_i, \boldsymbol{e}_j \right\rangle = \delta_{ij} = \begin{cases} 1, & \text{if } i = j, \\ 0, & \text{if } i \neq j \end{cases} \tag{C.1}$$

where $\left\langle \boldsymbol{e}_i, \boldsymbol{e}_j \right\rangle = \boldsymbol{e}_i \cdot \boldsymbol{e}_j$ is an inner product between $\boldsymbol{e}_i$ and $\boldsymbol{e}_j$, and δ_{ij} is Kronecker's delta (note that the inner product is defined later).

By using the basis $\{ \boldsymbol{e}_i \}$, coefficients v_i of a vector $\boldsymbol{v} = \sum v_i \boldsymbol{e}_i$ can be obtained as

$$v_i = \boldsymbol{v} \cdot \boldsymbol{e}_i . \tag{C.2}$$

The number n of the basis $\{ \boldsymbol{e}_i \}$ is referred to as the *dimension*, which is denoted as $n = \dim \mathcal{V}$. Note that if $m = \dim \mathcal{U}$ and $n = \dim \mathcal{V}$, the dimension of the set of all linear transformations $\mathcal{L}(\mathcal{U}, \mathcal{V})$ is $m \times n$; i.e., $\dim \mathcal{L}(\mathcal{U}, \mathcal{V}) = m \times n$.

---

[1] If the set is not linearly independent, it is linearly dependent. Note that $a_1 \boldsymbol{e}_1 + a_2 \boldsymbol{e}_2 + \cdots + a_n \boldsymbol{e}_n$ is referred to as a linear combination of $\{ \boldsymbol{e}_i \}$.

[2] For a finite dimensional case $(n < \infty)$, the above discussion is obvious. For an infinite dimensional case we need a further discussion of convergence of the coefficients $a_1, a_2, \cdots$ under a proper norm topology (cf. Appendix C.8).

C.4 Topological Spaces

We introduce a *topology* in a LVS in order to distinguish one element from others. The topology constitutes the most general class, and hierarchically a *metric*, a *norm* and an *inner product* are introduced.

1. Topological space $(X, \mathcal{T})$

   A class $\mathcal{T}$ of a set X[3] is referred to as a *topology* if the following are satisfied:

(t-1) $X \in \mathcal{T}$, and the empty set $\emptyset \in \mathcal{T}$.
(t-2) $G_1 \in \mathcal{T},\ \ G_2 \in \mathcal{T}\ \Rightarrow\ G_1 \cap G_2 \in \mathcal{T}$.
(t-3) For any class $\{ G_i \}$ that consists of elements of $\mathcal{T}$, we have $\cup_i G_i \in \mathcal{T}$.

2. Metric space (X, d)

(m-1) $d(x, y) = 0$ iff $x = y,\ \ \forall x, y \in X$
(m-2) $d(x, y) = d(y, x),\ \ \forall x, y \in X$
(m-3) $d(x, z) \leq d(x, y) + d(y, z),\ \ \forall x, y, z \in X$: triangular inequality
(m-4) $d(x, y) > 0,\ \ \forall x \neq y \in X$

3. Normed linear space $(X, \| \cdot \|)$

(n-1) $\| x \| \geq 0$ where $\| x \| = 0$ iff $x = 0$
(n-2) $\| \alpha x \| = |\alpha| \| x \|,\ \ \forall x \in X, \alpha$: a scalar number
(n-3) $\| x + y \| \leq \| x \| + \| y \|,\ \ \forall x, y \in X$: triangular inequality

4. Inner product space $(X, (\cdot, \cdot))$

(i-1) $(x, x) \geq 0$ where $(x, x) = 0$ iff $x = 0$
(i-2) $(x, y) = \overline{(y, x)}^*,\ \ \forall x, y \in X$ where $\overline{(y, x)}^*$ is the complex conjugate of (y, x).
(i-3) $(\lambda x, y) = \lambda^*(x, y),\ \ \forall x, y \in X, \lambda$: a scalar number where λ^* is the complex conjugate of λ.
(i-4) $(x, y + z) = (x, y) + (y, z),\ \ \forall x, y, z \in X$: Pythagoras' Theorem

Proof of $(X, d) \supset (X, \| \cdot \|)$ if $d(x, y) = \| x - y \|$:

(m-1) $d(x, y) = \| x - y \| = 0\ \Rightarrow\ x - y = 0$ i.e. $x = y$
(m-2) $d(x, y) = \| x - y \| = \| (-1)(y - x) \| = |-1| \| y - x \| = \| y - x \| = d(y, x)$
(m-3) $d(x, z) = \| x - z \| = \| x - y + y - z \| \leq \| x - y \| + \| y - z \| = d(x, y) + d(y, z)$
(m-4) $d(x, y) = \| x - y \| > 0,\ \ \forall x \neq y \in X$

Proof of $(X, \| \cdot \|) \supset (X, (\cdot, \cdot))$ if $\| x \|^2 = (x, x)$

[3]A class or family is a set of sets.

(n-1) Trivial
(n-2) $\| \alpha x \|^2 = (\alpha x, \alpha x) = \alpha^* \alpha (x, x) = |\alpha|^2 \| x \|^2$
(n-3) $\| x + y \|^2 = (x + y, x + y) = (x, x) + (y, x) + (x, y) + (y, y) \leq \| x \|^2 + 2\| x \| \| y \| + \| y \|$
where we applied the Schwartz inequality; i.e., $(x, y) \leq \| x \| \, \| y \|$.

Proof of the Schwartz inequality:
For $y \neq 0$, we have $\| x - \lambda y \|^2 = (x - \lambda y, x - \lambda y) \geq 0 \;\; \forall \lambda$

$$\Rightarrow \quad (x, x) - \lambda^*(y, x) - \lambda[(x, y) - \lambda^*(y, y)] \geq 0$$

Since λ is arbitrary, we can set $\lambda^* = (x, y)/(y, y)$, then the term [ ] of the above equation becomes null.

Thus, we have

$$\| x \|^2 - \frac{(y, x)\,(x, y)}{\| y \|^2} = \| x \|^2 - \frac{(x, y)^2}{\| y \|^2} \geq 0 \;\Rightarrow\; \| x \|^2 \, \| y \|^2 \geq (x, y)^2 \quad \blacksquare$$

C.5 Cauchy Sequence and Complete Space

A sequence $\{x_n\}$ in a metric space (X, d) is referred to as a *Cauchy sequence* or *fundamental sequence* if $d(x_m, x_n) \to 0$ for $m, n \to \infty$. A convergent sequence is Cauchy,[4] however the converse is not always true.

A metric space (X, d) is *complete* if any Cauchy sequence in (X, d) is convergent. The space of real numbers $\mathbb{R}$ is complete under $d(x, y) = |\, x - y \,|$, while the space of rational numbers $\mathbb{Q}$ is not complete under the same metric, since, e.g., $e = 1+1+1/2!+1/3!+\cdots$ (i.e., the l.h.s. term is an irrational number, while the r.h.s. terms are rational numbers).

An operation to add all the limits of Cauchy sequences to (X, d) is referred to as a *completion*. We recall that the space of real numbers $\mathbb{R}$ is a completion of the space of rational numbers $\mathbb{Q}$ under the uniform norm $| \cdot |$.

A complete normed space is referred to as a *Banach space*, and a complete inner product space is a *Hilbert space*.

C.6 Functionals and Dual Space

A set of all linear transformations $\mathcal{L}(\mathcal{U}, \mathbb{R})$ from a LVS $\mathcal{U} = \{X, +, *\}$ into the real number space $\mathbb{R}$ is referred to as a *dual space*, which is denoted as $\mathcal{U}^*$. An element

[4] A sequence $\{ x_m \}$ in (X, d) converges (strongly) to $x \in X$ if $d(x_m, x) \to 0$ for $n \to \infty$; i.e., $\forall \varepsilon > 0, \; \exists\, N > 0$ s.t. $d(x_n, x) < \varepsilon \;\, \forall n > N$. We have $d(x_m, x_n) \leq d(x_m, x) + d(x_n, x) \to 0$ for a convergent sequence, which implies that the convergent sequence is Cauchy.

f in $\mathcal{U}^*$ is called a *functional*. We have $f(\boldsymbol{u}) \in \mathbb{R}$ for $\boldsymbol{u} \in \mathcal{U}$, then we emphasize the fact that there exists a vector $\boldsymbol{f} \in \mathcal{U}^*$ and write

$$f(\boldsymbol{u}) = \langle \boldsymbol{f}, \boldsymbol{u} \rangle \in \mathbb{R} \tag{C.3}$$

where $\langle \cdot, \cdot \rangle : \mathcal{U}^* \times \mathcal{U} \to \mathbb{R}$ is a bilinear map such that

$$\langle \alpha \boldsymbol{f}_1 + \beta \boldsymbol{f}_2, \boldsymbol{u} \rangle = \alpha \langle \boldsymbol{f}_1, \boldsymbol{u} \rangle + \beta \langle \boldsymbol{f}_2, \boldsymbol{u} \rangle,$$
$$\langle \boldsymbol{f}, \alpha \boldsymbol{u}_1 + \beta \boldsymbol{u}_2 \rangle = \alpha \langle \boldsymbol{f}, \boldsymbol{u}_1 \rangle + \beta \langle \boldsymbol{f}, \boldsymbol{u}_2 \rangle.$$

Let $\mathcal{U}$ be a finite dimensional LVS and its dimension is n ($n = \dim \mathcal{U}$), then the dimension of $\mathcal{U}^*$ is also n ($n = \dim \mathcal{U}^*$).

If $\{ \boldsymbol{e}_i \}$ is a basis of $\mathcal{U}$ (referred to as a *covariant basis*), a basis of $\mathcal{U}^*$, denoted as $\{ \boldsymbol{e}^i \}$, can be given as

$$\langle \boldsymbol{e}^i, \boldsymbol{e}_j \rangle = \delta^i_j = \begin{cases} 1, & \text{if } i = j \\ 0, & \text{if } i \neq j. \end{cases} \tag{C.4}$$

$\{ \boldsymbol{e}^i \}$ is referred to as a *dual* or *contravariant basis* of $\{ \boldsymbol{e}_i \}$.

Including infinite dimensional cases, we can introduce a norm $\| \cdot \|_*$ on $\mathcal{U}^*$ as

$$\| \boldsymbol{f} \|_* = \sup \left\{ \frac{\langle \boldsymbol{f}, \boldsymbol{u} \rangle}{\| \boldsymbol{u} \|}, \boldsymbol{u} \neq \boldsymbol{0} \right\} = \sup_{\|\boldsymbol{v}\| \leq 1} \langle \boldsymbol{f}, \boldsymbol{v} \rangle \tag{C.5}$$

where $\| \cdot \|$ is a norm on $\mathcal{U}$.

C.7 Tensor Product and Tensor Space

Let $\mathcal{V}$ and $\mathcal{V}^*$ be a finite dimensional LVS and its dual space, which have a covariant basis $\{ \boldsymbol{e}_i \}_{i=1}^n$ (i.e., the basis of $\mathcal{V}$) and a contravariant basis $\{ \boldsymbol{e}^i \}_{i=1}^n$ (i.e., the basis of $\mathcal{V}^*$). Henceforth we use Einstein's summation convention.[5]

1. What is a tensor?

   If a bilinear map $\boldsymbol{\alpha}(\boldsymbol{u}, \boldsymbol{v})$ is defined for $\boldsymbol{u}, \boldsymbol{v} \in \mathcal{V}$, $\boldsymbol{\alpha}$ is referred to as a *second-order covariant tensor*. If $\boldsymbol{u} = u^i \boldsymbol{e}_i$, and $\boldsymbol{v} = v^j \boldsymbol{e}_j$, we have

$$\boldsymbol{\alpha}(\boldsymbol{u}, \boldsymbol{v}) = \boldsymbol{\alpha}(u^i \boldsymbol{e}_i, v^j \boldsymbol{e}_j) = u^i v^j \boldsymbol{\alpha}(\boldsymbol{e}_i, \boldsymbol{e}_j).$$

[5] Here we extend Einstein's summation convention for the three-dimensional space, which is shown in Appendix A.1. A vector in an n-dimensional space is represented as $\boldsymbol{u} = \sum_{i=1}^n u^i \boldsymbol{e}_i$, which can be denoted as $\boldsymbol{u} = u^i \boldsymbol{e}_i$ under the summation convention where i is a dummy index (appearing twice in a term), which implies summing from 1 to n. A free index appears once in a term, which implies 1 or 2, $\cdots$, or n.

We set $\boldsymbol{\alpha}(\boldsymbol{e}_i, \boldsymbol{e}_j) = \alpha_{ij}$, which represents coefficients of the second-order covariant tensor $\boldsymbol{\alpha}$.

2. Tensor product

For given $\boldsymbol{\xi}, \boldsymbol{\eta} \in \mathcal{V}^*$, we define a *tensor product* $\boldsymbol{\xi} \otimes \boldsymbol{\eta}$ as

$$\boldsymbol{\alpha}(\boldsymbol{u}, \boldsymbol{v}) = \xi(\boldsymbol{u}) \cdot \eta(\boldsymbol{v}) \equiv (\boldsymbol{\xi} \otimes \boldsymbol{\eta})(\boldsymbol{u}, \boldsymbol{v}) \quad \forall \boldsymbol{u}, \boldsymbol{v} \in \mathcal{V} \tag{C.6}$$

where $\boldsymbol{\alpha}$ is a corresponding covariant tensor. For $\boldsymbol{u} = u^i \boldsymbol{e}_i$, $\boldsymbol{v} = v^i \boldsymbol{e}_i$, $\boldsymbol{\xi} = \xi_i \boldsymbol{e}^i$, $\boldsymbol{\eta} = \eta_i \boldsymbol{e}^i$, we have

$$\xi(\boldsymbol{u}) \cdot \eta(\boldsymbol{v}) = \xi_i\, u^i\, (\eta_j\, v^j) = \xi_i\, \eta_j\, (u^i\, v^j)$$

where $\xi_i\, \eta_j$ represents coefficients of the covariant tensor $\boldsymbol{\alpha} = \boldsymbol{\xi} \otimes \boldsymbol{\eta}$.

If we use a contravariant basis $\{\boldsymbol{e}^i\}_{i=1}^n$, which provides the basis for $\boldsymbol{\xi}, \boldsymbol{\eta}$, a covariant tensor $\boldsymbol{\alpha}$ can be written as

$$\boldsymbol{\alpha} = \alpha_{ij}\, \boldsymbol{e}^i \otimes \boldsymbol{e}^j. \tag{C.7}$$

This shows that a covariant tensor is introduced on a contravariant basis $\{\boldsymbol{e}^i\}_{i=1}^n$ of the dual space, while a contravariant tensor $\boldsymbol{\alpha} = \alpha^{ij}\, \boldsymbol{e}_i \otimes \boldsymbol{e}_j$ is given on a covariant basis $\{\boldsymbol{e}_i\}_{i=1}^n$, which is defined by $\boldsymbol{\alpha}(\boldsymbol{\xi}, \boldsymbol{\eta})$ for $\boldsymbol{\xi}, \boldsymbol{\eta} \in \mathcal{V}^*$. Mixed tensors corresponding to $\boldsymbol{\alpha}(\boldsymbol{\xi}, \boldsymbol{u})$ and $\boldsymbol{\alpha}(\boldsymbol{u}, \boldsymbol{\xi})$ are similarly introduced.

3. Linear transformation and mixed tensor

Let $\boldsymbol{T} : \mathcal{V} \to \mathcal{V}$ $(\boldsymbol{u} \in \mathcal{V}, \boldsymbol{\xi} \in \mathcal{V}^*)$ be a linear transformation. There is a one-to-one correspondence between a mixed tensor $\boldsymbol{\alpha}$ and a linear transformation $\boldsymbol{T}$ such that

$$\boldsymbol{\alpha}(\boldsymbol{\xi}, \boldsymbol{u}) = \langle \boldsymbol{\xi}, \boldsymbol{T}(\boldsymbol{u}) \rangle = \xi_i\, T^i{}_j\, u^j \tag{C.8}$$

where $\langle \cdot, \cdot \rangle$ is a bilinear map. Then $\boldsymbol{T}$ can be written as

$$\boldsymbol{T} = T^i_j\, \boldsymbol{e}_i \otimes \boldsymbol{e}^j.$$

4. Tensor space

We can define a vector addition for two covariant tensors $\boldsymbol{\alpha}$ and $\boldsymbol{\beta}$, and a scalar multiplication between a scalar λ (for simplicity, a real number) and a covariant tensor $\boldsymbol{\alpha}$ as follows:

$$(\boldsymbol{\alpha} + \boldsymbol{\beta})(\boldsymbol{u}, \boldsymbol{v}) = \boldsymbol{\alpha}(\boldsymbol{u}, \boldsymbol{v}) + \boldsymbol{\beta}(\boldsymbol{u}, \boldsymbol{v}),$$
$$(\lambda\, \boldsymbol{\alpha})(\boldsymbol{u}, \boldsymbol{v}) = \lambda\, (\boldsymbol{\alpha}(\boldsymbol{u}, \boldsymbol{v}))\,.$$

Then a set of all covariant tensors is a LVS where a zero element and an inverse element are given as

$$\textbf{0-element}: \quad \mathbf{0}(\boldsymbol{u}, \boldsymbol{v}) = \mathbf{0},$$
$$\text{Inverse element}: \quad (-\boldsymbol{\alpha})(\boldsymbol{u}, \boldsymbol{v}) = -\boldsymbol{\alpha}(\boldsymbol{u}, \boldsymbol{v}).$$

Similarly, we can introduce a space of contravariant tensors or mixed tensors.

C.8 On Completeness of Function Spaces

We first show that the set of all continuous functions C$[a, b]$ is complete only under the uniform norm $\| x \|_\infty = \sup_{t \in T} | x(t) |$. Then we prove that the Fourier basis $\{\phi_n\}$ is complete for periodic functions in the Lebesgue integrable space $L^2[-L/2, L/2]$[6]; i.e., any function f in this space can be represented as $\sum_{n=-\infty}^{\infty} \hat{f}_n^* \phi_n = f$.

C.8.1 Sequence Space l^p

A vector $\boldsymbol{u} = (u_1, u_2, u_3, \ldots)$ belongs to the sequence space l^p if

$$\| \boldsymbol{u} \| = \left(\sum_{i}^{\infty} | u_i |^p \right)^{1/p} < \infty. \tag{C.9}$$

The sequence space l^p is a metric space since the metric axioms presented in Appendix C.4 are satisfied as follows:

(m-1) $d(x, y) = 0$ iff $x = y$: Trivial
(m-2) $d(x, y) = d(y, x)$: Trivial
(m-3) $d(x, z) \le d(x, y) + d(y, z)$: We will prove this in the following Step (a-1)–Step (a-4).
(m-4) $d(x, y) > 0$: Trivial

Step (a-1) We have the following auxiliary theorem:

$$\alpha \beta \le \frac{\alpha^p}{p} + \frac{\beta^q}{q} \qquad \forall \alpha, \beta > 0 \tag{C.10}$$

where p and q are real numbers, which satisfy

[6] A function f belongs to $L^p[\Omega]$ if

$$\| f \|_p < \infty, \qquad \| f \|_p = \left(\int_\Omega | f |^p dx \right)^{1/p}$$

where the integral is given in the sense of Lebesgue.

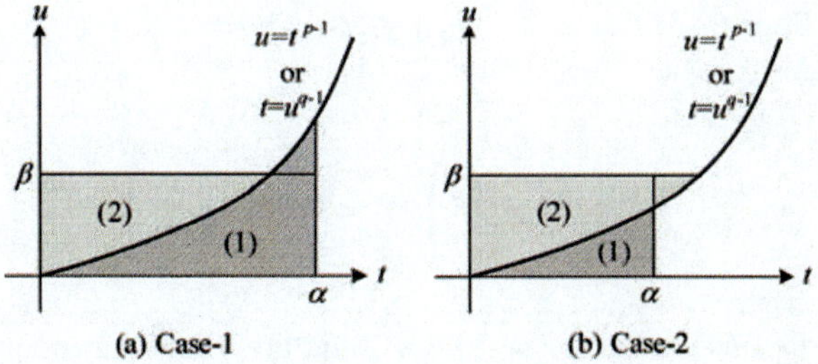

Fig. C.1 Auxiliary inequality

$$\frac{1}{p}+\frac{1}{q}=1 \quad \Rightarrow \quad (p-1)(q-1)=1. \tag{C.11}$$

Proof. Let $u = t^{p-1} \Rightarrow t = u^{q-1}$. Referring to Fig. C.1, we understand that the rectangular area $\alpha\beta$ is always smaller than the sum of the area (1) and the area (2). This results in

$$\alpha\beta \le \int_0^\alpha t^{p-1}\,dt + \int_0^\beta u^{q-1}\,du = \frac{\alpha^p}{p}+\frac{\beta^q}{q}. \qquad \blacksquare$$

Step (a-2) Let $x = (\xi_j) \in l^p$, $y = (\eta_j) \in l^q$. Then we have the following Hölder's inequality:

$$\sum_{j=1}^{\infty} |\, \xi_j \eta_j \,| \le \left(\sum_{j=1}^{\infty} |\, \xi_j \,|^p\right)^{1/p} \left(\sum_{j=1}^{\infty} |\, \eta_j \,|^q\right)^{1/q}. \tag{C.12}$$

Proof. We put

$$\tilde{\xi}_j = \frac{\xi_j}{(\sum |\, \xi_k \,|^p)^{1/p}}, \qquad \tilde{\eta}_j = \frac{\eta_j}{(\sum |\, \eta_k \,|^q)^{1/q}} \tag{C.13}$$

where $\sum |\, \tilde{\xi}_k \,|^p = 1$, $\sum |\, \tilde{\eta}_k \,|^q = 1$. From (C.10) we have

$$|\, \tilde{\xi}_j \tilde{\eta}_j \,| \le \frac{1}{p} |\, \tilde{\xi}_j \,|^p + \frac{1}{q} |\, \tilde{\eta}_j \,|^q.$$

By summing up with respect to the index j, we have

$$\sum_{j=1}^{\infty} |\, \tilde{\xi}_j \tilde{\eta}_j \,| \le \sum_{j=1}^{\infty} \frac{1}{p} |\, \tilde{\xi}_j \,|^p + \sum_{j=1}^{\infty} \frac{1}{q} |\, \tilde{\eta}_j \,|^q = \frac{1}{p}+\frac{1}{q} = 1.$$

By using (C.13), we obtain Hölder's inequality (C.12). Note that the inequality for $p = q = 2$ in (C.12) is referred to as the Cauchy-Schwartz inequality. $\blacksquare$

Step (a-3) Let $x = (\xi_j) \in l^p$, $y = (\eta_j) \in l^q$. Then we have the following Minkowski's inequality:

$$\left(\sum_{j=1}^{\infty} |\xi_j + \eta_j|^p\right)^{1/p} \leq \left(\sum_{j=1}^{\infty} |\xi_j|^p\right)^{1/p} + \left(\sum_{j=1}^{\infty} |\eta_j|^q\right)^{1/q}. \tag{C.14}$$

Proof. We put $\omega_j = \xi_j + \eta_j$, and the triangular inequality for real numbers ξ_j and η_j results in

$$|\omega_j|^p = |\xi_j + \eta_j|^p |\omega_j|^{p-1} \leq (|\xi_j| + |\eta_j|) |\omega_j|^{p-1}.$$

We add up j from 1 to a finite integer n, and we have

$$\sum_{j=1}^{n} |\omega_j|^p \leq \sum_{j=1}^{n} |\xi_j||\omega_j|^{p-1} + \sum_{j=1}^{n} |\eta_j||\omega_j|^{p-1}$$

$$\leq \left(\sum_{j=1}^{n} |\xi_j|^p\right)^{1/p} \left(\sum_{j=1}^{n} (|\omega_j|^{p-1})^q\right)^{1/q} + \left(\sum_{j=1}^{n} |\eta_j|^p\right)^{1/p} \left(\sum_{j=1}^{n} (|\omega_j|^{p-1})^q\right)^{1/q}$$

($\Leftarrow$ Hölder's inequality)

$$= \left[\left(\sum_{j=1}^{n} |\xi_j|^p\right)^{1/p} + \left(\sum_{j=1}^{n} |\eta_j|^p\right)^{1/p}\right] \left(\sum_{j=1}^{n} |\omega_j|^p\right)^{1/q}$$

where we used the relationship $(p-1)q = pq - q = p + q - q = p$. The above inequality can be rewritten as

$$\left(\sum_{j=1}^{n} |\xi_j + \eta_j|^p\right)^{1-1/q} \leq \left(\sum_{j=1}^{n} |\xi_j|^p\right)^{1/p} + \left(\sum_{j=1}^{n} |\eta_j|^p\right)^{1/p}.$$

Since $p = 1 - 1/q$, we have Minkowski's inequality (C.14) when $n \to \infty$. ■

Step (a-4) Let $x = (\xi_j)$, $y = (\eta_j)$, $z = (\zeta_j) \in l^p$. Then, from Minkowski's inequality (C.14) we have

$$d(x, y) = \left(\sum_{j=1}^{\infty} |\xi_j - \eta_j|^p\right)^{1/p} = \left(\sum_{j=1}^{\infty} |\xi_j - \zeta_j + \zeta_j - \eta_j|^p\right)^{1/p}$$

$$\leq \left(\sum_{j=1}^{\infty} |\xi_j - \zeta_j|^p\right)^{1/p} + \left(\sum_{j=1}^{\infty} |\zeta_j - \eta_j|^p\right)^{1/p}$$

$$= d(\boldsymbol{x}, z) + d(z, \boldsymbol{y}). \tag{C.15}$$

This result shows that l^p is a metric space. ■

C.8.2 Completeness of l^p

We prove here that the metric space l^p $(1 \leq p < \infty)$ is complete. Let $\boldsymbol{x}_m = (\xi_j^{(m)})$, $\boldsymbol{x}_n = (\xi_j^{(n)}) \in l^p$ be Cauchy sequences. Then for any $\varepsilon > 0$ there exists $N > 0$, which satisfies the following:

$$d(\boldsymbol{x}_m, \boldsymbol{x}_n) = \left(\sum_{j=1}^{\infty} |\, \xi_j^{(m)} - \xi_j^{(n)} \,|^p \right)^{1/p} < \varepsilon, \qquad \forall m, n > N. \tag{C.16}$$

This implies that for each jth component we have the inequality

$$|\, \xi_j^{(m)} - \xi_j^{(n)} \,| < \varepsilon, \qquad \forall m, n > N.$$

Therefore, for a fixed j, $(\xi_j^{(1)}, \xi_j^{(2)}, \cdots)$ forms a Cauchy sequence. Since the spaces of real numbers $\mathbb{R}$ and complex numbers $\mathbb{C}$ are complete, we have a limit ξ_j such that $\xi_j^{(m)} \to \xi_j$ for $m \to \infty$. For this limit ξ_j, we set $\boldsymbol{x} = (\xi_1, \xi_2, \cdots)$, and we prove that $\boldsymbol{x} \in l^p$ and $\boldsymbol{x}_m \to \boldsymbol{x}$.

From (C.16), for any $m, n > N$ we have

$$\sum_{j=1}^{k} |\, \xi_j^{(m)} - \xi_j^{(n)} \,|^p < \varepsilon^p, \qquad k = 1, 2, \cdots.$$

Let $n \to \infty$ $(\xi_j^{(n)} \to \xi)$, then for any $m > N$ we have

$$\sum_{j=1}^{k} |\, \xi_j^{(m)} - \xi_j \,|^p < \varepsilon^p, \qquad k = 1, 2, \cdots$$

Furthermore, let $k \to \infty$, then for any $m > N$ we obtain

$$\sum_{j=1}^{\infty} |\, \xi_j^{(m)} - \xi_j \,|^p < \varepsilon^p.$$

This shows that

$$\boldsymbol{x}_m - \boldsymbol{x} = (\xi_1^{(m)} - \xi_1,\, \xi_2^{(m)} - \xi_2,\, \cdots) \in l^p.$$

From Minkowski's inequality (C.14) we have

$$\| \boldsymbol{x} \| = \| \boldsymbol{x}_m + (\boldsymbol{x} - \boldsymbol{x}_m) \| \leq \left(\sum_{j=1}^{\infty} | \xi_j^{(m)} |^p \right)^{1/p} + \left(\sum_{j=1}^{\infty} | \xi_j^{(m)} - \xi |^p \right)^{1/p} < \infty.$$

This results in $\boldsymbol{x} \in l^p$. Therefore, l^p is complete.

It is not difficult to prove that the metric space l^∞ $(\| \boldsymbol{x} \|_\infty = \sup_j | \xi_j |)$[7] is also complete.

C.8.3 Completeness of C[a, b] Under the Norm $\| x \|_\infty$

Let C[a, b] be a set of all continuous functions defined on $T = [a, b]$.[8] We show that C[a, b] is complete under the norm $\| x \|_\infty = \sup_{t \in T} | x(t) |$.

Let $\{ x_m \}$ be a Cauchy sequence in C[a, b], then for a given $\varepsilon > 0$ there is $N > 0$ such that

$$d(x, y) = \| x(t) - y(t) \|_\infty = \sup_{t \in T} |x_m(t) - x_n(t) | < \varepsilon, \qquad \forall m, n > N. \quad \text{(C.17)}$$

Therefore, for a fixed $t_0 \in T$ we have

$$|x_m(t_0) - x_n(t_0) | < \varepsilon, \qquad \forall m, n > N.$$

Since the real number space $\mathbb{R}$ is complete, there exists the limit $x(t_0)$ such that $x_m(t_0) \to x(t_0)$ for $m \to \infty$. Therefore, for any $t \in T$ we have a value of the function x. We prove that the limit x belongs to C[a, b] and $x_m \to x$.

From (C.17), for $n \to \infty$ we have

$$\sup_{t \in T} |x_m(t) - x(t) | < \varepsilon, \qquad \forall m > N.$$

This shows that for a given $\varepsilon > 0$ we have the inequality

$$\sup |x_m(t) - x(t) | < \varepsilon, \qquad \forall m > N$$

[7] The notation 'sup' (supremum) shows the least upper bound and 'inf' (infimum) is the greatest lower bound.

[8] A function $x(t)$ is continuous at $t_0 \in T$ if for a given $\varepsilon > 0$, there exists $\delta > 0$ such that
$\| x(t) - x(t_0) \| < \varepsilon \quad \forall t \in T$ satisfying $|t - t_0 | < \delta$.

on any point $t \in T$. That is, $x(t)$ is continuous at any point t, which implies that $\{x_m(t)\}$ uniformly converges to $x(t)$.[9] In conclusion, under the norm $\| x(t) \|_\infty = \sup_{t \in T} | x(t) |$ the Cauchy sequence $\{x_m\}$ in $C[a, b]$ has a limit x, which is continuous, therefore $C[a, b]$ is complete. Note that since we choose the norm $\| x(t) \|_\infty = \sup_{t \in T} | x(t) |$, the sequence $\{x_m\}$ uniformly converges to x, while if we use, for example, a p-norm such as $\| x(t) \|_p = \left(\int_a^b | x(t) |^p \, dt \right)^{1/p}$, $(1 \le p < \infty)$, we cannot prove the completeness. For further discussion we need a knowledge of the Lebesgue integral.

Note C.1 (Uniform convergence and continuity of sequence of functions). We show that if a sequence of functions $\{x_m\}$ uniformly converges to x, the limit x is continuous.

Since $\{x_m\}$ uniformly converges to x, for a given $\varepsilon > 0$ there is $N > 0$ such that

$$| x_n(t) - x(t) | < \varepsilon \qquad \forall n > N \text{ and } \forall t \in T.$$

If x_n is continuous at $t = t_0$, there is $\delta > 0$ such that

$$| x_n(t) - x_(t_0) | < \varepsilon \qquad \forall t \in T \text{ satisfying } |t - t_0| < \delta.$$

For this $t \in T$ we have

$$| x(t) - x(t_0) | < | x(t) - x_n(t) | + | x_n(t) - x_n(t_0) | + | x_n(t) - x(t_0) |, < 3\varepsilon$$

which shows that x is continuous at $t_0 \in T$. Since $t_0 \in T$ is arbitrary, x is continuous at any point in T. ∎

C.8.4 Bessel's Inequality and Parseval's Equality

Let $\langle \cdot, \cdot \rangle$ be an inner product defined on a space X, and $\{\phi_n\}$ an orthonormal sequence in X. Then we have the following Bessel's inequality (Yoshida 1980):

$$\sum_{n=-\infty}^{\infty} | \langle f, \phi_n^* \rangle |^2 \le \| f \|^2. \tag{C.18}$$

---

[9] A sequence of functions $\{x_m\}$ uniformly converges to x if for a given $\varepsilon > 0$ there is $N > 0$ such that

$$\| x_n(t) - x(t) \| < \varepsilon \qquad \forall n > N \text{ and } \forall t \in T.$$

Proof. Let Y_N be a space, which is spanned by $\{\phi_n\}_{n=-N}^{N}$ ($N < \infty$ is a natural number):

$$Y_N = \left\{\alpha : \alpha = \textstyle\sum_{n=-N}^{N} a_n \phi_n,\ a_n \in \mathbb{R}\right\} \subset X.$$

That is, Y_N is the entirety of all linear combinations of $\{\phi_n\}$ (cf. Appendix C.3). Let $f \in X$ be given, and we introduce the following partial sum:

$$f_N(x) = \sum_{n=-N}^{N} \langle f, \phi_n^* \rangle \phi_n(x) = \sum_{n=-N}^{N} \hat{f}_n^* \phi(x) \quad \text{for } N : \text{fixed} \tag{C.19}$$

Let $h = f - f_N$, then we have

$$\begin{aligned}
\langle h, f_N^* \rangle &= \langle f, f_N^* \rangle - \langle f_N, f_N^* \rangle \\
&= \langle f, \textstyle\sum \langle f, \phi_n^* \rangle \phi_n \rangle - \langle f_N, f_N^* \rangle \\
&= \textstyle\sum \overline{\langle f, \phi_n^* \rangle}^* \langle f, \phi_n^* \rangle - \textstyle\sum \langle f, \phi_n^* \rangle \overline{\langle f, \phi_n^* \rangle}^* \langle \phi_n, \phi_n^* \rangle \\
&= 0,
\end{aligned}$$

since $\langle \phi_n, \phi_n^* \rangle = 1$, which shows that h is orthogonal to f_N. Therefore, from Pythagoras' theorem we obtain

$$\begin{aligned}
&\| f \|^2 = \| f_N \|^2 + \| h \|^2 \\
&\Rightarrow \quad \| h \|^2 = \| f \|^2 - \| f_N \|^2 = \| f \|^2 - \sum_{n=-N}^{N} |\langle f, \phi_n^* \rangle|^2 \ \geq\ 0 \\
&\Rightarrow \quad \sum_{n=-N}^{N} |\langle f, \phi_n^* \rangle|^2 \leq \| f \|^2 < \infty.
\end{aligned}$$

For any N we have $\| f \|^2 < \infty$. Thus for $N \to \infty$, we can obtain Bessel's inequality (C.18). ■

If we recall the Fourier analysis, we understand that $\hat{f}_n^* = \langle f, \phi_n^* \rangle$ are the Fourier coefficients. Note that Bessel's inequality (C.18) is reduced to an equality if we have

$$\langle f - \sum_{n=-\infty}^{\infty} \hat{f}_n^* \phi_n,\ f^* - \sum_{n=-\infty}^{\infty} \hat{f}_m \phi_m^* \rangle = \| f - \sum_{n=-\infty}^{\infty} \hat{f}_n^* \phi_n \|^2 = 0.$$

It is only in this case that the Fourier series is complete; i.e., $\sum_{n=-\infty}^{\infty} \hat{f}_n^* \phi_n = f$, then the Fourier basis $\{\phi_n\}$ can be complete in a function space such that an element is a periodic and continuous function belonging to the Lebesgue integrable space $L^2[-L/2, L/2]$. We will show this in the following discussion, though the above conditions can be relaxed.

If $\{ \phi_n \}$ is complete, we have for functions $f,\ g \in L^2[-L/2, L/2]$

$$\hat{f}_n^* = \langle f, \phi_n^* \rangle, \qquad \hat{g}_n^* = \langle g, \phi_n^* \rangle,$$

then from Parseval's identity (Parseval relation) we have

$$\langle f, g^* \rangle = \sum_{n=-\infty}^{\infty} \hat{f}_n^* \, \hat{g}_n. \tag{C.20}$$

The proof is as follows:

$$\left| \langle f, g^* \rangle - \sum_{n=-N}^{N} \hat{f}_n^* \, \hat{g}_n \right| = \left| \int f \left(g^* - \sum_{n=-N}^{N} \hat{g}_n \, \phi_n^* \right) dx \right|$$

$$\leq \| f \| \left(\int \left| g^* - \sum_{n=-N}^{N} \hat{g}_n \, \phi_n^* \right|^2 dx \right)^{1/2} \to 0 \quad \text{as } N \to \infty \quad \blacksquare$$

From Parseval's equality (C.20) we have $\sum_{n=-\infty}^{\infty} | \hat{f}_n^* |^2 < \infty$. Therefore, we obtain

$$\lim_{|n| \to \infty} | \hat{f}_n^* |^2 = 0$$

and the following equation is also satisfied:

$$\lim_{|n| \to \infty} \hat{f}_n^* = \lim_{n \to \infty} \int_{-L/2}^{L/2} f(x) \sin k_n x \, dx = \lim_{n \to \infty} \int_{-L/2}^{L/2} f(x) \cos k_n x \, dx = 0. \tag{C.21}$$

Based on the above discussions, it is understood that we must prove the completeness of the following Fourier basis in a function space:

$$\left\{ \cdots, \phi_{-1}, \phi_0, \phi_1, \cdots \right\} = \left\{ \phi_n(x) = e^{ik_n x} / \sqrt{L} \right\}_{n=-\infty}^{\infty}.$$

C.8.5 Completeness of Fourier Basis $\{\phi_n\}$

Let a function $f \in L^2[-L/2, L/2]$ be periodic ($f(-L/2) = f(L/2)$) and continuous. We write a partial sum of f under the Fourier basis $\{\phi_n\}$ as

$$f_N(x) = \sum_{n=-N}^{N} \langle f, \phi_n^* \rangle \phi_n(x) = \sum_{n=-N}^{N} \hat{f}_n^* \, \phi(x) = \sum_{n=-N}^{N} \frac{1}{L} \int_{-L/2}^{L/2} f(t) \, e^{ik_n(x-t)} \, dt \tag{C.22}$$

(cf. (C.19)). In addition, we assume that a derivative $f' = df/dx \in L^2[-L/2, L/2]$ is also periodic and continuous, so that at any point $x \in L^2[-L/2, L/2]$ for a given $h > 0$ there is $\delta > 0$ such that

$$\int_{x-\delta}^{x+\delta} \left| \frac{f(t+h) - f(t)}{h} \right|^2 dt < \infty. \tag{C.23}$$

Then, the partial sum f_N converges to f:

$$\lim_{N \to \infty} \| f_N - f \| = 0. \tag{C.24}$$

Proof. By using an integral transformation, the partial sum f_N given by (C.22) is written as

$$f_N(x) = \frac{1}{L} \int_{-L/2}^{L/2} D_N(x-t)\, f(t)\, dt \tag{C.25}$$

where

$$\begin{aligned} \frac{1}{L} \int_{-L/2}^{L/2} D_N(x-t)\, dt &\overset{(y=x-t)}{=} \frac{1}{L} \int_{x+L/2}^{x-L/2} D_N(y)\,(-dy) = \frac{1}{L} \int_{-L/2}^{L/2} D_N(t)\, dt \\ &= \frac{1}{L} \int_{-L/2}^{L/2} \sum_{n=-N}^{N} e^{ik_n t}\, dt = 1 \end{aligned} \tag{C.26}$$

($\because$ $e^{ik_n t} = 1$ for $k_n = 0$; otherwise, the integral is 0 due to the periodicity.)

$$\begin{aligned} D_N(t) = \sum_{n=-N}^{N} e^{ik_n t} &= e^{-ik_N t} \left(1 + e^{ik_1 t} + + e^{ik_2 t} + \cdots + e^{2ik_N t} \right) \\ &= e^{-ik_N t} \frac{1 - e^{i(2k_N + k_1)t}}{1 - e^{ik_1 t}} = \frac{\sin\{(k_N + k_1/2)t\}}{\sin(k_1 t/2)}. \end{aligned} \tag{C.27}$$

By using (C.26) and (C.27), we have

$$\begin{aligned} f_N(x) - f(x) &= \frac{1}{L} \int_{-L/2}^{L/2} D_N(t) \Big\{ f(x-t) - f(x) \Big\}\, dt \\ &= \frac{1}{L} \int_{-L/2}^{L/2} \sin\{(k_N + k_1/2)t\}\, g(t)\, dt \\ &= \frac{1}{L} \int_{-L/2}^{L/2} \sin k_N t \Big[\cos(k_1 t/2) \cdot g(t) \Big]\, dt \end{aligned}$$

$$+\frac{1}{L}\int_{-L/2}^{L/2} \cos k_N t \left[\sin(k_1 t/2) \cdot g(t)\right] dt \qquad \text{(C.28)}$$

where

$$g(t) = \begin{cases} \dfrac{f(x-t) - f(t)}{\sin(k_1 t/2)}, & \text{for } t \neq 0, \\ 0, & \text{for } t = 0. \end{cases} \qquad \text{(C.29)}$$

If $|t| \ll 1$, we can have a domain such that $\sin(k_1 t/2) \geq k_1 |t|/4$, and together with (C.23) we obtain $g \in L^2[-L/2, L/2]$. Thus, the terms $[\ \cdot\ ]$ of the r.h.s. of (C.28) also belong to $L^2[-L/2, L/2]$. By applying (C.21) to the above result, we have the final result (C.24) $\lim_{N\to\infty} \| f_N - f \| = 0$ as $N \to \infty$.

Appendix D
Classical and Chemical Thermodynamics

First, some fundamental concepts of thermodynamics are introduced and these will be discussed in greater detail in subsequent sections. In this Appendix we treat phenomena relevant to mechanics and temperature; other effects including chemical fields (cf. Appendix E) and electromagnetic effects are excluded. Further expositions are given by de Groot and Mazur (1962), Kestin (1979) and Kondepudi and Prigogine (1998).

D.1 Energy Conservation Law and Thermodynamic Systems

The total energy $\mathcal{E}$ consists of the kinetic energy $\mathcal{K}$ and the internal energy $\mathcal{U}$, and several types of energy fluxes contribute to the change in the internal energy $\mathcal{U}$ and the kinetic energy $\mathcal{K}$. The energy flux consists of the heat flux $d\mathcal{Q}$, the power $d\mathcal{W}$, which corresponds to a mechanical flux, the chemical energy flux $d\mathcal{C}$ due to the change of mass, called the mass-energy flux, and the electromagnetic energy flux. These fluxes induce a change of internal energy $d\mathcal{U}$. We now have the following *energy conservation law*:

$$d\mathcal{E} = d\mathcal{K} + d\mathcal{U} = d\mathcal{W} + d\mathcal{Q} + d\mathcal{C}. \tag{D.1}$$

The change of mass of this system can be denoted by $d\mathcal{M}$.

Each energy flux consists of an externally supplied part and an internally generated part. For example, the heat flux $d\mathcal{Q}$ is decomposed into the heat flow $d\mathcal{Q}^e$ supplied from outside the system and the internally generated heat flow $d\mathcal{Q}^i$; the mechanical power $d\mathcal{W}$ and the mass-energy flux $d\mathcal{C}$ are decomposed in the same manner:

$$d\mathcal{W} = d\mathcal{W}^e + d\mathcal{W}^i, \qquad d\mathcal{Q} = d\mathcal{Q}^e + d\mathcal{Q}^i, \qquad d\mathcal{C} = d\mathcal{C}^e + d\mathcal{C}^i. \tag{D.2}$$

Y. Ichikawa and A.P.S. Selvadurai, *Transport Phenomena in Porous Media*,
DOI 10.1007/978-3-642-25333-1, © Springer-Verlag Berlin Heidelberg 2012

First, we introduce the notation: $d\mathcal{Q}$ is the heat flux (i.e. the rate of heat supply), which consists of a part supplied from outside the system $d\mathcal{Q}^e$ and a part generated internally $d\mathcal{Q}^i$,

Let us define the following thermodynamical systems:

1. *Isolated system*: No interaction with the surroundings ($d\mathcal{W}^e = 0,\ d\mathcal{Q}^e = 0,\ d\mathcal{C}^e = 0,\ d\mathcal{M} = 0$). It should be noted that internally generated fluxes $d\mathcal{W}^i,\ d\mathcal{Q}^i,\ d\mathcal{C}^i$ do not always vanish.
2. *Closed system*: Energy is exchanged with the surroundings, while no mass is exchanged ($d\mathcal{W}^e \neq 0,\ d\mathcal{Q}^e \neq 0,\ d\mathcal{C}^e = 0,\ d\mathcal{M} = 0$).
3. *Adiabatic system*: Work is input in the system, while no heat energy is exchanged with the surroundings ($d\mathcal{W}^e \neq 0,\ d\mathcal{Q}^e = 0,\ d\mathcal{C}^e = 0,\ d\mathcal{M} = 0$).
4. *Open system*: Energy and material are exchanged with the surroundings ($d\mathcal{W}^e \neq 0,\ d\mathcal{Q}^e \neq 0,\ d\mathcal{C}^e \neq 0,\ d\mathcal{M} \neq 0$).

In order to describe the thermodynamical status of the system it is necessary to introduce both *extensive* and *intensive variables*, which form a scalar-valued 'energy' in their combination:

1. *Extensive variables*: Variables proportional to the amount of mass of the system, such as strain and entropy.
2. *Intensive variables*: Variables non-proportional to the amount of mass of the system, such as stress and temperature.

For example, a change of internal energy due to mechanical work is denoted by an inner-product of stress, an intensive variable, and strain increment under small strain conditions, which is an extensive variable: A change of internal energy due to heat supply is denoted by a product of temperature, an intensive variable, and an entropy increment, which is an extensive variable.

D.2 Existence of Entropy and Thermal Energy: First Part of the Second Law of Thermodynamics

There are two approaches for proving the existence of entropy in classical thermodynamics; these rely on either molecular-based statistical mechanics or classical thermodynamics. First we employ the latter approach because of its simplicity. It is not necessary to use either approach, because the conjugate property of temperature and entropy is important. In the sense of molecular physics, both the temperature and entropy are an aspect of kinetic movement and the energy level of molecules.

The concept of *entropy* is attributed to Carnot (1824) for his deep insight prior to the statement of the First Law of Thermodynamics (see Feynman et al. 1963, Vol. I, Chap. 44), although the introduction of the term *entropy* is due to Clausius in 1850 (Kestin 1979). Here the fundamental idea of entropy is examined on the basis of Carnot's concept. Then, based on the discussion of Prigogine (1967; Kondepudi

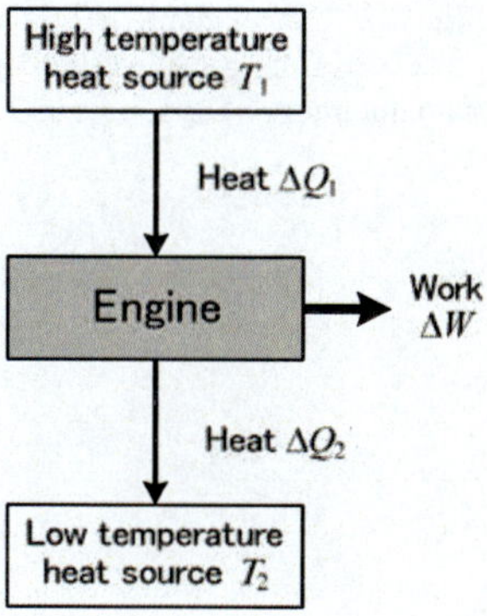

Fig. D.1 Heat engine

and Prigogine 1998), there exists an entropy production in the system due to an irreversible process, which results in the Second Law of Thermodynamics.

D.2.1 Carnot Cycle

Let us consider a heat engine or simply engine, shown schematically in Fig. D.1, which takes heat $\Delta \mathcal{Q}_1$ from the high-temperature source T_1, transfers a part of it into work $\Delta \mathcal{W}$, and delivers heat $\Delta \mathcal{Q}_2$ to the low temperature source T_2 (if we follow the notations of the First Law of Thermodynamics described in Sect. 3.2.4, $\Delta \mathcal{Q}_1,\ \Delta \mathcal{Q}_2,\ \Delta \mathcal{W}$ should be denoted as $d\,\mathcal{Q}_1,\ d\,\mathcal{Q}_2,\ d\mathcal{W}$, respectively; however, by keeping these variables under the 'finite values', we employ the notation with increments denoted by Δ. We apply the First Law to the working material of the heat engine. If there is no loss in the working material, the internal energy does not change and the transferred heat $\Delta \mathcal{Q}_2$ is just equal to the accepted heat $\Delta \mathcal{Q}_1$ less the work $\Delta \mathcal{W}$:

$$\Delta \mathcal{U} = \Delta \mathcal{Q}_1 - \Delta \mathcal{Q}_2 - \Delta \mathcal{W} = 0 \qquad \Rightarrow \qquad \Delta \mathcal{Q}_2 = \Delta \mathcal{Q}_1 - \Delta \mathcal{W}. \tag{D.3}$$

The efficiency of this engine is defined by

$$\eta = \frac{\Delta \mathcal{W}}{\Delta \mathcal{Q}_1} = \frac{\Delta \mathcal{Q}_1 - \Delta \mathcal{Q}_2}{\Delta \mathcal{Q}_1} = 1 - \frac{\Delta \mathcal{Q}_2}{\Delta \mathcal{Q}_1} \tag{D.4}$$

The ideal engine, that is, without loss due to friction etc., is referred to as the *reversible engine*. This reversible engine can be operated under $\Delta \mathcal{Q}_2 = 0$, therefore the efficiency is $\eta = 1$. Note that the working material of this reversible engine is not necessarily a perfect gas.

The Carnot cycle for this heat engine is shown in Fig. D.2a where the processes are represented as follows:

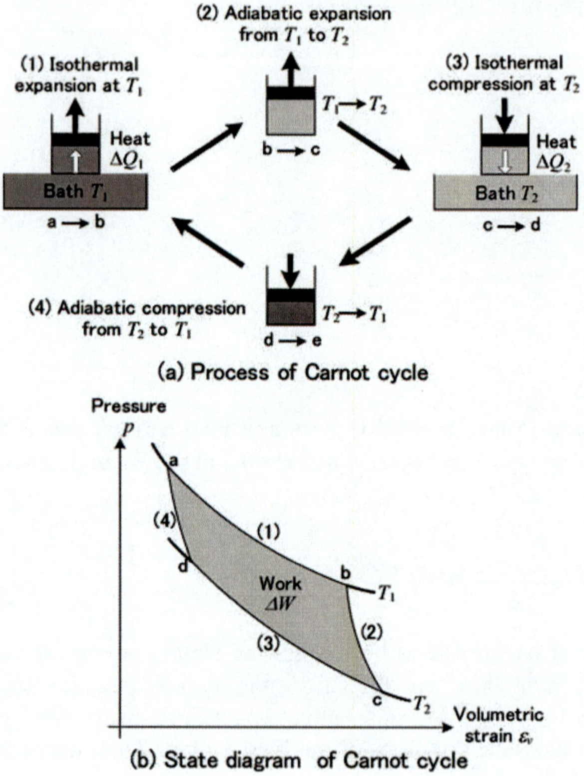

Fig. D.2 Carnot cycle (**a**) process of Carnot cycle (**b**) state diagram of Carnot cycle

1. *Isothermal expansion process*: Keeping the same temperature T_1 with a high temperature thermal bath, the piston is slowly pulled up to absorb heat ΔQ_{∞}.
2. *Adiabatic expansion process*: The engine is moved out of the bath, and the piston continues to be pulled up until the temperature reaches T_2. Note that $T_1 > T_2$.
3. *Isothermal compression process*: Keeping the same temperature T_2 with a low temperature thermal bath the piston is slowly pushed down to exhaust heat $\Delta Q_{\in}$.
4. *Adiabatic compression process*: The engine is moved out of the bath, and the piston continues to be pushed down until the temperature reaches T_1.

The total work during Carnot cycle a→b→c→d→a shown in Fig. D.2b is given by an integral of pressure p and volume change Δv:

$$\Delta W = -\oint p \, \Delta v \, ds \tag{D.5}$$

We observe that it is possible to operate the engine in complete cycles. The efficiency of the reversible engine under the Carnot cycle is $\eta = 1$.

We next consider a parallel operation of a reversible engine A and another engine B of an arbitrary efficiency (Fig. D.3). The engine B takes up heat ΔQ_1 from

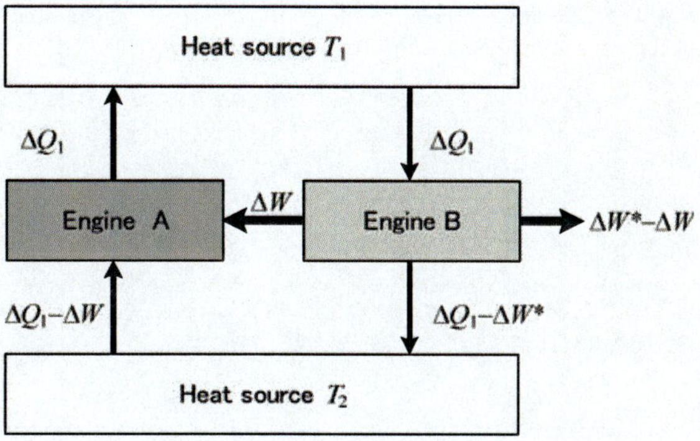

Fig. D.3 Reversible engine A being driven backwards by engine B (Feynman et al. 1963)

the high temperature heat source T_1 and performs work ΔW^* in which ΔW is transferred to engine A working in parallel. The assumption $\Delta W^* > \Delta W$ would indicate that engine B could perform external work $\Delta W^* - \Delta W$. Recall that the heat transferred from engine B to the low temperature heat source T_2 is $\Delta Q_1 - \Delta W^*$. Since engine A is reversible, it can be operated in the reverse Carnot cycle, taking $\Delta Q_1 - \Delta W$ from T_2 and transferring ΔQ_1 to T_1. If we observe the total system A+B, it works as a perpetual engine, taking heat from the high temperature heat source to perform the work $\Delta W^* - \Delta W$. Carnot concluded that such a situation is untenable. Thus we have

$$\Delta W^* \leq \Delta W \tag{D.6}$$

That is, no engine can work more efficiently than the reversible engine. In the above discussions we can exchange engine A with B, therefore the conclusion of Carnot is that the efficiency of all reversible engines is the same, and it depends only on T_1 and T_2. Thus the efficiency of a reversible engine can be defined as

$$\eta = \frac{\Delta W}{\Delta Q_1} = 1 - \frac{\Delta Q_2}{\Delta Q_1} = 1 - f(T_1, T_2) \tag{D.7}$$

where $f(T_1, T_2)$ is tentatively called the efficiency function.

Let us introduce a double Carnot cycle (1) and (2) as shown in Fig. D.4. The cycle (1) traverses a→b→c→f→a, and the efficiency function is

$$\frac{\Delta Q_2}{\Delta Q_1} = f(T_1, T_2).$$

The cycle (2) traverses f→c→d→e→f, and the efficiency function is

$$\frac{\Delta Q_3}{\Delta Q_2} = f(T_2, T_3).$$

If we consider the cycle (1) + (2), it traverses a→b→c→d→e→f→a, therefore the efficiency function is

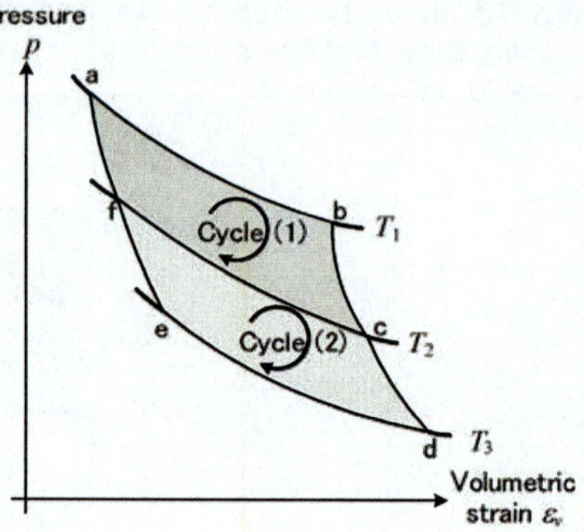

Fig. D.4 Double Carnot cycle

$$\frac{\Delta Q_3}{\Delta Q_1} = f(T_1, T_3).$$

In order to satisfy this relation we must have a function $F(T)$ such that

$$f(T_1, T_2) = \frac{F(T_2)}{F(T_1)},$$

then the efficiency function for the cycle (1) + (2) can be correctly described as

$$\frac{\Delta Q_3}{\Delta Q_1} = \frac{\Delta Q_2}{\Delta Q_1}\frac{\Delta Q_3}{\Delta Q_2} = \frac{F(T_2)}{F(T_1)}\frac{F(T_3)}{F(T_2)} = \frac{F(T_3)}{F(T_1)}.$$

Following Lord Kelvin (W. Thomson) we redefine the function $F(T)$ as the absolute temperature T ($F(T) = T$). The efficiency of the engine is now written as

$$\eta = 1 - \frac{\Delta Q_2}{\Delta Q_1} = 1 - \frac{T_2}{T_1}. \tag{D.8}$$

Alternatively (D.8) is

$$\frac{\Delta Q_1}{T_1} = \frac{\Delta Q_2}{T_2}.$$

For a more generalized cycle, the efficiency of the reversible engine can be written in an integral form as

$$\sum \frac{\Delta Q}{T} = 0 \quad \Rightarrow \quad \oint \frac{dQ}{T} = 0 \tag{D.9}$$

D.2.2 *Entropy as a State Variable*

Equation D.9 shows that in the "reversible process" we can introduce an increment of the state variable S, called the *entropy*, for the heat flux dQ by

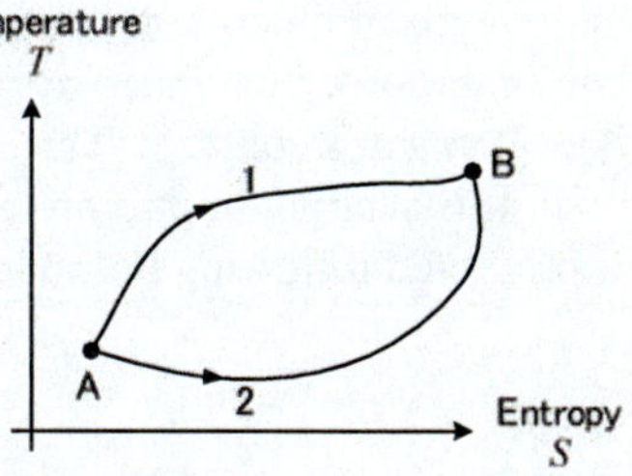

Fig. D.5 Integral path and entropy

$$dS = \frac{dQ}{T}. \tag{D.10}$$

As an example, let us consider two integral paths 1 and 2 from the initial state A to the final state B as shown in Fig. D.5. For a *reversible process* we require

$$\int_{A,1}^{B} \frac{dQ}{T} + \int_{B,2}^{A} \frac{dQ}{T} = 0 \quad \Rightarrow \quad \int_{A,1}^{B} \frac{dQ}{T} = \int_{A,2}^{B} \frac{dQ}{T}.$$

This implies that the difference in S between the state B and the state A is independent of the integration path:

$$S_B - S_A = \int_{A}^{B} \frac{dQ}{T}. \tag{D.11}$$

In conclusion, for a reversible process, the heat flux dQ input to the system is exchanged through a corresponding thermal internal energy increment $T\,dS$, and S is independent on the path of integration. This entropy increment can also be written as dS^e, since S^e is *exchangeable* with the surroundings and reversible. Then (D.10) can be written as

$$dQ = T\,dS^e. \tag{D.12}$$

The idea of exchangeable entropy is also supported by Boltzmann's theory based on statistical thermodynamics (see Appendix D.3). That is, Boltzmann's principle can be shown as

$$S^e = k_B \ln W \tag{D.13}$$

where k_B is Boltzmann's constant and W is the thermodynamic weight of configuration of the macroscopic state.

D.3 Interpretation of Entropy within the Framework of Statistical Thermodynamics: Boltzmann's Theory ♣

At the end of nineteenth century, based on a concept of probabilistic distribution of energy states, Ludwig Boltzmann (1844–1906) presented an innovative interpretation that the *entropy* S is the most feasible number of microscopic energy states under a given macroscopic environment; i.e., if W is the *thermodynamic weight* of configuration of the macroscopic state, the relation $\mathcal{S} = k_B \ln W$ holds, where k_B is Boltzmann's constant. This is referred to as *Boltzmann's Principle*.

In this section an interpretation of entropy based on statistical thermodynamics is presented following Schrödinger (1952) and Atkins and Paula (2010).

D.3.1 Statistical Representation of Molecular States and Boltzmann Distribution

The thermodynamical system treated is assumed to be closed and consists of an equal number of molecules (uniform population). The total number of molecules is N, and the total energy of this system is $\mathcal{E}$, which is the sum of energy of all molecules. The quantum mechanics theory indicates that the energy state of each molecule is discrete (see, e.g., McQuarrie and Simon 1997). Suppose an instantaneous configuration of the system occurs that corresponds to each energy level $\{\varepsilon_0, \varepsilon_1, \varepsilon_2, \ldots\}$ where there exist $\{n_0, n_1, n_2, \ldots\}$ molecules, respectively (the reason that the energy level starts with ε_0 is that it corresponds to the ground state). The total number of molecules and the total energy are kept constant:

$$\sum_{i=0} n_i = N, \qquad \sum_{i=0} n_i\, \varepsilon_i = \mathcal{E}. \tag{D.14}$$

Since all the molecules were originally the same, the number of combinations such that an instantaneous configuration is satisfied is given by

$$W = \frac{N!}{n_0!\, n_1!\, n_2! \cdots} \tag{D.15}$$

This W is referred to as the *weight of configurations*.

Taking the logarithm of (D.15) yields

$$\ln W = \ln N! - \sum_i \ln n_i!$$

Since N is an extremely large number, Stirling's approximation[1] can be applied to give

---

[1] Stirling's approximation can be proved as follows:

$$\ln x! = \ln 1 + \ln 2 + \ln 3 + \cdots \ln x = \sum_{i=1}^{x} \ln y_i \simeq \int_0^x \ln y\, dy = (y \ln y - y)\Big|_0^x = x \ln x - x.$$

$$\ln W \simeq N \ln N - \sum_i n_i \ln n_i. \tag{D.16}$$

In these combinations of W, the dominant configuration can be determined by maximizing W:

$$dW(n_i) = 0 \quad \Rightarrow \quad d(\ln W) = 0 \tag{D.17}$$

and the restrictions (D.14) are introduced via the Lagrangian multiplier method:

$$d(\ln W) = \sum_i \left(\frac{\partial(\ln W)}{\partial n_i} + \alpha - \beta \varepsilon_i \right) dn_i = 0$$

where α, β are Lagrangian multipliers. Thus we have

$$\frac{\partial(\ln W)}{\partial n_i} + \alpha - \beta \varepsilon_i = 0 \tag{D.18}$$

and substituting (D.16) into the above gives

$$-\ln N + \ln n_i = \alpha - \beta \varepsilon_i \quad \Rightarrow \quad \frac{n_i}{N} = e^{\alpha - \beta \varepsilon_i}$$

We sum this up to obtain $e^{\alpha} = 1/\sum_j e^{-\beta \varepsilon_j}$ since $\sum_i n_i = N$. Thus the following *Boltzmann distribution* for the molecular state $p_i = n_i/N$ is obtained:

$$p_i = \frac{1}{q} e^{-\beta \varepsilon_i}; \qquad q = \sum_j e^{-\beta \varepsilon_j} \tag{D.19}$$

where q is referred to as the *molecular partition function*, which indicates the measure of thermodynamically admissible states. Note that the Lagrangian multiplier β is shown later to be $\beta = 1/k_B T$.

D.3.2 Molecular Partition Function q

When the molecular partition function q is defined as (D.19)$_3$, each energy level $\{\varepsilon_0, \varepsilon_1, \varepsilon_2, \ldots\}$ corresponding to each state $\{n_0, n_1, n_2, \ldots\}$ is assumed to be different. However in practice there exist molecular states such that the energy levels become the same, therefore we can modify (D.19)$_2$ to

$$q = \sum g_j e^{-\beta \varepsilon_i} \tag{D.20}$$

where g_j is the partition coefficient which implies a g_j-fold degeneration in the energy level ε_i.

Assuming $\beta = 1/k_B T$, let us consider the case where the temperature approaches absolute zero degrees. We set the ground state ε_0 of energy as $T \to 0$. Since q is divergent if $\varepsilon_0 \neq 0$ as $T \to 0$, we have the conditions

$$\varepsilon_0 \equiv 0, \qquad \lim_{T \to 0} q = g_0. \tag{D.21}$$

On the other hand as $T \to \infty$, the combination of the energy states must be infinite:

$$\lim_{T \to \infty} q = \infty. \tag{D.22}$$

In this particular case, if the energy levels are distributed uniformly, we have

$$q = 1 + e^{-\beta\varepsilon} + e^{-2\beta\varepsilon} + e^{-3\beta\varepsilon} + \cdots = \frac{1}{1 - e^{-\beta\varepsilon}}, \tag{D.23}$$

and the Boltzmann distribution is calculated as

$$p_i = (1 - e^{-\beta\varepsilon})\, e^{-i\beta\varepsilon}. \tag{D.24}$$

If we apply the free particle solution E_n in a one-dimensional domain $(0, L)$, which will be shown in (D.47) to obtain the ground state energy $\varepsilon = E_1 = h^2/8mL^2$, the partition function can be calculated as

$$\varepsilon_n = (n^2 - 1)\varepsilon \quad \Rightarrow \quad q = \sum_{n=1}^{\infty} e^{-(n^2-1)\beta\varepsilon} \tag{D.25}$$

The sum in this equation is replaced by an integral form and we assume $n \gg 1$, therefore we obtain

$$q = \int_1^{\infty} e^{-(n^2-1)\beta\varepsilon}\, dn \simeq \int_0^{\infty} e^{-n^2\beta\varepsilon}\, dn.$$

We then set $n^2\beta\varepsilon = x^2 \Rightarrow dn = dx/\sqrt{\beta\varepsilon}$ and integrate the above to obtain

$$q = \frac{1}{\sqrt{\beta\varepsilon}} \cdot \frac{\sqrt{\pi}}{2} = \frac{1}{\Lambda} L, \qquad \Lambda = h\left(\frac{\beta}{2\pi m}\right)^{1/2} \tag{D.26}$$

where Λ is referred to as the thermal wave length. If we extend (D.26) into the three-dimensional domain $V = L_1 L_2 L_3$, we obtain

$$q = \frac{1}{\Lambda^3} V. \tag{D.27}$$

D.3.3 *Internal Energy and Entropy: Boltzmann's Formula*

Substituting the total energy of the system $\mathcal{E}$ of (D.14)$_2$ into the Boltzmann distribution (D.19) yields

$$\mathcal{E} = \frac{N}{q}\sum_i e^{-\beta\varepsilon_i} = -\frac{N}{q}\sum_i \frac{de^{-\beta\varepsilon_i}}{d\beta} = -N\frac{d(\ln q)}{d\beta}$$

where β is a function of the temperature. On the other hand, $\mathcal{E}$ is not only a function of the temperature but also the volume V of the system and other factors, therefore we write the above equation as

$$\mathcal{E} = -\left(N\frac{d(\ln q)}{d\beta}\right)_V. \tag{D.28}$$

Substituting the partition function (D.27) for the case of equi-difference energy, and differentiating this (recalling (D.26)$_2$), we obtain the following expression for the total energy:

$$\mathcal{E} = -\left(N\frac{d(\ln q)}{d\beta}\right)_V = \frac{3NV}{q\Lambda^4}\frac{\partial\Lambda}{\partial\beta} = \frac{3N}{2\beta}. \tag{D.29}$$

As described in Sect. 3.2.1, the total energy $\mathcal{E}$ consists of the kinetic energy $\mathcal{K}$ and the internal energy $\mathcal{U}$. However, in the system considered here, the change of the kinetic energy of molecules induces changes in the temperature and the volume, such that the kinetic energy is not treated directly, and the total energy is converted into internal energy. Thus we have

$$\mathcal{U} = \mathcal{E} = \sum_{i=0} n_i\,\varepsilon_i. \tag{D.30}$$

Here we should note that ε_0 corresponds to the energy at the ground state. Substituting this into (D.29), we obtain

$$\mathcal{U} = \frac{3N}{2\beta}. \tag{D.31}$$

On the other hand, as shown in (E.63) for a perfect gas, we have

$$\mathcal{U} = \frac{3}{2}n\,RT$$

where n is the total amount of substance. If we compare this with (D.31), we have

$$\beta = \frac{1}{k_B T} \tag{D.32}$$

where k_B is Boltzmann's constant, and $k_B = R/N_A$ for an ideal gas (cf. Appendix E.5).

As understood by (D.30) the change of internal energy $d\mathcal{U}$ is induced by a change of the energy state $d\varepsilon_i$ and/or by a change of the number of molecules dn_i belonging to that energy level. In quantum mechanics, it is known that the pressure applied induces $d\varepsilon_i$ and the heating induces dn_i. Now we consider the change due to heating

$$d\mathcal{U} = \sum_{i=0} \varepsilon_i \, dn_i = d\mathcal{Q} = T\, d\mathcal{S}^e,$$

and using (D.32) we have

$$d\mathcal{S}^e = \frac{d\mathcal{U}}{T} = k_B\, \beta \sum_{i=0} \varepsilon_i \, dn_i \tag{D.33}$$

Substituting (D.18) into this yields

$$d\mathcal{S}^e = k_B \sum_{i=0} \frac{\partial(\ln W)}{\partial n_i} dn_i + k_B\, \alpha \sum_{i=0} dn_i$$

Note that since we treat a closed system, $\sum_i dn_i = 0$. Thus we obtain the following *Boltzmann formula*:

$$d\mathcal{S}^e = k_B\, d(\ln W) \qquad \Rightarrow \qquad \mathcal{S}^e = k_B \ln W. \tag{D.34}$$

Substituting (D.16) into (D.34), we obtain

$$\mathcal{S}^e \simeq k_B \sum_i (n_i \ln N - n_i \ln n_i) = -N\, k_B \sum_i p_i \ln q. \tag{D.35}$$

Alternatively, from (D.19) we have

$$\ln p_i = -\beta \varepsilon_i - \ln q.$$

Substituting the above into (D.35), we obtain the Boltzmann formula of the molecular partition function as

$$\mathcal{S}^e = \frac{\mathcal{U}}{T} + N\, k_B \ln q \tag{D.36}$$

where we use the relation $\sum_i n_i \varepsilon_i = \mathcal{E} = \mathcal{U},\ k_B\, \beta = 1/T$.

Note D.1 (Schrödinger equation and the free particle problem). The Schrödinger equation specifies the quantum state of a particle: It is considered in the one-dimensional state for simplicity (for more general cases see, e.g., McQuarrie and Simon 1997, p. 73).

The one-dimensional wave equation can be written as

$$\frac{\partial^2 u}{\partial x^2} = \frac{1}{\mathfrak{v}^2}\frac{\partial^2 u}{\partial t^2} \tag{D.37}$$

where $\mathfrak{v}$ is the phase velocity. We consider harmonic motions with a circular frequency ω, and assume solutions of (D.37) of the form $u(x,t) = \psi(x)\cos\omega t$. This gives an ordinary differential equation in the spatial coordinate x for $\psi(x)$; i.e.,

$$\frac{d^2\psi}{dx^2} + \frac{4\pi^2}{\lambda^2}\psi(x) = 0 \tag{D.38}$$

where we set $\omega = 2\pi\nu$ and $\lambda\nu = \mathfrak{v}$ (ν is the frequency and λ is the wavelength).

In 1924 Louis de Broglie discovered that every particle of a mass m shows wave-particle duality. Let the linear momentum of the particle be $p = m\mathfrak{v}$, then the total energy is given by

$$E = \frac{p^2}{2m} + V(x) \tag{D.39}$$

where $V(x)$ is the potential energy. The linear momentum then becomes

$$p = \{2m(E - V)\}^{1/2}. \tag{D.40}$$

Alternatively, Einstein's relationship relates the wavelength of light λ to the linear momentum p according to $\lambda = h/p$ (where $h = 6.626 \times 10^{-34}$ [J·s] is the Planck constant). De Broglie applied this to the material wave of the mass m to obtain

$$\lambda = \frac{h}{p} = \frac{h}{m\mathfrak{v}} \tag{D.41}$$

The above is referred to as the de Broglie relation. Using (D.40) and (D.41) we have

$$\lambda = \frac{h}{\{2m(E-V)\}^{1/2}} \quad \Rightarrow \quad \frac{1}{\lambda^2} = \frac{1}{h^2}\{2m(E-V)\}$$

and substituting this into (D.38) we obtain the following *time-independent Schrödinger equation*:

$$\hat{H}\,\psi(x) = E\,\psi(x), \qquad \hat{H} = -\frac{\hbar}{2m}\frac{d^2}{dx^2} + V(x) \tag{D.42}$$

where $\hat{H}$ is Hamilton's operator and $\hbar = h/2\pi$.

We can extend this time-independent Schrödinger equation to obtain the *time-dependent Schrödinger equation*:

$$\hat{H}\,\Psi(x,t) = i\,\hbar\frac{\partial\,\Psi(x,t)}{\partial t} \tag{D.43}$$

($i = \sqrt{-1}$ is the imaginary unit). If we introduce the separation of variable $\Psi(x,t) = \psi(x) f(t)$ and set

$$E = \frac{1}{\psi(x)} \hat{H}\, \psi(x) = \frac{i\,\hbar}{f(t)} \frac{df(t)}{dt}$$

we obtain the following two ordinary differential equations:

$$\hat{H}\, \psi(x) = E\, \psi(x), \qquad \frac{df(t)}{dt} = -\frac{i}{\hbar} E\, f(t).$$

The first one is equivalent to the time-independent Schrödinger equation (D.42).

Let us solve the simplest problem, referred to as the free particle problem, which treats the motion of a particle whose potential energy can be ignored ($V(x) = 0$). The spatial domain is $(0, L)$, and the two-point boundary value problem is

$$\frac{d^2\psi}{dx^2} + \frac{2mE}{\hbar}\psi(x) = 0, \quad 0 \le x \le L \tag{D.44}$$

$$\text{BC:} \quad \psi(0) = \psi(L) = 0. \tag{D.45}$$

The general solution of this is

$$\psi(x) = A \cos kx + B \sin kx, \qquad k = \left(\frac{2mE}{\hbar}\right)^{1/2} = \left(\frac{4\pi m E}{h}\right)^{1/2} \tag{D.46}$$

and the boundary condition $\psi(0) = 0$ gives $A \equiv 0$, and the boundary condition $\psi(L) = 0$ gives the eigenvalues

$$kL = n\pi, \qquad n = 1, 2, \cdots$$

Substituting this into (D.46)$_2$, we obtain the quantized energy E_n for the free particle problem as

$$E_n = \frac{n^2 \pi h^2}{4mL^2}, \qquad n = 1, 2, \cdots \tag{D.47}$$

The three-dimensional time-independent Schrödinger equation is given by

$$\hat{H}\, \Psi(\boldsymbol{x}, t) = i\,\hbar \frac{\partial\, \Psi(\boldsymbol{x}, t)}{\partial t}, \qquad \hat{H} = -\frac{\hbar}{2m}\Delta + V(\boldsymbol{x}) \tag{D.48}$$

where $\Delta = \partial^2/\partial x_i \partial x_i$ is Laplace's operator. The time-dependent Schrödinger equation is similarly obtained as

$$\hat{H}\, \psi(\boldsymbol{x}) = E\, \psi(\boldsymbol{x}). \tag{D.49}$$

The quantized energy for the three-dimensional free particle problem ($V(\boldsymbol{x}) = 0$) can be calculated as

$$E_n = \frac{\pi h^2}{4m} \sum_{i=1}^{3} \left(\frac{n_i}{L_i} \right)^2, \qquad n_i = 1, 2, \cdots \qquad \blacksquare \tag{D.50}$$

D.4 Entropy Production: Second Part of the Second Law of Thermodynamics

The phenomenon that mechanical energy can be converted into heat through friction is a common observation. The Second Law of Thermodynamics states that in addition to such an energy exchange process, there exists an autonomous mechanism to increase the entropy due to the disturbance of the internal structure or order.

D.4.1 Second Law of Thermodynamics for the Irreversible Process

In Appendix D.1 we showed that in a reversible process the energy increment that is exchangeable with its surroundings is formed by temperature and entropy change (we referred to this as the first part of the Second Law). We now discuss the second part of the Second Law, which states that in an irreversible process the entropy increases.

Carnot (1824) concluded that in an irreversible process, the efficiency η of a heat engine is always less than unity, which gives the second part of the Second law. That is, in the irreversible process the rate of absorption of heat ΔQ_1 from the high temperature source that is converted to work becomes smaller than in a reversible process, and the heat output increases; therefore, from (D.4) we have

$$\eta = 1 - \frac{\Delta Q_2^{\text{irr}}}{\Delta Q_1} < 1 - \frac{T_2}{T_1} \qquad \Rightarrow \qquad \frac{\Delta Q_1}{T_1} < \frac{\Delta Q_2^{\text{irr}}}{T_2}$$

We rewrite this as

$$dS > \frac{dQ}{T}. \tag{D.51}$$

This is referred to as the *Clausius inequality*, and implies that in an irreversible process the entropy increases. The heat supply dQ is therefore not completely converted into the heat component of the internal energy $T\,dS$. This is the essential concept of the second part of the Second Law of Thermodynamics.

In an isolated system $dQ^e = 0$, and if we assume that the internal heat flux dQ^i also vanishes (i.e., $dQ^i = 0$ at the equilibrium state), we have $dS = 0$. This implies that S approaches the (local) maximum (since S is increasing). That is, at the equilibrium state of the isolated system, with no internal heat source, we have

$$dS = 0 \qquad \text{at} \quad S = S_{\text{max}} \tag{D.52}$$

and, in the course of an irreversible process, the entropy increases towards the above state: i.e.

$$dS > 0. \tag{D.53}$$

Note D.2 (Entropy increase in the isolated system). In the isolated system $dQ^e = 0$ as mentioned above, however, the flux dQ^i caused by some internal heat source is not always zero. Furthermore, it is not completely obvious whether or not an "irreversible process" which satisfies the inequality (D.51) exists in an isolated system (as mentioned in Appendix D.3, Boltzmann's interpretation of entropy $S^e = k_B \ln W$ also relates the exchangeable part of entropy S^e with the molecular configuration W). Any experiment to verify the inequality (D.51) for the isolated system poses an interesting *physical observation problem*, since it is almost impossible to measure the state of the isolated system by using a sensor 'inserted' into the system without introducing some disturbance.

At present it is not possible to prove that entropy increases, therefore we here follow classical thermodynamics and assume that in an irreversible process the inequality (D.51) is satisfied even for an isolated system. ■

For the general case with a heat source dQ the entropy change is divided into two paths, the autonomous internal change and that part affected by the external heat supply, attributed to Prigogine (1967), which will be discussed in the Appendix D.4.2. In this case the inequality (D.53) gives some of the entropy corresponding to the internal change.

D.4.2 Entropy Production

Let us reconsider the Second Law of Thermodynamics (cf. Kondepudi and Prigogine 1998). We partition the entropy change into dS^e due to the exchange of energy and/or material with the surroundings and dS^i due to the irreversible process in the system itself: i.e.

$$dS = dS^e + dS^i \tag{D.54}$$

where dS^i is the *entropy production*. Combining the inequality (D.51) for the irreversible process and the equality (D.10) for the reversible process we have the following inequality for all processes:

$$dS \geq \frac{dQ}{T}. \tag{D.55}$$

However, since

$$d\mathcal{S}^e = \frac{d\mathcal{Q}}{T} \tag{D.56}$$

the Second Law of Thermodynamics implies that for any system, either isolated, closed or open, the entropy production $d\mathcal{S}^i$ is non-negative:

$$\textit{Second Law of Thermodynamics:}\ d\mathcal{S} = d\mathcal{S}^e + d\mathcal{S}^i,\ d\mathcal{S}^e = \frac{d\mathcal{Q}}{T},\ d\mathcal{S}^i \geq 0. \tag{D.57}$$

The result (D.57) suggests (whether it is correct or not) that the system, even if it is isolated, involves an internal change. Considering the equilibrium state of an isolated system in which no internal heat source $d\mathcal{Q}^i$ exists, we have

$$\text{Equilibrium process in the isolated system:} \qquad d\mathcal{S} = 0. \tag{D.58}$$

Thus, in the irreversible process approaching the equilibrium state, the entropy is increasing ($d\mathcal{S} > 0$), while at the equilibrium state, it reaches a maximum. Note that for a closed or open system entropy can decrease in general if some energy flux is applied to the system.

Combining the First Law of Thermodynamics (3.40) and the Second Law of Thermodynamics discussed above, we can summarize the laws governing non-equilibrium thermodynamics as follows:

$$d\mathcal{E} = d\mathcal{K} + d\mathcal{U} = d\mathcal{W} + d\mathcal{Q} \tag{D.59}$$

$$d\mathcal{S} = d\mathcal{S}^i + d\mathcal{S}^e \tag{D.60}$$

$$d\mathcal{S}^i \geq 0, \qquad d\mathcal{S}^e = \frac{d\mathcal{Q}}{T}. \tag{D.61}$$

If Stokes' power formula is satisfied (cf. Sect. 3.2.4), the above formulae can then be modified as follows:

$$d\mathcal{K} + d\mathcal{U}_\sigma = d\mathcal{W} \quad \text{and} \quad d\mathcal{U} = d\mathcal{U}_\sigma + d\mathcal{Q} \tag{D.62}$$

$$d\mathcal{S} = d\mathcal{S}^i + d\mathcal{S}^e \tag{D.63}$$

$$d\mathcal{S}^i \geq 0, \qquad d\mathcal{S}^e = \frac{d\mathcal{Q}}{T}. \tag{D.64}$$

The closed system in which no material has been exchanged is schematically drawn as shown in Fig. D.6. The open system with material exchange can be configured in a similar manner.

Note D.3 (Internal dissipation and Clausius-Duhem inequality). The Second Law of Thermodynamics is considered in terms of the dissipation of internal structure. Referring to the definition of entropy (D.10) and its effects for a reversible process

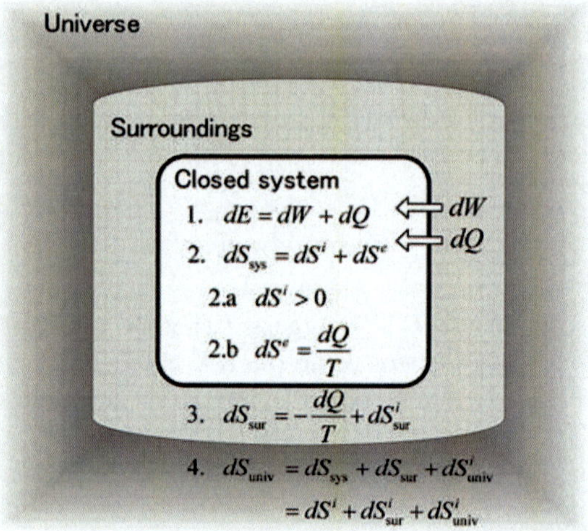

Fig. D.6 Dissipative structure for a closed system (cf. Stumm and Morgan 1996)

and recalling that $T\,dS$ corresponds to the change of internal energy and that $d\mathcal{Q}$ is the heat supply from the surroundings, we can define the following internal dissipation for any reversible and irreversible process;

$$d\mathcal{D} = T\,dS - d\mathcal{Q}. \tag{D.65}$$

Referring to (D.60) we understand that the Second Law requires that the internal dissipation should be non-negative: i.e.

$$d\mathcal{D} = T\,dS - d\mathcal{Q} = T\,dS^i \geq 0 \tag{D.66}$$

This is referred to as the *Clausius-Duhem inequality*.

It should be noted that the Clausius-Duhem inequality, giving the condition of internal dissipation, is exclusively satisfied for the equality part by the reversible process and for the pure inequality part by the irreversible process. The equality of the non-negative condition is not satisfied for the 'pure irreversible process'. ■

The above discussions treat the change of entropy, and do not give an absolute value of entropy. In 1918 Nernst showed that in its uniform state a material at absolute zero temperature has zero entropy (English version: Nernst 1969). This is referred to as the *Nernst-Planck theorem* or the *Third Law of Thermodynamics* (cf. Appendix D.3).

Example D.1 (Heat conduction problem). Let us consider an isolated system that consists of two parts, Cell 1 and Cell 2 (Fig. D.7). The temperature of Cell 1 is assumed to be T_1, and temperature of Cell 2 to be T_2 ($T_1 > T_2$). The heat flux flowing from Cell 1 to Cell 2 is given as $d\mathcal{Q}$. Since there is no heat exchange and no volume change in the isolated system, $dS^e = 0$ and $d\mathcal{W} = 0$. Thus the change of internal energy of each part is equal to the heat supply:

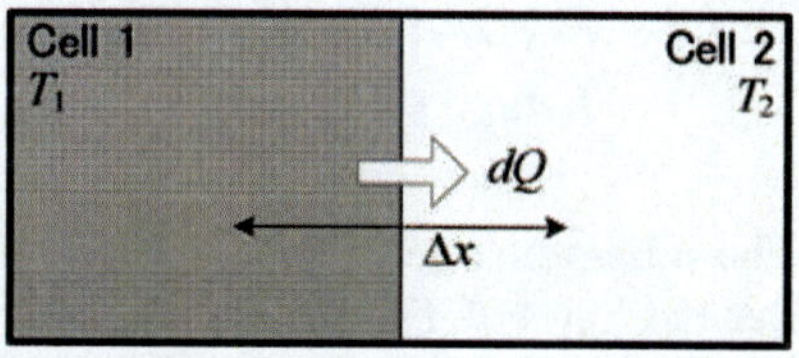

Fig. D.7 Heat transfer in an isolated system

$$d\mathcal{U}_1 = d\mathcal{Q}_1, \qquad d\mathcal{U}_2 = d\mathcal{Q}_2.$$

Applying the First Law of Thermodynamics to the total system we have $d\mathcal{Q} = d\mathcal{Q}_2 = -d\mathcal{Q}_1$. Thus the entropy production of the total system is

$$d\mathcal{S} = d\mathcal{S}^i = -\frac{d\mathcal{Q}}{T_1} + \frac{d\mathcal{Q}}{T_2} = \left(\frac{1}{T_2} - \frac{1}{T_1}\right) d\mathcal{Q}, \quad \text{or} \quad \frac{d\mathcal{S}^i}{dt} = \left(\frac{1}{T_2} - \frac{1}{T_1}\right) \frac{d\mathcal{Q}}{dt}. \tag{D.67}$$

In the classical irreversible thermodynamics (CIT) theory, the local equilibrium condition is applied, and T_1 and T_2 remain constant in each cell. This results in a strong inconsistency when applying CIT to the non-equilibrium problem. It should be noted that in our procedure the conditions of local equilibrium and constant temperature in each cell are not required, because (D.59)–(D.61) are applied directly.

The heat flux $q = d\mathcal{Q}/dt$ is related to the temperature gradient by *Fourier's law*

$$\boldsymbol{q} = -\boldsymbol{k}\nabla T \tag{D.68}$$

where $\boldsymbol{k}$ is the thermal conductivity tensor, which is a second-order tensor and is reduced to $K_{ij} = k\,\delta_{ij}$ if the material is isotropic (k is the thermal conductivity).

Let the 'thermal distance' between Cell 1 and Cell 2 be Δx. Thus, $dT/dx = (T_1 - T_2)/\Delta x$ is negative, and Fourier's law can be written as

$$\frac{d\mathcal{Q}}{dt} = k\frac{T_1 - T_2}{\Delta x}$$

where $k > 0$ is the thermal conductivity. Note that in a system where the subsystems are clearly separated with temperatures T_1 and T_2, we can assume that $d\mathcal{Q}/dt = k(T_1 - T_2)$. By using (D.67) the entropy production is given as

$$\frac{d\mathcal{S}^i}{dt} = \left(\frac{1}{T_2} - \frac{1}{T_1}\right) k\frac{T_1 - T_2}{\Delta x} = \frac{k\,(T_1 - T_2)^2}{T_1 T_2 \Delta x} \geq 0. \tag{D.69}$$

This implies that the non-negativeness of entropy production is equivalent to the non-negativeness of the thermal conductivity k. At the final equilibrium state the temperature of both subsystems becomes the same, and therefore the entropy production is zero. ■

D.4.3 Second Law of Thermodynamics and the Minimum Energy Principle

The minimum energy principle in the closed system is derived from the First and Second Laws of Thermodynamics. Since $d\mathcal{Q} = T d\mathcal{S}^e$, we have

$$d\mathcal{E} = T\,d\mathcal{S}^e + d\mathcal{W} = T\,d\mathcal{S} - T\,d\mathcal{S}^i + d\mathcal{W}. \tag{D.70}$$

We treat here a closed system, so that the volume does not change ($d\mathcal{W} = 0$). If the heat supply $d\mathcal{Q} = T d\mathcal{S}^e$ is controlled as the total entropy in the system is constant ($d\mathcal{S} = 0$), the total energy increment is

$$d\mathcal{E} = -T\,d\mathcal{S}^i \leq 0. \tag{D.71}$$

That is, in a system of constant entropy and constant volume the process moves towards the direction of minimum energy. This gives the *Principle of Minimum Energy* for the irreversible process, which is deduced from the Second Law of Thermodynamics.

D.4.4 Second Law of Thermodynamics in a Thermo-mechanical Continuum: Eulerian Description

Assume that, corresponding to the entropy $\mathcal{S}$ of a body, there exists an entropy density function s, measured per unit mass of a continuum, such that

$$\mathcal{S} = \int_\Omega \rho s(\boldsymbol{x}, t)\,dv. \tag{D.72}$$

As mentioned in Appendix D.4.2, the entropy change $d\mathcal{S}$ of a system consists of two parts in the form $d\mathcal{S} = d\mathcal{S}^e + d\mathcal{S}^i$, where $d\mathcal{S}^e$ is due to an energy exchange with the surroundings and $d\mathcal{S}^i$ is due to an internal irreversible process, so that the Second Law of Thermodynamics can be written as $d\mathcal{S}^i \geq 0$.

This can be applied to the continuum as understood from the definition of s (D.72). That is, the change of entropy ds is divided into two parts as $ds = ds^e + ds^i$, where ds^e is a result of energy exchange, and ds^i is the internal entropy production. For the reversible process $ds^i = 0$ and for the irreversible process $ds^i > 0$. Thus the Second Law of Thermodynamics for the continuum can be written as

Second Law of Thermodynamics: $$ds = ds^e + ds^i, \qquad ds^i \geq 0. \tag{D.73}$$

Note that CIT may be able to come to the same conclusion; however, it is important that we do not require the assumption of a local-equilibrium. Since the Second

Law is used here as a working hypothesis in the local form, we can introduce this inequality locally.

If no chemical processes are considered, the change of internal energy is caused by the deformation and the change of entropy ds^e, which is the exchangeable part with the surroundings. Thus we have

$$\frac{du}{dt} = \boldsymbol{\sigma}^{\ddagger} : \boldsymbol{D}^e + T\frac{ds^e}{dt}. \tag{D.74}$$

This is one of the most important results of non-equilibrium thermodynamics.

By substituting this result (D.74) into (D.73) under $ds = ds^e + ds^i$ we obtain the following entropy production inequality for the non-chemical process:

$$-\rho T\frac{ds^i}{dt} = \rho\left(\frac{du}{dt} - T\frac{ds}{dt}\right) - \rho\boldsymbol{\sigma}^{\ddagger} : \boldsymbol{D}^e \leq 0. \tag{D.75}$$

Substituting the dissipation equation (D.75) into (3.50), we obtain

$$\frac{ds}{dt}(\boldsymbol{x}, t) \geq \frac{1}{T}\boldsymbol{\sigma}^{\ddagger} : \boldsymbol{D}^i - \frac{1}{\rho T}\operatorname{div}\boldsymbol{q} + \frac{r}{T}. \tag{D.76}$$

This gives the Eulerian form of the *Second Law of Thermodynamics* for the continuum, also referred to as the *Clausius-Duhem inequality*. Note that the first term of the r.h.s. of (D.76) is a consequence of the internal dissipation in mechanical energy. It should also be noted that the equality is provided for reversible processes whereas the pure inequality is applicable to irreversible processes.

In many articles we find the following discussions on the temperature-independent process: Since (D.76) accounts for the dissipation, the equality is satisfied, and the dissipation term, excluding the thermal terms, must be positive ($\boldsymbol{\sigma}^{\ddagger} : \boldsymbol{D}^i \geq 0$). This might be true; however, here we state only that (D.76) is true as a result of the Second Law of Thermodynamics.

In a closed system if the kinetic energy density κ defined by $(3.19)_2$ and the internal energy density u are used, the inequality (D.71) is reduced to

$$d\kappa + du = -T\,ds^i \leq 0. \tag{D.77}$$

Thus the minimum energy principle is also given for the continuum.

Appendix E
Chemical Processes and Classical Thermodynamics

In most textbooks on chemical physics and thermodynamics, field variables are not given in the form of variables per unit mass or per unit volume but in the form of variables per mole. In this section we follow the basis of classical chemical physics.

E.1 Molar Description of Thermodynamic Functions

Both descriptions in continuum mechanics and in classical chemistry are listed as follows (note that since the inelastic volume change is small enough, the volume V is used instead of V^e for the conventional notations):

Field variables	*(Continuum mechanics)*	*(Molar description)*
Strain/volume	$\boldsymbol{\varepsilon}$	V
Stress/pressure	$\boldsymbol{\sigma}^{\ddagger}$	P
Exchangeable entropy	s^e	S^e
Temperature	T	T
Chemical potential	$\mu_\alpha^{\ddagger}$	μ_α
Concentration/amount of substance	c_α	n_α
Heat flux	dq	dQ

Here S^e is the exchangeable entropy in the molar description. Note that the internal energy per unit mass u [J/M] (M is the unit of mass) is related to the internal energy per unit substance U [J/mol] by $m\,u = U$ where m is the molecular weight. Most of the textbooks using the molar description employ the pressure p (compression positive) instead of stress $\boldsymbol{\sigma}$ (cf. Sect. 3.2.4). We here use the pressure P and volume V (expansion positive) in the molar descriptions.

Y. Ichikawa and A.P.S. Selvadurai, *Transport Phenomena in Porous Media*,
DOI 10.1007/978-3-642-25333-1, © Springer-Verlag Berlin Heidelberg 2012

1. Field of volume/pressure and entropy/temperature

Thermodynamic functions	*(Continuum mechanics)*	*(Molar description)*
Internal energy	$u(\boldsymbol{\varepsilon}^e, s^e)$	$U(V, S^e)$
Helmholtz free energy	$f(\boldsymbol{\varepsilon}, T)$	$F(V, T)$
Enthalpy	$h(\boldsymbol{\sigma}^{\ddagger}, s^e)$	$H(P, S^e)$
Gibbs free energy	$g(\boldsymbol{\sigma}^{\ddagger}, T)$	$G(P, T)$

Change of internal energy

$$dU(V, S^e) = -P\,dV + T\,dS^e, \quad P = -\frac{\partial U}{\partial V}, \quad T = \frac{\partial U}{\partial S^e}. \tag{E.1}$$

Helmholtz free energy

$$F(V, T) = U(V, S^e) - TS^e, \tag{E.2}$$

$$dF = dU - d(TS^e) = -P dV - S^e\,dT, \qquad P = -\frac{\partial F}{\partial V}, \quad S^e = -\frac{\partial F}{\partial T}. \tag{E.3}$$

Enthalpy

$$H(P, S^e) = U(V, S^e) + PV, \tag{E.4}$$

$$dH = dU - d(PV) = T\,dS^e + V\,dP, \qquad V = \frac{\partial H}{\partial P}, \quad T = \frac{\partial H}{\partial S^e}. \tag{E.5}$$

Gibbs free energy

$$G(P, T) = H(P, S^e) - TS^e = F(V, T) + PV = U(V, S^e) + PV - TS^e, \tag{E.6}$$

$$dG = dU + d(PV) - d(TS^e) = V\,dP - S^e\,dT, \quad V = \frac{\partial G}{\partial P}, \; S^e = -\frac{\partial G}{\partial T}. \tag{E.7}$$

2. Field with chemical process

Thermodynamic functions	*(Continuum mechanics)*	*(Molar description)*
Internal energy	$u(\boldsymbol{\varepsilon}, s, c_\alpha)$	$U(V, S^e, n_\alpha)$
Helmholtz free energy	$f(\boldsymbol{\varepsilon}, T, c_\alpha)$	$F(V, T, n_\alpha)$
Enthalpy	$h(\boldsymbol{\sigma}^{\ddagger}, s, c_\alpha)$	$H(P, S^e, n_\alpha)$
Gibbs free energy	$g(\boldsymbol{\sigma}^{\ddagger}, T, c_\alpha)$	$G(P, T, n_\alpha)$
Grand potential	$\phi(\boldsymbol{\varepsilon}, T, \mu_\alpha^{\ddagger})$	$\Phi(V, T, \mu_\alpha)$

Change of internal energy

$$dU(V, S^e, n_\alpha) = -P\,dV + T\,dS^e + \sum_\alpha \mu_\alpha\,dn_\alpha, \tag{E.8}$$

$$P = -\left.\frac{\partial U}{\partial V}\right|_{S^e, n_\alpha}, \quad T = \left.\frac{\partial U}{\partial S^e}\right|_{V, n_\alpha}, \quad \mu_\alpha = \left.\frac{\partial U}{\partial n_\alpha}\right|_{V, S^e} \tag{E.9}$$

Euler's equation

$$U = -PV + TS^e + \sum_\alpha \mu_\alpha\,n_\alpha \tag{E.10}$$

Gibbs–Duhem relation

$$-V\,dP + S^e\,dT + \sum_\alpha n_\alpha\,d\mu_\alpha = 0 \tag{E.11}$$

Helmholtz free energy

$$F(V, T, n_\alpha) = U(V, S^e, n_\alpha) - TS^e = -PV + \sum_\alpha \mu_\alpha\,n_\alpha \tag{E.12}$$

$$dF = -P\,dV - S^e\,dT + \sum_\alpha \mu_\alpha\,dn_\alpha, \tag{E.13}$$

$$P = -\left.\frac{\partial F}{\partial V}\right|_{T, n_\alpha}, \quad S^e = -\left.\frac{\partial F}{\partial T}\right|_{n_\alpha, V}, \quad \mu_\alpha = \left.\frac{\partial F}{\partial n_\alpha}\right|_{V, T} \tag{E.14}$$

Enthalpy

$$H(P, S^e, n_\alpha) = U(V, S^e, n_\alpha) + PV = TS^e + \sum_\alpha \mu_\alpha\,n_\alpha \tag{E.15}$$

$$dH = V\,dP + T\,dS^e + \sum_\alpha \mu_\alpha\,dn_\alpha, \tag{E.16}$$

$$V = \left.\frac{\partial H}{\partial P}\right|_{S^e, n_\alpha}, \quad T = \left.\frac{\partial H}{\partial S^e}\right|_{n_\alpha, P}, \quad \mu_\alpha = \left.\frac{\partial H}{\partial n_\alpha}\right|_{P, S^e} \tag{E.17}$$

Gibbs free energy

$$G(P, T, n_\alpha) = H(P, S^e, n_\alpha) - TS^e = U(V, S^e, n_\alpha) - TS^e + PV = \sum_\alpha \mu_\alpha\,n_\alpha \tag{E.18}$$

$$dG = V\,dP - S^e\,dT + \sum_\alpha \mu_\alpha\,dn_\alpha, \tag{E.19}$$

$$V = \left.\frac{\partial G}{\partial P}\right|_{T, n_\alpha}, \quad S^e = -\left.\frac{\partial G}{\partial T}\right|_{n_\alpha, P}, \quad \mu_\alpha = \left.\frac{\partial G}{\partial n_\alpha}\right|_{P, T} \tag{E.20}$$

Grand potential

$$\Phi(V,T,\mu_\alpha) = U(V,S^e,n_\alpha) - TS^e - \sum_\alpha \mu_\alpha n_\alpha = -PV \tag{E.21}$$

$$d\Phi = -P\,dV - S^e\,dT - \sum_\alpha n_\alpha\, d\mu_\alpha, \tag{E.22}$$

$$P = -\left.\frac{\partial\Phi}{\partial V}\right|_{T,\mu_\alpha}, \quad S^e = -\left.\frac{\partial\Phi}{\partial T}\right|_{\mu_\alpha,V}, \quad n_\alpha = -\left.\frac{\partial\Phi}{\partial\mu_\alpha}\right|_{V,T} \tag{E.23}$$

Maxwell's relations

$$\left.\frac{\partial P}{\partial S^e}\right|_{n_\alpha,V} = \left.\frac{\partial T}{\partial V}\right|_{S^e,n_\alpha}, \quad \left.\frac{\partial T}{\partial n_\alpha}\right|_{V,S^e} = \left.\frac{\partial \mu_\alpha}{\partial S^e}\right|_{n_\alpha,V}, \quad \left.\frac{\partial \mu_\alpha}{\partial V}\right|_{S^e,n_\alpha} = \left.\frac{\partial P}{\partial n_\alpha}\right|_{V,S^e}, \tag{E.24}$$

$$\left.\frac{\partial P}{\partial T}\right|_{n_\alpha,V} = -\left.\frac{\partial S^e}{\partial V}\right|_{T,n_\alpha}, \quad -\left.\frac{\partial S^e}{\partial n_\alpha}\right|_{V,T} = \left.\frac{\partial \mu_\alpha}{\partial T}\right|_{n_\alpha,V}, \quad \left.\frac{\partial \mu_\alpha}{\partial V}\right|_{T,n_\alpha} = \left.\frac{\partial P}{\partial n_\alpha}\right|_{V,T}, \tag{E.25}$$

$$-\left.\frac{\partial V}{\partial S^e}\right|_{n_\alpha,P} = \left.\frac{\partial T}{\partial P}\right|_{S^e,n_\alpha}, \quad \left.\frac{\partial T}{\partial n_\alpha}\right|_{P,S^e} = \left.\frac{\partial \mu_\alpha}{\partial S^e}\right|_{n_\alpha,P}, \quad \left.\frac{\partial \mu_\alpha}{\partial P}\right|_{S^e,n_\alpha} = -\left.\frac{\partial V}{\partial n_\alpha}\right|_{P,S^e}, \tag{E.26}$$

$$-\left.\frac{\partial V}{\partial T}\right|_{n_\alpha,P} = -\left.\frac{\partial S^e}{\partial P}\right|_{T,n_\alpha}, \quad -\left.\frac{\partial S^e}{\partial n_\alpha}\right|_{P,T} = \left.\frac{\partial \mu_\alpha}{\partial T}\right|_{n_\alpha,P}, \quad \left.\frac{\partial \mu_\alpha}{\partial P}\right|_{T,n_\alpha} = -\left.\frac{\partial V}{\partial n_\alpha}\right|_{P,T}, \tag{E.27}$$

$$\left.\frac{\partial P}{\partial T}\right|_{\mu_\alpha,V} = -\left.\frac{\partial S^e}{\partial V}\right|_{T,\mu_\alpha}, \quad -\left.\frac{\partial S^e}{\partial \mu_\alpha}\right|_{V,T} = -\left.\frac{\partial n_\alpha}{\partial T}\right|_{\mu_\alpha,V}, \quad -\left.\frac{\partial n_\alpha}{\partial V}\right|_{T,\mu_\alpha} = \left.\frac{\partial P}{\partial \mu_\alpha}\right|_{V,T}. \tag{E.28}$$

$$\left.\frac{\partial U}{\partial V}\right|_T = -T^2 \left.\frac{\partial (P/T)}{\partial T}\right|_V. \tag{E.29}$$

Gibbs-Helmholtz relation

$$\frac{\partial}{\partial T}\left(\frac{G}{T}\right) = -\frac{H}{T^2} \quad \text{or} \quad \frac{\partial}{\partial T}\left(\frac{\Delta G}{T}\right) = -\frac{\Delta H}{T^2} \tag{E.30}$$

E.2 Heat of Reaction and Change of Enthalpy

We consider the relation between the reaction heat and the change of enthalpy. As shown by r.h.s. of (E.15), $H(\sigma, S^e, n_\alpha)$ is the energy due to heat and mass exchange (cf. Note 3.11, p. 115). Thus the chemical processes accompanying reaction heat such as combustion, formation, fusion and vaporization are formulated by using the enthalpy change ΔH.

Let us consider a reaction at the standard state ($P^{\ominus} = 0.1$ MPa $= 1$ bar, $T^{\ominus} = 25°\text{C} = 298.15$ K):

$$a\text{A} + b\text{B} \quad \longrightarrow \quad c\text{C} + d\text{D} \tag{E.31}$$

Original system Reaction system

If the enthalpies of material α ($\alpha =$A, B, C, D) at the standard state are $H_{\alpha}^{\ominus}$, the change in enthalpy due to this reaction is given by

$$\Delta H^{\ominus} = (H_{\text{C}}^{\ominus} + H_{\text{D}}^{\ominus}) - (H_{\text{A}}^{\ominus} + H_{\text{B}}^{\ominus}). \tag{E.32}$$

It is common that values such as the change in enthalpy are obtained by the r.h.s. value of the reaction system subtracting the l.h.s. value of the original system. $\Delta H^{\ominus}$ is referred to as the *change in standard enthalpy*.

Let us generalize the reaction formula (E.31) for N_o substances of the original system and N_r substances of the reaction system by

$$\sum_{\alpha=1}^{N_o} a_{\alpha}\text{A}_{\alpha} \rightleftharpoons \sum_{\alpha=1}^{N_r} b_{\alpha}\text{B}_{\alpha}. \tag{E.33}$$

The change in enthalpy is

$$\Delta H^{\ominus} = \sum_{\alpha=1}^{N_r} b_{\alpha} H_{\text{B}_{\alpha}}^{\ominus} - \sum_{\alpha=1}^{N_o} a_{\alpha} H_{\text{A}_{\alpha}}^{\ominus}. \tag{E.34}$$

If an intermediate reaction is formed during the reaction from the original system A to the reaction system B such as

$$\text{A} \rightarrow \text{I} \rightarrow \text{II} \rightarrow \text{B},$$

the change in enthalpy can be calculated as

$$\Delta H_{\text{A}\to\text{B}}^{\ominus} = \Delta H_{\text{A}\to\text{I}}^{\ominus} + \Delta H_{\text{I}\to\text{II}}^{\ominus} + \Delta H_{\text{II}\to\text{B}}^{\ominus} \tag{E.35}$$

where we set $\Delta H_{\text{A}\to\text{I}}^{\ominus} = H_{\text{I}}^{\ominus} - H_{\text{A}}^{\ominus}$, etc. Equation E.35 is referred to as *Hess's law*.

There are many types of enthalpy changes, as listed in Table E.1; some typical enthalpy changes are described below.

The reaction heat of combustion under constant pressure is referred to as the combustion enthalpy. If the states before and after combustion are kept standard, this is known as the *standard enthalpy of combustion*, denoted by $\Delta_c H^{\ominus}$. For example, the reaction of the combustion of methane is written as

Table E.1 Changes of enthalpy (cf. Atkins and Paula 2010)

Transition	Process	Symbol[a]
Reaction	Reactants → products	$\Delta_r H^\ominus$
Formation	Elements → compound	$\Delta_f H^\ominus$
Combustion	Compound(g,l,s)+O_2(g)→CO_2(g),H_2O(l,g)	$\Delta_c H^\ominus$
Atomization	Molecule→atoms	$\Delta_{at} H^\ominus$
Ionization	$X(g) \to X^+(g) + e^-$	$\Delta_{ion} H^\ominus$
Cathode reaction[b]	$X(g) + e^- \to X^-(g)$	$\Delta_{eg} H^\ominus$
Activation	Reactants → activated complex	$\Delta^\ddagger H^\ominus$
Transition	Phase α → Phase β	$\Delta_{trs} H^\ominus$
Fusion	s→l	$\Delta_{fus} H^\ominus$
Vaporization	l→g	$\Delta_{vap} H^\ominus$
Sublimation	s→g	$\Delta_{sub} H^\ominus$
Mixing of fluids	Pure → mixture	$\Delta_{mix} H^\ominus$
Solution	Solute → solution	$\Delta_{sol} H^\ominus$
Hydration	$X^\pm(g) \to X^\pm(aq)$	$\Delta_{hyd} H^\ominus$

[a]Standard symbols of IUPAC (International Union of Pure and Applied Chemistry)
[b]Cathode reaction implies electron gain. $\Delta_{eg} H^\ominus$ denotes the electron gain

$$CH_4(g) + 2O_2(g) \longrightarrow CO_2(g) + H_2O(l), \qquad \Delta_c H^\ominus = -890.4\ \mathrm{kJ\ mol^{-1}},\ 25°C$$

where (g) denotes the state of gas and (l) denotes the state of liquid (and the state of solid is denoted as (s)). Since the above reaction is exothermic, the enthalpy is decreased if the reaction is toward the r.h.s., which implies that $\Delta_c H^\ominus$ is negative.

The reaction heat generated when producing a compound from simple elements is referred to as the formation enthalpy. If the states before and after formation are kept standard, this is known as the *standard enthalpy of formation*, denoted by $\Delta_f H^\ominus$. In Table E.2 some examples of the above enthalpies are shown.

The standard enthalpy of formation of methane can be calculated by using the standard enthalpies of formation of CO_2 and H_2 and the standard enthalpy of the combustion of methane. That is, since we have

$$\begin{array}{ll}
\text{(a) } C(s) + O_2(g) \longrightarrow CO_2(g) & \Delta_f H^\ominus(CO_2) = -393.51\ \mathrm{kJ\ mol^{-1}} \\
\text{(b) } 2H_2(g) + O_2(g) \longrightarrow 2H_2O(l) & 2\Delta_f H^\ominus(H_2O) = 2 \times (-285.85)\ \mathrm{kJ\ mol^{-1}} \\
\text{(c) } CH_4(g) + 2O_2(g) \longrightarrow CO_2(g) + H_2O(l) & \Delta_c H^\ominus = -890.4\ \mathrm{kJ\ mol^{-1}},
\end{array}$$

Hess's law is applied, as (a)+(b)−(c), to obtain the following thermo-chemical equation:

$$C(s) + 2H_2(g) \longrightarrow CH_4(g) \qquad \Delta_f H^\ominus(CH_4) = -74.81\ \mathrm{kJ\ mol^{-1}}.$$

Table E.2 Examples of thermodynamic data (298 K) (cf. Atkins and Paula 2010)

Compound (state)	Molar mass g/mol	$\Delta_f H^\ominus$ kJ/mol	$\Delta_f G^\ominus$ kJ/mol	$S^{e\ominus}$ J/mol K	C_p J/mol K
CO_2(g)	44.010	−393.51	−394.36	213.74	37.11
CH_4(g)	16.04	−74.81	−50.72	186.26	35.31
H_2O(l)	18.015	−285.83	−237.13	69.91	75.291
H_2O(g)	18.015	−241.82	−228.57	188.83	33.58
HCl(g)	36.46	−92.31	−95.30	186.91	29.12
$CaCO_3$(g)	100.09	−1206.9	−1128.8	92.9	81.88
NaCl(s)	58.44	−411.15	−384.14	72.13	50.50
SiO_2(s,α)	60.09	−910.94	−856.64	41.84	44.43
Fe_3O_4(s,magnetite)	231.54	−1118.4	−1015.4	146.4	143.43

E.3 Change of Entropy

The classical theory of thermodynamics suggests that the state of a system is governed not only by the First Law of Thermodynamics (energy conservation) but also by the Second Law of Thermodynamics (increase of entropy). That is, a reaction of materials in general not only tends to proceed toward a low level of energy, which implies that the reaction is exothermic but also tends toward disordering (i.e., toward $+\Delta S$). As mentioned in Sect. D.4 the change of entropy ΔS consists of the part ΔS^e exchangeable with surroundings and the part ΔS^i generated autonomously in the system ($\Delta S = \Delta S^e + \Delta S^i$), and, taking this into consideration, we will examine some chemical processes.

E.3.1 Progress of Chemical Reaction and Change of Entropy

In stoichiometry the l.h.s. terms are transferred to the r.h.s., so that

$$\sum_{\alpha=1}^{N} \nu_\alpha \mathrm{X}_\alpha = 0, \tag{E.36}$$

$$\nu_\alpha = \begin{cases} -a_\alpha & \text{for originals,} \\ b_\alpha & \text{for reactants,} \end{cases} \qquad \mathrm{X}_\alpha = \begin{cases} -\mathrm{A}_\alpha & \text{for originals,} \\ \mathrm{B}_\alpha & \text{for reactants,} \end{cases} \tag{E.37}$$

where $N = N_o + N_r$, and ν_α is the stoichiometric coefficient.

Since this reaction simultaneously proceeds for all elements X_α, the change in the amount of substance of X_α can be given by the following stoichiometric relation:

$$\frac{dn_\alpha}{\nu_\alpha} \equiv d\xi \qquad (!\alpha). \tag{E.38}$$

Here ξ is referred to as the extent of reaction, and the reaction rate is given as the rate of the extent of reaction:

$$w = \frac{d\xi}{dt}. \tag{E.39}$$

Let the amount of substance of a species X_α at the start of the reaction be $n_{\alpha 0}$, then the amount at the extent ξ is given by

$$n_\alpha = n_{\alpha 0} + \nu_\alpha \xi. \tag{E.40}$$

Since the amount at ξ is given by (E.40) and $n_{\alpha 0},\ \nu_\alpha$ are constant, the internal energy at ξ can be represented as

$$U(V, S^e, n_\alpha) \longrightarrow U(V, S^e, \xi) \tag{E.41}$$

where the extent of reaction is a state variable instead of the molar number. Note that the heat of reaction r can be calculated by the enthalpy $H(P, S^e, \xi)$ as

$$r = \left.\frac{\partial H}{\partial \xi}\right|_{P,T}. \tag{E.42}$$

Let us use $dS,\ dS^e,\ dS^i$ instead of $\Delta S,\ \Delta S^e,\ \Delta S^i$:

$$dS = dS^e + dS^i. \tag{E.43}$$

The change of internal energy is given by (E.8), therefore we have

$$dU(V, S^e, n_\alpha) = -P\,dV + T\,dS^e + \sum_\alpha \mu_\alpha\,dn_\alpha. \tag{E.44}$$

This gives

$$dS^e = \frac{dU + P\,dV}{T} - \frac{1}{T}\sum_\alpha \mu_\alpha\,dn_\alpha. \tag{E.45}$$

The Second Law of Thermodynamics is then

$$dS^i \geq 0. \tag{E.46}$$

E.3.2 Affinity and Direction of Reaction

The stoichiometric relation given by (E.38)

$$dn_\alpha = \nu_\alpha\,d\xi$$

is substituted into (E.45) to obtain

$$dS^e = \frac{dU + P\,dV}{T} + \frac{A}{T}d\xi \tag{E.47}$$

where

$$A = -\sum \mu_\alpha \nu_\alpha \tag{E.48}$$

is the *affinity*, which represents the driving force of the chemical reaction. By using the affinity, the change of internal energy given by (E.44) is represented as

$$dU(V, S^e, n_\alpha) = -P\,dV + T\,dS^e - Ad\xi. \tag{E.49}$$

Let us consider the following reaction:

$$\sum_{\alpha=1}^{N_o} a_\alpha \mathrm{A}_\alpha \rightleftharpoons \sum_{\alpha=1}^{N_r} b_\alpha \mathrm{B}_\alpha \qquad \Rightarrow \qquad \sum_{\alpha=1}^{N} \nu_\alpha \mathrm{X}_\alpha = 0. \tag{E.50}$$

At chemical equilibrium the total amount of mass is not changed, therefore the change of internal energy dU_c due to chemical reaction is null:

$$dU_c = \sum_\alpha \mu_\alpha\,dn_\alpha = -Ad\xi = 0.$$

Thus at chemical equilibrium we have

$$\frac{d\xi}{dt} = 0 \qquad \text{or} \qquad A = -\sum_\alpha \mu_\alpha \nu_\alpha = 0$$

The condition $w = d\xi/dt = 0$ implies a state of null reaction rate, which is not relevant. The non-obvious solution is obtained as $A = 0$, which implies equilibrium of the affinity. Thus, in general, the equilibrium condition is given by

$$A = -\sum_\alpha \mu_\alpha \nu_\alpha = 0. \tag{E.51}$$

The non-equilibrium reactions are classified according to the following four cases:

	$d\xi/dt$	A	dU_c
Case 1	> 0	> 0	< 0
Case 2	> 0	< 0	> 0
Case 3	< 0	> 0	< 0
Case 4	< 0	< 0	> 0

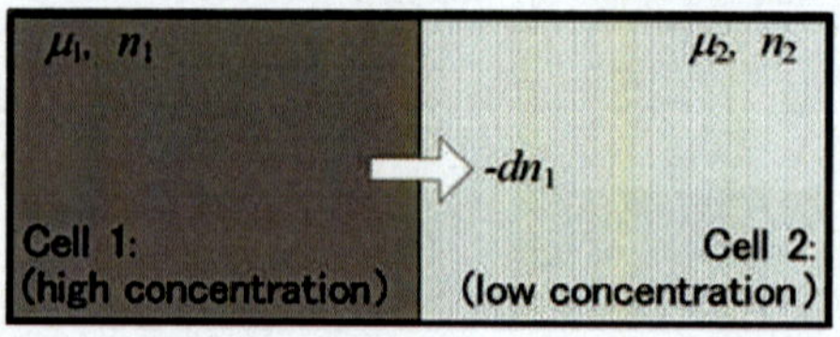

Fig. E.1 Diffusion problem of closed two-partitioned domain

Because of (E.50), the products increase and the originals decrease, therefore the reaction moves to the right (forward reaction) if $d\xi/dt > 0$. If products decrease and the originals increase, the reaction moves to the left (backward reaction) if $d\xi/dt < 0$. On the other hand, if $dU_c = -A\,d\xi/dt < 0$, the reaction is exothermic, and if $dU_c > 0$, the reaction is endothermic. Thus Case 1 is a forward exothermic reaction, Case 2 a forward endothermic reaction, Case 3 a backward exothermic reaction, and Case 4 a backward endothermic reaction.

E.3.3 Change of Entropy and the Diffusion Process

Let us suppose that two parts of substances (Part 1 and 2 of Fig. E.1) of different concentration are in contact with each other. We can prove that diffusion occurs from the higher concentration part (Part 1) to the lower concentration part (Part 2) because of the First Law of Thermodynamics. Here the total system is closed, the temperature is the same, T, the chemical potentials are μ_1 and μ_2 ($\mu_1 > \mu_2$), and the amounts of substance are n_1 and n_2, respectively. We set the stoichiometric coefficients of Part 1 and Part 2 as -1 and 1, respectively; thus the extent of reaction $d\xi$ is

$$d\xi = -dn_1 = dn_2 > 0.$$

Let $dU = dV = 0$, then by (E.45) we have

$$dS^e = -\frac{\mu_2 - \mu_1}{T}d\xi > 0. \tag{E.52}$$

That is, transport of the substance is autonomously caused, which implies 'diffusion' from the higher concentration to the lower concentration. It should be noted that this is a result of the First Law of Thermodynamics.

E.3.4 Changes of Temperature/Pressure/Phase and Entropy

Let us consider a closed system, $dn_\alpha = 0$. From (E.8) and (E.16) we have

$$\left.\frac{\partial S^e}{\partial U}\right|_V = T^{-1}, \qquad \left.\frac{\partial S^e}{\partial H}\right|_P = T^{-1},$$

and the changes of entropy under constant volume and constant pressure are calculated as follows:

$$\left.\frac{\partial S^e}{\partial T}\right|_V = \left.\frac{\partial S^e}{\partial U}\right|_V \left.\frac{\partial U}{\partial T}\right|_V = T^{-1}C_v, \qquad \left.\frac{\partial S^e}{\partial T}\right|_P = \left.\frac{\partial S^e}{\partial H}\right|_P \left.\frac{\partial H}{\partial T}\right|_P = T^{-1}C_p.$$

Thus if C_v, C_p are known as functions of temperature, the entropies that are exchangeable with the surroundings are given by

$$\text{Constant volume:} \quad S^e = \int_0^T C_v(T)\, d(\ln T), \tag{E.53}$$

$$\text{Constant pressure:} \quad S^e = \int_0^T C_p(T)\, d(\ln T). \tag{E.54}$$

Here we assume that, by virtue of the Third Law, the entropy is zero at 0 K.

When a substance undergoes a phase change, heat is adsorbed or discharged under constant temperature. In general, this change occurs under constant pressure, therefore $\Delta Q = \Delta H$. Table E.3 shows the temperature T_{fus} and change in enthalpy $\Delta_{\text{fus}} H$ for fusion, and the temperature T_{vap} and change in enthalpy $\Delta_{\text{vap}} H$ for vaporization for several substances; we can calculate the changes in entropy at fusion and vaporization by

$$\Delta_{\text{fus}} S^e = \frac{\Delta_{\text{fus}} H}{T_{\text{fus}}}, \qquad \Delta_{\text{vap}} S^e = \frac{\Delta_{\text{vap}} H}{T_{\text{vap}}}. \tag{E.55}$$

For example, the changes of entropy at fusion and vaporization of water are

$$\Delta_{\text{fus}} S^e(H_2O) = 21.995 \text{ J/K mol}, \qquad \Delta_{\text{vap}} S^e(H_2O) = 108.953 \text{ J/K mol}.$$

The change of entropy of water is schematically drawn as a function of temperature in Fig. E.2.

Table E.3 Change in enthalpy at fusion and vaporization

Substance	T_{fus} K	$\Delta_{\text{fus}} H$ kJ/mol	T_{vap} K	$\Delta_{\text{vap}} H$ kJ/mol
H_2O	273.15	6.008	373.15	40.656
CH_3OH	175.2	3.16	337.2	35.27
C_2H_5OH	156	4.60	351.4	38.56
CH_4	90.68	0.941	111.7	8.18
CO_2	217.0	8.33	194.6	25.23
N_2	63.15	0.719	77.35	5.586
O_2	54.36	0.444	90.18	6.820

Fig. E.2 Entropy change of water

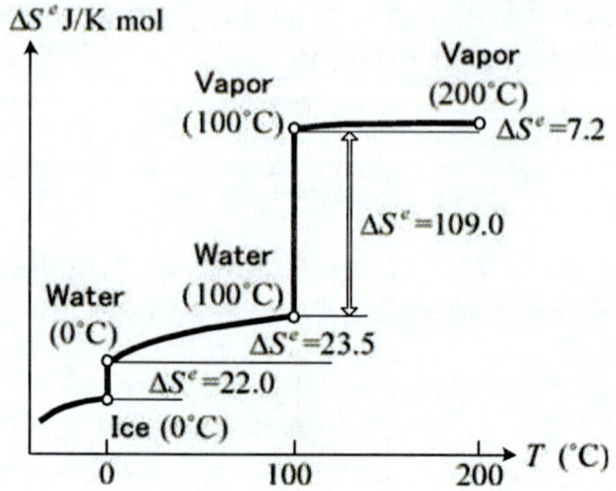

E.4 Change of Gibbs Free Energy and Chemical Potential

As understood from (E.19), the Gibbs free energy is the energy due to substances flowing in/out the system:

$$G(P,T,n_\alpha) = H(P,S^e,n_\alpha) - TS^e = \sum \mu_\alpha n_\alpha. \tag{E.56}$$

Gibbs' free energy is also a part of energy that is converted to work, but not converted to heat. We call the process where $\Delta G < 0$ the exergonic reaction, whereas $\Delta G > 0$ is the endergonic reaction ('ergon' implies work).

If the states before and after the reaction are kept standard and the change of entropy is $(\Delta S^e)^\ominus$ (change of standard entropy), the change in standard Gibbs free energy $(\Delta S^e)^\ominus$ is given by

$$\Delta G^\ominus = \Delta H^\ominus - T(\Delta S^e)^\ominus. \tag{E.57}$$

In particular, if the compound is generated from simple substances and the states before and after the reaction are kept standard, we designate $\Delta_f G^\ominus$, which is the *standard Gibbs free energy of formation*, as

$$\Delta_f G^\ominus = \Delta_f H^\ominus - T(\Delta_f S^e)^\ominus. \tag{E.58}$$

Some examples of thermodynamic data are given in Table E.2.

If we know the standard Gibbs energies of formation of each substance for the original system $\Delta_f G^\ominus_{A_\alpha}$ and the reaction system $\Delta_f G^\ominus_{B_\alpha}$, the change in standard Gibbs free energy $\Delta G^\ominus$ is calculated by

$$\Delta G^\ominus = \sum_{\alpha=1}^{N_r} b_\alpha \Delta_f G^\ominus_{B_\alpha} - \sum_{\alpha=1}^{N_o} a_\alpha \Delta_f G^\ominus_{A_\alpha}. \tag{E.59}$$

It is understood from (E.19) that the chemical potential μ_α is equivalent to the Gibbs free energy per unit mole. In other words we have

$$\mu_\alpha = \frac{\partial G}{\partial n_\alpha}. \tag{E.60}$$

Differentiating the Gibbs–Helmholtz relation (E.30) w.r.t. n_α and applying the above relation, we obtain

$$\frac{\partial}{\partial T}\left(\frac{\mu_\alpha}{T}\right) = -\frac{H_{\mathrm{m}\alpha}}{T^2}, \qquad H_{\mathrm{m}\alpha} = \frac{\partial H}{\partial n_\alpha} \tag{E.61}$$

where $H_{\mathrm{m}\alpha}$ is the partial molar enthalpy of species α (cf. Kondepudi and Prigogine 1998, Sect. 5.3, pp. 137 and Sect. 5.5, pp. 142–143).

E.5 Thermodynamics of Gas

The state quantity of a perfect gas can be decided by the following *state equation*:

$$\frac{PV}{T} = nR = Nk_B \tag{E.62}$$

where P is the pressure, V is the volume, T is the absolute temperature, n is the molar number, N is the number of particles of the system, $R = N_A k_B = 8.31$ $\mathrm{J\,K^{-1}\,mol^{-1}}$ is the gas constant, $N_A = 6.0221367 \times 10^{23}$ is Avogadro's number (the number of particles involved in one mole of substance), $k_B = R/N_A = 1.380658 \times 10^{-23}\ \mathrm{J\,K^{-1}}$ is Boltzmann's constant.

The temperature does not change if the perfect gas, enclosed in an adiabatic container, is released in a vacuum. This is because the internal energy U of the perfect gas is simply decided by the temperature, which is kinetic. We can prove this as follows: Assume that the perfect gas is contained in a cube with rigid walls of sides L. If the gas molecules undergo elastic collisions at the rigid wall in the x-direction and the mean velocity of the molecules is $\bar{v}_x$, the change of linear momentum ΔLM before and after a collision is $2m\,\bar{v}_x$, and the mean free path is $2L$, the mean number of collisions N_{col} per unit time against the wall is $\bar{v}_x/2L$. Thus the pressure of the wall area L^2 can be calculated by

$$P_x = \frac{\Delta \mathrm{LM} \times N_{col}}{L^2} = n\,N_A \times 2m\,\bar{v}_x \times \frac{\bar{v}_x}{2L} \times \frac{1}{L^2}.$$

The motion of the gas molecules is statistically homogeneous in all directions, and the mean velocity $\bar{v}$ is

$$\bar{v}_x^2 = \frac{1}{3}\bar{v}^2.$$

Set $P = P_x$, and we obtain

$$PV = \frac{1}{3} n\,N_A\,m\,\bar{v}^2$$

where $V = L^3$. Since the perfect gas satisfies the state equation (E.62), the kinetic energy of n mole of gas is calculated by

$$K = \frac{1}{2} n N_A \times m \bar{v}^2 = \frac{3}{2} nRT.$$

This shows that since the internal energy of the perfect gas is only kinetic energy, it is a function only of the temperature and we obtain

$$U = \frac{3}{2} nRT. \tag{E.63}$$

Recall that the kinetic energy of one molecule is given by

$$\frac{3}{2} k_B T.$$

Thus Boltzmann's constant is calculated as

$$k_B = \frac{R}{N_A} = \frac{8.314510}{6.0221367 \times 10^{23}} = 1.380658 \times 10^{-23} \mathrm{J\,K^{-1}}.$$

In a reversible process in a perfect gas defined by the state equation (E.62), we have

$$dQ = dU + PdV = C_v dT + \frac{nRT\,dV}{V}. \tag{E.64}$$

This implies that

$$dS^e = \frac{dQ}{T} = C_v \frac{dT}{T} + \frac{nR\,dV}{V}. \tag{E.65}$$

Thus the change of entropy between the state 1 and the state 2 can be written as

$$\Delta S^e = S_2^e - S_1^e = \int_1^2 C_v\, d(\ln T) + \int_1^2 nR\, d(\ln V).$$

Assuming that C_v is independent of the temperature, we have

$$\Delta S^e = C_v \ln \frac{T_2}{T_1} + nR \ln \frac{V_2}{V_1}. \tag{E.66}$$

For n mole of perfect gas, the Gibbs–Duhem relation (E.11) gives

$$d\mu = -S_{\mathrm{m}}^e\, dT + V_{\mathrm{m}}\, dP, \qquad S_{\mathrm{m}}^e = \frac{S^e}{n}, \quad V_{\mathrm{m}} = \frac{V}{n} \tag{E.67}$$

where S_{m}^e and V_{m} are the entropy per unit mole and the volume per unit mole, respectively. Let us assume that the chemical potential $\mu(P_0, T)$ is known for any pressure P_0 and temperature T, and we integrate (E.67) under $dT = 0$, to obtain

$$u(P, T) = \mu(P_0, T) + \int_{P_0}^{P} V_{\mathrm{m}}(P', T)\, dP'.$$

For the perfect gas $V_{\mathrm{m}} = RT/P$, and substituting this into the above equation, we obtain

$$\mu(P, T) = \mu^0(P_0, T) + RT \ln(P/P_0). \tag{E.68}$$

For the case of a real gas, we introduce a modification coefficient γ, which gives

$$\mu(f, T) = \mu^0 + RT \ln \gamma P/P_0 = \mu^0 + RT \ln f/P_0 \tag{E.69}$$

where $f = \gamma P$ is referred to as the fugacity.

Let us consider a mixture of perfect gases, which is also a perfect gas. In this case the total energy is given by the sum of each energy, therefore we have

$$\mu(P, T) = \sum_{\alpha} \mu_\alpha(P, T), \qquad \mu_\alpha(P, T) = \mu_\alpha^0(T) + RT \ln P_\alpha/P_0. \tag{E.70}$$

Let the molar fraction be $x_\alpha = n_\alpha/n$ and the partial pressure be $P_\alpha = x_\alpha P$, so that the chemical potential of species α is given by

$$\mu_\alpha(P, T, x_\alpha) = \bar{\mu}_\alpha(P, T) + RT \ln x_\alpha/P_0, \quad \bar{\mu}_\alpha(P, T) = \mu_\alpha^0(T) + RT \ln P/P_0 \tag{E.71}$$

where $\bar{\mu}_\alpha(P, T)$ is the chemical potential of pure gas of species α. For the mixture of real gases we can use the fugacity $f_\alpha = P_\alpha/P$ of species α to obtain

$$\mu_\alpha(P, T, f_\alpha) = \bar{\mu}_\alpha(P, T) + RT \ln f_\alpha/P_0. \tag{E.72}$$

E.6 Diffusion Behavior of Solutions without Inter-molecular Interaction

In this section we look at several diffusion problems of dilute solutions in which no inter-molecular interactions are considered.

E.6.1 Chemical Potential

The chemical potential of an ideal solution is given in the same form as that for a perfect gas (E.71):

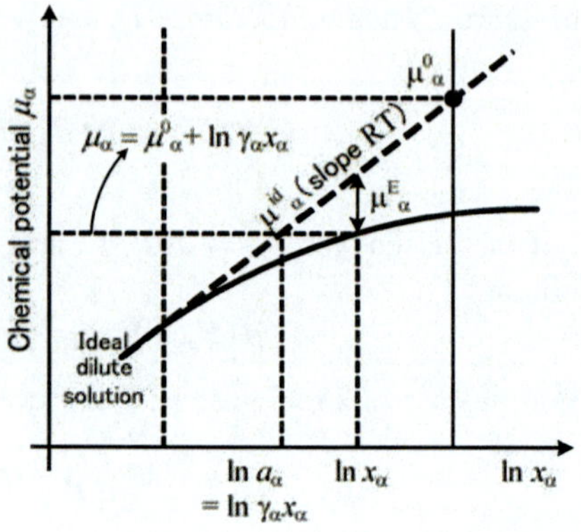

Fig. E.3 Chemical potential and concentration of solution

$$\mu^{id}_\alpha(P, T, x_\alpha) = \mu^0_\alpha(P, T) + RT \ln x_\alpha \tag{E.73}$$

where $\mu^0_\alpha(P, T)$ is the chemical potential at the standard state.

For the real solution we introduce the *activity* a_α as

$$a_\alpha = \gamma_\alpha x_\alpha \qquad (!\alpha) \tag{E.74}$$

and substituting this into x_α of (E.73), we obtain the following chemical potential:

$$\mu_\alpha(P, T, a_\alpha) = \mu^0_\alpha(P, T) + RT \ln \gamma_\alpha x_\alpha = \mu^0_\alpha(P, T) + RT \ln a_\alpha \qquad (!\alpha) \tag{E.75}$$

where γ_α is the *activity coefficient*. The difference between the real solution and the ideal one μ^E_α is referred to as the excess chemical potential:

$$\mu^E_\alpha = \mu_\alpha - \mu^{id}_\alpha = RT \ln \gamma_\alpha. \tag{E.76}$$

These are illustrated in Fig. E.3.

E.6.2 Diffusion Coefficient in an Ideal Solution: Stokes–Einstein Equation

Let there be a dilute solution in which molecules of the solute do not collide with each other, and suppose that the molecules of the solute move under viscous drag in a homogeneous and continuous solvent. Let us consider a solution of two elements, i.e., one solute and one solvent, and the radius of each molecule of a spherical shape be R_0. Then by Stokes' law of viscosity, the viscous drag is $f v_1 = (6\pi\eta R_0) v_1$ where η is the viscosity of the solvent, and f is the viscous coefficient. We can set this force equivalent to the negative gradient of the chemical potential of the solute:

$$-\nabla\mu_1 = f\boldsymbol{v}_1 = (6\pi\eta R_0)\boldsymbol{v}_1. \tag{E.77}$$

Let the concentrations of solute and solvent be c_1 and c_2, respectively. The solution is assumed to be sufficiently dilute, and it is approximated as an ideal solution:

$$\mu_1 = \mu_1^0 + RT\ln x_1 = \mu_1^0 + RT\ln\frac{c_1}{c_1+c_2} \simeq \mu_1^0 + RT\ln c_1 - RT\ln c_2. \tag{E.78}$$

Here $c_1 \ll c_2$, and we can assume that c_2 is constant. Differentiating this, we obtain

$$\nabla\mu_1 = \frac{RT}{c_1}\nabla c_1, \tag{E.79}$$

and substituting this into (E.77), we obtain the following flux of diffusing concentration:

$$\boldsymbol{q}_1^m = c_1\boldsymbol{v}_1 = -\frac{RT}{6\pi\,\eta\,R_0}\nabla c_1 \quad (= -D\nabla c_1) \tag{E.80}$$

where we assume the mean velocity $\boldsymbol{v}$ is zero. Then the diffusion coefficient D can be calculated as

$$D = \frac{RT}{6\pi\,\eta\,R_0}. \tag{E.81}$$

This is referred to as the *Stokes-Einstein equation*. However it does not give a good approximation for a real solution, therefore we can modify it as

$$D = \frac{RT}{6\pi\,\eta\,R_0}(1 + 1.5\phi_1 + \cdots) \tag{E.82}$$

where ϕ_1 is the volume fraction of the solute. Details can be found in Chap. 5, Cussler (1997).

E.6.3 Diffusion Coefficient of Solute without Inter-molecular Interaction

For the non-ideal solution of two elements the chemical potential can be written as

$$\mu_1 = \mu_1^0 + RT\ln\gamma_1 x_1 \simeq \mu_1^0 + RT\ln\gamma_1 c_1 - RT\ln\gamma_1 c_2. \tag{E.83}$$

Let the velocity of the solute be given by (E.77):

$$\boldsymbol{v}_1 = -\frac{1}{f}\nabla\mu_1 = -\frac{D_0}{RT}\nabla\mu_1 \tag{E.84}$$

where D_0 is a constant. Differentiating (E.83), we obtain the flux of the diffusing concentration as

$$q_1^m = c_1 v_1 = -c_1 D_0 \left(\frac{\nabla c_1}{c_1} + \frac{\nabla \gamma_1}{\gamma_1} \right) = -D_0 \left(1 + \frac{\partial \ln \gamma_1}{\partial \ln c_1} \right) \nabla c_1 \tag{E.85}$$

This gives *Fick's law*. Thus the diffusion coefficient can be evaluated as

$$D = D_0 \left(1 + \frac{\partial \ln \gamma_1}{\partial \ln c_1} \right). \tag{E.86}$$

E.7 Diffusion Process in an Electrolyte Solution

We discuss the diffusion behavior of an electrolyte solute in which the intermolecular interaction is strong.

E.7.1 Chemical Potential of Electrolyte Solute

In Sect. 3.2.4 the internal energy was treated as a sum of the mechanical and heat energies. Here, in addition, we discuss the effect of an electrostatic field ϕ on the change of internal energy dU_e under the molar description.

Let the electrical charge dQ move in the electrostatic field ϕ, then the change of internal energy is given by

$$dU_e = \phi \, dQ. \tag{E.87}$$

Note that ϕ is the difference from the reference charge. If an amount of dn_α of ionic species α with charge z_α is added to the electrolyte solute, then the change of internal energy is given as

$$dU_e = z_\alpha F \phi \, dn_\alpha \qquad (!\alpha)$$

where $F = 9.64853 \times 10^4$ C/mole is the Faraday constant. Thus the change of chemical potential due to the above charge is accounted for as $z_\alpha F \phi$;, thus the chemical potential μ_α defined by (E.75) becomes

$$\tilde{\mu}_\alpha = \mu_\alpha + z_\alpha F \phi. \tag{E.88}$$

$\tilde{\mu}_\alpha$ is referred to as *electrochemical potential*.

Note E.1 (Chemical potential in electrochemistry). In (E.88) the chemical potential is represented as the sum of the chemical potential μ_α and the change of electrostatic energy $z_\alpha F\phi$; however, this is a conventional treatment. In view of the physical situation and measurability, the contribution of these two phenomena to the chemical potential cannot be separated; since most of the electrochemical phenomena can be expressed by (E.88), we will conventionally use it here.

Since thermodynamic quantities such as the activity and activity coefficient for ionized species, give no physical background, they must be related to the original species by electrical neutrality. Let 1 mole of electrolyte A be ionized into ν_+ mole of cations with charge z_+ and ν_- mole of anions with charge z_-; then, from the electroneutrality principle we have

$$\nu_+ z_+ + \nu_- z_- = 0. \tag{E.89}$$

The chemical potential μ_A of the electrolyte A (electroneutral) consists of the sum of two ionized ions; however, assuming a constant charge in the solution, the electrostatic terms $z_\alpha F\phi$ are canceled out, and we have

$$\mu_A = \nu_+\tilde{\mu}_+ + \nu_-\tilde{\mu}_- = \nu_+\mu_+ + \nu_-\mu_-, \qquad \mu_A^0 = \nu_+\mu_+^0 + \nu_-\mu_-^0. \tag{E.90}$$

Let the chemical potential be written by (E.75), then the chemical potential μ_A and activity a_A of the species A can be calculated as

$$\mu_A = \nu_+(\mu_+^0 + RT\ln a_+) + \nu_-(\mu_-^0 + RT\ln a_-) = \mu_A^0 + RT\ln a_A \tag{E.91}$$

$$a_A = a_+^{\nu_+} a_-^{\nu_-}. \tag{E.92}$$

Let the molar fractions of electrolyte A, cation and anion be x_A, x_+, x_-, respectively, giving

$$x_+ = \nu_+ x_A, \quad x_- = \nu_- x_A.$$

Set $\nu = \nu_+ + \nu_-$ to obtain the activity a_A as

$$a_A = \left(\nu_+^{\nu_+}\nu_-^{\nu_-}\right)\left(\gamma_+^{\nu_+}\gamma_-^{\nu_-}\right) x_A^{\nu} \tag{E.93}$$

where γ_+ and γ_- are the activity coefficients of the cation and anion, respectively. The mean activity $a_\pm$, mean activity coefficient $\gamma_\pm$ and mean ionic molar fraction $x_\pm$ of the electrolyte A are defined by

$$a_\pm = (a_A)^{1/\nu}, \quad \gamma_\pm = \left(\gamma_+^{\nu_+}\gamma_-^{\nu_-}\right)^{1/\nu}, \quad x_\pm = \left(\nu_+^{\nu_+}\nu_-^{\nu_-}\right)^{1/\nu} x_A, \tag{E.94}$$

and (E.93) can be written as

$$a_\pm = \gamma_\pm x_\pm \tag{E.95}$$

■

E.7.2 Electrostatic Field due to Distributed Ions and Activity Coefficient: Debye–Hückel Theory

The electrostatic field ϕ must be known in order to identify the electrochemical potential; however, this is almost impossible. In order to avoid this difficulty, in electrochemical engineering the electrostatic field is determined by assuming the distribution of ions (Tamamushi 1991).

Let the concentration number n_α (i.e., the number of species per unit volume) of the ion α yield Boltzmann's distribution[1]:

$$n_\alpha = n_\alpha^0 \exp\left(-\frac{z_\alpha e\phi}{k_B T}\right) \tag{E.96}$$

where n_α^0 is the number concentration at $\phi = 0$, $e = 1.60217733 \times 10^{19}$ C is the elementary charge and k_B is Boltzmann's constant. The charge due to n_α is $\rho_\alpha = z_\alpha e n_\alpha$, and the charge ρ at a point p can be given by summing up all the distributed charges:

$$\rho = \sum_\alpha \rho_\alpha = \sum_\alpha z_\alpha e\, n_\alpha^0 \exp\left(-\frac{z_\alpha e\phi}{k_B T}\right). \tag{E.97}$$

Let $\boldsymbol{E}$ be the electric field. Since $\nabla \wedge \boldsymbol{E} = \boldsymbol{0}$ for the electrostatic field, we have the electromagnetic potential ϕ such that $\boldsymbol{E} = -\nabla\phi$ (recalling $\nabla \wedge \nabla\phi \equiv \boldsymbol{0}$). Let the electric flux density be $\boldsymbol{D} = \varepsilon \boldsymbol{E}$ ($\varepsilon = \varepsilon_0 \varepsilon_r$, ε the dielectric constant, ε_0 the dielectric constant in vacuum, ε_r the relative dielectric constant), then the principle of conservation of charge (i.e., Coulomb's law) $\nabla \cdot \boldsymbol{D} = \rho$ gives

$$\Delta\phi = -\frac{\rho}{\varepsilon} \tag{E.98}$$

($\Delta = \nabla \cdot \nabla$ is Laplace's operator). Assuming that the electrostatic energy is significantly lower than the heat energy ($z_\alpha e\phi / k_B T \ll 1$), a Taylor expansion can be applied to the r.h.s. of (E.98):

$$-\frac{e}{\varepsilon}\sum_\alpha z_\alpha e n_\alpha^0 \exp\left(-\frac{z_\alpha e\phi}{k_B T}\right) \simeq -\frac{e}{\varepsilon}\sum_\alpha z_\alpha n_\alpha^0 \left(1 - \frac{z_\alpha e\phi}{k_B T}\right) = \frac{e^2}{\varepsilon k_B T}\sum_\alpha n_\alpha^0 z_\alpha^2 \phi.$$

Here we apply the electroneutrality principle and set $\sum_\alpha z_\alpha e n_\alpha^0 = 0$. Then (E.98) is approximated as

[1] It is an assumption that each ion is a point charge and the distribution occurs according to the Boltzmann rule. Note that the number concentration n_α is related to the volume molar concentration c_α shown in Table 3.1 as $n_\alpha = N_A c_\alpha$, where N_A is the Avogadro's number.

$$\Delta\phi = \kappa^2\phi \tag{E.99}$$

$$\kappa^2 = \frac{e^2}{\varepsilon k_B T}\sum_\alpha n_\alpha^0 z_\alpha^2 = \frac{2N_A^2 e^2}{\varepsilon RT} I, \qquad I = \frac{1}{2}\sum_\alpha c_\alpha z_\alpha^2, \quad c_\alpha = \frac{n_\alpha^0}{N_A} \tag{E.100}$$

where κ^{-1} is the Debye radius and I is the ionic strength.

Let us consider a spherical polar coordinate system around the charge ρ. In this system the differential equation (E.99) becomes

$$\frac{1}{r^2}\frac{d}{dr}\left(r^2\frac{d\phi}{dr}\right) = \kappa^2\phi. \tag{E.101}$$

If $\phi \to 0$ as $r \to 0$, the solution of (E.101) is given by

$$\phi = A\frac{\mathrm{e}^{-\kappa r}}{r}$$

where A is a constant. Substituting this into (E.98) and (E.99), we obtain

$$\rho = -A\varepsilon\kappa^2\frac{\mathrm{e}^{-\kappa r}}{r}. \tag{E.102}$$

Let the mean ionic diameter be a and assume that no other ion can come into the space inside the radius $a/2$, then from the electroneutrality principle we obtain

$$\int_a^\infty 4\pi r^2\rho\, dr = -z_\alpha e.$$

Next, substituting the above result in (E.102), we calculate A as

$$A = \frac{z_\alpha e}{4\pi\varepsilon}\cdot\frac{\mathrm{e}^{\kappa a}}{1+\kappa a}.$$

The electrostatic potential ϕ is then obtained as

$$\phi(r) = \frac{z_\alpha e}{4\pi\varepsilon}\cdot\frac{\mathrm{e}^{\kappa a}}{1+\kappa a}\frac{\mathrm{e}^{-\kappa r}}{r}. \tag{E.103}$$

In the dilute solution, $\kappa a \ll 1$, therefore ϕ is given by

$$\phi(r) = \frac{z_\alpha e}{4\pi\varepsilon}\cdot\frac{\mathrm{e}^{-\kappa r}}{r}. \tag{E.104}$$

This potential involves the following electric potential, which is the effect of the central ion α:

$$\phi_\alpha(r) = \frac{z_\alpha e}{4\pi\varepsilon} \cdot \frac{1}{r}, \tag{E.105}$$

and we exclude this to obtain the remaining part

$$\phi_r(r) = \phi(r) - \phi_\alpha(r) = -\frac{z_\alpha e}{4\pi\varepsilon} \cdot \frac{1 - \mathrm{e}^{-\kappa r}}{r} \tag{E.106}$$

as the outer potential, which is the effect of the ions distributed outside of the central charge. If $r \ll \kappa^{-1}$, the outer potential becomes

$$\phi_r(r) = -\frac{z_\alpha e \kappa}{4\pi\varepsilon}. \tag{E.107}$$

This is obtained as $r \to 0$ for the r.h.s. of (E.106).

Let us evaluate the activity coefficient by using the Debye-Hückel theory: Recall that the excess chemical potential μ_α^{E} is given by (E.76):

$$\mu_\alpha^{\mathrm{E}} = RT \ln \gamma_\alpha. \tag{E.108}$$

We can assume that this discrepancy from the ideal solution is mainly caused by the electrostatic interaction between ions. Let the molar fraction of ionic species α with charge z_α be x_α, the internal energy of the single ion be $u_e(x_\alpha)$, and the internal energy at the molar fraction zero (i.e., infinitely dilute condition) be $u_e(x_\alpha = 0) = u_e^0$. Then we have

$$\mu_\alpha^{\mathrm{E}} = N_A \left(u_e(x_\alpha) - u_e^0 \right) \tag{E.109}$$

where N_A is Avogadro's number. When one ionic particle of species α is added, the internal energy is calculated by using the electric potential ϕ of (E.104) as

$$u_e(x_\alpha) = \int_0^{z_\alpha e} \phi \, dQ$$

where $Q = z_\alpha e$. On the other hand, in the case of infinite dilution, $\phi_r = 0$, therefore we have

$$u_e^0 = \int_0^{z_\alpha e} \phi_\alpha \, dQ.$$

Thus μ_α^{E} can be evaluated as

$$\mu_\alpha^{\mathrm{E}} = N_A \int_0^{z_\alpha e} \phi_r \, dQ. \tag{E.110}$$

That is, the excess chemical potential can be estimated as the effect of the outer potential.

Substituting (E.107) into (E.110) and integrating, we have

$$\mu_\alpha^{\mathrm{E}} = -N_A \frac{z_\alpha^2 e^2 \kappa}{8\pi\varepsilon}, \tag{E.111}$$

and we obtain $\ln \gamma_\alpha$ from (E.108):

$$\ln \gamma_\alpha = -\frac{z_\alpha^2 e^2 \kappa}{8\pi\varepsilon k_B T} = -A z_\alpha^2 \sqrt{I}, \qquad A = \frac{(2N_A)^{1/2} e^3}{8\pi(\varepsilon k_B)^{3/2}} \cdot \frac{1}{T^{3/2}} \tag{E.112}$$

where (E.100) is used. From this equation, the mean ionic activity coefficient $\gamma_\pm$ of a strong electrolyte solution with a charge z_1 of cations and a charge z_2 of anions can be calculated as

$$\ln \gamma_\pm = -A|z_a z_2|\sqrt{I}. \tag{E.113}$$

For a dilute solution of a strong electrolyte, this gives the limiting Debye–Hückel law.

The results (E.112) and (E.113) are calculated using the approximated potential (E.104). This approximation can be improved if we apply (E.103) in which the assumption of a dilute solution is avoided, giving the following results:

$$\ln \gamma_\alpha = -\frac{A z_\alpha^2 \sqrt{I}}{1 + Ba\sqrt{I}}, \qquad \ln \gamma_\pm = -\frac{A|z_a z_2|\sqrt{I}}{1 + Ba\sqrt{I}}, \qquad B = \left(\frac{2e^2 L}{k_B \varepsilon}\right)^{1/2} \frac{1}{\sqrt{T}}. \tag{E.114}$$

E.7.3 Diffusion of Ionic Species in an Electrolyte Solution

As shown in Sects. E.6.2 and E.6.3, the diffusion property of an electrically neutral species can be denoted by the gradient of the chemical potential. Similarly, the transport and diffusion of ionic species in an electrolyte solution are designated by the gradient of the electrochemical potential (E.88).

Recall that the coefficient f of (E.77) represents the viscous drag. The inverse $u = 1/f$ may give the mobility of the ionic species in the solution. Then the particle velocity of species α is defined by

$$\boldsymbol{v}_\alpha = -u_\alpha \nabla \tilde{\mu}_\alpha = -u_\alpha \left(\nabla \mu_\alpha + z_\alpha F \nabla \phi\right). \qquad (!\alpha). \tag{E.115}$$

The first and second terms of the r.h.s. of (E.115) represent diffusivity and electric migration, respectively.

Let 1 mole of electrolyte A be ionized into ν_+ mole of cations with a charge of z_+ and ν_- mole of anions with a charge of z_-, and both ions flow at the same velocity $\boldsymbol{v}_A$. Then we have

$$\boldsymbol{v}_A = -u_+ \left(\nabla \mu_+ + z_+ F \nabla \phi\right) = -u_- \left(\nabla \mu_- + z_- F \nabla \phi\right) \tag{E.116}$$

where μ_+ and μ_- are the chemical potentials of the cation and anion, respectively, and u_+ and u_- are their motilities. In these equations the electrophoresis terms are eliminated and we use the electroneutrality principle $\nu_+ z_+ + \nu_- z_- = 0$, which gives

$$\boldsymbol{v}_A = -\frac{u_+ u_-}{u_+ z_+ - u_- z_-}(z_+ \nabla\mu_- - z_- \mu_+) = -\frac{u_+ u_-}{u_+ z_+ - u_- z_-}\nabla\mu_A \tag{E.117}$$

where we have used the relation

$$\mu_A = \nu_+ \mu_+ + \nu_- \mu_- . \tag{E.118}$$

On the other hand, let $\boldsymbol{q}_A^m$ be the flux of the diffusing concentration and D_A be the diffusivity. Then Fick's law can be written as

$$\boldsymbol{q}_A^m = c_A \boldsymbol{v}_A = -D_A \nabla c_A \qquad (!A), \tag{E.119}$$

and together with (E.117) we obtain

$$D_A = \frac{u_+ u_-}{u_+ z_+ - u_- z_-} c_A \frac{\partial \mu_A}{\partial c_A} \qquad (!A). \tag{E.120}$$

E.7.4 Electric Conduction in an Electrolyte Solution

The mobility of ions in a solution implies the generation of an electric current. Let $\boldsymbol{i}_\alpha$ be the current density due to the mobility of ionic species α. Then we have

$$\boldsymbol{i}_\alpha = z_\alpha F \boldsymbol{v}_\alpha = -z_\alpha F c_\alpha u_\alpha \nabla\tilde{\mu}_\alpha = \boldsymbol{i}_\alpha^{\,\mathrm{dif}} + \boldsymbol{i}_\alpha^{\,\mathrm{mig}}, \tag{E.121}$$

$$\boldsymbol{i}_\alpha^{\,\mathrm{dif}} = -z_\alpha F c_\alpha u_\alpha \nabla\mu_\alpha, \quad \boldsymbol{i}_\alpha^{\,\mathrm{mig}} = -z_\alpha^2 F^2 c_\alpha u_\alpha \nabla\phi \tag{E.122}$$

where $\boldsymbol{i}_\alpha^{\,\mathrm{dif}}$ and $\boldsymbol{i}_\alpha^{\,\mathrm{mig}}$ correspond to the terms due to diffusion and electric migration, respectively. By summing (E.121) the total current can be estimated as

$$\boldsymbol{i} = \sum_\alpha \boldsymbol{i}_\alpha . \tag{E.123}$$

The electric field $\boldsymbol{E}$ is related to the current density $\boldsymbol{i}_\alpha$ by

$$\boldsymbol{i}_\alpha = \kappa_\alpha \boldsymbol{E} \tag{E.124}$$

where κ_α is the conductivity of the ionic species α. Since for $\boldsymbol{i}_\alpha^{\,\mathrm{mig}}$ we have $\boldsymbol{E} = -\nabla\phi$, κ_α is given by

$$\kappa_\alpha = z_\alpha^2 F^2 c_\alpha u_\alpha \qquad (!\alpha). \tag{E.125}$$

The conductivity κ_α is a function of the concentration c_α, and we define the molar conductivity by

$$\lambda_\alpha = \frac{\kappa_\alpha}{c_\alpha} = z_\alpha^2 F^2 u_\alpha \qquad (!\alpha) \tag{E.126}$$

where the unit of the molar conductivity λ_α is $\mathrm{S\,m^2\,mole^{-1}}$. Thus for the partial current (E.124) and for the total current we have

$$\boldsymbol{i}_\alpha = \kappa_\alpha \boldsymbol{E} = \lambda_\alpha c_\alpha \boldsymbol{E}, \qquad \boldsymbol{i} = \kappa \boldsymbol{E}, \qquad \kappa = \sum_\alpha \kappa_\alpha = \sum_\alpha \lambda_\alpha c_\alpha \qquad (!\alpha). \tag{E.127}$$

The ratio of the total current $\boldsymbol{i}$ in the solute to a partial current $\boldsymbol{i}_\alpha$ of species α is referred to as the transport number or the Hittorf number:

$$\boldsymbol{i}_\alpha = t_\alpha \boldsymbol{i} \qquad \Rightarrow \qquad t_\alpha = \frac{\lambda_\alpha c_\alpha}{\sum \lambda_\alpha c_\alpha} \qquad (!\alpha). \tag{E.128}$$

For example, in a solution with solvent S in which one electrolyte solute A is dissolved, the transport numbers of cation t_+ and of anion t_- are given by

$$\begin{aligned} t_+ &= \frac{\lambda_+ c_+}{\lambda_+ c_+ + \lambda_- c_-} = \frac{\lambda_+/z_+}{\lambda_+/z_+ - \lambda_-/z_-}, \\ t_- &= \frac{\lambda_- c_-}{\lambda_+ c_+ + \lambda_- c_-} = -\frac{\lambda_-/z_-}{\lambda_+/z_+ - \lambda_-/z_-} \end{aligned} \tag{E.129}$$

because of the electrical neutrality $z_+ c_+ + z_- c_- = 0$. The conductivities of solute (electrolyte A+solvent S) and of electrolyte A are

$$\kappa = \kappa_S + \kappa_A, \qquad \kappa_A = c_+ \lambda_+ + c_- \lambda_-.$$

Let 1 mole of electrolyte A be ionized completely into ν_+ mole of cations with charge z_+ and ν_- mole of anions with charge z_- (complete dissociation), then the concentration of each ion is

$$c_+ = \nu_+ c_A, \qquad c_- = \nu_- c_A.$$

By applying the electroneutrality principle $z_+ \nu_+ + z_- \nu_- = 0$, the conductivity of the completely dissociated solution is calculated as

$$\kappa_A = c_A (\nu_+ \lambda_+ + \nu_- \lambda_-) = \nu_+ z_+ c_A \left(\frac{\lambda_+}{z_+} - \frac{\lambda_-}{z_-} \right). \tag{E.130}$$

Thus the molar conductivity is

$$\Lambda_A = \frac{\kappa_A}{c_A} = \nu_+ z_+ \left(\frac{\lambda_+}{z_+} - \frac{\lambda_-}{z_-} \right). \tag{E.131}$$

We next treat an incompletely dissociated solution due to, for example, the formation of solvation shells because of ion-pairs. Let 1 mole of this weak electrolyte MA be dissolved into ν_+ mole of cations with a charge z_+ and ν_- mole of anions with a charge z_- under the dissociation constant K, and suppose that the solution reaches a dissociation equilibrium. Then we have

$$\begin{array}{cccc} \mathrm{MA} & \rightleftharpoons \; \nu_+\mathrm{M}^{z+} & + & \nu_-\mathrm{M}^{z-} \\ (1-K)c_{MA} & K(\nu_+\, c_{MA}) & & K(\nu_-\, c_{MA}). \end{array}$$

Since the concentrations of cation and anion are $c_+ = K(\nu_+\, c_{MA})$ and $c_- = K(\nu_-\, c_{MA})$, respectively, the conductivity of the electrolyte MA is calculated as

$$\kappa_{MA} = K\, c_{MA}\, (\nu_+\lambda_+ + \nu_-\lambda_-) = K\, \nu_+ z_+ c_{MA} \left(\frac{\lambda_+}{z_+} - \frac{\lambda_-}{z_-} \right) \qquad \text{(E.132)}$$

where from the electroneutrality principle $K(z_+\nu_+) + K(z_-\nu_-) = 0 \Rightarrow z_+\nu_+ + z_-\nu_- = 0$. The molar conductivity is given by

$$\Lambda_{MA} = \frac{\kappa_{MA}}{c_{MA}} = K\, \nu_+ z_+ \left(\frac{\lambda_+}{z_+} - \frac{\lambda_-}{z_-} \right). \qquad \text{(E.133)}$$

By using the molar conductivities $\lambda_+,\ \lambda_-$ we can evaluate the diffusivity D_A of (E.120). The motilities $u_+,\ u_-$ are related to the molar conductivity by

$$u_+ = \lambda_+/z_+^2 F^2, \quad u_- = \lambda_-/z_-^2 F^2,$$

then the diffusivity can be evaluated by measuring the molar conductivity as

$$D_A = \frac{\lambda_+\lambda_-}{F^2 z_+ z_- (\lambda_+ z_- - \lambda_- z_+)} c_A \frac{\partial \mu_A}{\partial c_A} \qquad (!A). \qquad \text{(E.134)}$$

E.8 Chemical Equilibrium and the Equilibrium Constant

Let us denote the reaction formula under chemical equilibrium by the following stoichiometric relation:

$$\sum_{\alpha=1}^{N} \nu_\alpha \mathrm{X}_\alpha = 0. \qquad \text{(E.135)}$$

The extent of reaction ξ is uniform as defined by (E.38), and it is written by

$$\nu_\alpha d\xi = dn_\alpha. \qquad \text{(E.136)}$$

On the other hand, from (E.56) the Gibbs free energy $G(P,T,n_\alpha) = \sum \mu_\alpha n_\alpha$ represents the chemically induced energy. When the reaction reaches chemical equilibrium, it has apparently ceased because the forward reaction is equal to the backward reaction, and the Gibbs free energy G attains a minimum:

$$dG = \sum_\alpha \mu_\alpha \, dn_\alpha = 0. \tag{E.137}$$

Substituting (E.136) into (E.137) and eliminating $d\xi$, we obtain

$$\sum_\alpha \nu_\alpha \, \mu_\alpha = 0. \tag{E.138}$$

In a solution, as shown by (E.75), the chemical potential is represented as $\mu_\alpha(P, T, a_\alpha) = \mu_\alpha^0(P, T) + RT \ln a_\alpha$. We substitute this into (E.138), and obtain the standard Gibbs free energy in the reaction, $\Delta G^\ominus$, as

$$\Delta G^\ominus = \sum_\alpha \nu_\alpha \, \mu_\alpha^0 = -RT \sum_\alpha \ln a_\alpha^{\nu_\alpha} = -RT \ln \prod\nolimits_\alpha a_\alpha^{\nu_\alpha} \tag{E.139}$$

where $\prod_\alpha a_\alpha^{\nu_\alpha} = a_\alpha^{\nu_1} \cdot a_\alpha^{\nu_2} \cdot a_\alpha^{\nu_3} \cdots$. Then the equilibrium constant K is defined by

$$K \equiv e^{-\Delta G^\ominus / RT} = \prod\nolimits_\alpha a_\alpha^{\nu_\alpha}. \tag{E.140}$$

As given in (E.57), $\Delta G^\ominus = \Delta H^\ominus - T(\Delta S^e)^\ominus$, therefore we have

$$K = e^{-\Delta H^\ominus / RT} e^{(\Delta S^e)^\ominus / R}. \tag{E.141}$$

If the solution is dilute, $\gamma_\alpha = 1 \Rightarrow a_\alpha = \gamma_\alpha x_\alpha = x_\alpha$, and using (E.140) the equilibrium constant becomes

$$K = \prod\nolimits_\alpha x_\alpha^{\nu_\alpha} \tag{E.142}$$

These relations are known as the *Law of Mass Action*.[2]

The chemical potential defined by (E.70) and (E.72) can be employed for treating the process of the reaction of a gas, but we will not discuss it here.

---

[2]It is understood from (E.142) that the equilibrium constant K gives the relation between the mass before the reaction and the mass after the reaction. For example, if the reaction

$$\mathrm{A} + \mathrm{B} \rightleftharpoons \mathrm{C}$$

is of the first order ($\nu_\alpha = 1$) for every element, using (E.142) we obtain

$$K = \frac{[\mathrm{C}]}{[\mathrm{A}][\mathrm{B}]}.$$

E.9 Phenomenological Theory of Non-equilibrium Chemical Reaction Processes

In terms of modern quantum chemistry, a reaction is a complex process related to bonding and debonding of electron clouds. It is not possible to write this reaction using a simple formula, and phenomenological descriptions are commonly used that correspond to each chemical process. In this Section we summarize the phenomenological reaction theory (see, e.g., Seki 1997, pp. 322–331).

E.9.1 Reaction Velocity and Order of Reaction

As in the previous section, the stoichiometric relation (E.135) and the extent of reaction ξ given by (E.136) are used. As described in (E.39) the reaction rate w is denoted by using ξ as

$$w = \frac{d\xi}{dt}. \tag{E.143}$$

Let the molar concentration of species X_α be $[X_\alpha]$, then the reaction rate of a non-equilibrium reaction is commonly and phenomenologically represented by

$$w = \frac{1}{\nu_\alpha}\frac{d[X_\alpha]}{dt} = k\prod_\alpha [X_\alpha]^{p_\alpha}, \qquad p = \sum_\alpha p_\alpha \tag{E.144}$$

where p_α is the order of reaction of species X_α and p is the overall order of reaction. In (E.144) k is the reaction rate constant, which generally depends on temperature (described later). As implied by (E.144) the equation of reaction is generally nonlinear.

Example E.1 (The simplest first order reaction). Consider the simplest reaction equation

$$\mathrm{A} \longrightarrow \mathrm{B} \tag{E.145}$$

where the overall order of reaction is unity. Let the molar concentration of reactant A consumed by this reaction at a time t be x, and the initial amount of A be a. Then the concentration of A at time t is $(a–x)$, and the amount of reaction product B is x. Since the reaction is of order one, the differential equation giving the concentration of A is

$$-\frac{d(a-x)}{dt} = \frac{dx}{dt} = k(a-x). \tag{E.146}$$

Solving this under the initial condition $x(t = 0) = a$, we obtain the concentration of A as

$$[\mathrm{A}] = a - x = ae^{-kt}. \tag{E.147}$$

The time required for the concentration of reactant A to become half is referred to as the half-life denoted as $t_{1/2}$. The half-life of the above reaction is $t_{1/2} = \ln 2/k$ which is independent of the initial concentration. ■

Example E.2 (Second-order reaction). Suppose that a second-order reaction is given by the following reaction equation:

$$\mathrm{A} + \mathrm{B} \longrightarrow \mathrm{C}. \tag{E.148}$$

Let the molar concentration of reactant A at time t be x. The initial concentrations of A and B are a and b, respectively. At time t they becomes $(a-x)$ and $(b-x)$. Then the equation of the overall reaction of order two (each reaction order is one, both for A and B) can be written as

$$\frac{dx}{dt} = k(a-x)(b-x). \tag{E.149}$$

If the initial concentrations of A and B are the same ($a = b$), the solution of (E.149) is obtained as

$$\frac{1}{a-x} - \frac{1}{a} = kt. \tag{E.150}$$

The half-life is $t_{1/2} = 1/kt$ which is inversely proportional to the initial concentration.

If the initial concentration of A is far less than that of B ($a \ll b$), we have $b–x \simeq b$ since x never exceeds a and $b \gg x$. Thus (E.149) is approximated as

$$\frac{dx}{dt} \simeq kb(a-x) = k'(a-x), \qquad k' = kb. \tag{E.151}$$

Note that this is of the same form as the first order reaction equation (E.146). k' is referred to as the pseudo-first order rate constant. ■

E.9.2 Elementary Reaction and Complex Reaction

Chemical reactions are, in general, complex combinations of simple elementary reactions. The molecular number of reactants of the elementary reaction, which are given as the l.h.s. terms of a reaction formula is one, two and three, and these are called unimolecular, bimolecular and termolecular reactions, respectively. The order of reaction is one, two and three corresponding to each case. It is not always true that the reaction of order one is the unimolecular reaction, and this is not always true similarly for other cases of the reaction order.

Example E.3 (Unimolecular reaction with two elementary processes). Consider the following two elementary processes in a unimolecular reaction:

$$\begin{aligned} &A \xrightarrow{k_1} B \\ &B \xrightarrow{k_2} C \end{aligned} \tag{E.152}$$

where k_1 and k_2 are the reaction rate constants for each elementary reaction. Let the amount of substance of A consumed until time t be x, and the initial concentration be a. As shown in Example E.1 the concentration of the first elementary process is given by

$$[A] = a - x = a\, e^{-k_1 t}, \tag{E.153}$$

and if the concentration of B is denoted as $y = [B]$, the balance of mass of B for both processes is given by

$$\frac{dy}{dt} = k_1(a - x) - k_2 y.$$

Substituting (E.153) into this gives the differential equation with respect to the concentration y:

$$\frac{dy}{dt} + k_2 y = k_1 a\, e^{-k_1 t}. \tag{E.154}$$

Since the initial condition is $y = 0$, the solution of (E.154) for the case $k_1 \neq k_2$ is obtained as

$$y = \frac{a k_1}{k_2 - k_1}\left(e^{-k_1 t} - e^{-k_2 t}\right). \tag{E.155}$$

$y(t)$ reaches the maximum when $t_{\max} = \ln(k_1/k_2)/(k_1 - k_2)$. Note that the concentration $z = [C]$ of the substance C is calculated by $x + y + z = a$.

If $k_1 \gg k_2$, the first process is rate-determining, and if $k_2 \gg k_1$, the second is rate-determining. ■

Example E.4 (Reaction producing an intermediate). Suppose that two elementary processes of reactions produce an intermediate C under an equilibrium reaction:

$$A + B \underset{k_{-1}}{\overset{k_1}{\rightleftharpoons}} C \xrightarrow{k_2} D \tag{E.156}$$

where the reaction rate constants of the forward reaction and of the backward reaction for the first reaction are k_1 and k_{-1}, respectively, and the reaction rate constant of the second reaction is k_2.

The intermediate C is produced constantly, which gives the condition

$$\frac{d[C]}{dt} = k_1[A][B] - k_{-1}[C] - k_2[C] = 0.$$

Thus we have

$$[C] = K[A][B], \qquad K = \frac{[C]}{[A][B]} = \frac{k_1}{k_{-1} + k_2} \tag{E.157}$$

where K is the equilibrium constant. On the other hand, from the second reaction we have

$$\frac{d[\mathrm{D}]}{dt} = k_2[\mathrm{C}] = k[\mathrm{A}][\mathrm{B}], \qquad k = k_2 K = \frac{k_1 k_2}{k_{-1} + k_2}. \tag{E.158}$$

The order of the overall reaction to produce D from A and B is two, and its reaction rate constant is $k = k_2 K$. ■

Example E.5 (Enzyme reaction). In biochemistry the enzyme reaction is important, and it is represented in a similar manner as in Example E.4. Let an enzyme be E, a substrate be S, an intermediate complex be ES, and a product be P, then the reaction system can be written by

$$\mathrm{S} + \mathrm{E} \underset{k_{-1}}{\overset{k_1}{\rightleftharpoons}} \mathrm{ES} \overset{k_2}{\longrightarrow} \mathrm{E} + \mathrm{P}. \tag{E.159}$$

This is referred to as the *Michaelis-Menten structure*. All the reaction equations of this system can be represented by

$$\frac{d[\mathrm{S}]}{dt} = -k_1[\mathrm{S}][\mathrm{E}] + k_{-1}[\mathrm{ES}], \tag{E.160}$$

$$\frac{d[\mathrm{E}]}{dt} = -k_1[\mathrm{S}][\mathrm{E}] + (k_{-1} + k_2)[\mathrm{ES}], \tag{E.161}$$

$$\frac{d[\mathrm{P}]}{dt} = k_2[\mathrm{ES}], \tag{E.162}$$

$$\frac{d[\mathrm{ES}]}{dt} = k_1[\mathrm{S}][\mathrm{E}] - (k_{-1} + k_2)[\mathrm{ES}] = 0. \tag{E.163}$$

In (E.163), we have assumed that the intermediate complex is produced by an equilibrium reaction. From (E.163) we obtain

$$[\mathrm{ES}] = K[\mathrm{E}][\mathrm{S}], \qquad K = \frac{[\mathrm{ES}]}{[\mathrm{E}][\mathrm{S}]} = \frac{k_1}{k_{-1} + k_2}. \tag{E.164}$$

From the view point of the overall reaction, we define the reaction velocity as $v = d[\mathrm{P}]/dt = -d[\mathrm{S}]/dt$. Then using (E.160) and (E.164) we have

$$v = -\frac{d[\mathrm{S}]}{dt} = \frac{d[\mathrm{P}]}{dt} = \left(\frac{k_1}{K} + k_{-1}\right)[\mathrm{ES}]. \tag{E.165}$$

Since the enzyme is a type of catalyst, the amount of substance is not changed before or after the reaction. Let $[\mathrm{E}]_0$ be the initial amount of substance of the enzyme, so that

$$[\mathrm{E}] + [\mathrm{ES}] = [\mathrm{E}]_0 = \text{constant}. \tag{E.166}$$

Substituting (E.164) into (E.166), we obtain

$$[\mathrm{E}] = \frac{[\mathrm{E}]_0}{1 + K[\mathrm{S}]}. \tag{E.167}$$

We substitute (E.167) into (E.165), and finally the reaction rate of Michaelis-Menten structure is determined as

$$v = -\frac{d[\mathrm{S}]}{dt} = \frac{d[\mathrm{P}]}{dt} = \frac{k_2[\mathrm{E}]_0[\mathrm{S}]}{K_M + [\mathrm{S}]}, \qquad K_M = \frac{k_{-1} + k_2}{k_1}. \tag{E.168}$$

This is referred to as the Michaelis-Menten equation, and K_M is the Michaelis constant.

In (E.168), let $[\mathrm{S}] \to \infty$, then we have

$$v \to k_2[\mathrm{E}]_0. \tag{E.169}$$

This implies that in a region with a high concentration of the substrate S, the resolution rate does not increase even if we increase the concentration of enzyme.

■

E.9.3 Temperature Dependence of Reaction Rate: Arrhenius Equation

The reaction rate constant k is known to be temperature dependent, which is determined experimentally by

$$k = A\,e^{-E_a/RT} \tag{E.170}$$

where R is the universal gas constant, A is referred to as the frequency factor, E_a is the activation energy, corresponding to the energy gap between the transition state and the final product state. The result (E.170) is known as the Arrhenius equation. Note that in this process the reaction starts with the original state, passes to the transition state and reaches the product state. In Fig. E.4 the change of potential energy is schematically illustrated according to the reaction extent. In the sense of molecular statistics, the reaction extent signifies the extent of atomic rearrangement of the reaction system.

Taking a natural logarithm of (E.170), we obtain

$$\ln k = \ln A - \frac{E_a}{RT}. \tag{E.171}$$

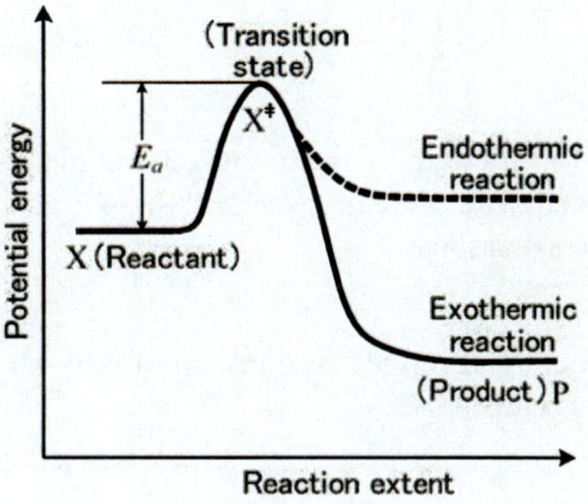

Fig. E.4 Reaction coordinate: reaction extent and change of potential energy

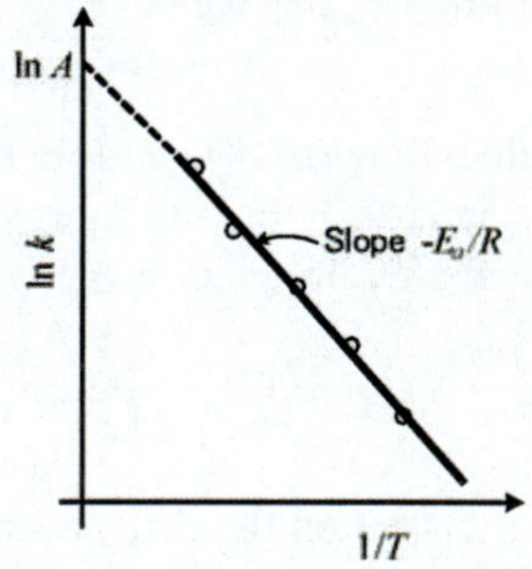

Fig. E.5 Arrhenius plot

Thus from the results of an experiment, we plot $\ln k$ versus $1/T$ to obtain A as the intercept of the ordinate, and E_a from the slope (Arrhenius plot: Fig. E.5). Note that the molecular-based interpretation of k is given by the transition state theory.

E.9.4 Transition State Theory

The process shown in Example E.4, where the intermediate is produced under the equilibrium state, is strongly related to the transition state theory, and the intermediate corresponds to the activated complex of the transition state theory.

Consider a bimolecular reaction to produce P from the reactants A and B through the transition state in which the activated complex $AB^\ddagger$, corresponding to an intermediate, is generated (the notation ‡ implies the transition state):

$$A + B \overset{K^\ddagger}{\rightleftharpoons} AB^\ddagger \overset{k^\ddagger}{\longrightarrow} P. \tag{E.172}$$

The reaction $A+B \rightleftharpoons AB^\ddagger$ is under equilibrium, and we set the equilibrium constant as $K^\ddagger$. From (E.157) we have

$$K^{\ddagger} = \frac{[\mathrm{AB}^{\ddagger}]}{[\mathrm{A}][\mathrm{B}]}. \tag{E.173}$$

The lifetime of the activated complex $\mathrm{AB}^{\ddagger}$ is relatively short (less than 10^{-13} s), therefore we can regard A+B→P as one process and estimate its reaction rate constant k as

$$\mathrm{A} + \mathrm{B} \xrightarrow{k} \mathrm{P}. \tag{E.174}$$

Let the reaction rate constant of $\mathrm{AB}^{\ddagger}$→P be $k^{\ddagger}$, then the production rate of P can be written as

$$\frac{d[\mathrm{P}]}{dt} = k[\mathrm{A}][\mathrm{B}] = k^{\ddagger}[\mathrm{AB}^{\ddagger}]. \tag{E.175}$$

From (E.158) we have

$$k = k^{\ddagger} K^{\ddagger}. \tag{E.176}$$

Recalling (E.141) and (E.142), the equilibrium constant $K^{\ddagger}$ is represented using the standard Gibbs free energy in the reaction $\Delta G^{\ominus\ddagger}$ of the transition state, the standard enthalpy of reaction $\Delta H^{\ominus\ddagger}$ and the standard entropy of the reaction $(\Delta S^{e})^{\ominus\ddagger}$ as

$$K^{\ddagger} = e^{-\Delta G^{\ominus\ddagger}/RT} = e^{-\Delta H^{\ominus\ddagger}/RT} e^{(\Delta S^{e})^{\ominus\ddagger}/R}. \tag{E.177}$$

Based on the discussions of the molecular path in the space of potential energy, Eyring showed that the reaction rate constant $k^{\ddagger}$ is obtained as

$$k^{\ddagger} = \frac{k_B T}{h} \tag{E.178}$$

(details are found in Chap. 4 of Glasstone et al. 1941; Barrow 1996, Sect. 16.6). Here $k_B = R/N_A$ is Boltzmann's constant, and $h = 6.63 \times 10^{-34}$ J·s is Planck's constant. Thus the reaction rate constant of (E.176) is represented as

$$k = \kappa \frac{k_B T}{h} e^{-\Delta G^{\ominus\ddagger}/RT} = \left(\kappa \frac{k_B T}{h} e^{(\Delta S^{e})^{\ominus\ddagger}/R}\right) e^{-\Delta H^{\ominus\ddagger}/RT}. \tag{E.179}$$

Here the transmission coefficient κ is introduced because of the uncertainty of reaction ($\kappa < 1$).

References

Akagi T (1980) Study to specify viscoelastic constants. Report on Scientific Research Funds of Ministry of Education (in Japanese)

Allen MP, Tildesley DJ (1987) Computer simulation of liquids. Oxford Science Publications, New York

Atkins P, de Paula J (2010) Physical chemistry, 9th edn. Oxford University Press, New York

Atkinson JH (1993) An introduction to the mechanics of soils and foundations: through critical state soil mechanics. McGraw-Hill, New York

Atkinson JH, Bransby PL (1978) The mechanics of soils: an introduction to critical state soil mechanics. McGraw-Hill, New York

Baeyens B, Bradbury M (1997) A mechanistic description of Ni and Zn sorption on Na-montmorillonite part I: titration and sorption measurement. J Contam Hydrol 27:199–222

Bakhvalov N, Panasenko G (1984) Homogenization: averaging processes in periodic media. Nauka (English trans: Kluwer Academic, Dordrecht, 1989)

Barber JR (2002) Elasticity. Kluwer Academic, Dordrecht

Barrow GM (1996) Physical chemistry, 6th edn. McGraw-Hill, New York

Bear J, Bachmat Y (1990) Introduction to modeling transport phenomena in porous media. Kluwer Academic, Dordrecht

Bear J, Verruijt A (1987) Modeling groundwater flow and pollution. Reidel, Dordrecht

Bird RB, Stewart WE, Lightfoot EN (2002) Transport phenomena, 2nd edn. Wiley, New York

Bowen R (1976) Theory of mixture. In: Eringen AC (ed) Continuum physics, vol III. Academic, New York, pp 1–127

Buchdahl HA (1966) The concepts of classical thermodynamics. Cambridge Univesity Press, London

Carnot S (1824) Réflections sur la Puissance Motrice de Feu, found in the book ed. by Fox R, Librarie Philosophique (1978) (English trans: Mendoza E (ed) (1960): Reflections on the Motive Power of Fire and Others). Dover, New York

Carslaw HS, Jaeger JC (1959) Conduction of heat in solids. Oxford University Press, Oxford

Chadwick P (1976) Continuum mecahnics: concise theory and problems. Allen and Unwin, London

Chapman N, McCombie C (2003) Principles and standards for the disposal of long-lived radioactive wastes. Elsevier, Oxford

Christensen RM (2003) Theory of viscoelasticity, 2nd edn. Academic, New York

Coleman BD (1964) Thermodynamics of materials with memory. Arch Ration Mech Anal 17:1–47

Coussy O (1995) Mechanics of porous continua. Wiley, Chichester

Cussler EL (1997) Diffusion: mass transfer in fluid systems, 2nd edn. Cambridge University Press, Cambridge

Darve F (ed) (1990) Geomaterials: constitutive equations and modelling. Elsevier, London

Y. Ichikawa and A.P.S. Selvadurai, *Transport Phenomena in Porous Media*,
DOI 10.1007/978-3-642-25333-1, © Springer-Verlag Berlin Heidelberg 2012

Davis RO, Selvadurai APS (1996) Elasticity and geomechanics. Cambridge University Press, Cambridge

Davis RO, Selvadurai APS (2002) Plasticity and geomechanics. Cambridge University Press, Cambridge

Desai CS, Siriwardane HJ (1984) Constitutive laws for engineering materials. Prentice-Hall, Englewood Cliffs

de Groot SR, Mazur P (1962) Non-equilibrium thermodynamics. North Holland/Dover, Amsterdam (1984)

Donat J (1929) Wisserkraft und Wisserwirtschaft, 25 (English trans: An Article on the Permeability of Sand. Bureau of Reclamation, Denver, 1933)

Dormieux L, Kondo D, Ulm FJ (2006) Microporomechanics. Wiley, Chichester

Drew DA, Passman SL (1998) Theory of multicomponent fluids. Springer, New York

Ene HI, Polisevski D (1987) Thermal flows in porous media. Reidel, Dordrecht

Ericksen JL (1991) Introduction to the thermodynamics of solids. Chapman and Hall, London

Feynman RP, Leighton RB, Sands M (1963) The Feynman lectures on physics, vol 1. Addison-Wesley, Reading

Frankel T (2004) The geometry of physics: an introduction, 2nd edn. Cambridge University Press, Cambridge

Freundlich CGL (1926) Colloid and capillary chemistry. Methuen, London

Friedman A (1982) Foundations of modern analysis. Dover, New York

Fu MH, Zhang ZZ, Low PF (1990) Changes in the properties of a montmorillonite-water system during the adsorption and desorption of water hysteresis. Clays Clay Miner 38:485–492

Fujii K, Nakano M (1984) Chemical potential of water adsorbed to bentonite. Trans Japanese Soc Irrigation Drainage Rural Eng 112:43–53 (in Japanese)

Fujii N (2002) On diffusion problem in bentonite and homogenization analysis. Doctral thesis, Nagoya University (in Japanese)

Fung YC (1969) A first course in continuum mechanics. Prentice-Hall, Englewood Cliffs

Gibson RE, Lo KY (1961) A theory of consolidation for soils exhibiting secondary consolidation. Pub. No. 41, Norwegian Geotechnical Institute, Oslo

Gibson RE, England GL, Hussey MJL (1967) The theory of one-dimensional consolidation of saturated clays: 1. Finite non-linear consolidation of thin homogeneous layers. Geotechnique 17:261–273

Gillham RW, Robin MJL, Dytynyshyn DJ (1983) Diffusive transport of strontium-85 in sand-bentonite mixtures, AECL-6838. Atomic Energy of Canada Ltd. Whiteshell Nuclear Research Establishment, Pinawa

Glasstone S, Laidler KJ, Eyring HE (1941) The theory of rate processes: The kinetics of chemical reactions, viscosity, diffusion and electrochemical phenomena. McGraw-Hill, New York

Greenberg MD (1998) Advanced engineering mathematics. Prentice-Hall, Upper Saddle River

Gurtin ME (1972) The linear theory of elasticity. In: Flugge S (ed) Mechanics of solids II, Encyclopedia of physics, VIa/2. Springer, Berlin, pp 1–295

Gurtin ME, Sternberg E (1962) On the linear theory of viscoelasticity. Arch Ration Mech Anal 11(4):291–356

Hassani S (1999) Mathematical physics: a modern introduction to its foundations. Springer, New York

Hazen A (1892) Some physical properties of sands and gravels, with special reference to their use in filtration. 24th Annual Rep., Massachusetts State Board of Health, Pub. Doc. No. 34, 539–556

Hendricks DW (1972) Sorption in flow through porous media. In: Bear J (ed) Transport phenomena in porous media. Elsevier, Amsterdam, pp 384–392

Hill R (1978) Aspects of invariance in solid mechanics. In: Yih C-S (ed) Advances in applied mechanics, vol 18. Academic, New York, pp 1–73

Houlsby GT, Schofield AN (eds) (1993) Predictive soil mechanics. Thomas Telford, London

Hornung U (ed) (1997) Homogenization and porous media. Springer, New York

Hughes TJR (1987) The finite element method. Prentice-Hall, Englewood Cliffs

Ichikawa Y (1990) Introduction to the finite element method in geotechnical engineering. Nikkagiren-shuppan, Tokyo (in Japanese)
Ichikawa Y, Kawamura K, Nakano M, Kitayama K, Kawamura H (1998) Unified molecular dynamics/homogenization analysis for water flow in bentonite. In: Proceedings of the 98 international high-level radioactive waste management conference, American Nuclear Society, Las Vegas, May 1998, pp 422–428
Ichikawa Y, Kawamura K, Fujii N, Theramast N (2002) Molecular dynamics and multiscale homogenization analysis of seepage/diffusion problem in bentonite clay. Int J Numerical Methods Eng 55:1–40
Ichikawa Y, Kawamura K, Theramast N, Kitayama K (2004) Secondary and tertial consolidation of bentonite clay: consolidation test, molecular dynamics simulation and multiscale homogenization analysis. Mech Mater 36:487–513
Ito M et al (1993) Analysis of mineral composition of bentonite. PNC[1] Report, TN8430 93-003 (in Japanese)
JNC[2] (1999) H12: Project to establish the scientific and technical basis for HLW disposal in Japan. JNC Report, TN1400 99-020
Kachanov LM (2005) Fundamentals of the theory of plasticity (trans: from Russian by Konyaeva M). Dover, Mineola
Kamon M, Yasuahara M, Kida D (1985) Volume change of soil. Soils Found Japanese Geotech Soc 33(11):75–84 (in Japanese)
Kawamura K (1990) Molecular simulations using perosonal computer. Kaibun-do, Tokyo (in Japanese)
Kawamura K, Ichikawa Y (2001) Physical properties of clay minerals and water: by means of molecular dynamics simulations. Bull Earthquake Res Inst Univ Tokyo 76:311–321
Keener JP (1988) Principles of applied mathematics: transformation and approximation. Addison-Wesley, Redwood City
Kestin J (1979) A course in thermodynamics, Revised printing, vols 1 and 2. Hemisphere, Washington
Kitagawa H (1987) Mechanics in elasto-plasticity: fundamental theories for nonlinear analysis. Shokabo, Tokyo (in Japanese)
Klitzsche CS, Weistroffer K, Elshasly EM (1976) Grundwasser der Zentralsahara: Fossile Vorrate. Geol Rundschau 65:264
Kondepudi D, Prigogine I (1998) Modern thermodynamics. Wiley, Chichester
Kozaki T, Sato Y, Nakajima M, Kato H, Sato S, Ohashi H (1999) Effect of particle size on the diffusion behavior of some radionuclides in compacted bentonite. J Nucl Mater 270:265–272
Kozeny J (1927) Ueber kapillare leitung des wassers im boden. Sitzungsber Akad Wiss Wien 136(2a):271–306
Kubota K, Kono I, Uno H (1976) Permeability – approach to design. Kajima Institute Publishing, Tokyo (in Japanese)
Langmuir I (1916) Chemical reaction at low temperatures. J Am Chem Soc 37:1139
Langmuir I (1918) The adsorption of gases on plane surfaces of glass, mica and platinum. J Am Chem Soc 40:1161–1403
Lapidus L, Amundson NR (1952) Mathematics of adsorption in beds VI. The effect of longitudal diffusion in ion exchange and chromatographic columns. J Phys Chem 56:984–988
Leigh DC (1968) Nonlinear continuum mechanics. McGraw-Hill, New York
Lewis RW, Schrefler BA (2000) The finite element method in the static and dynamic deformation and consolidation of porous media. Wiley, Chichester

[1]PNC: Power Reactor and Nuclear Fuel Development Co. Currently, Japan Atomic Energy Agency (JAEA).

[2]JNC: Japan Nuclear Cycle Development Institute. Currently, Japan Atomic Energy Agency (JAEA).

Little RW (1973) Elasticity. Prentice-Hall, Englewood Cliffs
Lubarda VA (2000) Elastoplasticity theory, CRC Press, Boca Raton
Lubarda VA (2004) Constitutive theories based on the multiplicative decomposition of deformation gradient: Thermo-elasticity elastoplasticity and biomechanics. Appl Mech Rev 57:95–108
Lubliner J (1990) Plasticity theory. Collier-Macmillan, New York
McQuarrie DA, Simon JD (1997) Physical chemistry. University Science Books, Sausalito
Malvern LE (1969) Introduction to the mechanics of a continuous medium. Prentice-Hall, Englewood Cliffs
Mase GE (1970) Theory and problems of continuum mechanics. McGraw-Hill, New York
Milton GW (2002) The theory of composites. Cambridge University Press, Cambridge
Mitchell JK (1993) Fundamentals of soil behavior, 2nd edn. Wiley, Hoboken
Miyahara K, Ashida T, Kohara Y, Yusa Y, Sasaki N (1991) Effect of bulk density on diffusion for cesium in compacted sodium bentonite. Radiochimica Acta 52/53:293–297
Moore WJ (1972) Physical chemistry, 4th edn. Prentice-Hall, Englewood Cliffs
Nakano M (1991) Material transporting soils. University Tokyo Press, Tokyo (in Japanese)
Nernst W (1918) Die theoretischen und experimentellen Grundlagen des neuen Wrmesatzes, Knapp (English trans: The new heat theorem, 1926/Dover, New York, 1969)
Oka F (1995) Exercises of soil mechanics. Morikita, Tokyo (in Japanese)
Oscarson DW, Hume HB (1994) Diffusion of 14C in dense saturated bentonite under steady-state conditions. Trans Porous Media 14:73–84
Owen DRJ, Hinton E (1980) Finite elements in plasticity. Pineridge, Swansea
Parry RHG (ed) (1971) Stress-strain behaviour of soils. G.T. Foulis, Henley-on-Thames
Pietruszczak S (2010) Fundamentals of plasticity in geomechanics. CRC, Boca Raton
Prigogine I (1967) Etude thermodynamique des processus irreversible, 4th edn. Desoer, Liège
Protter MH, Morrey CB (1977) A first course in real analysis. Springer, New York
Pusch R (1994) Waste disposal in rock. Elsevier, Amsterdam
Pusch R, Karnland O, Hökmark H (1990) GMM – A general microstructural model for qualitative and quantitative studies of smectite clays. Clay Technology AB/Lund, Sweden
Raniecki B (1976) An introduction to the concepts of applied thermoplasticity, Lecture Note. Materials Center, Royal Institute of Technology, Stockholm
Raniecki B, Nguyen HV (2005) Favorable theoretical framework for pressure-sensitive isotropic elastic-plastic damaging bodies under finite strains. In: Proceeding of the symposium elasto-plasticity for K. Hashiguchi Retirement Aniv., Kyushu University, Feb. 2005, pp 115–134
Rasilainen K (1997) Matrix diffusion model: in situ tests using natural analogues, pp 1–81, Technical Research Centre of Finland, VTT Publications 331, Espo
Riley KF, Hobson MP, Bence SJ (1998) Mathematical methods for physics and engineering. Cambridge University Press, Cambridge
Sanchez-Palencia E (1980) Non-homogeneous media and vibration theory. Springer, Heidelberg
Sato H (1998) A study on nuclide migration in buffer materials and rocks for geological disposal of radioactive waste. PNC Technical Report, TN8410 97-202 (in Japanese)
Sato H, Ashida T, Kohara Y, Yui M, Sasaki N (1992) Effect of dry density on diffusion of some radionuclides in compacted sodium bentonite. J Nucl Sci Tech 29(9):873–882
Sato H, Shibutani T (1994) Study on adsorption and diffusion mechanism of nuclides in buffer material and geosphere. PNC Technical Report, 91, 71–88 (in Japanese)
Schiffman RL, Ladd CC, Chen ATF (1964) The secondary consolidation of clay. Rensselaer Polytechnic Institute, Troy
Schofield AN, Wroth CP (1968) Critical state soil mechanics. McGraw-Hill, London
Schrödinger E (1952) Statistical thermodynamics, 2nd edn. Cambridge University Press, Cambridge (Dover, 1989)
Seki K (1997) Physical chemistry. Iwanami Shoten, Tokyo (in Japanese)
Selvadurai APS (2000a) Partial differential equations in mechanics, vol 1. Fundamentals, Laplace's equation, diffusion equation, wave equation. Springer, Berlin
Selvadurai APS (2000b) Partial differential equations in mechanics, vol 2. The biharmonic equation, Poisson's equation. Springer, Berlin

Selvadurai APS, Yu Q (2006) On the indentations of polymeric membrane, Proc Roy Soc Math Phys Ser A 462:189–209
Selvadurai APS (2007) The analytical method in geomechanics. Appl Mech Rev 60:87–106
Singh A, Mitchell JK (1968) General stress-strain-time function for soils. Proceedings of the american society of civil engineers (ASCE), This is a journal of ASCE, which is published in 1968. SM1, University of California, Berkeley, 21–46
Slattery JC (1999) Advanced transport phenomena. Cambridge University Press, Cambridge
Spencer AJM (2004) Continuum mechanics. Dover, Mineola
Spiegel MR (1971) Advanced mathematics for engineers and scientists. McGraw-Hill, New York
Stumm W, Morgan J (1996) Aquatic chemistry. Wiley, New York
Tamamushi R (1991) Electrochemistry, 2nd edn. Tokyo Kagaku-dojin, Tokyo (in Japanese)
Taylor DW (1948) Fundamentals of soil mechanics. Wiley, New York
Terzaghi K, Peck RB (1964) Soil mechanics in engineering practice. Wiley, New York
Theramast N (2003) A study of seepage and consolidation of smectic bentonite. Doctoral Thesis, Nagoya University
Torquato S (2002) Random heterogeneous materials: microstructure and macroscopic properties. Springer, New York
Truesdell C, Noll W (1965) The nonlinear field theory of mechanics. In: Flugge S (ed) Encyclopedia of physics, III/3. Springer, Berlin
Truesdell C, Toupin R (1960) Classical field theories. In: Flugge S (ed) Encyclopedia of physics, III/1. Springer, Berlin, pp 226–793
Ueda A (1990) Computer simulations: atomic motions in a macro-system. Asakura-shoten, Tokyo (in Japanese)
Yamamoto M (1972) Deformation of materials. Seibundo-Shinkousha, Tokyo (in Japanese)
Yamamoto Y (1987) Historical review of thermodynamical concepts. Gendai-Suugakusha, Kyoto (in Japanese)
Yoshida K (1980) Functional analysis, 6th edn. Springer, Berlin
Zienkiewicz OC, Chan AHC, Pastor M, Schrefler BA, Shiomi T (1999) Computational geomechanics: with special reference to earthquake engineering. Wiley, Chichester
Zunker F (1930) Das Verhalten des Bodens Zum Wasser. Handbuch der Bodenlehre 6:66–220

Index

Absorption, 178
Acceleration, 14
Activated complex, 371
Activation energy, 370
Adiabatic system, 318
Adsorption, 162, 178
  coefficient, 179
  isotherm, 177
    equilibrium, 178
    nonequilibrium, 178
  physical, 177
Affinity, 132, 347
After-effect function, 57
Almansi strain, 20
Apparent diffusion coefficient, 257
Approximation, 152
Arrhenius plot, 371
Avogadro's number, 351

Base vectors, 285
Basis, 301
Bauschinger effect, 69
Bentonite, 267
Bessel's inequality, 311
Best approximation, 156
Bilinear form, 141, 144
Biochemistry, 369
Body force, 30
Boltzmann distribution, 325
Boltzmann formula, 328
Boltzmann's constant, 324, 351
Boltzmann's Principle, 324
Bulk modulus, 51

Cam clay model, 71, 196
  modified, 204
  original, 198, 203
Carnot cycle, 319
Cauchy sequence, 303
Cauchy stress, 32
Cauchy's lemma, 31
Cayley-Hamilton's theorem, 47
Centrifugal force, 18
Change in standard enthalpy, 343
Change in standard Gibbs free energy, 350
Change of standard entropy, 350
Characteristic function, 210, 235, 236, 246
  velocity and pressure, 217
Chemical potential, 111
Chemisorption, 177
Clausius-Duhem inequality, 334, 337
Clausius inequality, 331
Closed system, 318, 333
Complementary energy, 103
Completeness, 309
Compression index, 200
Configuration, 12
Conjugate, 98
  stress, 86
Conservation of moment of linear momentum, 35
Consolidation, 268
  equation, Biot, 187
  1D finite strain, 271
  finite strain, 187
Constrictivity, 175
Continuity equation, 30
Continuum mechanics, 1, 11
Contravariant basis, 304
Contravariant derivative, 42
Convected derivative, 42
Convective term, 14
Convolution integral, 147
Coordinate transformation, 285

Y. Ichikawa and A.P.S. Selvadurai, *Transport Phenomena in Porous Media*,
DOI 10.1007/978-3-642-25333-1, © Springer-Verlag Berlin Heidelberg 2012

Coriolis force, 18
Corotational derivative, 43
Covariant basis, 304
Covariant derivative, 42
Creep, 61
  curve, 64
  function, 61, 64
Current configuration, 12

Darcy's law, 167, 185
  homogenized, 218
Datum head, 185
Deformation gradient, 19
Degree of consolidation, 278
Delta function, 148
Desorption, 178
Deviatoric plastic strain increment, 68
Deviatoric strain, 50
Deviatoric stress, 50, 68
Deviatoric tensor, 47
Diffusion
  coefficient, 166, 174
  equation, 126, 162, 165
  experiment, 257
  velocity, 125
Diffusive body force, 130
Diffusive stress, 130
Direction cosines, 286
Directional derivative, 145
Dissipation function, 109
Dissipative energy law, 88
Distribution coefficient, 165, 177
Divergence theorem, 292
Drained test, 196
Drucker's stability postulate, 72
Dual space, 303
Dummy index, 283

Effective diffusion coefficient, 175, 257
Effective stress, Terzaghi, 187
Efficiency, engine, 319
Elastic strain increment, 68
Elasto-plastic constitutive law, 205
Electrochemical potential, 356
Elevation head, 219
Embedded coordinate system, 23
Energy
  conservation law, 317
  flux, 85
  functional, 142, 145
  minimization, 142
Energy-conjugate, 87

Engineered barrier system: EBS, 267
Engineering shear strain, 52
Enthalpy, 102, 106, 119
  molar description, 340
Entropy, 322, 323
  increase, 332
  per unit mass, 336
  production, 332
Enzyme reaction, 369
Equilibrium adsorption isotherm
  Freundlich, 181
  Langmuir, 178
  Temkin, 181
Equivalent plastic strain increment, 75
Equivalent stress, 75
Euler's equation, 113
Euler's theorem, 108
Euler stress, 34
Eulerian description, 13
Eulerian equation of motion, 33
Eulerian spin, 45
Eulerian strain, 20
Eulerian triad, 22
Excess pore pressure, 187
Extensive variable, 318
Extent of reaction, 346
External work, 83

Fick's law, 166, 174
Finite Element Analysis, 191
First Law of Thermodynamics, 82, 88
First Piola-Kirchhoff stress, 33
Flow rule, plasticity, 72
Flow theory, plasticity, 67
Formation factor, 175
Fourier basis, 313
Fourier's law, 335
Fractional coverage, adsorption, 178
Frame indifference, 16
Free index, 283
Froude number, 184
Fugacity, 353
Functional, 304
Fundamental Identity, 129

Galilean principle of relativity, 10
Galilean transformation, 10, 11
Gas constant, 351
Gâteaux differential, 98
Gauss' double integral theorem, 292
Gauss' theorem, 291
Generalized derivative, 148
Generalized Eulerian strain measure, 22

Generalized Kelvin-Voigt model, 62
Generalized Lagrangian strain measure, 22
Generalized Lagrangian stress, 87
Generalized Maxwell model, 60
Generalized strain measure, 21, 86
Geological disposal, 267
Gibbs-Duhem relation, 101, 113
Gibbs free energy, 102, 106, 119
  molar description, 340
Gibbs-Helmholtz relation, 117
Gibbs' phase rule, 118
Grand potential, 120
  molar description, 340
Gravity, 31
Green strain, 20
Green's theorem, 292

Hagen-Poiseulle flow, 170
Hardening law, 68
Hardening model
  anisotropic, 69
  isotropic, 69
  kinematic, 69
Hardening parameter, 68, 73
Hardening rule, 203, 204
Heat capacity under constant pressure, 104
Heat capacity under constant volume, 104
Heat conduction
  equation, 104
  problem, 334
Heat engine, 319
Heat flux, 88, 89
Heat of reaction, 346
Heat supply, 88
Helmholtz free energy, 102, 106, 119
  molar description, 340
Helmholtz relation, 117
Hereditary integral, 56
Hess's law, 343
Hittorf number, 363
Homogeneous function, 75, 108
Homogenization analysis, 3, 207
  porous media flow, 213
Hookean solid, 51
Hvoslev surface, 196
Hydraulic conductivity, 186
Hydrostatic axis, 50
Hyperbolic PDE, 150
Hyperelastic material, 102
Hyperelastic, finite strain, 107

Incompressibility condition, 30
Incompressible, 56
Inertial frame, 9
Initial strain, 52
Initial stress, 52
Inner product space, 302
Integrability, 100
  condition, 297
Intensive variable, 318
Internal dissipation, 334
Internal energy, 99, 118
  molar description, 340
  per unit mass, 89
  reversible, 88
Interpolation, 151
  functions, 151
Invariant, 47
Isolated system, 318, 331, 333

Jacobian, 25
Jump, 28

Kelvin-Voigt model, 61
Kinematic viscosity, 108
Kozeny-Carman formula, 174
Kronecker's delta, 284

Lagrangian description, 13
Lagrangian equation of motion, 35
Lagrangian multiplier method, 146
Lagrangian spin, 45
Lagrangian strain, 20
Lagrangian strong form, 147
Lagrangian triads, 21
Lagrangian weak form, 147
Lamé's constant, 51, 104
Langmuir, 178
  plot, 179
Laplace transformation, 59
Law of action and reaction, 9
Law of conservation of linear momentum, 9
Law of inertia, 9
Least squares method, 152
Left Cauchy-Green tensor, 20
Left form, 14
Legendre polynomials, 151
Legendre transformation, 98
Linear momentum, 30
Linear vector space: LVS, 300
Lode's angle, 49
  for stress, 68
Logarithmic creep law, 67
Lower convected rate, 42

Macro-phenomenological approach, 2
Macroscale equation, 210, 247
Macroscale seepage equation, 237
Mass-averaged velocity, 125
Mass concentration, 112
Mass conservation law, 30
Mass density, 29, 111
Mass-energy flux, 317
Mass flux, 126
Material time derivative, 13
Matric diffusion, 165
Maxwell model, 58
Maxwell's relations, 116
Mean tensor, 47
Mechanical dispersion, 176
Melan's formula, 74
Metric space, 302
Michaelis-Menten structure, 369
Micro-inhomogeneous material, 1
Microscale equation, 210, 246
  Stokes' flow, 217
Microscale incompressibity, 236
Microscale Stokes' equation, 235
Miscible solution, 129
Molar fraction, 112
Molecular diffusion coefficient, 175
Molecular dynamics, 2
Molecular partition function, 325
Molecular weight, 111
Moment of linear momentum, 35
Moment of tensor, 47
Monotonicity, 99
Multiscale porous medium, 230

Nanson's formula, 27
Navier's equation, 53
Navier-Stokes equations, 55
Nernst-Planck theorem, 334
Newtonian mechanics, 9
Newton's second law, 30
Nominal stress, 34
  rate, 37
Nonequilibrium adsorption isotherm
  Freundlich, 182
  Langmuir, 182
Norm, 48
Normalized measure of Cauchy stress, 38
Normally consolidated clay, 196
Normed linear space, 302
Norton's power law, 67

Objectivity, 41
Open system, 318
Orthonormal, 16
Outer product, 287
Over-consolidated clay, 197
Over-consolidation ratio, 197

Parabolic PDE, 149
Parseval's equality, 311
Partial molar enthalpy, 351
Partial pressure, 353
Péclet number
  diffusion, 184
  temperature, 184
Penalized strong form, 146
Penalized weak form, 146
Penalty method, 146
Perfect gas, 351
Periodic boundary, 215
Periodic function, 209
Periodic microstructure, 207
Permeability, 167
Perturbation, 215
Phase change, 348
Physisorption, 177
Π-plane, 50
Plane strain, 52
Plane stress, 52
Plastic strain increment, 68
Poisson's ratio, 51, 104
Polar decomposition, 20
Pore pressure, 186
Porosity, 157
Positive definiteness, 156
Power input, 83
Prager's consistency condition, 71
Pressure head, 219
Principal stretch, 21
Principle of determinism, 41
Principle of frame indifference, 41
Principle of local action, 41
Principle of Minimum Energy, 336
Principle of the energy minimization, 145

Rate-of-deformation tensor, 23
Rayleigh-Ritz method, 156
Reference configuration, 12
Relaxation function, 58
Relaxation spectrum, 58, 59, 63, 64
Representative elementary volume: REV, 157
Response function, 58
Retardation coefficient, 166
Retardation spectrum, 61, 63
Retardation time, 62

Reversible engine, 319
Reversible process, 322
Reynolds number, 184
Reynolds' transport theorem, 26
Right Cauchy-Green tensor, 20
Right form, 14
Roscoe surface, 196
Rotation tensor, 20
Rotation vector, 18

Schrödinger equation, 328
Secondary consolidation, 268
Second Law of Thermodynamics, 331
Second Piola-Kirchhoff stress, 34
Seepage equation, 186
    homogenized, 219
Shear modulus, 51
Similitude law, 183
Singular surface, 28
Small strain, 20
Solute coefficient, 176
Sorption, 178
Specific description, 38
Spherical decomposition, 49
Spin tensor, 23
Standard enthalpy of combustion, 343
Standard enthalpy of formation, 344
Standard state, 343
State boundary surface, 196
State equation, 351
Step function, 148
Stirling's approximation, 324
Stoichiometric coefficient, 345
Stokes' equation, 214
Stokes' law, 55
Stokes' power formula, 82, 83
    Lagrangian, 88
Stokes' theorem, 294
Strain energy, 84, 103
    specific, 84
Strain hardening, 73
    rule, 69
Stress, 32
    relaxation, 58
Stress–dilatancy relation, 201
Stretch tensor, 20, 23
Strong form, 139
Strouhal number, 184
    diffusion, 184
Summation convention, 283
Superposition principle, 57
Swelling index, 200

Tensor, 304
Terzaghi, 187
Thermal conduction tensor, 335
Thermal conductivity, 335
Thermal expansion coefficient, 52
Thermal wave length, 326
Thermodynamical system, 318
Third Law of Thermodynamics, 334
Through-diffusion test, 248
Topology, 302
Tortuosity, 175
Total head, 219
Total Lagrangian form, 36
Total mechanical energy, 85
Traction, 30, 31
Transport number, 363

Undrained test, 196
Unit cell, 208
Updated Lagrangian form, 36
Upper convected rate, 42

Van der Waals force, 178
Vector product, 287
Velocity gradient, 23
Velocity of the material point, 12
Virtual displacement, 144
Virtual strain, 144
Virtual work equation, 140, 144
Viscosity, 54
Void ratio, 71, 280
Volume molar fraction, 111
Volumetric plastic strain increment, 68
Volumetric strain, 50
Volumetric stress, 50
Volumetric tensor, 47

Weak form, 140
Weight of configurations, 324
Work hardening, 73
    rule, 69

Yield function, 68
Young's modulus, 51, 104

Zaremba-Jaumann rate, 44